| Hydrogen $^1$H 1.008 | Helium $^2$He 4.003 |

| Boron $^5$B 10.81 | Carbon $^6$C 12.011 | Nitrogen $^7$N 14.01 | Oxygen $^8$O 16.00 | Fluorine $^9$F 19.00 | Neon $^{10}$Ne 20.18 |

| Aluminum $^{13}$Al 26.98 | Silicon $^{14}$Si 28.09 | Phosphorus $^{15}$P 30.97 | Sulfur $^{16}$S 32.06 | Chlorine $^{17}$Cl 35.45 | Argon $^{18}$Ar 39.95 |

| Nickel $^{28}$Ni 58.71 | Copper $^{29}$Cu 63.55 | Zinc $^{30}$Zn 65.37 | Gallium $^{31}$Ga 69.72 | Germanium $^{32}$Ge 72.59 | Arsenic $^{33}$As 74.92 | Selenium $^{34}$Se 78.96 | Bromine $^{35}$Br 79.90 | Krypton $^{36}$Kr 83.80 |

| Palladium $^{46}$Pd 106.4 | Silver $^{47}$Ag 107.87 | Cadmium $^{48}$Cd 112.40 | Indium $^{49}$In 114.82 | Tin $^{50}$Sn 118.69 | Antimony $^{51}$Sb 121.75 | Tellurium $^{52}$Te 127.60 | Iodine $^{53}$I 126.90 | Xenon $^{54}$Xe 131.30 |

| Platinum $^{78}$Pt 195.09 | Gold $^{79}$Au 196.97 | Mercury $^{80}$Hg 200.59 | Thallium $^{81}$Tl 204.37 | Lead $^{82}$Pb 207.19 | Bismuth $^{83}$Bi 208.98 | Polonium $^{84}$Po (209) | Astatine $^{85}$At (210) | Radon $^{86}$Rn (222) |

| Gadolinium $^{64}$Gd 157.25 | Terbium $^{65}$Tb 158.93 | Dysprosium $^{66}$Dy 162.50 | Holmium $^{67}$Ho 164.93 | Erbium $^{68}$Er 167.26 | Thulium $^{69}$Tm 168.93 | Ytterbium $^{70}$Yb 173.04 | Lutetium $^{71}$Lu 174.97 |

| Curium $^{96}$Cm (247) | Berkelium $^{97}$Bk (249) | Californium $^{98}$Cf (249) | Einsteinium $^{99}$Es (254) | Fermium $^{100}$Fm (257) | Mendelevium $^{101}$Md (258) | Nobelium $^{102}$No (259) | Lawrencium $^{103}$Lr (260) |

# SOILS

## AN INTRODUCTION

# SOILS
## AN INTRODUCTION

*Fifth Edition*

## Michael J. Singer

*Professor of Soil Science*
*University of California, Davis*

## Donald N. Munns

*Late Professor of Soil Science*
*University of California, Davis*

Prentice Hall

Upper Saddle River, New Jersey 07458

**Library of Congress Cataloging-in-Publication Data**

Singer, Michael J. (Michael John), 1945-
    Soils : an introduction / Michael J. Singer, Donald N. Munns.--5th ed.
      p. cm.
    Includes bibliographical references (p. ).
    ISBN 0-13-027825-4
    1. Soil science. 2. Soils. 3. Plant-soil relationships. 4. Soil management. I. Munns,
Donald N. (Donald Neville), 1931-2000 II. Title.

S591.S555 2002
631.4--dc21                                                  00-064984

**Executive Editor:** Debbie Yarnell
**Assistant Editor:** Kimberly Yehle
**Production Editor:** Lori Dalberg, Carlisle Publishers Services
**Production Liaison:** Eileen M. O'Sullivan
**Director of Manufacturing and Production:** Bruce Johnson
**Managing Editor:** Mary Carnis
**Manufacturing Buyer:** Ed O'Dougherty
**Senior Design Coordinator:** Miguel Ortiz
**Cover Designer:** Joseph Sengotta
**Cover Photo:** Bruce Heinemann, PhotoDisc, Inc.
**Marketing Manager:** Jimmy Stephens
**Interior Design:** Carlisle Communications, Ltd.
**Composition:** Carlisle Communications, Ltd.
**Printing and Binding:** Courier Westford

Prentice-Hall International (UK) Limited, *London*
Prentice-Hall of Australia Pty. Limited, *Sydney*
Prentice-Hall Canada Inc., *Toronto*
Prentice-Hall Hispanoamericana, S.A., *Mexico*
Prentice-Hall of India Private Limited, *New Delhi*
Prentice-Hall of Japan, Inc., *Tokyo*
Prentice-Hall Singapore Pte. Ltd.
Editora Prentice-Hall do Brasil, Ltda., *Rio de Janeiro*

10 9 8 7 6 5 4 3 2 1
ISBN 0-13-027825-4

This edition of *Soils: An Introduction* is dedicated to my colleague and friend Don Munns, whose inspired teaching, quiet sense of humor, masterful skill as an artist, and helpful editing made him a joy to work with. Don was stricken with cancer during the preparation of this edition and died on March 16, 2000. He will be greatly missed.

# CONTENTS

# SOIL AND ENVIRONMENT

# PREFACE

*Soils: An Introduction* has been revised to include concepts and information new since the fourth edition in 1999. This edition, like the first four, is written to be understood by all students, but it has sufficient depth to serve as a text for soil science majors and as a reference for anyone interested in soil. We have expanded the environmental emphasis in most chapters to highlight the role of soils in nonagricultural uses.

Why be interested in soil? Soil is one of the "ultimate" resources, like water and air. There are no replacements for soil for growing plants for food and fiber (cotton, pine, and redwood trees), for raising cattle and sheep, for anchoring the foundations of our homes, for building golf courses, and for burying our garbage. Soils, a close friend once said, are "the excited skin of the earth" in which geology, biology, chemistry, and physics combine. With knowledge we can manage this slowly renewable resource so that it can serve us and provide us with our requirements for life, while we in turn should leave it intact for future generations.

Soil science is changing, and we wanted a text that would make the current concepts easy to understand. We therefore have excluded unnecessary jargon and have defined necessary terms both in the text and in a glossary. We have left out or deemphasized obsolete concepts that seem no longer worth the trouble they cause and have used chapter supplements to deliver valuable detail without cluttering the main text. We have tried to keep our language direct, to stress principles, and to avoid trivia. We have also included many original diagrams and pictures that we hope will inform and interest when the words falter.

Repetition aids learning. There is wisdom in the old advice to tell what you are going to tell, then tell it, then tell what you told. We begin each chapter with an outline and an overview. Further summaries of the main points appear in chapter summaries, in summary diagrams and tables, or at the ends of sections within the chapter. Finally, each chapter has questions to encourage thought on the main points.

We begin by naming parts—mineral particles, organic matter, organisms, pores, water—and explain how they relate to one another to form soil and with plants and microbes to form an ecosystem. We then relate this complex soil body to its larger environment by discussing soil origins, classification, and interpretation. This is a logical order of treatment, but, like any other order, it will not suit all readers, and so they should be able to change the order or read only some of the chapters. Accordingly, we have repeated important ideas to help make each chapter less dependent on the others.

We have continued to use our "building the pedon" concept introduced in the first edition. We introduce the pedon concept in Chapter 1 as an empty rectangle and develop it in each subsequent chapter. As inorganic and organic matter, water, and pores are added, the empty rectangle takes on the characteristics of a "real" soil. Our goal is to have every student understand the parts that contribute to the whole soil individual and then appreciate how the parts function together.

Our guiding philosophy is that soils are interesting and important parts of natural systems, and we hope that the users of this book develop an understanding and appreciation of soils from these pages.

Special thanks are given to Mary Savina and Jim White for their valuable suggestions on the new edition.

Michael J. Singer
Donald N. Munns

# SOILS

## AN INTRODUCTION

# 1

# INTRODUCTION

## OVERVIEW

*Soils are complex biogeochemical materials on which plants may grow. They have structural and biological properties that distinguish them from the rocks and sediments from which they normally originate. They are dynamic ecological systems, providing plants with support, water, nutrients, and air. They support all ecosystems on land including large populations of microorganisms that recycle the materials of life. They provide the entire human population with food, fiber, water, building materials, and sites for construction and waste disposal. We rely on them to protect groundwater by filtering toxic chemicals and disease organisms from wastewater. Understanding soils and managing them well are essential to human welfare.*

*Farmers, ranchers, engineers, foresters, gardeners, and ecologists think of soil in different ways and for different purposes (Figure 1–1). The purpose of this book is to help you think of soil in all these ways and understand what soils are and do. To understand a soil system, one must first know the components, and much that follows will be naming parts and describing processes. You will need to be thinking about how these parts and processes fit into the whole.*

*Be aware of the vast range of sizes of the different parts that make up the whole soil; note the scales in many of the diagrams. We use metric units throughout this book. A conversion table for metric (SI) and English units is given on the inside back cover. SI is an abbreviation of* Système International d'Unités.

**FIGURE 1–1** People's concept of soil depends on their direct experience of it. Farmers, gardeners, and agronomists think of it primarily as a medium for plant growth; engineers and builders, as a foundation or construction material; environmentalists, as a pollutant source or sink; conservationists, as something to preserve; and parents and housekeepers, as dirt. Soil scientists study soil as a fascinating system.

## 1.1  SOIL

### 1.1.1  What Is Soil?

*Soil* as a general term usually denotes the unconsolidated, thin, variable layer of mineral and organic material, usually biologically active, that covers most of the earth's land surface. Dictionaries often describe soil as the loose surface material in which plants grow. This text is intended to add details that such definitions hide, to name soil components and processes, and to help you assemble them into a working concept of soil and understand its many variations. Soil variation is so important that we will often write of "soils," or "a soil" in particular, rather than "soil" in general.

In different contexts, the expression "a soil" can mean either (1) a particular stuff, soil material, or blend of components, or (2) a particular three-dimensionally structured body of soil materials occurring naturally within a landscape. The first meaning of "a soil" as a somewhat complex material is the one most often intended by engineers, greenskeepers, horticulturists, and even soil scientists in discussions of soil chemistry, physics, or microbiology. The second, more formal sense of "a soil" as a natural body in a landscape is the one most often intended by soil scientists talking of the development, behavior, and classification of soils in their natural environment as parts of ecological systems.

*Ecology* is the study of interactions of organisms and the environments in which they live. A soil is both an ecosystem in itself and part of a larger ecosystem. Soils are fundamental to terrestrial ecosystems and ecological processes and often influence aquatic ecosystems nearby or downstream. Too often, we think of soil only in the context of food-crop production. Soils are part of all ecosystems on land, be they rural or urban, managed or undisturbed.

In these ecological contexts, "a soil" means a particular three-dimensional example of the larger body of material we call soil, separately defined for purposes of characterization, study, and interpretation; it is part of the landscape in which it naturally exists; and it has characteristic layers with specific properties such as color, mineral and organic content, hardness, and thickness. If we dig it up, bulldoze it into a pile, or otherwise disturb it, it becomes just "soil material." This natural soil body must be large enough to contain all the physical, chemical, biological, and morphological properties necessary to describe it and distinguish it from other soils. A minimum volume, with a surface area of 1 square meter (1 m$^2$) and depth of 1.5 m, unless rock is encountered, is a **pedon.** We will say more about the pedon concept later.

## 1.1.2 How Do Soils Form?

A soil develops from some starting material such as consolidated rock or unconsolidated material deposited by wind or water. The starting material is the **parent material** for a soil. The parent material consists of specific minerals of different sizes or plant material of different plant types. A soil that originates from consolidated rock or unconsolidated material inherits the mineral types found there. In addition, over time, the original minerals are dissolved and new minerals form and accumulate in the soil. This process of dissolution and reformation of new minerals is part of the complex set of processes called **weathering.**

Soil formation begins after unconsolidated parent material is deposited on a stable landscape or after bedrock has been exposed at the earth's surface. The rate of soil formation depends on the climate, es-

pecially the annual amount of precipitation and the average annual temperature. It also depends on the kind and amount of vegetation and the numbers, kinds, and activity of the other organisms that live on and in the parent material. The organisms that colonize the parent material help to convert it to soil.

Two other factors affect the rate of soil formation and the kind of soil that is created. These are topography and time. *Topography,* the slope of the land from the horizontal and the compass direction the slope faces, affects the amount of precipitation that enters the parent material or soil and the soil temperature. These influence the kinds and numbers of organisms that live on and in the soil. All of the processes take time. *Time zero* is the time at which a soil begins to form from parent material. Depending on the environment, thousands or tens of thousands of years may be necessary before the parent material is converted to a highly weathered and developed soil (Figure 1–2).

Parent material, climate, vegetation, topography, and time are the factors of soil formation. These factors differ in subtle and complex ways over the surface of the earth to create an infinite array of soils. Each soil, in turn, influences the success or failure of our use of the soil, whether we use it for agriculture, urban engineering, recreation, or waste disposal. These factors of soil formation are discussed in detail in Chapter 12.

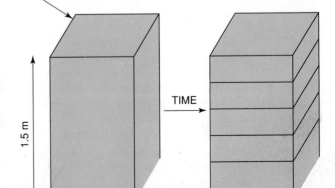

**FIGURE 1–2** Over time, under the influence of climate, organisms, and topography, processes of soil formation change undifferentiated parent material into a vertically differentiated soil.

The remainder of this chapter begins the process of naming parts of a soil and describing in more detail a few of the concepts introduced in this section.

## 1.2  MINERALS AND ORGANIC MATTER

Soil has two kinds of solid components: inorganic minerals derived from weathering rocks, and organic materials derived from plants and microorganisms. Rocks change when exposed to water, air, and organisms. These changes are called *weathering,* which refers to the separation of the rock's individual mineral particles and their alteration or destruction and resynthesis to form new minerals. Sometimes wind and rain erode the weathering material and deposit it somewhere else. But whether it is transported or left in place, the weathering rock becomes the parent material in which mineral soils form.

The minerals of soils and rocks are nearly always distinct crystalline substances: mostly compounds of oxygen, silicon, and aluminum, sometimes with appreciable amounts of iron, calcium, potassium, or magnesium. Soil may contain both unweathered minerals residual from the rock and the minerals that weathering produces. The principal weathering products are the abundant clay minerals that occur as small particles that cohere to one another and adhere to other particles.

Organic matter in soil is living and dead, large and recognizable, or small and detectable only with sophisticated instruments. Living in the soil are organisms of many kinds: plants, fungi, filamentous and single-celled **bacteria** and **algae,** and small animals from **protozoa** to worms, insects, and mammals. The plants, bacteria, and fungi are especially important to soil processes, including soil formation. Higher plants are photosynthetic and make the organic compounds that feed the other organisms and us. The bacteria and fungi are the destroyers, decayers, and decomposers of organic matter. During the decay of plant tissue, large complex molecules form that remain in the soil as resistant by-products. These organic colloids are **humus,** which is responsible for the brown and black colors of the top layers of some soils. It is produced from plant material by microbes operating under the influence of the soil's special environment. Humus is important because it has properties of water retention, nutrient retention, and cohesion that complement those of clay.

Humus and clay are said to be *surface active,* meaning that much happens on their surfaces. They are **colloids,** materials in the form of fine particles in the size range from 1 nanometer to 1 micrometer (1 nm $= 10^{-9}$ m; 1 $\mu$m $= 10^{-6}$ m). The smallness of colloidal particles confers enormous surface area per unit mass (e.g., 80 m$^2$/g or more for fine clays). Clay minerals cohere to each other and adhere to larger mineral particles. Their surfaces adsorb and hold water, organic compounds, plant nutrient ions, and toxic ions.

## 1.3  SIZE AND ORGANIZATION OF PARTICLES

Most soils consist of mineral particles of different sizes: large particles, called *gravel;* smaller ones, called *sand;* still smaller ones, called *silt;* and submicroscopic ones, called *clay.* The proportions of each size fraction combine to determine the soil's texture, be it coarse (gravelly or sandy), intermediate (loamy), or fine (clayey).

Soil **texture** refers to particle sizes, whereas soil **structure** refers to particle arrangements. In most soils, individual particles are bound together into larger, complex structural units or **aggregates.** The most important binding agents are clay and humus. Structure distinguishes most soils from weathered rock, loose sand or dust, a lump of clay, or a brick (Figure 1–3). The size, form, and strength of the aggregates vary. A structureless (nonaggregated) soil may be either an incoherent heap of particles in which no particles are stuck together or a single compacted mass in which all the particles are stuck together. Structure favorable to plants is common in uncultivated soils of forest and grassland but is often degraded in cultivated soils. Structure affects soil behavior mainly because it affects the spaces, called **pores** or **voids,** within and between the aggregates.

## 1.4  SOIL PORES

Soil is a three-phase system: solid, liquid, and gas in intimate contact. The solid mineral particles and organic coatings are the soil's skeleton. The pores between particles and aggregates control the soil's

**FIGURE 1–3**   The pile of loose sugar is structureless. The grains are touching but are not organized. A single sugar cube is an aggregate of individual grains. The pile of cubes represents a blocky structure.

ventilation, water intake, water storage, and drainage. The sizes and shapes of the pores, the total pore space, and the continuity of interconnected pores are all important. Small pores, for instance, hold water well, but large interconnected pores are needed for water and air to move freely into and out of the soil.

The gas phase, the **soil air,** is the channel for the movement (diffusion) of oxygen and other gases. It is the soil's connection with earth's atmosphere. The liquid phase, the **soil water** or **soil solution,** is the solvent in which many reactions occur, and it supplies water and dissolved nutrients to plants and microorganisms. Even in soils that seem dry to the touch, a water film covers the solid surface and fills the smallest pores. If the soil gets wetter, the films thicken and the larger pores fill. When even the largest pores are filled with water, the soil is said to be **waterlogged** or **saturated,** a state in which water and dissolved substances move freely, but gases move so slowly that oxygen normally becomes deficient.

## 1.5   SURFACE REACTIONS AND TRANSFER PROCESSES

Many soil processes occur at the surfaces where the soil solution meets solid, air, or living cells. *Surface reactions* include the **adsorption** of both ions and neutral molecules from solution onto the soil colloids. These reactions allow plant nutrients, toxins, and other pollutants to be largely held rather than leaching freely as water passes through the soil. Adsorption processes are reversible; thus, a reserve of adsorbed nutrients can be mobilized by **desorption** as nutrients

become depleted from the soil solution. Other surface reactions are the exchange of gases between soil air and soil solution, the crystallization of new solids from solution, and the dissolution of soil minerals (weathering). Surface reactions are coupled with transfer processes; they modify each other.

*Transfer processes* move heat, water, gases, dissolved substances, and even particulate solids in soil and the soil-plant-atmosphere system. These movements vary in scale and speed. The **diffusion** of oxygen or a nutrient ion to a root from adjacent soil might involve millimeters and minutes. Heat flows measurably from the soil surface to the subsoil daily and back toward the surface at night under normal weather conditions. Water flows downward from the soil surface during rain or irrigation; this movement slows over several days once the rain or irrigation stops and then reverses as the soil dries. Diffusion of adsorbed solutes may take weeks or months to become measurable. Perhaps the slowest significant transfer process in soils is the creep of clay particles from topsoil to subsoil, which can take thousands of years to become apparent.

## 1.6   SOIL AS AN ECOSYSTEM

Soils are ecosystems. Some of the main soil processes are biological. Roots probe, split, and squeeze the mineral particles. Roots inhabit soil, gain nourishment, remove water, encourage microbial growth, and influence their environment by sloughing off tissue and exuding chemicals. Dead leaves, roots, and stems return organic matter for eventual decay, releasing inorganic nutrients

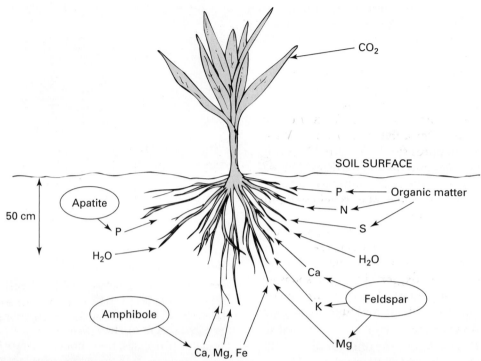

**FIGURE 1–4**　Plant nutrients come from the weathering (dissolution) of soil minerals such as apatite, amphibole, and feldspar. Nitrogen, phosphorus, and sulfur are part of organic matter and are frequently in short supply in soil.

for use again. Much carbon is locked up in soil organic matter. The decay of this organic matter by soil microorganisms and the release of carbon dioxide and ions are the largest microbial processes in soil. Other essential biological processes are the transformations of nitrogen, phosphorus, and sulfur compounds and the mobilization of iron. Roots and microbes together virtually control the composition of soil air.

Most plants need soil to hold them up to the sun. Roots exploit the soil and spread vertically and horizontally, holding the plant upright and extracting nutrients and water. The nature of the **soil fabric,** particularly the density, affects root extension. **Soil density** is a measure of the relative amount of pores (voids) and solid particles. The fewer pores there are per unit volume, the higher is the density, and the more difficult it is for roots to extend into the soil.

Plants need **nutrient elements** to grow. They get carbon, oxygen, and hydrogen from air and water, but all the other nutrients come from the soil. Those nutrient elements needed in the largest quantities are carbon (C), hydrogen (H), nitrogen (N), phosphorus (P), sulfur (S), potassium (K), calcium (Ca), and magnesium (Mg). Required in smaller amounts are **micronutrients,** including iron (Fe), manganese

(Mn), zinc (Zn), copper (Cu), molybdenum (Mo), boron (B), and chlorine (Cl). Elements often found at levels toxic to plants are aluminum (Al), hydrogen ($H^+$), sodium (Na), chlorine (Cl), and boron (B).

Each nutrient element has one or more sources in soil (Figure 1–4). Nitrogen, for example, is mainly present in humus. Only small amounts of N are found in inorganic forms at any time, but there is a continuous release of new inorganic N from humus due to microbial decomposition. Like many other nutrients, N moves in the soil and changes from one form to another. The series of changes from organic N to inorganic and back again is called the *nitrogen cycle,* one of the most important sets of processes in nature. The other nutrients also have cycles, each different.

None of the nutrient cycles is closed. Some losses are caused by leaching, erosion, harvesting, and the escape of gases to the atmosphere. Inputs or gains include fertilizer application, absorption of nitrogen from the air by specialized soil microbes, and release of essential elements from minerals as they weather.

Nutrient (and toxin) retention and movement, as well as root growth, determine much of a soil's fertility. Most important in holding nutrients are the different clay minerals and humus. Their surfaces can chemically retain nutrients in forms that plants can use. Very often a nutrient is held too tightly or moves too slowly, and so it is not available to plants in sufficient amounts. Deficiencies or excesses can determine the nature of vegetation and the need for fertilization in agriculture, as can the soil's ability to hold water in a state available to plants.

Soils also hold the water that plants use. Soil pores fill with water after a rain or irrigation. Water flows through these pores slowly, impeded by attraction between the water and the soil particles. This *adhesion* between water and clay minerals becomes stronger as the soil dries. Some water is held so tightly that plants cannot remove it.

If soil pores do not contain water, they contain air. Soil air is similar to the air we breathe, except it is usually more humid and higher in carbon dioxide ($CO_2$). Plant roots need oxygen, and if the soil is waterlogged for too long, most plants will drown. Waterlogged soils of swamps, bogs, and tundra carry vegetation adapted to oxygen shortage.

## 1.7  SOIL MORPHOLOGY

**Soils are three-dimensional**  Everyone is familiar with the horizontal dimensions of soil as area, but not everyone is aware of its third dimension, depth. That could mean digging! This third dimension contains many of the soil's most interesting and important properties. At some depth below the surface, the soil changes to parent material and unweathered geological material. The change may be either gradual or abrupt. Soil scientists call the three-dimensional soil body a *pedon,* and Figure 1–5 shows its basic form.

This particular soil individual is 1 m square at the surface and 1.5 m deep, about the smallest size a pedon can be to contain all the properties we need in order to classify and study the soil individual. The upper part of the pedon, most influenced by rainfall, roots, and microorganisms, is the **solum.** The part below the solum is usually the parent material from which the solum developed. At times, the solum may sit directly on the geological material.

**Soil horizons**  A soil is not uniform with depth but consists of layers called **horizons.** In Figure 1–6, we have added five horizons to the empty pedon. Horizons are identified by letter and number codes, the **master horizons** being the O, A, E, B, C, and R horizons (Table 1–1). At the surface, we normally find one or

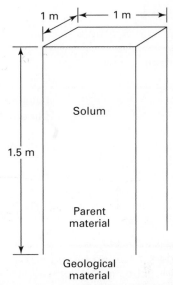

**FIGURE 1–5**  The pedon is a three-dimensional body that contains the minimum volume of earth necessary to contain all the properties of a soil individual. This pedon is the minimum size.

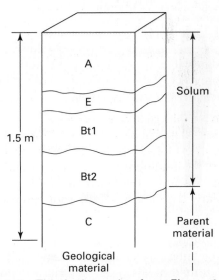

**FIGURE 1–6**  This is the pedon from Figure 1–5. We have added horizons to give it some "character." Clay has accumulated in the subsoil. These horizons are designated Bt. In the book we will continue to build this pedon until it looks like a soil you might see if you were to dig a hole somewhere. We will also classify this pedon and describe its properties for potential uses.

**TABLE 1–1**

Master Horizon Symbols

| Master horizon symbols | Horizon property or characteristic |
|---|---|
| O | Layers dominated by organic material |
| A | Mineral horizons formed at the surface or below an O horizon and containing accumulated decomposed organic matter |
| E | Mineral horizon in which the main feature is the loss of silicate clay, Fe, or Al, leaving a concentration of resistant sand and silt particles |
| B | Horizons formed below an A, E, or O horizon and dominated by the obliteration of the original rock structure and the accumulation of silicate clay, Fe, Al, humus, carbonate, gypsum, or Si |
| C | Horizons excluding hard bedrock and little affected by soil genesis |
| R | Hard bedrock such as basalt, granite, or sandstone |

Source: *USDA Soil Survey Manual*, Chapter 3. Soil Survey Staff, 1993. Soil Survey Manual, Supt. of documents, US Govt. Printing Office, Pittsburg, PA. http://www.statlab.iastate.edu/soils/ssm/gen_cont.html

more A horizons of high biological activity, where humus accumulates to darken the soil. In addition, the A horizon is a zone of leaching: As rainwater enters and moves down, it dissolves soluble components and picks up colloidal particles. This load of dissolved and suspended material moves slowly downward into the lower subsoil horizons. The E horizon is one from which particularly large amounts of constituents such as clay and iron have been removed. E represents the word **eluvial.**

Material that has left the A or E horizon may accumulate in the B horizon, the zone of accumulation (*illuviation*). Some may pass below the B horizon. Still more material may leave the pedon entirely and find its way into shallow or deep groundwater. Water is the solvent that does most of the weathering and all of the leaching in soils.

Earlier we said that rocks weather to form soil. Consolidated, hard rock is given the letter R in the horizon classification system. Materials that are **unconsolidated**—not hard and solid like rock—may also be soil parent material. Such parent material is given the letter C. Examples of C material are beach sand, windblown silt (*loess*), alluvium deposited by rivers, and glacial till deposited by glacial ice.

## 1.8   SOIL FORMATION

No single property clearly distinguishes all soils from all rocks. We have said that soils are natural, three-dimensional, vertically differentiated portions of the earth's surface. But some rocks also have these properties. Soils, however, are generally softer, less dense, and less consolidated than are rocks. Some soft rocks and hard soils, however, are exceptions. The ability to grow plants was once a part of nonengineers' definition of soil until the surface material of the moon was sampled and called soil (Figure 1–7). Now, space exploration has extended to Mars, where Martian "soil" has been remotely sampled and analyzed. If we accept this moon and Martian material as soil, we cannot specify that plants must be growing on material for it to be called soil. But we can specify that soil have the potential to grow plants. Thus, moon soil, if brought to earth and given water, would meet this definition.

What really differentiates soil from rock, or solum from parent material, is the set of processes and factors that produced the soil. **Igneous rocks** form when an intensely hot liquid magma cools; **sedimentary rocks** form when sediments are pressed and consolidated; and **metamorphic rocks** result when preexisting rocks are transformed by intense pressure, heat, or injection. None of these processes forms soil. Rather, soils form through a combination of chemical, physical, and biological processes acting on rocks, sediments, or other geological materials. These are processes that loosen, reduce in size, differentiate, and change the starting material.

**FIGURE 1–7** The moon's surface. Is it soil? (Photo: NASA)

**Soil-forming processes** and the kind of soil that results are controlled by five **soil-forming factors:** parent material, climate, organisms, topography, and time. Knowing the effects of these soil-forming factors helps explain what kinds of soil are found throughout the world. These factors account for the productive alluvial and deep black soils in the world's breadbasket areas, the lime-rich salty soils of many desert areas, the acid-leached soils of cool humid forests, and the bright red-and-yellow acidic soils of the humid tropics.

## 1.9  SOIL CLASSIFICATION

It is easy to imagine the existence of an almost infinite variety of soils. The five soil-forming factors can combine in almost endless ways. In addition, soil formation never stops; there is always the slow addition, removal, and cycling of constituents. Indeed, there are over 20,000 kinds of soil individuals in the United States alone and many more thousands around the world. It is clearly impossible to remember the names of all these soils, much less their important properties and uses. For these reasons, soils information is organized by classification systems. Several systems are in use for different purposes and in different countries. One of the most generally useful, in the United States and elsewhere, is *Soil Taxonomy,* developed by the U.S. Department of Agriculture (USDA).

### 1.9.1  Soil Taxonomy

In soil taxonomy, as in plant taxonomy, there are several levels of generalization at which plants or soils can be classified. Soil taxonomy's 12 **soil orders** are designed to contain all the soils of the world. Each order is a broad grouping of similar soils separated by particular morphological (visible) characteristics. Within these broad groups are more specific groups—suborders, great groups, subgroups, and families.

The Entisols and Inceptisols are the least-developed soils (Table 1–2). Aridisols are separated from the other soil orders on the basis of climate. Histosols (organic soils) are separated from the other orders on the basis of the amount of organic matter they contain. Mollisols are primarily mineral soils that are also distinguished by the organic matter in the solum, but they have much less organic matter than do the Histosols. Vertisols have a high clay content, shrink when dry, and swell when wet. Ultisols are soils of warm, wet climates that are highly leached and generally infertile. The Alfisols, Oxisols, and Spodosols are separated from the other orders and from one another on the basis of their subsoil horizons. Andisols are soils formed from volcanic ash. The Gelisols, the newest soil order, are soils of cold climates that contain permanently frozen layers.

At first glance these taxonomic names may look awkward and unpronounceable, but much information is packed into each word. The words themselves have Greek or Latin roots, plus additional syllables indicating features of the soil. Table 1–2 summarizes some of the orders' basic characteristics.

### 1.9.2  Soil Series

At lower levels in the taxonomic system, distinctions are based on climate, thickness of the profile, acidity, mineralogy, and other properties of the soil and the environment. At the most detailed level is the *soil series,* which consists of a central concept, a typical pedon, and a carefully defined range of properties, such as kinds of horizons, numbers and thicknesses of horizons, color, and texture.

For example, the Yolo series is defined as having certain horizons, each with a specified color, structure, and texture, as well as other properties (Figure 1–8). For a soil to be in the Yolo series, it must have all the properties of the **modal** Yolo soil and be within each property's acceptable range. The central

## TABLE 1–2

Definitions of the Soil Orders

| Order name | Diagnostic characteristic |
| --- | --- |
| Entisol | Simple soils, no subsoil diagnostic horizons |
| Inceptisol | Soils with minimum development, little or no subsoil clay accumulation |
| Aridisol | Soils of hot, dry regions |
| Vertisol | No subsoil diagnostic horizons, much clay |
| Mollisol | Thick, soft, dark mineral soil of grasslands |
| Alfisol | Subsoil accumulation of clay, not strongly leached |
| Spodosol | Subsoil accumulation of iron |
| Ultisol | Subsoil accumulation of clay, strongly leached |
| Oxisol | Extreme leaching; Fe, Al oxides, and quartz left |
| Histosol | Dark organic soil, little mineral matter |
| Andisol | Mineral soil formed on volcanic ash parent material |
| Gelisol | Soils containing permanent ice |

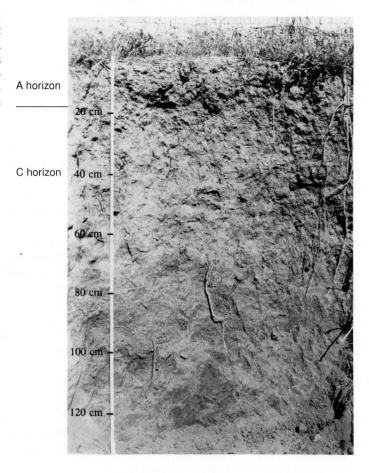

**FIGURE 1–8** This is a soil profile, a two-dimensional slice of soil. It is a young, simple profile with A and C horizons. The A horizon is darker than the C horizon because of humus accumulation. It has aggregates bound together by the humus.

A horizon

C horizon

20 cm

40 cm

60 cm

80 cm

100 cm

120 cm

or modal Yolo pedon has two horizons, A and C, and is 100 cm thick, with a range in thickness from 80 to 120 cm. A soil like the Yolo in every other property but only 70 cm thick or thicker than 120 cm would not be Yolo. In this way, an infinite variety of individuals can be categorized into units that, though numerous, are understandable and useful.

## 1.10 SOIL MANAGEMENT

Soil is the material that supports what we build, treats our waste, and purifies our water. Soils influence virtually all our environmental problems. For example, plants and lakes are damaged by acid rain, but many soils react with acid rain, reducing its acidity and protecting water quality. Soils frequently replace water as the disposal site for garbage and sewage. Soil management includes selecting the proper sites for waste disposal to ensure good water quality under the waste disposal site. All our uses of soils alter the soils, just as soil properties affect the success or failure of our use. Proper management matches the soil with appropriate uses that will minimize damage and maximize benefits.

**Soil surveys** Information about soils is found in **soil survey** reports (Figure 1–9), which are made by soil scientists who walk through fields digging holes, examining pedons, and describing horizons. The information they collect on soils and the landscapes where soils are found helps in the proper use of the soils and the land. One application of soil science is the interpretation of soil survey information for land-use planning, which requires discovering those properties that are critical to a particular use and that will be difficult to modify economically. For example, for crop production, the USDA Natural Resources Conservation Service (NRCS) developed an interpretive classification of soils called the Land Capability Classification. The land capability rating rests heavily on relief (slope angle), climate, and physical soil properties that are difficult to improve, but it usually ignores the plant nutrient supply, which can be improved with fertilizers.

Soil may seem nearly indestructible and in very large supply, but it can be eroded, compacted, salinized, polluted, or covered with roads, houses, and parking lots. As you can see from Table 1–3, only about 10 percent of the world's land area is **arable**

**FIGURE 1–9** This soil survey report contains information about the distribution of soils in Yolo County and the chemical, physical, and morphological properties of these soils.

(able to be tilled for crops) or in permanent crops such as orchards, plantations, or vineyards. The remaining area is too steep, too cold, too hot, too wet, or too dry for the major cultivated crops.

Compared with the total land area of the world and the United States, the irrigated land area is small. There were approximately 267,727,000 hectares (ha) of irrigated land in the world in 1997 and about 21,400,000 ha, or nearly 8 percent, in the United States. Soil maps and soil survey reports indicate where soils with different properties are located.

**Soil erosion** Soil management is a group of methods for maximizing the benefits of soil, be they yields of corn or strong foundations that will not crack. Use of a soil changes the soil. Although some changes are beneficial, many are not. In particular, any use that disturbs soil cover (usually plants) or compacts the soil aggravates erosion. For example, when soils on

**TABLE 1–3**

World and U.S. Land Use, 1977 and 1997

| Land Use | World area (1,000 ha) | | U.S. area (1,000 ha) | |
|---|---|---|---|---|
| | 1977 | 1997 | 1977 | 1997 |
| Arable plus permanent crops | 1,400,117 | 1,510,442 | 188,293 | 179,000[a] |
| Arable | 1,311,065 | 1,379,114 | 186,552 | 176,950[a] |
| Permanent | 89,052 | 131,328 | 1,741 | 2,050[a] |
| Nonarable plus nonpermanent | 11,628,055 | 11,537,965 | 769,018 | 736,912[a] |

Source: Food and Agriculture Organization of the United Nations. 1999. *FAO production yearbook.* Vol. 52. FAO Statistics Series No. 148. Rome, Italy: FAO.
[a]Estimated from the Food and Agriculture Organization (FAO) of the United Nations. Note how land use has changed over 20 years. Land that was considered nonarable is now being cultivated. This has led to land degradation through erosion and salinization.

**FIGURE 1–10**   Erosion removed soil from the edge of the field, reducing the potential productivity of the field. The soil lost is deposited elsewhere as sediment, one of the nation's major nonpoint pollutants.

hillsides are cultivated for crops, they become more susceptible to wind and water erosion (Figure 1–10). **Erosion** removes mineral particles, organic matter, and nutrients from the soil, reducing its thickness and water-holding capacity. Eroded soil then may become a pollutant in streams and reservoirs. The time required to form new soil is so long that from a human viewpoint, soil lost through erosion is lost forever.

**Soil compaction**   Any weight or other physical force on the soil pushes the soil particles together, a process called soil **compaction,** which impairs aeration, water flow through soil, and root growth. Sediments in the ocean and in deep alluvial soils compact naturally, and the extreme result is sedimentary rock. Part of soil management is knowing when and how to best use soil; it should not be cultivated or trafficked when wet enough to be plastic and easily compressed.

**Salinity** In arid and semiarid regions, too much or too little irrigation water can lead to an increase in soluble salts, reducing plant growth. Salts such as chlorides, sulfates, and bicarbonates of Na, Ca, and Mg accumulate as water evaporates from the soil. When this problem is severe, only the most salt-tolerant species can grow.

**Soil pollution** Other chemicals besides salt can **pollute** the soil, and the improper use of pesticides or excess fertilization can inhibit or exclude plant growth. Using soil to treat wastes from animals or humans can lead to on-site pollution by adding unwanted chemicals or desirable chemicals in excess. Chemicals can be leached from the soil to become water pollutants. Industrialized society produces a wide variety of chemical and radioactive wastes, most of which are disposed of on land. Here, proper soil management includes maintaining a variety of organic and inorganic ions and compounds at levels that are acceptable in the particular soil and place.

## 1.11 SUMMARY

Soils are part of ecosystems and are themselves ecosystems. The soil individual is the vertically differentiated pedon. Each horizon consists of three phases: solids, liquids, and gases. The kinds, num-

bers, arrangement, and properties of horizons determine the soil classification.

Soils have different capabilities for different uses. Some are excellent for farming; others are too steep or shallow. Some are excellent for most uses but are difficult to retain for agriculture because their development for competing uses is profitable and attractive. Part of soil management is selecting the correct use or uses for each soil.

Soil management requires comprehending the soil system (all three phases and their interactions) and its part in larger systems. For agriculture, gardens, greenhouses, range, and forest, the larger system must include plants, water, and climate, as well as the soil and its microorganisms. Environmental concerns require attention to streams, the surrounding soil, and the groundwater beneath.

## QUESTIONS

1. Define the terms *mineral, colloidal, humus, soil texture, soil structure, adsorption, leaching, pedon, profile, horizon, parent material, soil-forming factors, soil orders, soil series, soil density, soil survey, soil compaction, soil erosion,* and *salinity.*
2. Define soil in your own words.
3. Why is it difficult to define soil?
4. What is the difference between texture and structure? Why do they matter?
5. Which is easier to change in a soil, texture or structure?
6. What is arable land?
7. Why is nonarable land valuable?
8. What do plants, microbes, and soils do for each other?
9. What properties do clay and humus share that influence both soil fertility and human health?
10. Name five ways soil plays a role in your life.
11. How would the world be different if there was no soil?

# 2

# SOLIDS AND PORES

## OVERVIEW

Soils are composed of solids that surround holes. The solids are mostly inorganic, but they may also include living organisms and dead organic matter in various stages of decay. Individual mineral particles consist of many different sizes. Particle size is a continuous variable that is divided into discrete classes, or soil separates, named sand, silt, and clay.

Individual particles are held together by clay and organic matter to form aggregates of many sizes and shapes, and the arrangement of these particles is the soil's structure. Between the individual particles and between the aggregates are holes. The holes are pores or voids. They are always full of gas, liquid, or some of each in soils. They are never empty. Pores are the passageways for fluids and gas to flow into, through, and out of soils. They control the rate at which fluid flows through soil, whether the fluid is irrigation water or sewage.

The mineral particles vary in kind, some inherited from the parent material and others formed in the soil. Their size and chemical, physical, and mineralogical characteris-

*tics contribute to the soil's overall character. The most important minerals in soils are the layer silicates of the clay fraction, formed of sheets of silicon, aluminum, and oxygen atoms. Individual sheets are stacked one upon the other. Clay minerals have a large surface area per unit weight and are electrically charged, which makes them highly chemically active in soils.*

*Organic matter is the living and dead organic carbon–containing material in soil and on the soil surface.*

*Humus is the well-decomposed, chemically active part of organic matter that contributes to dark soil color. Humus has a high surface area and charge per unit weight. Although it is typically present in much smaller amounts than are clay minerals, humus is important in determining water-holding capacity, permeability, aggregate stability, and soil behavior such as consistence.*

The pedon introduced in Figure 1–6 was an empty rectangle divided into layers called horizons. Figure 2–1 is the same pedon, now partially filled with mineral and organic material. The A horizon is darker than the remainder of the pedon because it has accumulated more humus than the lower horizons. More humus accumulated because most of the roots and microorganisms that produce humus are near the soil surface, not uniformly distributed throughout the pedon. The pedon is only partially filled with mineral and organic particles because a soil is typically 40 to 50 percent holes. Many of the holes (pores or voids) are too small to see without a microscope, while others may be large enough to swallow a shovel.

## 2.1 PARTICLE SIZES

Size is a continuous variable (Figure 2–2). Although dividing particle size into discrete size ranges is arbitrary, it is also useful. Individual mineral particles range in size from boulders that are meters in diameter to clay particles that are less than 0.002 mm (less than 2 μm) in diameter. Particles so different in size

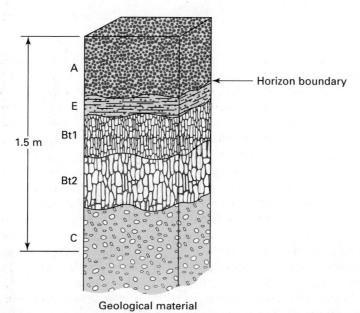

**FIGURE 2–1** Each horizon of our pedon has morphological, chemical, or physical characteristics that distinguish it from other horizons in the pedon.

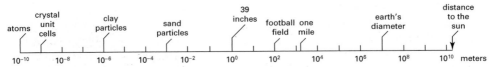

**FIGURE 2–2**   The sizes of some familiar and not-so-familiar objects should help orient you to the range of sizes in nature.

also have different properties. The terms used to divide the continuous size variable, diameter, into discrete units are considered next.

## 2.1.1   The Coarse Fraction

The first division in the size of inorganic particles is made at a 2-mm diameter, separating the *coarse fraction* (greater than 2 mm) from the *fine-earth fraction* (less than 2 mm). The fine-earth fraction controls the properties of most soils because it has the most surface area and is the most chemically and biologically active. The coarse fraction influences some important soil properties. For example, gravel 3 cm in diameter occupies space but does not contribute space for air or water in the soil because it has no pores. Thus, the volume of soil occupied by the coarse fraction reduces the soil's capacity to hold water. The coarse fraction also can hinder cultivation and damage machinery (Figure 2–3).

## 2.1.2   The Fine-Earth Fraction

The fine-earth fraction has been divided into three main **size separates** by the U.S. Department of Agriculture: **sand** (2.00 to 0.05 mm diameter), **silt** (0.05 to 0.002 mm diameter), and **clay** (less than 0.002 mm diameter). These are sometimes further subdivided, as shown in Table 2–1.

Mineral properties change radically with decreasing size below about 1 to 5 μm (0.005 mm). Particles between 1 and 1,000 nm (nanometers, $10^{-9}$ m) are colloidal particles. Fog and smoke are common examples of colloidal particles mixed with air. The *colloid* fraction includes any mineral as long as it is of colloid size. Clay minerals are the most abundant of the clay-sized minerals and include the layer aluminosilicates, allophanes, and oxides (see Section 2.3).

The mineral colloids are the most important mineral fraction of the soil because of their physical and chemical reactivity. *Clay* has three meanings in soil science, and they are easy to confuse, because the specific meaning depends on the context. Clay is (1) a size fraction of the soil, (2) a texture class name, and (3) a class of minerals.

The third meaning of the word *clay* may be the most important. *Clay* describes a group of minerals with special properties that make them important to soils' chemical and physical properties, even when they are present only in small amounts. Section 2.3 describes the nature and properties of the major soil minerals, including clay minerals. (The origin of clay minerals is discussed in Chapter 12.)

## 2.1.3   Soil Texture

Few soils consist of inorganic particles of a single size class. Although some soils are nearly 100 percent sand or 100 percent clay, they are rare. Usually soils are a mixture of sand, silt, and clay, whose relative proportions determine the soil's texture.

A textural class is a defined range of particle size distributions with similar behavior and management needs. The U.S. Department of Agriculture (USDA) soil textural triangle was designed so that any combination of particle sizes could be included within a textural class (Figure 2–4). Each corner of the textural triangle represents 100 percent of a size fraction: sand, silt, or clay. Within the triangle are areas that represent the allowable combinations of the three size separates—sand, silt, and clay—for each textural class name. For example, a sandy loam (A) may have no more than 20 percent clay or 85 percent sand or 50 percent silt (Figure 2–5).

*Loam* and *clay* are textural class names familiar to most people but often misused. Loam and clay each have a specific range of sand, silt, and clay, just as sandy loam does (Figure 2–4). A clay texture must have at least 40 percent clay and may have as much as 40 percent silt or 45 percent sand. A loam may have

(A)

(B)

**FIGURE 2–3** (A) This stone-covered field cannot be cultivated without first removing the stones. Coarse fragments such as these determine the agricultural productivity of the field. (B) This machine is removing stones from a New Hampshire field. (Photo: H. Mount, U.S. Department of Agriculture, Natural Resources Conservation Service.)

## TABLE 2–1

Size Fractions of Soil Particles

| Fraction name | Diameter (mm) |
|---|---|
| Sand | 2.00 to 0.05 |
| Very coarse | 2.00 to 1.00 |
| Coarse | 1.00 to 0.50 |
| Medium | 0.50 to 0.25 |
| Fine | 0.25 to 0.10 |
| Very fine | 0.10 to 0.05 |
| Silt | 0.05 to 0.002 |
| Coarse | 0.05 to 0.02 |
| Medium | 0.02 to 0.005 |
| Fine | 0.005 to 0.002 |
| Clay | <0.002 |
| Coarse | 0.002 to 0.0002 |
| Fine | <0.0002 |

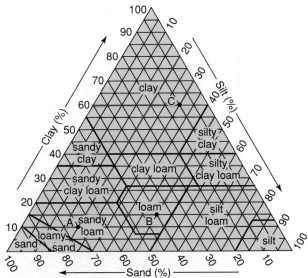

**FIGURE 2–5** Three soil textures are shown on the USDA texture triangle. They are sandy loam (A), loam (B), and clay (C). Particle size distribution data are given in Table 2–2. To use the triangle to determine the texture class name, read horizontally across from the left side along the line that represents the clay content. Next, read diagonally down from the right along the silt percentage. The two lines intersect in the texture class name area. The sand percentage, read diagonally from the bottom, intersects at the same point.

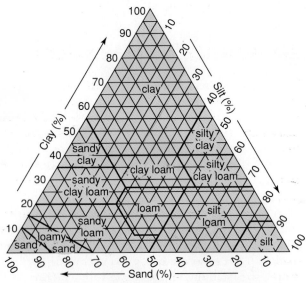

**FIGURE 2–4** The USDA texture triangle. The orientation of the numbers on each axis of the triangle indicate how to read the triangle to determine the texture class name. For example, silt content is read diagonally from right to left starting at the percentage under "silt." Sand is read diagonally from right to left starting at the bottom of the triangle. Examples are shown in Figure 2–5. (From *USDA-NRCS Handbook of Soil Survey.* Soil Survey Staff, Natural Resources Conservation Service, 1999. National Soil Survey Handbook, Title 430-VI. Washington, D.C. U.S. Government Printing Office Exhibit 618-8 http://www.statlab.iastate.edu/soils/nssh/)

from 7 to 27 percent clay, from 28 to 50 percent silt, and from 23 to 52 percent sand.

Two additional methods for describing a soil sample's particle size distribution are the particle size distribution curve and the bar graph (see Supplement 2–1).

### 2.1.4  Particle Size Analysis

How is particle size determined? First, the coarse particles (>2 mm) are separated from the fine particles (<2 mm). Then the less-than-2-mm "fine earth" is treated to separate the individual particles. Humus, which helps individual particles stick together, is removed by treating the sample with a strong oxidizing agent such as hydrogen peroxide. The sample is then mixed in water with sodium hexametaphosphate, a dispersing agent, to **disperse** (separate) the particles.

**FIGURE 2–6** Tools for measuring particle size distribution. The nest of sieves is used to separate size fractions larger than 0.05 mm from those smaller. The hydrometer is used to measure silt and clay amounts in a soil suspension.

ical relationship between particle size and settling time is Stokes' law. It is shown in Supplement 2–2. The hydrometer (Figure 2–6) measures the density of the soil–water suspension. From measurement of the density at appropriate times, we can calculate the amount of silt or clay remaining in suspension and the particle size distribution of the sample.

A less quantitative but common method of determining soil texture is the "feel" method. A moist soil sample is kneaded until aggregates are crushed and is then squeezed between the thumb and forefinger. The amount of sand, silt, and clay in the sample is estimated from the sample's roughness (grittiness), smoothness, and stickiness (Figure 2–7). The sand contributes to the roughness, and the silt, which is the size of baking-flour particles, contributes to the smoothness. The clay, which is cohesive, makes the sample sticky when wet. Guidelines for determining texture by feel are given in Table 2–3.

## 2.2 PARTICLE ARRANGEMENT

We now have a pedon (Figure 2–1) that contains particles of different sizes and different compositions. It is rare for these particles, which are often called **primary particles** (and sometimes called *soil separates*), to be separate from other particles. Except in dunes and on beaches, the sand particles are usually attached to silt and clay particles, forming groups of particles called aggregates, or peds.

*Aggregate* and *ped* are synonymous terms referring to groups of primary particles held together by soil-stabilizing agents such as clay, humus, and iron oxides. Aggregates are described according to their shape, size, stability, and the ease with which we can see them in soil. When most soil particles are aggregated, the soil has structure. Structure is an important morphological characteristic of soils because it controls the size and number of pores associated with the aggregates. The kind of structure is determined by the kind of aggregates.

### 2.2.1 Aggregate Shape

The simplest shape of an aggregate is a sphere. Soils with mostly spherical aggregates have a **granular** structure (Figure 2–8). Granular structure is common in surface A horizons. **Blocky** structure consists of aggregates shaped like cubes or short rectangles. If

Once the particles have been dispersed, the sand is separated from the silt and clay by pouring the suspension through a sieve whose openings are 0.050 mm. The sands, which remain on the sieve, may be dried and separated into various size fractions by passing them through a nest of sieves with different-sized openings (Figure 2–6). The weight of the sample remaining on each sieve after a specific amount of shaking is a measure of the content of that size fraction in the soil (Table 2–2).

The silt and clay, too small to sieve, are separated by means of **sedimentation** in a tall cylinder of water mixed with a dispersing agent. Particles settle at rates proportional to their diameter. The mathemat-

**TABLE 2–2**
Particle Size Distribution Data for the Three Soil Textures Shown in
Figure 2–5

| Size fraction (mm) | Percentage of sample smaller than each size fraction | | |
| | Sandy loam | Loam | Clay |
| --- | --- | --- | --- |
| 2.0 | 100 | 100 | 100 |
| 1.0 | 50 | 90 | 100 |
| Sand   0.5 | 40 | 85 | 99 |
| 0.25 | 35 | 75 | 98 |
| 0.10 | 32 | 70 | 95 |
| Silt    0.05 | 30 | 60 | 90 |
| 0.02 | 18 | 40 | 85 |
| 0.005 | 15 | 30 | 80 |
| 0.002 | 10 | 15 | 60 |
| Clay   0.001 | 2 | 5 | 20 |

See Supplement 2–1 for a discussion of these data. The table shows percentages finer
than the given size (diameter). The sandy loam has 30 percent finer than 0.05, or 70
percent sand. It has 10 percent finer than 0.002 mm = 10 percent clay and 30 to 10 percent
= 20 percent silt content.

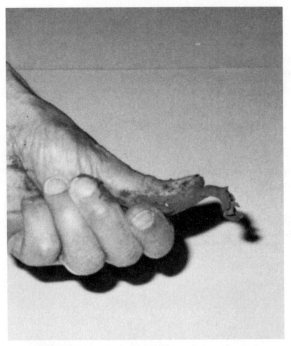

**FIGURE 2–7**  Example of a "ribbon" made
when determining soil texture. The sample
has 6 percent sand, 34 percent silt, and 60
percent clay.

the aggregates have rounded corners, they are **sub-angular blocky,** and if the corners are not rounded, they are **angular blocky.** Blocky structure is often found in B horizons. Aggregates that are longer than they are wide are prisms, and the soil structure is **prismatic.** Prisms have angular edges and tops. When the tops and sides are rounded, the aggregates are columns, and the structure is **columnar.** In some soils, particularly in compacted horizons and in E horizons, the aggregates are thin, flat, and platelike, and the structure is **platy.**

Some soils have no structure and thus are *structureless*. Structureless soils are divided into two groups: massive and single grained. A massive structureless condition means that all of the particles are stuck together with no aggregates and no natural cleavage planes. The entire pedon may be one coherent mass of material. By contrast, sand dunes and beach sands are structureless and single grained, each particle acting as an individual.

### 2.2.2  Aggregate Size

Table 2–4 shows the size ranges for the different structure types and the standard names given to the sizes by the USDA.

**TABLE 2–3**

Guidelines for Estimating Texture by Feel

| Texture | Guidelines |
|---|---|
| Sand and loamy sand | Individual grains easily seen and felt; when squeezed moist, forms a cast that crumbles when touched |
| Sandy loam | When squeezed moist, forms a cast that can be gently handled |
| Loam | Somewhat gritty feel but fairly smooth; when squeezed moist, forms a cast that can be freely handled without breaking |
| Silty loam | Soft and floury when dry; forms a cast when dry or moist, but when squeezed between thumb and forefinger, will not ribbon when moist |
| Clay loam | Forms a thin ribbon that barely sustains its own weight; moist soil is plastic and forms a cast that can be handled |
| Clay | Sticky and plastic when wet; forms a strong ribbon |

Source: *USDA-NRCS Handbook of Soil Survey Investigations, field procedures* I-2.5-1. Web site: http://www.statlab.iastate.edu/soils/nssh/. Soil Survey Staff, Natural Resources Conservation Service, National Soil Survey Handbook, Title 430-VI. Washington, D.C., U.S. Government Printing Office, September 1999.

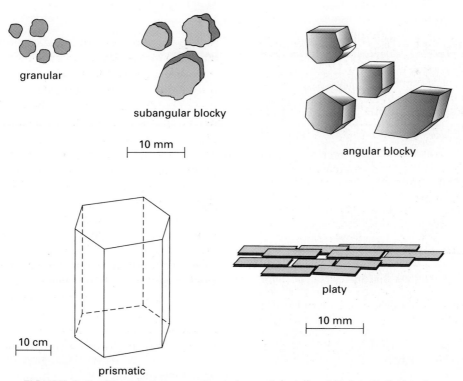

**FIGURE 2–8** Structure types: Granular, subangular blocky, and angular blocky are common ped shapes. Prismatic and platy are less common. Note the size and shape of each ped type.

## SOIL AND ENVIRONMENT
### The Relevance of Soil Texture, Structure, and Depth

Soil texture (particle size distribution), structure (particle arrangement), and depth of pedon all influence the flow and quality of underground and surface water and the environmental behavior of hazardous chemicals. These three soil properties largely control the land's ability to

- Take in, retain, and release water
- Maintain plant cover and resist erosion
- Retain, destroy, or leak hazardous materials from spills, waste disposals, pesticides, and fertilizers

The extensive charged surfaces of clay minerals and humus give soils the ability to retain toxic compounds, as well as water and plant nutrients, which would otherwise readily leave the soil and pass into groundwater. So fine-textured soils, with high clay content, usually retain nutrients and toxins better than do coarse-textured soils. Likewise, retention tends to increase with increasing soil thickness. But few soils effectively retain loosely held ions such as nitrate and chloride, which accordingly move readily to groundwater.

In all but the coarsest soils, structure (particle arrangement) determines the size and continuity of pore spaces between the particles and aggregates. These pores control the flow of water, petroleum, or any other fluid into and through the soil.

Soil structure, unlike texture, is readily altered by management, especially at the ground surface, with both positive and negative effects on fluid flow rates. For instance, drought, cultivation, overgrazing, clearing, and traffic—by animals, people, or machines—can lead to compaction and crust formation, which impede the entry of water. Water that does not enter the soil profile runs off the surface, carrying eroded soil and attached substances and microbes. Soil erosion is a major source of nonpoint pollution of lakes and waterways, especially with mineral sediments, humus, phosphate, and other pollutants that are so well retained that they seldom appear in underground waters. Nonpoint pollution, in contrast with point pollution, has a diffuse source.

## TABLE 2–4
Names and Sizes of Structure Types (millimeters)

| Size class | Diameter of granules | Thickness of plates | Diameter of blocks | Diameter of prisms and columns |
|---|---|---|---|---|
| Very fine[a] | <1 | <1 | <5 | <10 |
| Fine[a] | 1–2 | 1–2 | 5–10 | 10–20 |
| Medium | 2–5 | 2–5 | 10–20 | 20–50 |
| Coarse[a] | 5–10 | 5–10 | 20–50 | 50–100 |
| Very coarse[a] | >10 | >10 | >50 | >100 |

Source: Soil Survey Staff. 1993. Soil Survey Manual. Supt. of Documents, U.S. Government Printing Office, Pittsburg, P.A. Table 3–13
http://www.statlab.iastate.edu/soils/ssm/gen_cont.html
[a]Platy structure is described as thin or thick rather than fine or coarse.

### 2.2.3 Aggregate Grade

**Structure strength** or grade refers to (1) how easy it is to see the structure in the pedon, (2) how many of the primary particles are aggregated, and (3) how strongly the primary particles are held together in the aggregates.

There are three visual **structure grades.** *Weak structure* is difficult to see in a horizon. When the horizon is poked with a knife, the particles that fall out are mostly not aggregated and the aggregates present break with little handling. *Moderate structure* can be seen in the horizon, and most of the material removed from the horizon is aggregated. *Strong struc-*

*ture* can easily be seen in the pedon; most of the primary particles are aggregated, and the aggregates do not readily fall apart when handled. When a soil's properties are described in the field, the structure is described for each horizon. Structure differences help define horizons.

### 2.2.4  Mode of Formation

When the moist primary particles of sand, silt, and clay size are squeezed together, they often stick, usually weakly. This is the beginning of structure formation.

In soils, swelling and shrinking, freezing and thawing repeatedly push particles together and pull them apart. With some bonding (by dried clay, for example), the developing units may become difficult to pull apart, and the result is an aggregate.

Other natural processes that help form structure are the growth of roots and the burrowing of small animals. As a root or earthworm extends into soil, it pushes primary particles to the side, mixing them with gums and compressing them. Animals such as earthworms also ingest soil particles, mix them with organic products, and defecate the castings.

Aggregates that hold together are stable aggregates. The stability of aggregates usually refers to their resistance to destruction by water. To measure it, weighed aggregates, on a sieve, are repeatedly dunked in water; the fraction remaining after 3 minutes is a conventional measure of stability.

Stabilizing agents include silicate clay, organic materials, and oxides of Fe and Al. Soils lacking these constituents have weak structure or none. The charged clay surfaces can interact with each other and with sand and silt particles to form aggregates. Oxides of Fe and Al also help link particles. Some oxides have positive charge and bond strongly to particles with negative charge. Other oxides are not charged but build up tough coatings that connect other particles. Large organic molecules also form bridges between inorganic particles, either with electrostatic bonds or simply by linking particles like a net. Polysaccharides—long-chain sugar polymers—decompose quickly in soil. But more resistant organic polymers associated with metal cations such as iron are persistent and responsible for much aggregate stability. On a larger scale, fungi and roots link particles in the same way. In A horizons, the stable aggregates, soil structure, and dark color all depend on humus.

Figure 2–9 outlines a conceptual model in which large aggregates (>2 mm in diameter) are made of smaller aggregates, mostly 20–250 μm in diameter. These are stabilized by a network of roots and persistent organic materials derived from roots, fungi, and bacteria. Aggregates of this size are in turn composed of particles 2–20 μm in diameter, also held together by persistent organics. Below 2 μm, electrostatic interactions and organic compounds bind particles.

## 2.3  SOIL MINERALS

### 2.3.1  General Properties

Oxygen and silicon are the two most abundant elements in the earth's crust—46 and 28 percent by weight, respectively. These two elements, along with lesser amounts of others, combine in different ways to form an array of minerals, differing in composition and crystallinity.

Composition describes the kinds of atoms in the mineral; crystallinity describes the degree and manner of organization of the atoms. Most minerals are highly crystalline; their atoms are arranged in a definite order that repeats extensively in three dimensions. But some are amorphous short-range order minerals that lack a definite long-range atomic pattern. The properties of a mineral can depend greatly on both its composition and the kind and degree of crystallinity.

The kinds of minerals in a soil influence its fertility, its physical properties, and the direction and rates of its further development. The minerals in the soil weather—that is, undergo further changes in composition and crystallinity over time (see Chapter 12).

**Silicate** minerals are the main constituents of all particle size classes in soil. The silicon atoms in silicate minerals can be regarded as positively charged (+4) cations, and the oxygen atoms can be regarded as anions with a negative charge (−2). The silicon and oxygen atoms combine to form a unit called a *silicon tetrahedron*. It is called a **tetrahedron** because it has four sides, like a pyramid (Figure 2–10). The silicon tetrahedron is the main building block of common minerals because silicon and oxygen are abundant and their ionic sizes allow the four oxygen atoms to fit nicely around the central silicon atom and to bond firmly.

FIGURE 2–9 Model of aggregate organization with major binding agents indicated. (From Tisdall, J. M., and J. M. Oades. 1982. Organic matter and water stable aggregates. *J. Soil Sci.* 33:141–163.)

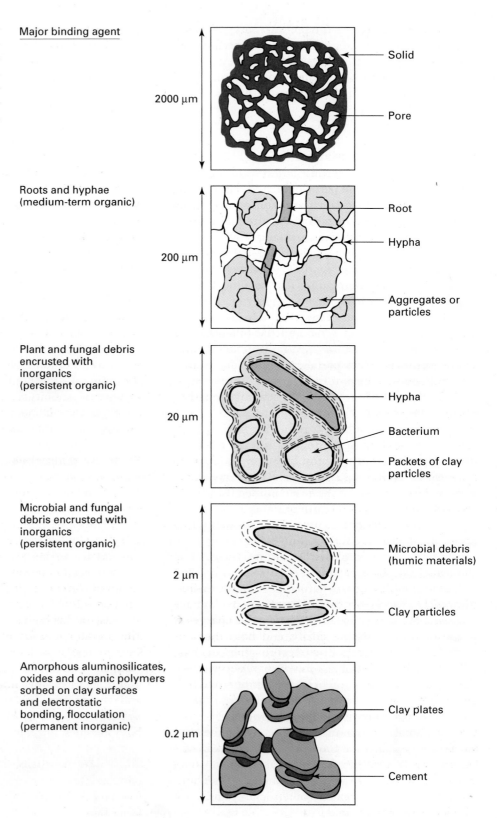

Major binding agent

2000 μm — Solid, Pore

Roots and hyphae
(medium-term organic)

200 μm — Root, Hypha, Aggregates or particles

Plant and fungal debris
encrusted with
inorganics
(persistent organic)

20 μm — Hypha, Bacterium, Packets of clay particles

Microbial and fungal
debris encrusted with
inorganics
(persistent organic)

2 μm — Microbial debris (humic materials), Clay particles

Amorphous aluminosilicates,
oxides and organic polymers
sorbed on clay surfaces
and electrostatic
bonding, flocculation
(permanent inorganic)

0.2 μm — Clay plates, Cement

The $SiO_4$ grouping is negatively charged. If it is to be stable, it must be electrically neutral. Depending on the type of mineral, neutrality is achieved by the sharing of oxygen ions among tetrahedral units and by the incorporation of "accessory" cations in addition to silicon.

**Accessory cations** and oxygen sharing have important consequences. The accessory cations, mostly $Fe^{3+}$, $Fe^{2+}$, $Ca^{2+}$, $Mg^{2+}$, $Na^+$, and $K^+$, influence weatherability and soil fertility and lead to the property of cation exchange (see Chapter 3). The sharing of oxygen ions in different ways and different degrees produces very different mineral structures, consisting of single tetrahedra chains, layers, or frameworks of tightly linked tetrahedra, (Figure 2–10).

## 2.3.2  Silicate Minerals of Sand and Silt

Silicate minerals, those that contain the element silicon (Si), are abundant in rocks, and the most resistant of them also dominate the sand and silt in soils. The simplest structure of a silicate mineral is a single tetrahedron and its accessory cation (Figure 2–10). Olivine, $(Mg,Fe)_2SiO_4$, is a mineral with this structure. Olivine can vary from pure $Fe_2SiO_4$ to pure $Mg_2SiO_4$. Because of its structure and composition, olivine is the most weatherable silicate mineral in most rocks and in soils. Thus, there is little olivine in soils, unless the soils are very young and have been formed from fresh, unweathered rocks. Yet zircon $(ZrSiO_4)$, with similar structure, is one of the minerals most resistant to chemical weathering.

In most silicates, the tetrahedra are not single, as in olivine, but joined through shared oxygen atoms, to form complex crystal structures (Figure 2–10). Most include accessory cations. In pyroxenes and amphiboles, Mg and Fe balance the negative charges of single or double silicate chains and hold them together in linear crystals. These ferromagnesian chain silicates, formed in cooling molten rock, persist in the sand and silt of many soils that have undergone only slight weathering.

With further oxygen sharing (Figure 2–10), silica tetrahedra form the basic sheet structures of the **layer silicates.** These include the clay silicates. Readily visible examples are the **mica** minerals, of which **muscovite,** white mica, and **biotite,** black mica, are common. These are the thin sheets that pack together as "books" of mica flakes and that were once used as isin-glass windows for stoves and wagons. The mica minerals often weather directly into clay minerals. If you look at the formulas for the micas and for the clay minerals in Supplement 2–3, you will see the similarities.

The most complex silicate minerals are those that have complete three-dimensional framework silicate structures in which all the oxygen atoms are shared among tetrahedra (Figure 2–10). Quartz and the feldspars are the most important examples of minerals with a continuous silicate structure. **Quartz** is $SiO_2$. All the bonds are Si–O, and there is no need for accessory cations to balance the charge; therefore it is hard and durable. Much of the sand in the world is quartz because it weathers slowly. Many soils start with considerable quartz because it is an important rock-building mineral. Quartz persists even in intensely weathered soils.

The basic structure of the **feldspar** mineral group is also three-dimensional. However, in addition to Si, the feldspars have $Al^{3+}$ substituted in some tetrahedra, balanced by $Na^+$, $K^+$, or $Ca^{2+}$ as accessory cations. The two most common groups of feldspar minerals are orthoclase (K feldspar) and plagioclase (Na, Ca feldspar). Feldspars are more weatherable than quartz because of the K, Na, and Ca needed to balance the charge. They are a source of these plant nutrients.

**Nonsilicate minerals**   Nonsilicate minerals do not have Si as part of their structure. Common salts of Na and Ca are soil minerals, and in areas where evaporation is greater than precipitation, salts such as $NaCl$, $Na_2SO_4$, $CaSO_4$ (gypsum), and $CaCO_3$ (calcite) or $MgCO_3$ can accumulate. These can also affect plant growth (see Chapters 10 and 11). Soils formed from limestone inherit Ca from the parent material. Calcite, gypsum, and the other simple salts are very soluble in water, compared with the silicates. Because of this high solubility, they are important to the overall chemistry of the soils in which they are found. These salts sometimes form visible concretions in soil but are more often found as clay-sized (less than 0.002 mm) minerals. Apatite is another key nonsilicate because it is one of the few minerals in soil to provide phosphorus (P) for plants.

## 2.3.3  Minerals of the Clay Fraction

General properties   Clay minerals have some common properties and important differences:

| Arrangement of SiO$_4$ tetrahedra (central Si$^{4+}$ not shown) | Unit composition | Mineral example |
|---|---|---|
| 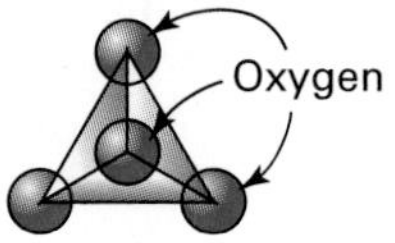 Oxygen | $(SiO_4)^{4-}$ | Olivine, $(Mg, Fe)_2SiO_4$ |
| 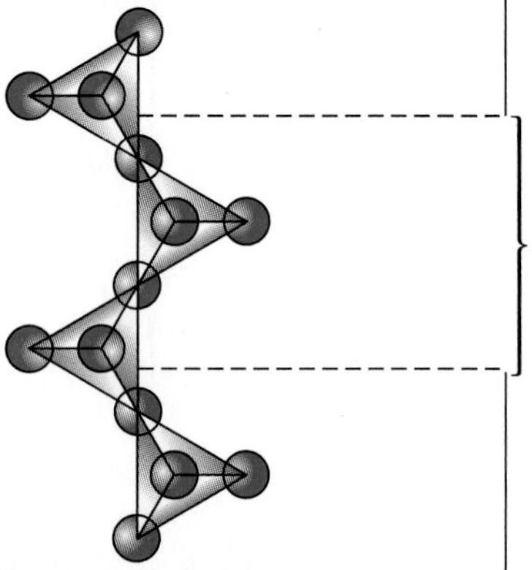 | $(Si_2O_6)^{4-}$ | Pyroxene e.g. Enstatite, $MgSiO_3$ |
| 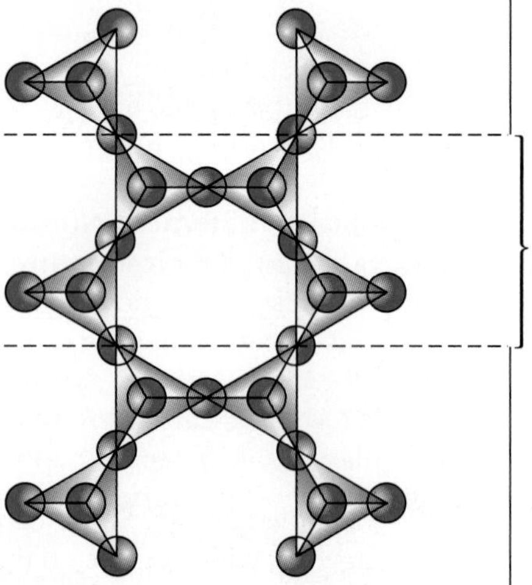 | $(Si_4O_{11})^{6-}$ | Amphibole e.g. Anthophyllite, $Mg_7Si_8O_{22}(OH)_2$ |

**FIGURE 2–10** Structures of some silicate minerals. The simplest structure is the single tetrahedron. Tetrahedra join to form single chains. Single chains combine to form double chains, and double chains join together to form sheets. Finally, all oxygens are shared among tetrahedra in three dimensions. The negative charge of the unit cell is balanced by accessory cations (not shown). (From Klein, C., and C. S. Hurlbut. 1993. *Manual of mineralogy.* 23rd ed. New York: Wiley.

| Arrangement of SiO₄ tetrahedra (central Si⁴⁺ not shown) | Unit composition | Mineral example |
|---|---|---|
| 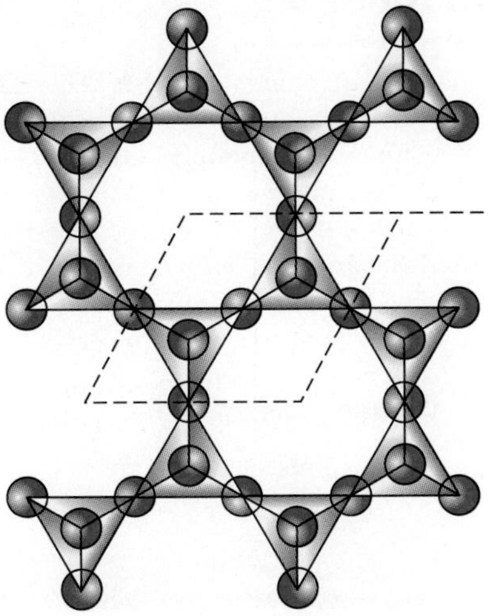 | $(Si_2O_5)^{2-}$ | Mica e.g. Phlogopite, $KMg_3(AlSi_3O_{10})(OH)_2$ |
| 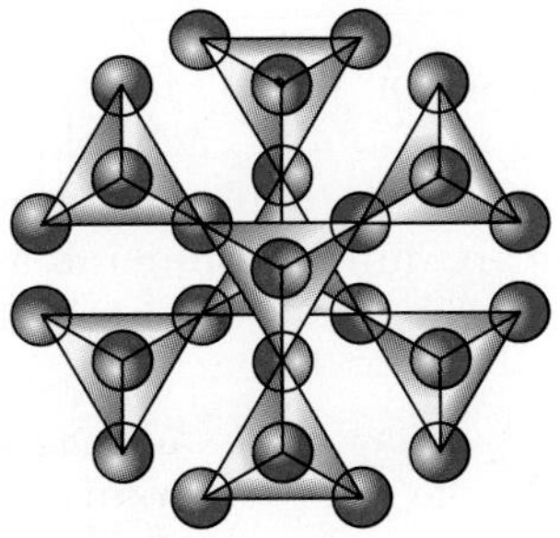 | $(SiO_2)^0$ | High cristobalite, $SiO_2$ |

**FIGURE 2–10** *Continued*

1. They are colloidal ($<2\ \mu m$), with a resulting large surface.
2. They are platy or flaky microcrystals (with some needle-shaped or amorphous exceptions), reflecting their layered crystal structures. The shape and size explain clay's slipperiness and plasticity when wet and the tendencies of clay particles to stack and stick together in loose "domains" and aggregates, to coat larger particles, and to line pores.
3. Most clay crystals have a built-in negative electrical charge, balanced by accessory cations, which at the crystal surface can interchange with cations from the surrounding soil solution. Types of clay differ in the amount of built-in negative charge. Some clays (and humus) also have variable surface charge depending on the pH of the surrounding solution, and this can become positive. Regardless of the source, surface charge must be balanced by oppositely charged ions near the surface.
4. All clays sorb water on their surface, but some also allow water to enter between the layers within the crystal. These expanding-lattice clays, and the soils they dominate, shrink and swell markedly depending on water content.

These commonalities and differences can mostly be explained by the chemical structure and composition of the minerals.

## BOX 2–1
## Isomorphous Substitution and Charge in Minerals

As a mineral crystal forms, ions present in large concentrations may substitute for others, if they can occupy similar space so as not to disrupt the growing crystal's structure. Thus $Al^{3+}$ can substitute for $Si^{4+}$ in oxygen tetrahedra, and $Fe^{3+}$, $Fe^{2+}$, and $Mg^{2+}$ can substitute for $Al^{3+}$ in oxygen octahedra. This is **isomorphous substitution.** It can have interesting consequences:

1. Substitution creates a range of mineral species. Because they form under different conditions, their presence in a soil can help us to reconstruct the soil's history.
2. Substitution usually alters the crystal's electrical charge. Once the crystal is formed, its constituents

do not change until it is weathered. Hence the charge built into the crystal lattice by substitution is a fixed, "permanent" charge.

3. Substitution usually results in a net negative charge that must be balanced by accessory cations, also built into the lattice. However, in the layered microcrystals of aluminosilicate clays, much of the accessory cations are at the crystal surface, where they are easily replaced or exchanged with other cations from the soil solution.

Discussions of exchangeable cations will recur in Chapters 3, 9, 10, 11, and 12.

**Aluminosilicate clays**   Most clay minerals are **aluminosilicates,** the major elements in their structure being oxygen, silicon, and aluminum.

Aluminum is the third most abundant element in the earth's crust (8.1 percent) after oxygen and silicon. Aluminum has three missing electrons ($Al^{3+}$) and attracts six oxygen atoms, forming an eight-sided octahedron (Figure 2–11). The silicon tetrahedron and aluminum octahedron are the basic building blocks of the clay minerals.

The silicon tetrahedra join in sheets by sharing oxygen atoms (Figure 2–11), and the aluminum octahedra do the same. These tetrahedral and octahedral sheets can be sandwiched together to form layers, which can be stacked on top of one another to form the crystals of the different clay minerals. The sheets are held together by the oxygen atoms shared between the sheets. The aluminosilicate clays, like the mica minerals, are called layer silicates.

Aluminosilicate clay minerals are divided into groups based on the number of sheets of Si tetrahedra and Al octahedra in their layers. These groups are the **kandites,** which have one silicon sheet and one aluminum sheet; the **smectites,** which have two silicon sheets for every aluminum sheet; **vermiculite,** which is also a 2:1, Si:Al mineral but which has properties different from those of the smectites; and **chlorite,** which has two silicon sheets and one aluminum sheet plus a sheet of magnesium atoms between adjacent silicon sheets (Table 2–5). This may seem confusing now, but as each mineral is explained and as you study the figures, it should become clear why these are different minerals and why the differences are pertinent to the behavior of soils.

*Kandites (kaolinite group)*   The simplest of the silicate clay mineral groups is the kandites. These 1:1 minerals have one silicon tetrahedral sheet connected to one aluminum octahedral sheet. Kaolinite, one member of the kandite family, has an equal number of aluminum and silicon atoms that correspond to the one sheet of aluminum atoms and one sheet of silicon atoms (see Figure 2–11). There also are oxygens and **hydroxyls.** Several questions need to be answered. Why are there so few atoms in this formula if there are sheets of silicon and aluminum atoms? Do four atoms make a sheet? Where do the hydroxyl molecules fit in the structure? The formula in Table 2–5 is for a unit cell of the kaolinite molecule. A **unit cell** is the minimum unit that is repeated in orderly fashion in three dimensions to form the crystal structure. In a single crystal of kaolinite, there may be millions of unit cells linked together horizontally and vertically. The hydroxyls are needed to balance the structure electrically. A detailed kaolinite structure is shown in Supplement 2–3.

A single layer of kaolinite is only 0.71 nm (nanometers, $10^{-9}$m) thick, but it may be many

**(A) ATOMS AND BUILDING BLOCKS**

central small "cations"–Si, Al, etc ● ◦ larger coordinating oxygen and/or OH

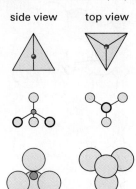

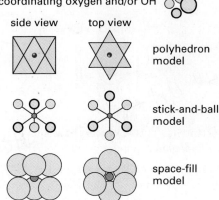

| side view | top view | | side view | top view | |
|---|---|---|---|---|---|

polyhedron model

stick-and-ball model

space-fill model

**tetrahedron**
4 oxygen atoms (or OH ions)
around central Si (or Al)

**octahedron**
6 oxygen atoms and/or OH ions
around central Al (or Mg, Fe . . .)

**(B) SHEETS**

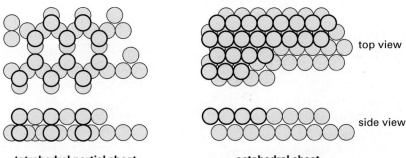

top view

side view

**tetrahedral partial sheet**
(part of silicate unit layer)

**octahedral sheet**
(part of silicate unit layer)
(whole unit layer of hydrous oxides)

**FIGURE 2–11** Summary of aluminosilicate clay structures. (A) Building blocks: Oxygen, OH, or $H_2O$—each 0.3 nm diameter—coordinate around smaller atoms of Si and Al, forming the two basic building blocks: the Si–O tetrahedron and the Al–O, OH octahedron. These units are represented in three ways: as polyhedra, as stick-and-ball drawings showing positions of atom centers and bonds, or as space-fill (sphere-packing) drawings indicating volumes filled by oxygen electron shells. (Parentheses—(Al), (Mg, Fe)—indicate possible isomorphous substitutions.) (B) Sheet structures: These are formed by Si–O tetrahedra, each sharing three of their oxygens, or by octahedra sharing all six of their OH or O. Sheets combine to form layers.

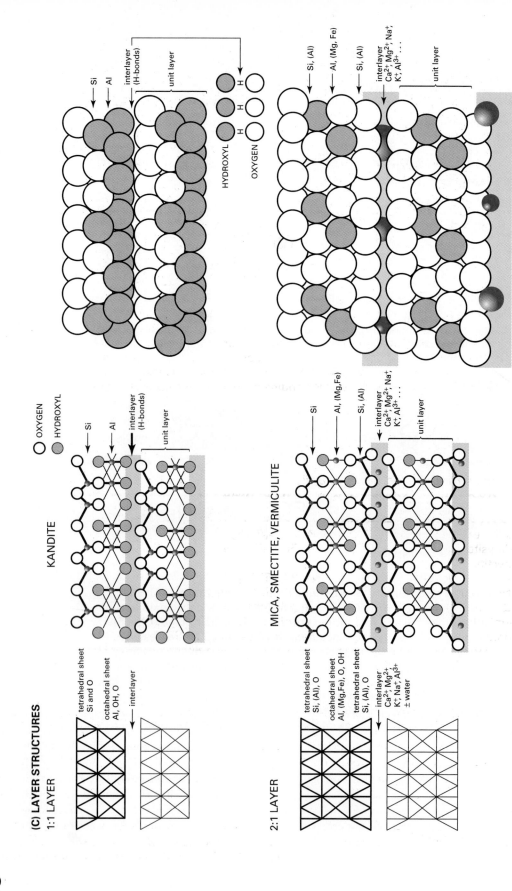

**(C) LAYER STRUCTURES**

**1:1 LAYER**

KANDITE

○ OXYGEN
● HYDROXYL

tetrahedral sheet
Si and O
octahedral sheet
Al, OH, O

interlayer

Si
Al
interlayer
(H-bonds)
unit layer

HYDROXYL
OXYGEN

H  H
H
H  H

Si
Al
interlayer
(H-bonds)
unit layer

**2:1 LAYER**

MICA, SMECTITE, VERMICULITE

tetrahedral sheet
Si, (Al), O
octahedral sheet
Al, (Mg,Fe), O, OH
tetrahedral sheet
Si, (Al), O
interlayer
Ca²⁺ Mg²⁺,
K⁺; Na⁺, Al³⁺
± water

Si
Al, (Mg,Fe)
Si, (Al)
interlayer
Ca²⁺ Mg²⁺; Na⁺,
K⁺; Al³⁺ ...

Si, (Al)
Al, (Mg, Fe)
Si, (Al)
interlayer
Ca²⁺ Mg²⁺; Na⁺,
K⁺; Al³⁺ ...
unit layer

**FIGURE 2–11** *Continued.* (C) Layer structures: The two basic types, 1:1 and 2:1, are shown. Each is represented (left to right) as polyhedral, stick-and-ball, and space-fill drawings, each depicting a side view of two unit layers and the interlayer space between them. The 1:1 layer, about 0.7 nm ($0.7 \times 10^{-3}$ μm) thick, has one Al–O, OH octahedral sheet laminated with one tetrahedral Si–O sheet. Scores or hundreds of these layers, stacked and hydrogen bonded, form a crystal particle of kandite clay. The 2:1 layer, 1.4 nm thick, has one octahedral sheet sandwiched between two tetrahedral sheets, commonly with much isomorphous substitution. Crystals of 2:1 layer silicates require accessory cations to neutralize negative charges and bond the layers. Different kinds of 2:1 minerals—smectites, vermiculite, micas—have different kinds of substitution and interlayers.

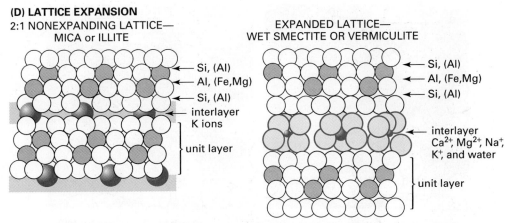

**(D) LATTICE EXPANSION**

2:1 NONEXPANDING LATTICE—
MICA or ILLITE

EXPANDED LATTICE—
WET SMECTITE OR VERMICULITE

← Si, (Al)
← Al, (Fe,Mg)
← Si, (Al)
← interlayer K ions
unit layer

← Si, (Al)
← Al, (Fe,Mg)
← Si, (Al)
← interlayer Ca²⁺, Mg²⁺, Na⁺, K⁺, and water
unit layer

**FIGURE 2–11** *Continued.* (D) Lattice expansion/swelling in 2:1 clays: Smectite and vermiculite swell when wet because water can enter their interlayer spaces, hydrating the interlayer cations. Soils rich in these clays shrink, swell, and crack markedly with wetting and drying. Mica does not expand because its dominant interlayer cation, $K^+$, coordinates perfectly with cavities on the opposing Si–O sheets, locking the layers together. Kandites do not expand because of H bonding between layers (likewise chlorite—see text).

## TABLE 2–5

Formulas for Selected Clay Minerals

| Mineral | Formula |
|---|---|
| Kaolinite | $Al_4Si_4O_{10}(OH)_8$ |
| Halloysite | $Al_4Si_4O_{10}(OH)_8 \cdot 4H_2O$ |
| Smectite | |
|    Montmorillonite | $(Al_3Mg)Si_8O_{20}(OH)_4$ |
|    Nontronite | $Fe_4(Si_7Al)O_{20}(OH)_4$ |
|    Saponite | $Mg_6(Si_7Al)O_{20}(OH)_4$ |
| Vermiculite | $Mg(Al, Fe, Mg_4)(Al_2Si_6)O_{20}(OH)_4 \cdot nH_2O$ |
| Chlorite | $Mg_6(OH)_{12} \cdot (Al, Mg_5)(Al_2Si_6)O_{20}(OH)_4$ |

nanometers long or wide. Nearly 3,000 of these 0.71-nm units are needed to make a particle of 0.002 mm clay (Figure 2–12). What holds these layers together to make crystals? For kaolinite, hydroxyl groups on the face of one unit layer share hydrogen atoms with the oxygen atoms on the face of the adjacent layer. Although they are individually weak, the billions of these hydrogen bonds acting together hold the layers together tightly, not permitting water and cations to enter the interlayer spaces in kaolinite.

Kandites are a group or family of clay minerals that includes nacrite, dickite, and halloysite, as well as kaolinite. They differ in composition, with different degrees of substitution of Al for Si and of Fe or Mg for Al.

Halloysite is another kandite. It has one sheet of silicon tetrahedra linked to one sheet of aluminum octahedra, but in the interlayer space between the layers of Si and Al is one layer (sometimes two layers) of water molecules. (Because of the water, a single halloysite layer is thicker than a kaolinite layer—

**FIGURE 2–12**   Kaolinite flakes. (From Dixon, J. B. In *Minerals in Soil Environments,* ed. J. B. Dixon and S. B. Weed, Chapter 10.)

**FIGURE 2–13**   Electron photomicrograph of halloysite crystals. (Photo: Kenneth Towe, Smithsonian Institution, Department of Paleobiology.)

1.025 nm rather than 0.71 nm.) The "sheets" roll to form distinctive tube-shaped crystals (Figure 2–13).

*Smectites*  The smectites are 2:1 layer aluminosilicates. Each aluminum octahedral sheet is sandwiched between two silicon tetrahedral sheets to make a unit layer. Montmorillonite is the most

prominent member of the smectite group. Isomorphous substitution of $Al^{3+}$ for $Si^{4+}$ in the tetrahedral sheet and $Fe^{2+}$, $Fe^{3+}$, $Mg^{2+}$, and other cations for $Al^{3+}$ in the octahedral sheet produces the other members of the group, such as nontronite (Fe substitution), saponite (Mg substitution), and hectorite (Li substitution).

## BOX 2–2
## Aluminosilicate Clays

### Kandites

- 1:1 layer silicates
- 1 Si tetrahedral sheet
- 1 Al octahedral sheet
- Little isomorphous substitution
- Interlayer closed to water and cations
- Not very reactive
- Little shrink-swell

### Smectites and Vermiculite

- 2:1 layer silicates
- 2 Si tetrahedral sheets
- 1 Al octahedral sheet
- Much isomorphous substitution
- Interlayer open to water and cations
- Very reactive
- Much shrinking when dry and swelling when wet

Compared with kandites, smectites have twice as many Si atoms in the unit layer, twice as many O atoms, and half the number of OH atoms. Like halloysite, the smectites have water molecules in the interlayer, but in smectites the number of water molecules varies, making the smectite unit layer variable in thickness. Smectite spacing may vary from 0.96 to 1.8 nm or more.

The variable spacing implies that smectites can change volume, and they can. A soil that has montmorillonite or vermiculite as the dominant clay shrinks when dry and swells when wet because the clay crystal can expand and contract as water enters and leaves the interlayer spaces. Both water and cations can enter the interlayer space in the smectites but not in the kandites. The reason is found in the structure. In the 2:1 clays, Si tetrahedra face each other across the interlayer space. The oxygen atoms at the surface of the unit cells face each other and do not engage in hydrogen bonding. This lack of hydrogen bonding means that the unit layers are held together weakly and that water can enter between them. The number of water molecules and the kind of cations associated with the water determine the layer's thickness.

Swelling is also influenced by the location of isomorphous substitution. If you add the positive and negative charges in kaolinite or pyrophyllite, the balance is zero (Table 2–6). With smectites, such as montmorillonite, the substitution of an $Al^{3+}$ for a $Si^{4+}$ in the tetrahedral sheet and a $Mg^{2+}$ for an $Al^{3+}$ in the octahedral sheet results in a net negative charge of $-2$ per layer (for a review of charge, see Supplement 2–4). These charges are satisfied by interlayer accessory cations. If the source of negative charge is predominantly in the octahedral sheet, it will be farther from the surface of the mineral than if it were in the tetrahedral sheet. The strength of attraction to the charge decreases rapidly with distance; thus a charge originating in the octahedral sheet is less capable of holding the units together across the interlayer space than is a charge originating in the tetrahedral sheet.

Why isomorphous substitution occurs in the smectite clays and not in the kandites and why the smectites are 2:1 rather than 1:1 layer clays are still matters of speculation. However, one explanation is that smectites form under conditions in which much silicon, more than twice as much as aluminum, and many cations are present. Kandites, on the other hand, form in soils in which the ratio of Si to Al is closer to 1:1, and few cations are available to substitute into the clay structure or balance the charges that would result from the substitute ions. Climatic conditions that favor leaching of cations favor kandite formation.

*Vermiculite* Like the smectites, vermiculite is a 2:1 expanding clay mineral (Supplement 2–3). Substitution of $Al^{3+}$ for $Si^{4+}$ in the tetrahedral sheet and of $Fe^{3+}$, $Fe^{2+}$, and $Mg^{2+}$ for $Al^{3+}$ in the octahedral sheet is extensive. The interlayer cations in vermiculite commonly include complexes of Mg and Al with OH and water. The layer spacing of vermiculite can

## TABLE 2–6

Charge Balance for a Unit Layer of Kaolinite and Montmorillonite

| Kaolinite | | | | Montmorillonite | | | |
|---|---|---|---|---|---|---|---|
| Number of atoms | Atom | Atomic charge | Layer charge | Number of atoms | Atom | Atomic charge | Layer charge |
| 4 | Al | +3 | +12 | 4 | Al | +3 | +12 |
| 4 | Si | +4 | +16 | 1 | Mg | +2 | +2 |
| 10 | O | −2 | −20 | 7 | Si | +4 | +28 |
| 8 | OH | −1 | −8 | 20 | O | −2 | −40 |
| | | | | 4 | OH | −1 | −4 |
| | | Balance | 0 | | | Balance | −2 |

range from 1.0 to 1.5 nm or more. Vermiculite expands less and takes less water into the interlayer than smectite, because more of the substitution is in the tetrahedral Si layer than in the octahedral layer. This puts the negative charge and the interlayer cations closer together, so hydrogen bonding holds layers together more tightly.

*Illite*    Illite, another 2:1 clay mineral, is sometimes called hydrous mica because it is believed to be a weathering product of mica. There is at least one nonexpanding clay mineral with potassium (K) in the interlayer, as in mica. The interlayer K contributes to the stability and nonexpanding nature of illite and mica (because it coordinates perfectly with O atoms in the tetrahedral sheets).

*Chlorite*    Chlorite is a 2:1:1 nonexpanding mineral. A magnesium hydroxide sheet—$Mg_6(OH)_{12}$ brucite—occupies the interlayer space between 2:1 sheets (Supplement 2–3). The 2:1 silicon and aluminum structure is similar to the 2:1 structure in the smectites and vermiculite, the major difference being the brucite sheet. Aluminum, $Fe^{3+}$, and $Fe^{2+}$ can substitute for $Mg^{2+}$ in the brucite sheet, giving it a net positive charge that promotes the formation of strong bonds between the layers. Chlorite has a 1.4-nm-thick layer.

**Nonsilicate clays**    Some soils contain an abundance of clay-sized minerals that lack silicon but have some of the properties of silicate clays. These are **sesquioxides** or **hydrous oxides** of iron and aluminum. They may be well-crystallized minerals, com-

pletely noncrystalline (amorphous), or poorly crystalline. The crystalline types have some properties in common with the silicate clays. The amorphous varieties often have high cation or anion exchange properties. Silica is also found as amorphous compounds in soils. The amorphous minerals are sometimes given the general name **allophane,** which is thought to be a gel (jellylike substance) in soils.

The crystalline hydrous oxides of iron ($Fe_2O_3$, hematite, and FeOOH, goethite) and aluminum ($Al(OH)_3$, gibbsite, and AlOOH, boehmite) are common soil minerals. The iron minerals give soils and rusting iron their yellow, red, and dark reddish brown colors. They are made of octahedral building blocks with O or OH coordinated to the $Fe^{3+}$ or $Al^{3+}$. The octahedra share oxygen atoms to form sheets that are stacked together and held with hydrogen bonds. Oxides contribute to stabilizing soil structure and they are charged.

Hydrous oxides, whether crystalline or amorphous, do not get their charge from isomorphous substitution but from surface **protonation** and **deprotonation.** Hydrogen ions ($H^+$) are added to the structure (protonation) or are released (deprotonation), depending on the concentration of $H^+$ in the soil solution. These minerals have a **variable charge,** or a pH-dependent charge. An example of the reaction is as follows. The lines to the Al indicate that it is part of an octahedral sheet.

$$\diagdown AlO^- + H^+ \leftharpoondown \diagdown AlOH + H^+ \leftharpoondown \diagdown AlOH_2^+$$
$$\diagup \text{negative} \quad \rightarrow \diagup \text{neutral} \quad \rightarrow \diagup \text{positive}$$

pH decreasing $\rightarrow$
Hydrogen ion concentration increasing $\rightarrow$

# BOX 2–3
# Fixed and Variable Charge

There are two major sources of charge in soil colloids:

1. Fixed charge caused by the isomorphous substitution of one cation for another in a clay mineral's structure (also called permanent or lattice charge).

2. Variable charge due to protonation and deprotonation (pH-dependent charge).

Fixed charge predominates in 2:1 silicates; pH-dependent charge predominates in hydrous oxides of Fe and Al and in humus. A minor source of charge is at edges of crystals.

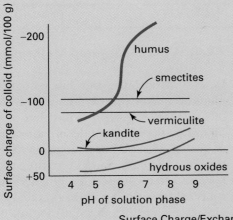

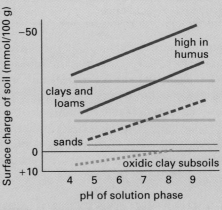

Surface Charge/Exchange Capacities of Colloids and Soils

Layer aluminosilicates have a small proportion of variable charge because of the OH at the edges. All the negative charge on humus is variable. Oxides and humus are positively charged in some very acidic soils and help retain anions.

**Important clay properties**  Two properties of clay minerals make them important to soils: charge and surface area (see Table 2–7). The net effect of the presence of edges, isomorphous substitution, and variable charge is that clay minerals have a negative (and sometimes positive) charge. Electrically charged substances do not exist alone for long in nature, and positively charged materials (cations, organic compounds, charged molecules) are attracted to clays to balance the negative charge. These positively charged materials are not held too tightly, however, but become the exchangeable cations and compounds discussed in the next chapter. The capacity of clays to hold and release (or exchange) what they are

holding for other positively charged ions is an important property of soils.

The importance of mineral and organic surfaces cannot be overemphasized. It is on the surfaces where nutrients, toxins, and water are adsorbed and later released. The cation exchange capacity and amount of shrinking and swelling of clays are partly functions of the surface area of the clay minerals. Surfaces provide niches for soil organisms and are particularly important for bacteria, most of which do not live freely in soil solution. Surfaces influence soil strength and the transport of fluids, solids, and dissolved constituents through the soil. In general, the more surface area there is in a given volume, the more numerous and the faster the reactions will be. The surface area per unit mass increases enormously as the particle size decreases. Consider Figure 2–14. A cube 1 cm on each side has a total surface area of 6 $cm^2$. The volume occupied by the cube is 1 $cm^3$. If the cube were divided into eight cubes, the volume

**TABLE 2–7**

Surface Area and Charge of Some Clay Minerals and Humus

| Mineral | Cation exchange capacity ($cmol_c$/ kg) | Surface area ($10^3$ $m^2$/ kg) |
|---|---|---|
| Kaolinite | 1–10 | 10–20 |
| Chlorite | 20–40 | 70–150 |
| Mica | 20–40 | 70–120 |
| Montmorillonite | 80–120 | 600–800 |
| Vermiculite | 120–150 | 600–800 |
| Humus | 150–300 | 900 |

and mass would remain the same, but the surface area would double to 12 cm$^2$. This could be repeated many times. Clay particles, which are only nanometers thick and tens of nanometers wide, have enormous surface-to-mass ratios (Table 2–7). Much of the reason smectites have more surface charge than micas or kandites is that they normally form much smaller crystals.

## 2.4  SOIL ORGANIC MATTER AND HUMUS

Most soils have less than 5 percent organic matter by weight, but this small amount profoundly affects many soil properties. Carbon, the essential element of organic compounds, combines with itself, hydrogen, nitrogen, phosphorus, and other elements to form an immense array of compounds. Structures are described later, in Supplement 8–1. Soil organic matter can be described as consisting of six important groups of materials:

1. Plants
2. Plant roots
3. Dead and decaying residues of plants (litter on the soil surface)
4. Decay agents, mainly microorganisms but including a wide array of soil microorganisms and soil fauna
5. Colloidal decayed residues
6. Water-soluble organic compounds

All of this is broadly referred to as soil organic matter (SOM), though often we exclude plants themselves except for small roots.

### 2.4.1  Coarse Organic Materials

Dead and decaying but still recognizable organic material makes up the Oi and Oe horizons at the soil surface. A horizon consisting of mostly unrecognizable decayed and decaying organic material is an Oa horizon. Aboveground plants and roots contribute particulate material to the soil surface as litter and to the subsoil as sloughed cells and dead roots. Much of this carbon is cycled back to the atmosphere as $CO_2$ through a multitude of decay processes (see Chapter 8). This particulate organic material may be subdivided into two fractions. Fragments that are sand sized or larger and that may be separated from inorganic components by sieving are macroorganic matter. The light fraction is organic matter that can be separated from inorganic material by floating soil suspensions on water. Inorganic particles sink, and the light fraction floats.

Charcoal is an inert form of SOM that marks locations of fires but is of little importance in nutrient cycles or soil carbon dynamics.

### 2.4.2  Colloidal Fraction

The most biologically active part of the SOM is the microbial fraction. The most chemically reactive part is the colloidal fraction, which often includes microbial biomass simply because it is difficult to separate the two. A working definition of humus is the organic materials that remain in soil after removal of macroorganic and dissolved organic matter.

Humus consists of four fractions that are operationally defined. That is to say, there is a large array of components that together are called humus, and,

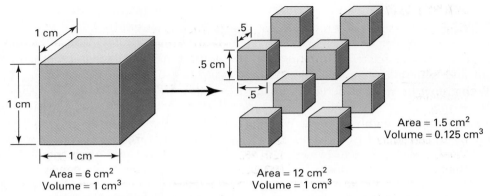

**FIGURE 2–14**  Surface area increases as the number of particles in a volume increases; that is, a 1-cm³ volume of soil filled with clay has a much larger total surface area than does the same volume filled with sand. In this example, a simple cube of 1 cm³ has the same total volume and weight as the eight smaller cubes do, but the total surface area of the eight smaller cubes is twice that of the single cube.

for convenience, this array is divided into four parts based on how they are extracted. The first is a part consisting of identifiable organic structures such as sugars, proteins, polysaccharides, amino acids, fats, waxes, other lipids, and lignin.

Cellulose is a polymer of the simple sugar glucose. It is the most abundant constituent of plant tissue and a source of food for the soil microorganisms. Hemicellulose is the second most abundant material (by weight) in plant tissue, and lignin, a phenolic polymer, is another major constituent of plant tissue and thus a major part of SOM.

A second part is humic acid, which consists of organic materials that are soluble in alkaline solution but insoluble on acidification of the alkaline extracts. A third fraction is fulvic acid. Fulvic acid compounds are soluble in alkaline solution and remain in solution when the alkaline extract is acidified. The fourth fraction is humin. Humin is the organic fraction that is insoluble in an alkaline solution.

*Humus* decomposes slowly and colors soils brown or black. Some of the products of organic matter decomposition are soluble in water and disappear quickly. Other compounds remain. Clays sometimes combine with organic compounds to form new compounds of large molecular weight that decompose very slowly. Humus, the stable colloidal organic matter, is the organic glue that helps hold mineral particles together to form aggregates.

Humus has two properties in common with clay: It is highly charged (Box 2–3), and it has a large surface area per unit mass (Table 2–7). Thus it is very reactive in soils. Humus molecules cannot be recognized with the eye, because they have undergone many chemical and physical changes from their original state (a leaf, a tree trunk, a root).

Humus is both a product synthesized (created) by microorganisms and a product of the breakdown or decomposition of other organic compounds. It is in a dynamic state, always changing. The amount of humus in soil depends on the balance between the rate of addition of organic matter and the rate of its decomposition.

Humus molecules coat mineral grains and help hold them together as aggregates. The many branches and chains contain "active groups" such as COOH and OH that can gain or lose hydrogen ions depending on the soil pH. Loss of a hydrogen ion from the COOH or OH results in a negative charge; addition of H may result in a positive charge. This pH-dependent charge of the organic matter attracts cations and anions from the soil solution, just as the permanent charge of silicate clays attracts cations. Finally, the large size and branched, linear shape of some of the molecules contribute to the large surface area and high activity of humus and its ability to stabilize aggregates. For a more detailed discussion of these substances, see Chapter 8.

## 2.5 SOIL COLOR AND CONSISTENCE

Two other properties of soils related to the organic matter content and mineralogy are soil color and consistence. Both properties are considered essential parts of a soil description.

Most minerals are not highly colored. When they are coated with humus and iron oxides, they take on the colors of humus (black or brown) and iron oxides and hydroxides (red and yellow). Carbonates of calcium and magnesium color the soil white. The color plates illustrate the wide range of colors created by the combination of humus and oxides. A soil's color is described with a number and letter code that describes its hue, value, and chroma (e.g., 10YR 4/4). *Hue* names the dominant wavelength range—that is, the spectral (rainbow) color. Hue in soils is normally in the red-orange-yellow range, from 10R (100 percent red) to 5Y (75 percent yellow and 25 percent red) (see Figure 2–15). A common hue is 10YR (50 percent yellow and 50 percent red). *Value* is the relative lightness of color (as it would appear in gray tones in a black-and-white photo), with 0 being black and 10 white.

*Chroma* is the relative purity, intensity, or strength of the hue. A chroma of 0 is pure gray, white, or black (depending on value) and theoretically can be 20 (as in a sodium fog light). In soils, the chroma rarely exceeds 8.

*Consistence* is a measure of the soil's cohesion (how it sticks to itself) and adhesion (how it sticks to other things) and resistance to deformation (changing shape) or rupture (breaking when dry). It is described at three moisture contents—dry, moist, and wet. *Dry* soil consistence ranges from loose for sand and loamy sand textures, in which the individual grains do not stick together, to extremely hard for textures that form structural units that cannot be broken in two hands. *Moist* consistence ranges from loose to extremely firm, and *wet* consistence is described as stickiness and plasticity. Stickiness is the amount of adhesion of the soil to the fingers, and plasticity is the ability of the soil to retain a shape when rolled or shaped. A key factor besides water content in influencing soil consistence is the clay content. More clay increases the hardness of dry soil, the firmness of moist soil, and the stickiness and plasticity of wet soil. Smectite clays are stickier and more plastic than kandites are.

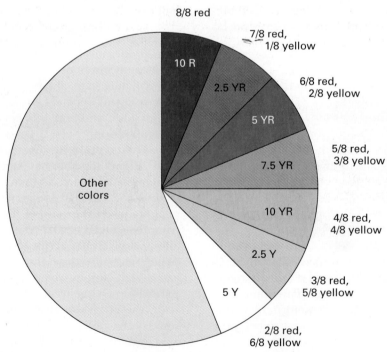

**FIGURE 2–15** A color wheel showing the most common hues (spectral wavelengths) found in soils.

## 2.6  PORES

At the beginning of this chapter, we stated that soil is a three-phase system containing solids, liquids, and gases. Thus far, we have described the solid portion but have left the liquid and gas phases for Chapter 3. The gases and liquids occupy the pores, or spaces, between and within the aggregates. Pores are the pathways through which water and air flow. They come in all sizes and shapes and may be oriented in any direction.

The pores between aggregates are *interaggregate* pores, and those inside aggregates are *intraaggregate* pores. The interaggregate pores (Figure 2–16) are usually larger and straighter than the intraaggregate pores, so they carry more water more quickly. Some pores go nowhere and transmit neither water nor air. These are **vesicular pores.**

Pores are often compared with thin tubes, though in fact they are more like thin tangled strands of damaged macaroni. Most pores are tortuous, not straight, and anything flowing through them must flow slowly. If the pores are small, it flows even slower.

Pores may vary in size from large cracks or animal burrows to microscopically small spaces between sand, silt, and clay particles. Pore sizes have been arbitrarily placed into different groups (Table 2–8). The macropores and mesopores are also termed biopores, shrinkage pores, and interaggregate pores. The term *biopore* is used because these pores are

formed as earthworms, ants, termites, other soil fauna, and roots move through soil. Roots create pores and enlarge existing pores as they grow. Living roots block pores, making them ineffective for water transport, but as they die and decay the pores remain.

Pores also form between structural units, hence the name shrinkage or interaggregate pores. These pores are a function of how particles and aggregates are packed together. Removal of water from soil by roots enhances shrinking and formation of shrinkage pores. These are often the most important pores for water movement in soils with much clay. Also included within the categories of macropores and mesopores are the large intraaggregate pores.

The volume of pores >30 μm and their continuity determine to a large extent the flow of water, solutes, and air in soils. When the soil is saturated, water flow occurs principally in these pores. These pores drain quickly, and the water is replaced by air. Pores in this size range are sensitive to management. They collapse when exposed to stresses on the soil surface such as loading by machinery or repeated human traffic.

The micropores and ultramicropores (Table 2–8) are the pores most responsible for storage of water that is available to plants. They also provide habitat for microorganisms and soil fauna. They are less affected by traffic on the soil surface than are the larger pores.

Cryptopores are the smallest pores. Because of their size, they are the last to fill with water and the

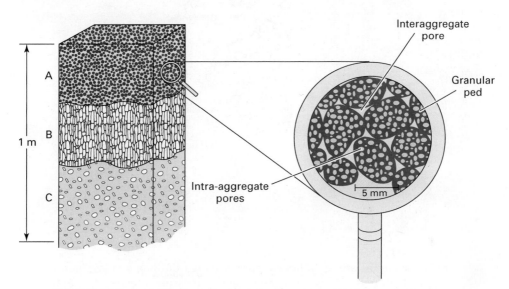

**FIGURE 2–16**  In this example, the A horizon has a porous granular structure, with many pores between peds and within peds.

## TABLE 2-8

Pore Size Classification

| Class | Class limits equivalent diameter (μm) |
|-------|---------------------------------------|
| Macropores | >75 |
| Mesopores | 30–75 |
| Micropores | 5–30 |
| Ultramicropores | 0.1–5 |
| Cryptopores | <0.1 |

Source: Soil Science Society of America, 1997.

last to empty. This water is unavailable to most plants, and flow through these pores is very slow. Little biological activity takes place in these pores. They are too small for roots and most microorganisms to penetrate. Among all the pore sizes, soil management least affects cryptopores.

Tillage, traffic, wetting and drying, freezing and thawing, root growth, and animal burrowing exert stresses on pores that influence the total volume, pore size distribution, tortuosity, and continuity of pores. These four pore properties determine the amount of water held by soils for plants and the rate at which water and air flow into and through the soil.

Most soils have pores of all these sizes, and thus average pore size is not a very useful concept. Rather, it is useful to know the pore size distribution in a particular soil. Pore size distribution (Figure 2–17) varies with texture, aggregate size and type, and bulk density. This is why air and water flow rapidly

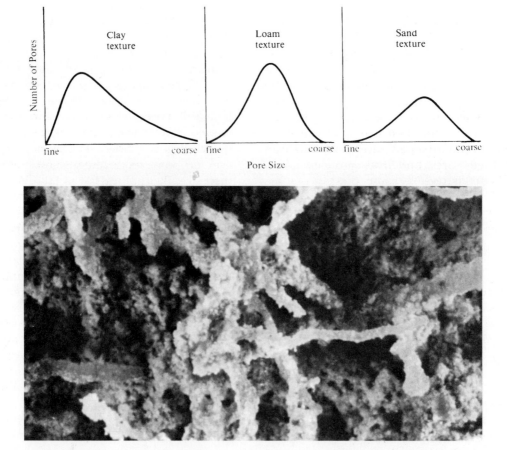

## FIGURE 2-17

Top: These curves are hypothetical pore size distributions for clay, loam, and sand texture horizons. Note the similarity between these curves and the particle size distribution bar graph. Bottom: Resin cast of tubular macropores in the Bv horizon of an Alfisol. (Photo courtesy of R. Tippkotter. Reprinted with the permission of Elsevier Science Publishers B. V., Amsterdam.)

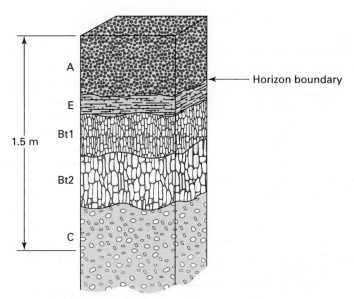

**FIGURE 2–18** Different horizons have different structures: in this case, granular (A), platy (E), blocky (B), and massive (C).

through sandy soils and slowly through soils high in clay content. Sandy soils have mainly large pores; clay soils, unless well structured, have mainly fine pores. Sandy and clay textured soils also have different capacities to retain water against the pull of gravity (Chapter 5).

Our pedon now has some form and substance. We have added mineral and organic matter to the pedon from Chapter 1. In Figure 2–18, only about 55 percent of the pedon is full of mineral and organic material; the remaining is pore space. Notice, too, that the surface and subsurface horizons are not randomly filled with material but that they both have structure. The structure is granular in the surface horizon, platy in the E, blocky in the B, and massive (structureless) in the C.

## 2.7 SUMMARY

Mineral soils are a framework of mineral particles of different sizes, the **soil matrix,** surrounding pores of different sizes. The mineral particles are usually held together in larger units called aggregates by humus, clay, and iron oxides. Aggregates take many common shapes such as granular and blocky, and they have different degrees of stability.

Soil pores transmit air and water and are formed by roots and animals that move through the soil and by cracking while drying or freezing. As particles are squeezed together to form aggregates, the spaces left behind are the pores. Pores have many shapes and sizes and are rarely straight.

The minerals in soils are mainly silicates, silica tetrahedra being the main building blocks of the silicate minerals. Clay minerals are aluminosilicates composed of alternating sheets of silica and aluminum tetrahedra and octahedra. Clays are important to soils because they have large surface areas and are charged.

Humus refers to a wide array of colloidal organic compounds formed from the microbial decomposition of plants and from plant and animal exudates. It has properties similar to those of clay, such as high surface area and charge, but its main building blocks are carbon, hydrogen, oxygen, nitrogen, phosphorus, and sulfur.

## QUESTIONS

1. What is clay?
2. Why are clay minerals important to soils and us?
3. What is isomorphous substitution?

4. What textural class does each of the following particle size distributions describe?

| Soil | Sand (%) | Silt (%) | Clay (%) |
|---|---|---|---|
| a | 10 | 20 | 70 |
| b | 70 | 10 | 20 |
| c | 30 | 30 | 40 |
| d | 40 | 40 | 20 |
| e | 8 | 42 | 50 |
| f | 60 | 15 | 25 |

5. Draw cumulative size curves with the following data and name each sample's textural class:

### Percentage of Soil Sample in Each Size Range

| Size range (mm) | Sample number | | | | |
|---|---|---|---|---|---|
| | 1 | 2 | 3 | 4 | 5 |
| 2–1 | 5 | 30 | 10 | 30 | 25 |
| 1–0.5 | 5 | 1 | 2 | 10 | 10 |
| 0.5–0.25 | 10 | 2 | 1 | 10 | 5 |
| 0.25–0.10 | 7 | 1 | 2 | 5 | 10 |
| 0.10–0.05 | 3 | 1 | 2 | 3 | 15 |
| 0.05–0.025 | 20 | 20 | 3 | 2 | 10 |
| 0.025–0.01 | 10 | 10 | 5 | 3 | 6 |
| 0.01–0.002 | 10 | 30 | 10 | 2 | 7 |
| 0.002–0.001 | 10 | 3 | 25 | 5 | 2 |
| <0.001 | 20 | 2 | 40 | 30 | 10 |

6. How are kaolinite and montmorillonite different? The same?
7. What is soil organic matter?
8. What is nonsilicate clay?
9. How does structure influence porosity?
10. What pore characteristics are important besides pore volume?

## FURTHER READING

Dixon, J. B., and S. B. Weed, eds. 1989. *Minerals in soil environments.* Madison, Wis.: American Society of Agronomy.

Kay, B. D., and D. A. Angers. 2000. Soil structure. In *Handbook of soil science,* ed. M. E. Sumner. New York: CRC Press. A229–A276.

McBride, M. P. 1994. *Environmental chemistry of soils.* New York: Oxford University Press.

Soil Science Society of America. 1997. *Glossary of Soil Science Terms.* Madison, Wis.: Soil Science Society of America.

Sparks, D. L. 1995. *Environmental soil chemistry.* San Diego: Academic Press.

Tan, K. H. 1993. *Principles of soil chemistry.* New York: Dekker.

## WEB RESOURCES

For more information on USDA texture, structure, color, and other descriptive terms, see the *National Soil Survey Handbook* Web site at *http://www.statlb. iastate.edu/soils/nssh*

For great three-dimensional crystal models, check out the virtual museum of soil minerals at *http://www.soils.wisc.edu/virtual_museum*

If you want to look at pictures of soils, check out the Canadian soils at *http://quarles.unbc.edu/nres/soc/ ggroup/ggroups.html* and soils from the United States at *http://www.statlab.iastate.edu/soils/photogal*

## SUPPLEMENT 2–1 Additional Particle Size Classifications and Methods to Display Data

Not all scientists use the USDA system of particle size classification. For example, highway engineers use a system developed by the American Association of State Highway and Transportation Officials (AASHTO). In the AASHTO system, silt is separated from sand at 0.075 mm and from clay at 0.005 mm rather than at 0.05 and 0.002 mm, as in the USDA system. The International Society of Soil Science uses four size separates: coarse sand (2.0 to 0.2 mm), fine sand (0.2 to 0.02 mm), silt (0.02 to 0.002 mm), and clay (<0.002 mm). Selection of the size limits used to separate sand from silt and silt from clay is not entirely arbitrary; in part it is for convenience and is based on the particles' behavior. For example, size affects the particles' packing behavior and cohesion (stickiness). Engineers have found that the particles in the USDA's very fine sand size behave more like silt when they build roads, and so for them there is a practical reason for raising silt's upper size limit.

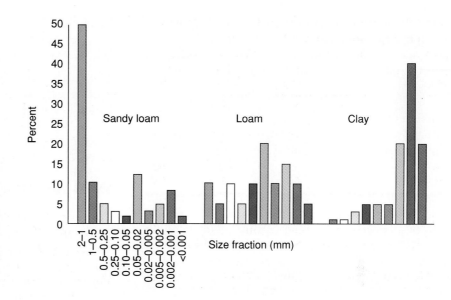

**FIGURE 2–19** Data from Table 2–2 are displayed here as bars on a graph. The percentage by weight of each size separate is shown by a separate bar. Sizes shown for the sandy loam apply to the loam and clay textures and are in the same order from coarse sand to fine clay.

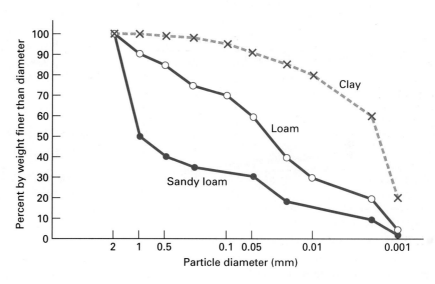

**FIGURE 2–20** Data from Table 2–2 are displayed as a cumulative size distribution curve. It is continuous because you can read off the curve the percentage of the sample that is smaller than any particle diameter. To use the figure, find the size fraction of interest and read vertically upward to the curve of interest. Next read horizontally to the left to determine the percentage by weight finer than the original fraction.

Figure 2–19 illustrates how the data in Table 2–2 are displayed in a bar graph, showing the percentage by weight for each fraction in a vertical bar. Figure 2–20 is a cumulative size distribution curve, plotted on semilog paper to fit the entire range of particle sizes onto one graph. Any point on the graph represents the percentage of the total sample that is smaller than the diameter on the horizontal axis.

All the curves are for the data in Table 2–2. All three lines begin at 100 percent (2 mm), because there are no particles larger than 2 mm in these soils. Twenty percent of the particles in the clay soil are

0.001 mm diameter or smaller, 60 percent of the particles are less than 0.002 mm, and 40 percent of the particles are smaller than 0.002 mm and larger than 0.001 mm.

## SUPPLEMENT 2–2  Stokes' Law

Stokes' law is the theoretical basis for particle size analysis. The maximum or terminal velocity of a spherical particle settling in a fluid under the force of gravity is proportional to the square of the particle's radius and the density of the fluid. The downward force ($F_D$) acting on the particle due to gravity is given by Equation 1, where $4/3\pi r^3$ is the volume of a sphere of radius $r$; $g$ is the acceleration due to gravity; $\rho_s$ is the particle density; and $\rho_F$ is the fluid density.

$$F_D = (4/3)\pi r^3(\rho_s - \rho_F)g \tag{1}$$

Friction ($F_F$) resists the downward force, according to Equation 2, where $\eta$ is fluid viscosity and $r$ and $u$ are the particle radius and velocity, respectively.

$$F_F = 6\pi\eta r u \tag{2}$$

The terminal velocity is reached when the gravitational and frictional forces are equal. The particle falls at a constant rate once terminal velocity, $u_T$, is reached (Equation 3). If we assume that $u_T$ is reached instantly and $F_D$ is set equal to $F_F$, then the time necessary for a particle to fall through a distance, $h$, can be calculated by using Equation 4. In addition, we can calculate the minimum diameter of particles that will settle beyond a given distance, $h$, in time, $t$, by means of Equation 5. Note that diameter ($d$) replaces radius ($r$) in Equations 3 through 5.

$$u_t = \frac{d^2 g(\rho_s - \rho_f)}{9h} \tag{3}$$

$$t = \frac{9h\eta}{d^2 g(\rho_s - \rho_f)} \tag{4}$$

$$d = \sqrt{\frac{9h\eta}{tg(\rho_s - \rho_f)}} \tag{5}$$

Both the hydrometer and pipette methods of particle size analysis depend on these relationships.

## SUPPLEMENT 2–3  Mineral Formulas and Structures

Hornblende is the most common of the double-chain amphibole group. Because of the wide range of chemical composition of this group, it is easier to give a general, rather than a specific, representation of the chemical formula. A common general formula for the amphiboles is:

$$(Ca, Na, K, Mn)_2(Mg, Fe, Al, Ti, Mn, Cr, Li, Zn)_3$$
$$(Si, Al)_8 O_{22}(OH)_2$$

Muscovite's has the general formula at the top of the right column, with varying amounts of substitution in the octahedral (oct.) and tetrahedral (tet.) layers.

$$K_2Al_4\underbrace{(Si_6Al_2)}_{}O_{20}(OH, F)_4$$
$$\text{oct.} \quad \text{tet.}$$

The K fits in the interlayer, linking layers and preventing expansion. When mica weathers, the K released is a major source of K for plants in many soils.

Trioctahedral biotite is structurally the same as dioctahedral muscovite, but magnesium and iron replace all or part of the Al in the octahedral sheet. A dioctahedral mineral has two out of every three octahedral positions filled with a cation, and a trioctahedral mineral has all octahedral positions filled with a cation. Biotite weathers more easily than does muscovite, in part because it is trioctahedral. The chemical formula for a typical biotite is:

$$K_2\underbrace{(Mg, Fe)_{4-6}(Fe, Al, Ti)_{0-2}}_{\text{oct.}}\underbrace{(Si_6, Al_2)}_{\text{tet.}}O_{20}(OH, F)_{2-4}$$

This formula is complicated because of the large amount of isomorphous substitution in the Al octahedral sheet ($Mg$, $Fe^{2+}$, $Fe^{3+}$, and $Ti^{3+}$) and in the Si tetrahedral sheets (Al). Many of these elements are important plant nutrients.

Orthoclase (K feldspar) has the general formula:

$$KAlSi_3O_8$$

Plagioclase is a group of feldspar minerals with similar structures and composition that vary from pure albite (Na-variety) to pure anorthite (Ca-variety).

$$NaAlSi_3O_8$$

$$CaAl_2Si_2O_8$$

Plagioclase can have any combination of the two cations Na and Ca in its structure.

Clay structures are shown in Figures 2–21 through 2–25.

## SUPPLEMENT 2–4  Charge in Ions

Mineral structures and the reactions that occur at their surfaces depend on the fundamental properties of atoms. Each atom is composed of a center, the **nucleus,** composed of positively charged particles (**protons**) and neutral particles (**neutrons**). Modern high-energy physics has shown that these "primary" particles consist of even smaller particles. The charge on a nucleus is equal to the number of protons because the neutrons

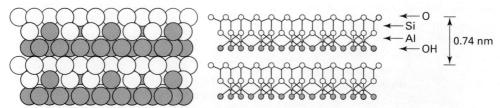

**FIGURE 2–21** Kaolinite is a 1:1 aluminosilicate mineral consisting of one tetrahedral sheet and one octahedral sheet. Layers are stacked one upon another to form particles of the mineral. (From Schulze, D. G. In *Minerals in soil environments,* ed. J. B. Dixon and S. B. Weed, Chapter 1.)

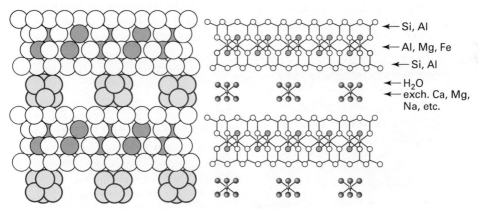

**FIGURE 2–22** Vermiculite and smectite are 2:1 minerals. Smectite is the family name for minerals with different amounts of $Fe^{2+}$, $Fe^{3+}$, and $Mg^{2+}$ substitution for $Al^{3+}$ in the octahedral layer. The layer charge for a unit cell of smectite ($-0.6$ to $-1.2$) is less negative than that for a unit cell of vermiculite ($-1.2$ to $-1.8$), which helps account for its greater capacity to shrink and swell. The layer charge is balanced by exchangeable cations that surround the minerals. (From Schulze, D. G. In *Minerals in soil environments,* ed. J. B. Dixon and S. B. Weed, Chapter 1.)

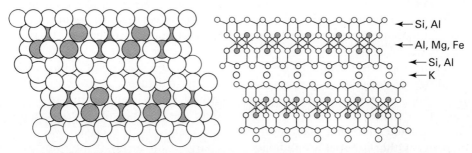

**FIGURE 2–23** Mica has nonexchangeable $K^+$ in the interlayer that balances a unit cell layer charge of $-2$. The potassium helps to hold the units together, resulting in no shrinking and swelling and lower surface area compared with smectite and vermiculite. (From Schulze, D. G. In *Minerals in soil environments,* ed. J. B. Dixon and S. B. Weed, Chapter 1.)

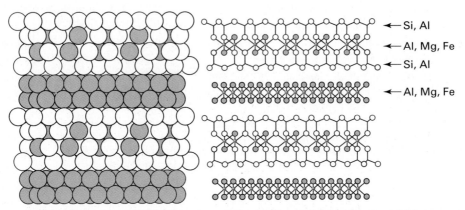

**FIGURE 2–24**   Chlorite has a layer of $Mg_6(OH)_{12}$ in the interlayer, which retards shrinking and swelling. (From Schulze, D. G. In *Minerals in soil environments,* ed. J. B. Dixon and S. B. Weed, Chapter 1.)

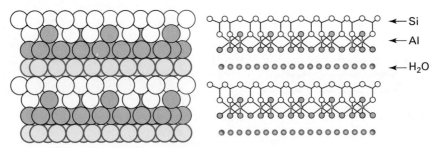

**FIGURE 2–25**   Halloysite is a 1:1 aluminosilicate clay like kaolinite, but it has one or two layers of water in the interlayer. (From Schulze, D. G. In *Minerals in soil environments,* ed. J. B. Dixon and S. B. Weed, Chapter 1.)

have no charge. The protons and neutrons in the nucleus are held tightly together. The *atomic mass* of the nucleus is within a fraction of a percent of the sum of the masses of protons and neutrons. The number of protons is the *atomic number.* Atoms of the same number (number of protons) but different mass (protons plus neutrons) are *isotopes.* Consult the inside front cover to see the atomic mass and atomic number of each element. Note that the atomic mass is not an integer. For example, the mass of hydrogen is 1.008. This results from the fact that hydrogen has several isotopes and the atomic mass of hydrogen is an average mass based on the atomic mass and abundance of each isotope.

Some isotopes are stable, meaning that they do not decay spontaneously. Other isotopes are radioactive, meaning they do decay spontaneously. The *half-life* of a radioactive nucleus is a measure of its stability or its tendency toward spontaneous decay. The half-life is the amount of time it takes for half of the nuclei to decay.

Surrounding the tight cluster of the nucleus is a cloud of one or more negatively charged particles (**electrons**). If the number of protons and electrons is equal, the atom will have no net charge. If the number of protons is more than the number of electrons, the atom will be positively charged. This is a cation. If the number of protons is less than the number of electrons, the atom will carry a negative charge. This is an anion. Atoms acquire charge by gaining or losing electrons.

In complex combinations of atoms, such as a crystal structure, electrons are shared among several atoms. If the overall number of electrons is greater than the number of protons among atoms sharing the electrons, the crystal will have a net negative charge. Conversely, if the number of electrons is insufficient to balance all of the protons, the crystal will have a net positive charge. In some compounds, particularly organic matter, the complexity is such that part of the molecule has a positive charge and part has a negative charge.

# 3

# LIQUIDS AND GASES

## OVERVIEW

*In Chapter 2 our pedon was shown as a three-dimensional body consisting of solid inorganic and organic material and pores. The pores are never empty. They contain air or water or both, and they are the pathways through which water and air move through the soil. This chapter discusses the composition of the air and water and the interactions between dissolved substances and the soil's solids.*

*The three phases are in intimate contact and interact with one another continuously unless frozen. One of the most important interactions between the solid phase and the solution phase is cation exchange. Cation exchange is a process in which positively charged ions in the soil solution attach themselves loosely to the solid phase, which has a net negative charge. At the same time, other cations on the solids move away from the solids and enter the bulk solution. Soils hold water and cations, two ingredients vital to plant growth. Likewise, other surface reactions hold many different kinds of soluble materials, including nutrient anions, heavy metals, and organics.*

## 3.1   SOIL AIR

### 3.1.1   Composition

Except when soils are contaminated, the composition of soil air is similar, but not identical, to the air we breathe. It is similar because some soil pores are open to the atmosphere. Atmospheric gases enter and move throughout the soil mass because many soil pores are interconnected. It is not identical because movement is slow and soil organisms (including roots) affect the composition of the soil air (Table 3–1).

Microorganisms such as bacteria and fungi and macroorganisms such as plant roots take oxygen from the soil air and release carbon dioxide into it. This results in the soil air's reduced oxygen content and its increased carbon dioxide content relative to the atmospheric content of these two gases.

When soil pores are not full of water, which is most of the time, aerobic organisms that require free oxygen can live. However, for short times during a rainstorm or irrigation, or for several days or weeks during the year, most soil pores may be full of water

**TABLE 3–1**
Average Composition of Soil Air and Atmosphere

| Component | Soil air (%) | Atmosphere (%) |
|---|---|---|
| $N_2$ | 79.2 | 79.0 |
| $O_2$ | 20.6 | 20.9 |
| $CO_2$ | 0.25 | 0.03 |

Source: Russel, E. J., and A. Appleyard. 1915. The atmosphere of the soil, its composition, and causes of variation. *Journal of Agricultural Science* 7:1–48.

and little or no air is in the pores. This is saturation. Extreme cases of saturation, such as in permanently wet swamps, occur when air is rarely in the soil pores (Figure 3–1). Some soil horizons and most rocks are **impermeable;** that is, they do not let water penetrate. Impermeable horizons, or bedrock near the soil surface, are often responsible for saturated soils.

The atmosphere of contaminated soils and permanently saturated soils may be very different from the composition of the atmosphere. Air within con-

**FIGURE 3–1**   Characteristics of poorly drained soils are dark surface horizons rich in organic matter and soil pores filled with water. Here the soil is saturated below 50 cm.

taminated soils may contain volatile organic compounds. The soil atmosphere's composition depends on the kind of contaminant, its solubility in water, and its volatility. When the composition of soil air is low in oxygen for protracted periods, the soil atmosphere contains the products of anaerobic microbial processes. These include methane ($CH_4$), nitrous oxide ($N_2O$), and hydrogen sulfide ($H_2S$). The "rotten-egg smell" of some poorly drained soils is due to gases such as hydrogen sulfide. Anaerobic conditions are not good for most common cultivated plants.

Besides being higher in $CO_2$ and lower in $O_2$ than atmospheric air, soil air is also almost always saturated with water vapor. **Relative humidity** represents the amount of water held in the atmosphere at any temperature relative to the maximum amount that the atmosphere can hold at the same temperature. The relative humidity of the atmosphere varies from very low in deserts to very high in humid climates. Unlike the atmosphere, soil air is almost always between 99 percent and 100 percent relative humidity.

### 3.1.2    Air Movement

Two mechanisms are responsible for the movement of gases in soil. The most important is diffusion, but some gas moves by means of mass flow. *Diffusion* occurs when molecules of a particular gas move from a zone of high **concentration** to a zone of low concentration (Figure 3–2). This continues until there is an equal concentration of that gas throughout the soil and atmosphere. Because this **equilibrium** condition rarely occurs, gases are constantly diffusing through, into, and out of the soil. Differences in $O_2$ and $CO_2$ concentrations in the soil and atmospheric air normally cause a net flow of $O_2$ into the soil air from the atmosphere and a net flow of $CO_2$ out of the soil air into the atmosphere.

The rate at which molecules diffuse through the soil depends on the soil's water content, size and number of pores, kind of molecule, and temperature. The total porosity is more important to the rate of diffusion than is the pore size distribution, because pores are very large compared with the short distances that gas molecules travel before hitting another molecule. Pore **continuity,** the length of interconnected pores, also affects diffusion, which is slower through short, disconnected pores than through lengthy, interconnected pores.

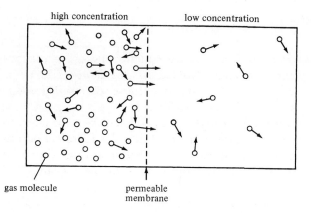

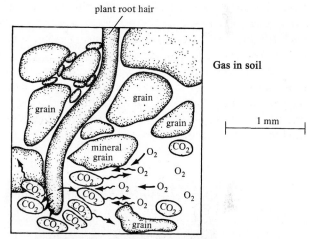

FIGURE 3–2 Diffusion occurs because gas molecules move from places of high concentration to places of lower concentration, as illustrated by the gas in the box. Plant roots lower $O_2$ and raise $CO_2$ concentrations nearby as they grow. Because of the difference in gas concentrations near the root and in the soil atmosphere away from the root, $CO_2$ diffuses from the root, and $O_2$ diffuses toward it.

The *mass flow* of gas is believed to be much less important than diffusion. It occurs when all gas molecules move from a zone of higher to lower pressure. Mass flow occurs when (1) the soil temperature changes, thereby changing the velocity of all gas molecules; (2) the atmospheric pressure changes; (3) plant roots extract water, and air flows into empty pores; (4) the wind blows over the soil surface, thereby increasing the evaporative loss; and (5) soil is being flooded (Figure 3–3).

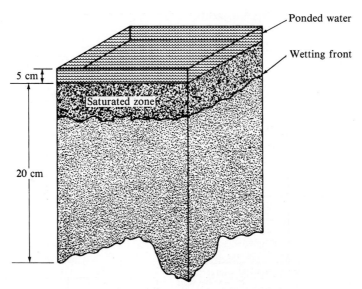

**FIGURE 3–3** The pedon in this figure is flooded. As the water enters the soil, it forces all the gas molecules in the pores downward in front of the wetting front. This is called mass flow.

Air flow through dry pores is much more rapid than that through water-filled pores. Plants and microorganisms that need oxygen die when the soil is saturated for any length of time, because in saturated soils the rate of $O_2$ replenishment is too slow (see Chapter 8).

## 3.2  AIR–WATER RELATIONSHIPS

Air and liquid compete for the same pore space. The total amount of air in a soil depends on two factors: the amount of pore space and the amount of liquid in the pores. A simple equation describes this:

$$V_A = V_P - V_W \tag{1}$$

In Equation 1, $V_A$ represents the volume of pores filled with air, $V_P$ the total soil volume filled by pores, and $V_W$ the volume filled with water.

### 3.2.1  Porosity

A hypothetical "average" medium-textured soil is 50 percent solids and 50 percent pores. The pores vary in size, shape, and **tortuosity** (straightness or curvi-

ness). The pores in this ideal system are shown in Figure 3–4 as being 50 percent air and 50 percent water filled. The total mass, $M_T$, is the sum of the mass of air, water, and solid; the total volume, $V_T$, is the volume occupied by the air, $V_A$, water, $V_W$, and solid, $V_S$; and the total volume of pores is $V_P$. Finally, the **porosity** ($P_T$) is defined in Equation 2:

$$P_T = V_P / V_T \tag{2}$$

In words, Equation 2 states that the total porosity is equal to the volume occupied by the pores divided by the total soil volume. The total volume of all the soil that exists is never measured. Rather, this equation and the others refer to the mass and volume relationships in a single soil sample.

Equation 2 can be written in another way:

$$P_T = \frac{V_A + V_W}{V_A + V_W + V_S} \tag{3}$$

Equation 3 shows that the total volume of pores, $V_P$, is equal to the volume of the air plus the volume of water, and the total volume of the soil is equal to the volume of air, water, and solids. These simple relationships can be used to calculate some important soil characteristics. Soil porosities ($P_T$) generally range from 30 to 60 percent for most soils.

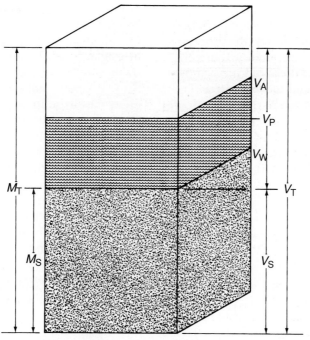

**FIGURE 3–4** This illustrates the "ideal" soil volume occupied by pores and solids and the symbols used in the text. The mass of solids greatly exceeds the mass of air and water because the solid density is much greater than water or air density.

## 3.2.2 Bulk Density and Void Ratio

**Bulk density** is often measured as an indicator of compaction and a basis for calculating porosity. **Void ratio** ($e$) is used to express the relationship between the soil's solid and pore phases:

$$e = (V_A + V_W)/V_S = V_P/(V_T - V_P) \qquad (4)$$

In Equation 4, the void ratio is shown to be a ratio of the volume of air and water to the total volume of solids. Two other values can be determined from the relationships in Figure 3–4: specific gravity and dry bulk density. **Specific gravity** is also referred to as *mineral* or *particle density* and is the mass-to-volume ratio of a mineral relative to water at 4°C, which is the temperature at which the density of water is 1.000 Mg/m$^3$ or g/cm$^3$. Bulk density, or apparent density, is the soil mass ($M_S$) divided by the soil volume ($VT$). Supplement 3–5 has details of how to measure and calculate bulk density. Bulk density is always less than the specific gravity because it includes pores, all of which have zero mass. Equations 5 and 6 show that specific gravity, $\rho_S$ (rho), is equal to the particle mass ($M_P$) divided by the particle volume ($V_S$) and that bulk density ($\rho_B$) is the soil mass ($M_S$) divided by the soil volume ($V_T$):

$$\rho_S = M_P/V_S \qquad (5)$$

$$\rho_B = M_S/V_T \qquad (6)$$

## 3.3 SOIL WATER

### 3.3.1 Characteristics of Water

Water is common, unusual, and interesting. For example, a covered glass filled with ice cubes and water has water molecules in three phases—solid, liquid, and gas. In soils, water is most commonly found in its liquid and gas phases.

A water molecule has an asymmetrical shape, with the oxygen atom at the center and the two hydrogen atoms to the side (Figure 3–5). The hydrogen atoms are bonded to the oxygen atom covalently, and each hydrogen atom shares its single electron with the oxygen atom. The molecule is electrically neutral, having no excess positive or negative charge. However, the electrons (which carry the negative charge) are not uniformly distributed around the molecule. The oxygen end of the molecule is more negative, and the hydrogen end more positive (Figure 3–5). Because of the uneven distribution of charge, the water molecule is said to be **polar.**

Polarity is important to water and to soils. Water molecules attach themselves to one another to form clusters that are larger than single $H_2O$ molecules.

The hydrogen (positive) end of one water molecule bonds to the oxygen (negative) end of an adjacent water molecule. This kind of polar bond is called a **hydrogen bond.** Polarity results in **cohesion,** the attraction of one water molecule to another (Figure 3–6). Likewise, **adhesion** is the attraction of water molecules to solid surfaces. Hydrogen bonding between water molecules produces **surface tension,** because water molecules attract one another more strongly than do air molecules (Figure 3–7).

Cations such as $Na^+$, $Ca^{2+}$, $Mg^{2+}$, and $K^+$ are common in the soil solution and on the charged surfaces of clays and humus. They are attracted to the oxygen (more negative) end of the water molecule. When a cation is associated with water molecules through this attraction, it is **hydrated,** the normal state of cations in solution (Figure 3–8). The number of water molecules with which any cation is associated is determined primarily by the ion charge and size.

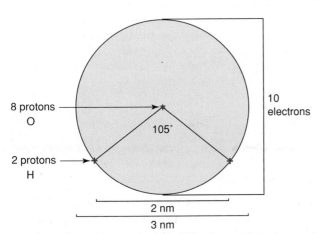

**FIGURE 3–5** A water molecule showing positions of protons associated with oxygen and hydrogen nuclei and a likely set of transient positions for electrons. The electrons are constrained to orbits traversing both top and bottom hemispheres, but the positive charges are distinctly off-center due to the placement of the two protons (H atoms) at the lower periphery. This arrangement makes the molecule polar.

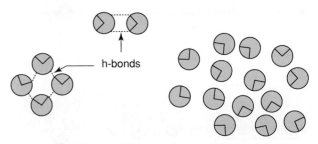

**FIGURE 3–6** "Water clusters" illustrating hydrogen bonding among water molecules. This hydrogen bonding is responsible for many of the interesting and important properties of water.

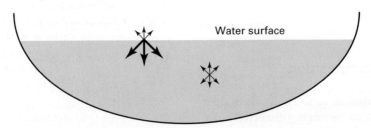

**FIGURE 3–7** Surface tension occurs because water molecules are more strongly attracted to one another than to air molecules. The arrow lengths represent the strength of molecular attraction among water molecules and between water and air molecules.

## SOIL AND ENVIRONMENT
### Soil Density

Increasing soil density improves the suitability of soils for engineering purposes such as highway construction and house foundations. Increased density decreases soil porosity, reduces the maximum soil water content, and minimizes the hazards of freeze-thaw and shrink-swell that damage structures. But for many other uses, increasing soil density is detrimental. For example, human foot traffic across lawns increases natural soil density to the point that the lawn cannot grow. The same problem occurs on golf course greens, recreation and camp areas, and playgrounds and in parks, forests, and wildlands subject to uncontrolled traffic by people, recreational vehicles, and logging machinery. The same decrease in soil porosity that is beneficial for structures is detrimental to plants and increases the potential for floods and soil erosion. A reduction in porosity reduces the rate at which water can enter soil, thus increasing the proportion of water flowing over the soil surface. Instead of the soil holding water and slowly releasing it to streams, the water flowing over the soil surface reaches streams quickly and may increase flooding potential. This same water, flowing over the soil surface rather than through the soil, may detach and carry soil particles into streams. Once the soil reaches the stream, it is sediment that is detrimental to water quality.

In the subsoil, the greater the density, the more difficult it is to remediate soil through pollutant extraction because the permeability decreases as density increases. One method of soil remediation is through stimulation of soil microorganisms to use pollutants as substrate (food). Nutrients to stimulate microorganism growth may be injected into the subsoil or below the soil to the contaminated volume. The greater the density, the more difficult it is to distribute the nutrients.

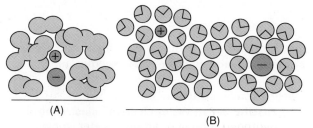

**FIGURE 3–8** Hydration of ions. (A) Ions in a nonpolar solvent are free to approach and form a solid. (B) Ions in water (a polar solvent) hydrate, orienting water molecules into "hydration shells." Ionic substances are more soluble in polar than in nonpolar solvents.

### 3.3.2 Pore Water Composition

Water in the soil is referred to as the soil solution. It is not pure, but contains **solutes** (dissolved substances), including cations and anions, organic and inorganic compounds, and gases. Frequently, small suspended particles such as clay are in the soil solution. The dominant soil solution cations and anions are $Ca^{2+}$, $Mg^{2+}$, $K^+$, $Cl^-$, $NO_3^-$, $SO_4^{2-}$, $H_2PO_4^-$, $HPO_4^{2-}$, and $HCO_3^-$. Also found in varying amounts are $H^+$, $Na^+$, $NH_4^+$, $Mn^{2+}$, and $Al^{3+}$ (Figure 3–9).

The amount of each constituent depends on the kinds and amounts of the minerals and organic matter in the soil's solid phase, the kind and activity of the vegetation, and the source of water. For example, frequent high-volume rains washing through the soil quickly produce a soil solution composition more dilute and quite different from a soil solution that is replenished infrequently.

Before "fresh" water enters the soil, it often comes into contact with vegetation. Water may remove some of the cations or anions in the vegetation by means of **leaching**. Leaching occurs when moving water contacts soluble compounds and the compounds or ions dissolve into it. Leaching redistributes cations and anions through the soil. Water also enters the soil and reacts with the soil's solid components, some of which dissolve readily and others slowly. Dissolution and leaching are subjects of soil genesis and soil fertility and are discussed in subsequent chapters. In any case, the soil's solid phase contributes solutes to the solution, which interact with the solid phase as the soil solution travels through the soil. The soil solution constantly changes whether or not new water is added to the soil. The soil solution

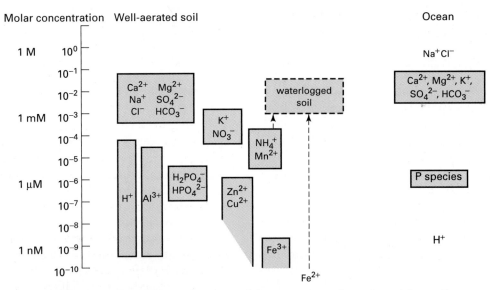

**FIGURE 3–9** The expected range of ion concentrations found in well-aerated soils, waterlogged soils, and the ocean. Note the large range of $H^+$ and $Al^{3+}$ concentrations in well-aerated soils. Also note the large increase in $Fe^{2+}$, $Mn^{2+}$, and $NH_4^+$ concentrations (denoted by the arrows) when a well-aerated soil becomes waterlogged. Some metals (Al, Fe, Cu, Zn . . . ) normally occur in solution, mainly in complexed forms.

can be regarded as integrating or communicating with the entire soil profile, carrying solids, cations, and anions from one part of the profile to another.

The composition of the solution is influenced not only by vegetation and soil minerals but also by the climate and season. In cool, humid climates such as those in the northeastern United States, the soil solution may be constantly renewed by precipitation. In cold winter months there is little biological activity to influence the composition of the soil solution. In the humid tropics, there is a continual replenishment of the soil solution, as well as warm temperatures that ensure nearly continuous biological activity. These conditions influence both average and seasonal soil solution composition. In a cool, humid climate the soil solution composition differs between the winter and summer seasons, whereas the humid tropical soil solution is more constant throughout the year.

What difference does the soil solution make? The soil solution is where plant roots and soil microbes obtain nutrients. Few nutrients are taken directly from the soil's solid phase, and so if the solution phase is deficient in a nutrient, the plant will suffer. Some soil solution may be leached beyond the root zone into groundwater, where its constituents may be pollutants. Anions, such as nitrate $(NO_3)^-$, move readily through the soil into aquifers.

**Determining solution composition**   Soil solution chemistry cannot easily be measured in place. It is simpler to extract a sample of the soil solution and determine the constituents in the laboratory. Two methods are frequently used. In one, a sample of soil from the horizon of interest is saturated with water in the laboratory, and the soil solution (referred to as the **saturation extract**) is removed for analysis (Figure 3–10). Another method is used to sample the soil solution directly in the field. Porous disks or cups called *suction lysimeters* (Figure 3–11) are installed in a horizon of interest. A weak suction is applied to the sampling device, and the soil solution is withdrawn into a sample container. Both methods are successful only at high soil water contents.

### 3.3.3   Soil Water Amounts

The amount of soil solution affects the composition of the soil solution, the rate at which water flows through the soil, and the availability of water for plants (Chapter 5).

**Saturation**  Figure 3–12 shows two different states of the soil solution. In Figure 3–12A the soil pores are saturated. A small volume of gas is dissolved in the solution phase. Only a few of the very small, "dead-end" pores are filled with air, because their openings are so small that water cannot push in and force the air out. If any more water were to enter from the soil surface or from a water table, it would have to displace water already in the pores.

**Oven dry**  Figure 3–12B illustrates the condition of soil dried at 105°C for 24 hours. Almost all the water in the pores has evaporated, although small amounts

of water do remain, tightly held by clay and organic matter. The weight of the soil in this condition is the *oven-dry weight*. The oven-dry weight of soil is used as the standard basis for expressing many measurements in soil science.

A soil may contain any amount of water between its oven-dry state and its saturated state, though few soils in nature are found at either extreme. The surfaces of desert soils may reach very high temperatures and be nearly oven dry. Tidal flats, swamps, meadows, and rice paddies are waterlogged for long times, but other soils are saturated only briefly during floods or intense rain.

**Intermediate moisture states**  **Field capacity** and **permanent wilting point** represent the approximate upper and lower limits of water that is held in soils and is available for plants. This water is sometimes referred to as **plant-available moisture.** The actual numerical values of field capacity and permanent wilting point are different for different soils, depending on the soils' particle size distribution, volume of pores, and pore size distribution.

At saturation, water can flow so rapidly out of the soil through the macropores that little of it is useful for plant growth. At a point when most of the largest pores have been emptied of the soil solution, the flow of water out of the soil slows. This point is the field capacity and is shown in Figure 3–13A. At field capacity, the micropores remain full of water, many of the mesopores are partially filled with water, and most of the large pores are nearly empty.

Toward the dry end of the wetness spectrum is the permanent wilting point (Figure 3–13B). At this moisture content, most plants cannot withdraw water from the soil. If the soil remains for any length of time

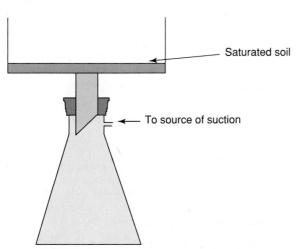

**FIGURE 3–10**  Suction is applied to the saturated soil in the Büchner funnel to extract soil solution, which is analyzed for its constituent soluble ions.

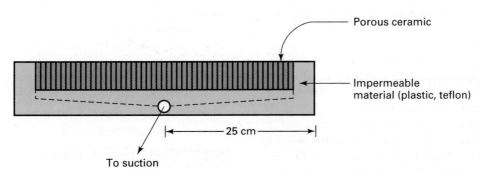

**FIGURE 3–11**  A suction lysimeter is a porous ceramic disk coated with impermeable material on the sides and bottom. Suction pulls the soil solution through the disk into a collection bottle.

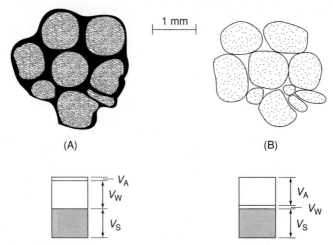

**FIGURE 3–12** (A) Saturation occurs when the soil pores are nearly completely full of water. (B) Oven-dry soil has no visible water at this scale. Any water remaining is in thin films on soil particles. Note the different volumes of air and water in the two moisture states. Note also the water in aggregates as well as around aggregates.

at or below the permanent wilting point, plants will permanently wilt and die. The films of water on the mineral particles are much thinner at the permanent wilting point than at the field capacity (Figure 3–13A), and the large and medium pores are completely empty except for these thin films. A few of the "dead-end" pores may remain filled. The water is now held so tightly by the mineral and organic particles that most plants fail to extract it fast enough to survive.

Between the field capacity and the permanent wilting point, a plant may be under stress; that is, it may not be supplied with enough water from the soil. However, if more water is given to the plant, it will regain its **turgor** and continue to grow. Many plants that live in arid environments are able to extract water from the soil at water contents below the conventional permanent wilting point; thus, the concept has to be used carefully.

## 3.4  SOLID–LIQUID INTERACTIONS

The solid (inorganic and organic particles), liquid (soil solution and its constituents), and gas (soil air) phases frequently interact. Among the most important interactions are those that occur between the solid and liquid phases on inorganic and organic particle surfaces. One of the best understood is the **electrostatic** interaction between a cation (positively charged) or anion (negatively charged) and the charged solid surface. This is **ion exchange.** Others are more specific bondings of solutes and surfaces, dissolution of the solid phase in the solution phase, precipitation of solids, the weak sorption of large organic molecules, phase partitioning, and pore retention.

### 3.4.1  Ion Exchange

Clay particles and organic matter have negative surface charge; that is, they have extra electrons in their structure. This arises from isomorphous substitution and broken edges in clays and from loss of $H^+$ from acid groups in humus. Clay and humus are the main constituents of the exchange complex. Likewise, some colloids may develop positive charge sites. In nature, electrical neutrality must be maintained. In soils, electrical neutrality is attained when negatively charged clay and humus particles attract positively charged ions.

This **electrostatic** or **coulombic** attraction is generally weaker than covalent bonding, and it is less specific, or less selective between ions of similar charge. So it allows the ready exchange of one cation for another (cation exchange) and one anion for another (anion exchange). Ion exchange oc-

Cation exchange is important because the **exchangeable ions** (those held on the exchange complex) are (1) available to plants, supplementing the small quantity in solution, and (2) retained in soils and not lost with leaching water.

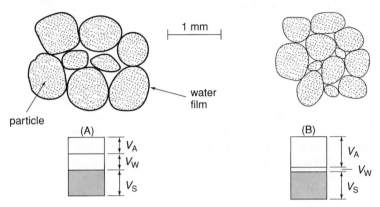

**FIGURE 3–13**   (A) Field capacity. Large pores are filled with air, and many small pores are filled with solution. This is the nearly ideal situation described in Figure 3–4. (B) Permanent wilting point. Most pores are empty, and the remaining water is tightly held as thin films on surfaces.

curs in soil when ions in solution exchange places with ions held on the **exchange complex** (the soil's clays and humus).

**Cation exchange**   Figure 3–14 depicts the exchange of $Ca^{2+}$ and $Mg^{2+}$ ions in the presence of other cations at a clay surface. Unless retarded by slow diffusion, exchange reactions rapidly reach equilibrium.

There are rules or preferences that govern cation exchange equilibria. First, cations with a small hydrated radius (Table 3–2) tend to be held more tightly and to be displaced from the exchange complex less easily than large cations; they can get closer to the surface of the exchange complex. Second, highly charged cations (Table 3–2) tend to be held more tightly than cations with less charge. The lyotropic series (Equation 7) summarizes these combined effects of size and charge; the general order of preference for electrostatic adsorption of the common cations is as follows:

$$Al^{3+} > Ca^{2+} > Mg^{2+} > K^+, NH_4^+ > Na^+ > Li^+ \quad (7)$$

The degree of exchange also depends on the relative concentrations of the competing cations in the bulk solution. The tendency of a cation species to be adsorbed increases if it is present in the bulk solution at a large concentration in relation to competing species. When we add calcium sulfate (gypsum) to soil, increasing the concentration of dissolved Ca, we expect to see more Ca ions on the exchange complex and fewer of the other ions. (In the solution we will see more of all the ions.) Weakly attracted ions such as K and Na can displace Ca and Al if they are concentrated, as in saltwater or in KCl added as fertilizer.

Hydrogen ion in cation exchange is a special case. It competes poorly with other cations because of its single charge, its large hydrated radius (Table 3–2), and its low concentrations even in the most acidic soil solutions (0.1 mmol at pH 4). Thus $H^+$ is nearly always a minor fraction of the total exchangeable cations. Nevertheless, soil colloids with pH-dependent surface charge adsorb $H^+$ in large quantity when the soil is acidified and release $H^+$ when the pH is raised (Figure 3–15). This adsorption is selective for $H^+$ and, like other specific adsorption reactions (see Section 3.4.2), does not depend simply on electrostatic attraction.

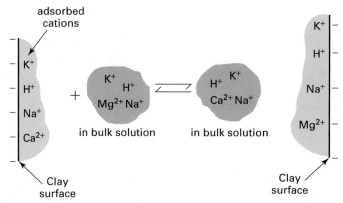

**FIGURE 3–14**  Cation exchange occurs when a loosely held cation in soil solution very near a charged surface changes places with a cation from the bulk soil solution. Here a $Mg^{2+}$ in the bulk solution is exchanged with a $Ca^{2+}$ at the clay surface. The reaction is reversible.

## TABLE 3–2

Selected Ions and Ion Characteristics

| Ion | Charge | Radius (nm) | | Atomic weight[a] (g) | Equivalent weight[b] (g) |
| | | Nonhydrated | Hydrated | | |
|---|---|---|---|---|---|
| Al | +3 | 0.050 | — | 27 | 9 |
| Ca | +2 | 0.099 | 0.300 | 40 | 20 |
| Mg | +2 | 0.065 | 0.400 | 24 | 12 |
| Na | +1 | 0.095 | 0.215 | 23 | 23 |
| K | +1 | 0.133 | 0.155 | 39 | 39 |
| H | +1 | — | 0.455 | 1 | 1 |
| O | −2 | 0.145 | — | 16 | 8 |

[a]The mass of 1 mol of atoms.
[b]The mass necessary to exchange for 1 mol of charge, which is equivalent to 1 mol of hydrogen.

The concentration of hydrogen ions in soils is significant in many ways (Chapter 11). Supplement 3–3 gives introductory background on pH.

**Cation exchange capacity**  Soils have a **cation exchange capacity (CEC),** the total amount of exchangeable cations that can be held by a given mass of soil. In cation exchange, each positive charge is balanced by an equal negative charge. Table 3–2 shows that the cations are singly, doubly, triply, or quadruply charged, and so one $Al^{3+}$ ion can replace three $K^+$ ions. Because $Al^{3+}$ is missing three electrons, it can balance three of the extra electrons

(negative charges) on the clay. An aluminum cation is equivalent to three potassium cations.

A useful unit to express exchange capacities and quantities of ions being exchanged is the **equivalent,** or $mol_c$, the quantity of ion that supplies 1 mol of charge ($mol_c$). The milliequivalent, or $mmol_c$, is $10^{-3}$ equivalent.

The *equivalent weight* of a cation or anion is its atomic weight divided by its charge. For example, the equivalent weight of a divalent ion such as calcium is the atomic weight divided by 2. One mole of an element is its atomic weight. For monovalent ions, equivalent weight, or $mol_c$, is equal to atomic weight. Table 3–2

$$R{-}COOH \underset{+H^+}{\overset{-H^+}{\rightleftarrows}} RCOO^- \qquad \text{Humus}$$

$$Fe(OH)_2{}^+ \underset{+H^+}{\overset{-H^+}{\rightleftarrows}} FeOOH \underset{+H^+}{\overset{-H^+}{\rightleftarrows}} FeOO^- \qquad \text{Sesquioxide}$$

**FIGURE 3–15** The charge on some humus particles and on nonsilicate clays is often determined by the pH of the soil solution. On the left side of these equations is the protonated species, and on the right is the deprotonated species. Deprotonation results in a net negative charge. Protonation adds an extra $H^+$ to the particle, reducing the negative charge or creating a net positive charge.

gives the equivalent weights of cations and anions commonly found in soils.

The cation exchange capacities of some soil horizons from the United States are given in Table 3–3; the data are for some of the profiles illustrated in Color Plates 1 through 10 (Chapter 12). The type and amount of clay and organic matter affect the total cation exchange capacity. Soils with dominantly smectite clays have a higher exchange capacity than those that have kaolinitic clays, because smectite clays have more isomorphous substitution than kaolinite. Similarly, soils with much humus have a higher cation exchange capacity than those with the same amounts and types of clay but less humus. In soils with significant amounts of humus or hydrous oxide clays, CEC increases with increasing pH. Measuring CEC is considered in Supplement 3–1, and example CEC calculations are given in Supplement 3–2 at the end of this chapter.

The pH-dependent charge is essential to the CEC of humus, which arises mainly from the dissociation of active groups such as carboxyls (Figure 3–15).

**Anion exchange** Anion exchange is associated with hydrous oxides of Fe, Al, and Mn and with octahedral groups of the silicate clays. The OH groups of the minerals accept protons (hydrogen ions) in acidic environments (Figure 3–15). When this happens the minerals develop a net positive (pH-dependent) surface charge and can attract anions (**anion sorption**).

Most soils have very little anion exchange capacity, but it does affect retention and movement of anionic pesticides such as 2,4–D and accounts for a degree of retention of $SO_4{}^{2-}$, $NO_3{}^-$, and $Cl^-$ in some acid oxidic soils. Ion size and charge have similar influence in simple anion exchange to that in cation exchange. In nearly all cases, simple electrostatic anion exchange in soils is overshadowed by stronger adsorption processes specific for particular anions.

### 3.4.2 Specific Adsorption Reactions

A diversity of specific reactions lead to strong retention of specific solutes, including the following important nutrients, toxins, and pollutants: phosphate, arsenate, selenate, molybdate, borate, chromate, and fluoride. Electrostatic attraction is involved but may

## TABLE 3–3
Characteristics of Some Soil Horizons from the Soils Illustrated in the Color Plates

| Soil series | Sample location | Horizon | Major clay | $OC^a$ (%) | Clay (%) | CEC $(cmol_c/kg)$ |
|---|---|---|---|---|---|---|
| Exum | NC | Bt1 | Kaolinite | 0.14 | 19.7 | 6.3 |
| Hazleton | PA | Ap | Mixed[b] | 1.70 | 10.6 | 12.7 |
| | | C1 | Kaolinite | 0.05 | 5.1 | 4.4 |
| Houston | TX | A2 | Smectite | 0.84 | 59.4 | 48.7 |
| Nicollet | MN | Bw1 | Mixed | 0.40 | 33.6 | 30.2 |
| Tavares | FL | A1 | Kaolinite | 1.20 | 2.0 | 6.0 |
| Vilas | WI | E | Smectite | 0.62 | 2.5 | 2.6 |

[a]OC, organic carbon content.
[b]Mixed, no single clay type dominant.

be less important than electron sharing and coordination (how well things fit). The terminology—**specific adsorption,** chemisorption, **ligand exchange,** coprecipitation, surface complexation, **surface chelation,** and so on—can become confusing and is not very relevant here.

**Covalent bonding *to organic matter*** Covalent bonding, a sharing of electrons, retains phosphate, sulfate, and some organic chemicals such as benzylamines. Release largely depends on microbial decay of the organic matter.

**Specific anion sorption** The retention of phosphate and similar anions on particular clay minerals is important to soil fertility, water quality and waste disposal, and geochemistry. Unlike simple anion exchange, specific phosphate adsorption occurs on negative and neutral surfaces as well as positive ones. The anion is adsorbed by becoming bound into the surfaces of minerals that have OH bound to Fe or Al, such as the hydrous oxides, hydroxides, or amorphous minerals and, to some degree, the edges of silicate clays such as the kandites (Figure 3–16A). Anions such as $H_2PO_4^-$, $HPO_4^{2-}$, $H_3SiO_4^-$, and $SO_4^{2-}$ can replace the mineral's surface OH groups that are bound to Fe or Al, with the anion's own O or OH fitting into the mineral's surface structure (Figure 3–16A).

The process dominates the geochemistry and soil chemistry of phosphate and molybdate. It holds P in soil, sometimes tightly enough to cause plant deficiency. It is the main reason that P deficiency limits algal growth in most lakes, rivers, and oceans, so that accidental inputs of P can greatly disturb these ecosystems by permitting increased growth. Closely similar reactions are involved in the retention of borate, selenate, and arsenate, and somewhat related reactions can bind basic organic chemicals such as amines.

**Specific metal sorption by oxide minerals and humus** Most polyvalent (more than one charge) metals, including aluminum, iron, nickel, cobalt, copper, zinc, cadmium, and mercury, are immobilized by precipitation and by specific adsorption on surfaces of manganese and iron oxides. This immobilization is so effective that only traces of these metals occur as free or exchangeable cations. The process of sorption on metal oxides, involving coordination and electron sharing, is suggested by Figure

3–16B. Some of the polyvalent metals also form ring-structured surface chelate complexes with closely spaced negative groups on humus (Figure 3–16C). In alkaline soils rich in organic matter, this process can so immobilize zinc that plants barely grow. Soluble complexes with small organic anions, unlike the insoluble complexes formed with humus, may be the main form in which these polyvalent metals become mobile in soils.

**Precipitation and dissolution** When a mineral dissolves—during weathering, for example—ions from its surface dissociate from the solid, hydrate, and go into solution. This **dissolution** process tends to continue until the concentrations of dissolved ions reach an equilibrium point of saturation (defined by the temperature and the particular solid substance's *solubility product*). Soil solutions tend to be at or near saturation in regard to each major mineral in the clay fraction. Any increase in concentrations—for instance, by evaporation of water—may induce **precipitation** of fresh mineral onto existing surfaces. This might be considered another form of specific adsorption.

### 3.4.3 Weak Sorption Reactions

There is a spectrum of weak, unspecific processes that bind organic molecules (amino acids, pesticides, enzymes, humus, and so on) and water to clays and bind hydrophobic/lipophilic substances (oils, paraffins, and nonpolar pesticides) to humus.

**London forces and polar bonding** Collectively called **van der Waal's forces,** these very weak, short-range forces become additive when large molecules get intimate. **London forces** are attributed to resonances between transient dipoles or electron vibrations in adjacent atoms. The somewhat stronger hydrogen bonding is attributed to partial sharing of electrons between permanent dipoles, which are groups or molecules with an uneven distribution of charge, such as OH groups. This can also be represented as sharing of H between O atoms. It makes water viscous and sticky and helps water bond to soil colloids (Figure 3–17) (Section 3.3). Because water is everywhere on surfaces in moist soil, hydrogen bonding acts as a glue holding clays, organic colloids, larger particles, and hydrophilic solutes together.

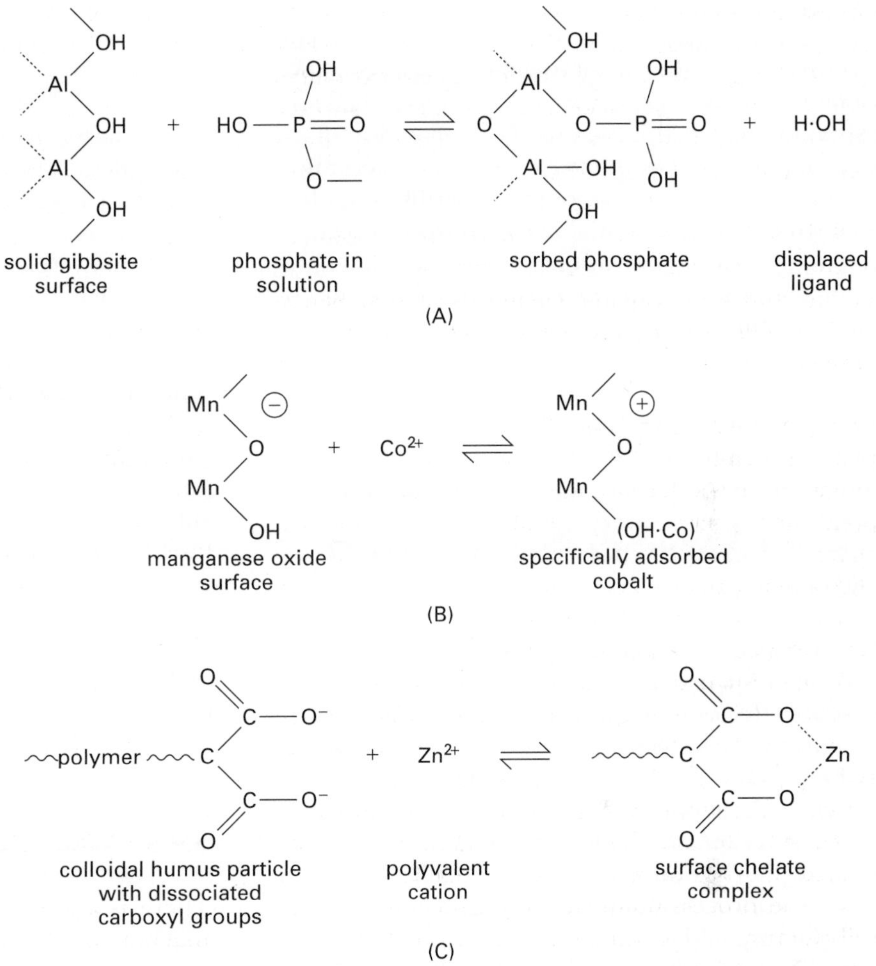

**FIGURE 3–16** Some representative specific adsorption reactions. (A) Adsorption of phosphate on aluminum hydroxide (ligand exchange). (B) Adsorption of cobalt on manganese oxide. (C) Adsorption of zinc on humus (surface chelation).

solid gibbsite surface        phosphate in solution        sorbed phosphate        displaced ligand

(A)

manganese oxide surface        specifically adsorbed cobalt

(B)

colloidal humus particle with dissociated carboxyl groups        polyvalent cation        surface chelate complex

(C)

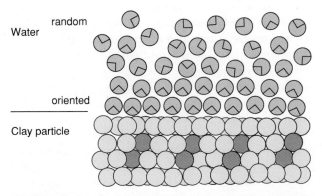

**FIGURE 3–17** Water molecules near a clay surface. Note that the molecules nearest the surface are oriented with their more positive side toward the negatively charged clay. Those in the bulk solution are randomly oriented.

The weak forces also supplement the ionic bonding of large organic cations or anions to clays or humus. Examples include the retention of some pesticides and of proteins exuded by microbes to act as external enzymes in the breakdown of complex organics.

**Phase partitioning**  Soils can retain even nonpolar, hydrophobic, lipophilic organic compounds such as alkanes, alkenes, aromatics, and chlorinated hydrocarbons of petroleum, pesticides, industrial solvents, and insulating fluids. In doing so, they restrict the spread of spills and leakages and partially protect groundwater. Clay adsorbs these compounds, but with little effect unless the soil is dry, very clayey, and high in smectite or vermiculite. Humus is much more sorptive. *Sorption* is used here to include adsorption and absorption, because the exact nature of phase partitioning is complex and not fully understood. The nonpolar or weakly polar organic substances move from the highly polar soil water to the less polar sorptive solid phase within humus. This separation of organic molecules from the soil solution to an accepting solid phase is an example of phase partitioning. It is analogous to the industrial and analytical process of "stripping" or extracting organic substances from water by shaking with an organic solvent. Properties of the contaminant molecule such as size, shape, ionization tendencies, and water solubility affect its tendency to be sorbed into humus. And the drier the soil, the more readily the organic molecules can expel water from the humus and replace it.

## 3.5 ORGANIC CHEMICALS AND THEIR ADSORPTION

The adsorption of manufactured organic industrial and agricultural chemicals in soils reduces their immediate effects on organisms (targeted or not), their volatilization into the air, leaching into groundwater, and entry into food chains. It slows their decay and increases their persistence in the soil. They are retained in or on soil solids by the same processes described in Section 3.4. The particular sorption process and its strength depend on the molecule's size, polarity, and ionization.

The molecules of organic electrolytes ionize in water and therefore adsorb strongly on oppositely charged surfaces. But even uncharged molecules can have polarity—uneven distribution of the density of electrons—because different atoms have different electronegativity, or tendency to attract electrons. (Atoms with small size and many protons are most electroneg-

ative.) The electronegativity of common elements in organic substances ranks approximately thus:

$$H, P < C, S < N, Cl < O < Br, F$$

Bonds between atoms of similar electronegativity have little polarity (e.g., H–C and H–S) or none (e.g., H–H and C–C). But bonds such as H–N, H–O, or C–O are distinctly polar (the O or N has a partial negative charge relative to the H or C). Molecules are highly polar if they contain prominent, strongly polar bonds.

Polar and ionized solutes associate readily with other ions, charged clay surfaces, and polar solvents such as water. All are hydrophilic. Nonpolar (lipophilic) solutes are squeezed into whatever nonpolar or weakly polar environment might be available. Octanol, $CH_3(CH_2)_7OH$, because of its long C–H chain, is nonpolar relative to ethanol, $CH_3CH_2OH$. Ethanol and water, both distinctly polar, mutually dissolve in all proportions with minimal shaking; but no matter how much you shake a mixture of octanol and water, they always separate into two phases (slightly wet octanol and slightly alcoholic water). A solute shaken with octanol and water partitions between the two phases according to its own degree of polarity. The *octanol coefficient* is high for nonpolar substances and low for electrolytes and highly polar substances.

### 3.5.1 Nonpolar Substances

Examples of nonpolar substances include hydrocarbons such as trimethyl-pentane (isooctane in petrol fields) (1), cyclohexane (2), benzene (3), toluene (4), naphthalene (5), and phenanthrene (6).

These are weakly retained in soils by sorption ("partitioning") into humus, more so if humus is abundant, the soil is dry, and the substance's octanol partition coefficient is high. Otherwise, they move freely, unless they are thick oils and waxes (e.g., very large hydrocarbon molecules).

### 3.5.2 Polar Nonionic Substances

Examples of polar nonionic substances include aniline (7), phenol (8), pyridine (9), chlorobenzenes (10), dichlorodiphenyl trichloroethane (DDT) (11), and polychlorinated biphenyls (PCBs) (12).

Such substances adsorb weakly on surfaces of clay and humus, probably orientating themselves with water molecules, OH groups, and charge sites at the solid–water interface. Organic electrolytes are discussed in Supplement 3–6.

## 3.6 ADSORPTION ISOTHERMS

Each soil material and sediment has, depending on its colloid components, a particular capacity to retain a particular solute ion or substance. All but the sandiest soils have significant sorption capacity. In no case is it unlimited or an absolute quantity. In fact, as more and more of a solute accumulates in the adsorbed state, it is held more loosely and its free concentration increases in the solution phase.

Soil and environmental chemists use a notion of *retention capacity,* a graph or index showing how the free concentration and the adsorbed concentration relate for the particular solute and soil material. For example, if large additions of copper produce very small increases of free copper in the soil solution, the soil is said to have a large copper-sorption capacity. Such *sorption isotherms* or indices are measured experimentally. They help to predict the retention and movement of plant nutrients, toxins, and pollutant chemicals and the effects of leaching and disposal practices.

Figure 3–18A is an example showing that a particular hydrous iron oxide holds aminobenzoate more strongly than nitrobenzoate. In this case, the isotherms are straight lines, characterized simply by a single sorption index (the slope of the line, which is about five times larger for the aminobenzoate).

Isotherms are more commonly curved; single indices are then approximations good for only a small range of concentrations.

Figure 3–18B shows phosphate adsorption isotherms for the A horizons of three tropical soils. The large differences are due to clay mineralogy; the alfisol has mixed clays, the oxisol has kandite and hydrous Al and Fe oxides, and the andisol has amorphous clays with extreme sorption capacity. Correcting the andisol's phosphate deficiency would be exorbitantly expensive; this soil would be ideal for "filtering" phosphate and other oxyanions from wastewater.

## 3.7 PORE RETENTION

The small pores and vesicular pores in a soil retain solutes, albeit incompletely, regardless of adsorption. For instance, most soils do not adsorb nitrate, chloride, bicarbonate, and sulfate. Yet some of these

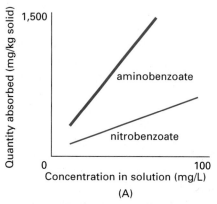
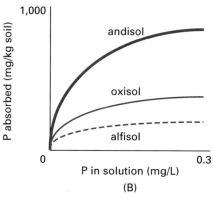

FIGURE 3–18    Examples of sorption isotherms. (A) Adsorption of two organic solutes on a hydrous iron oxide. (B) Adsorption of phosphate on three topsoils with different sorptive capacities (high to very high).

anions and their accompanying cations normally remain in solution, even in the A horizon, despite heavy rain or irrigation. This happens because water does not flow uniformly through the whole pore system. Most of it travels through the largest, best-connected pores (such as cracks, root channels, and interaggregate voids); it bypasses the smaller pores, and only a portion of their solutes diffuse out fast enough to join the moving stream and escape.

## 3.8 SUMMARY

Water is vital to life; it flows into and through soil and is held by soils for use by plants. The physical characteristics of soils, such as the number and sizes of pores, greatly influence the amount of water that can flow through soils in a time period and the amount that can be held in soils against the pull of gravity. Water vapor is always present in soils. Water flows through the soil pores. It is held by soils because it is a polar molecule and is attracted to the surfaces of clays and organic matter.

The solid and liquid phases interact in soils. One of the most important interactions in soils is cation exchange, in which positively charged ions are attracted to the negatively charged surfaces of clays and humus, where they are loosely held. These cations are a major source of plant nutrients. Hydrogen ions are abundant in many soils, and the concentration of hydrogen ions is controlled by the interaction between the soil's solid and liquid phases. In turn, hydrogen ion concentration influences many reactions, including weathering and the availability of plant nutrients.

## QUESTIONS

1. What is the meaning of *relative humidity, diffusion, mass flow, porosity, bulk density, polarity, hydrated, cation, anion, field capacity, saturation, ion exchange,* and *hydrogen bonding?*
2. Why does the composition of the soil air change from winter to spring?
3. What is available water?
4. How much water is in a clay soil, compared with a sandy soil, at the field capacity and the permanent wilting point?
5. How does cation exchange affect plant growth?
6. What is an equivalent?
7. Would a soil high in clay but low in organic matter have a higher cation exchange capacity than would a soil low in clay and high in organic matter? Why? How could you test such soils to find out whether you were correct?
8. What is the difference between surface chelation and cation exchange?
9. Describe a soil you might use as a filter bed to remove (a) phosphate and (b) benzene from water.
10. A 100-cm$^3$ soil sample is saturated. The water content $V_W$ is 45 cm$^3$. What is the volume of pores?

11. The soil in Question 10 begins to drain. Calculate the volume of air at the following water contents: 70, 50, and 30 percent.
12. What is the fraction of the total volume filled with air and fraction volume filled with water at each of the moisture contents in Question 11?
13. What is the porosity for soil samples with the following volume of pores ($V_P$) and total volume ($V_T$)?

| $V_P$ | $V_T$ |
|-------|-------|
| 30 | 60 |
| 20 | 50 |
| 40 | 70 |

14. Calculate the bulk density of a 110-cm$^3$ soil sample with an oven-dry weight of 250 g.
15. Calculate the bulk density of a 100-cm$^3$ soil sample that is 12 percent water (mass basis) and weighs 160 g.

## FURTHER READING

Bohn, H., B. McNeal, and G. O'Connor. 1985. *Soil chemistry*. 2d ed. New York: Wiley.

McBride, M. P. 1994. *Environmental chemistry of soils.* New York: Oxford University Press.

Scow, K. M., and C. R. Johnson. 1997. Effect of sorption on biodegradation of soil pollutants. *Advances in Agronomy* 58:1–56.

Sparks, D. L. 1995. *Environmental soil chemistry.* San Diego: Academic Press.

Tan, K. H. 1993. *Principles of soil chemistry.* New York: Dekker.

Xu, S., G. Sheng, and S. A. Boyd. 1997. Use of organoclays in pollution abatement. *Advances in Agronomy* 59:25–62.

## SUPPLEMENT 3–1   Measuring CEC

The measurement of cation exchange capacity takes advantage of the easy replacement of one cation for another on clay and humus surfaces. Cation exchange capacity is measured in the laboratory by two common methods. In the first, a soil sample is saturated with a selected cation, often ammonium ($NH_4^+$). Cations on the exchange complex are replaced by ammonium ions by mass action, because the number of ammonium ions is large compared with the number of any other ion. Excess ammonium is washed away, usually with alcohol, and a solution of some other salt is used to exchange and remove the exchangeable $NH_4^+$ ions from the soil. The amount of ammonium ion removed with this second solution is measured; it is equal to the CEC of the soil sample. One mole of charge (mol$_c$) of ammonium has a mass of 18 g, and 1 mmol of charge has a mass of 18 mg. If a 5-g soil sample retains 27 mg of ammonium, then it has a cation exchange capacity of 27 mg/18 mg/mmol of charge = 1.5 mmol of charge. Cation exchange capacity may be reported as millimoles of charge per 100 g of soil. The sample that held 27 mg of ammonium was 5 g, or 1/20 of 100 g. Multiply 1.5 mmol of charge per 5 g by 20 to find that 100 g of soil has a cation exchange capacity of 30 mmol of charge per 100 g of soil.

In the second method, a solution containing a single cation is used to remove the cations from the soil in the same way as in the first method. However, instead of the amount of the single cation retained by the soil, we measure the sum of the mole of the Ca, Na, Mg, K, and H washed from the soil. The cation exchange capacity measured by the two methods should be nearly the same. The second method is useful because it indicates both the composition of the exchangeable cations and the total amount. The two methods give different results if the soil has significant pH-dependent charge or a large content of soluble salts.

## SUPPLEMENT 3–2   Example CEC Calculations

Calculation of cation exchange capacity requires conversion of mass to moles of charge and moles of charge back to mass. The conversions are done most easily if units are used consistently and carefully. The example given in Supplement 3–1 shows one such conversion.

What is the cation exchange capacity of a 100-g soil sample that contains 160 mg $Ca^{2+}$, 23 mg $Na^+$, 117 mg $K^+$, and 96 mg $Mg^{2+}$? One mmol$_c$ of Ca is 20 mg, of Na is 23 mg, of K is 39 mg, and of Mg is 12 mg. Note that 1 mmol$_c$ of Ca and Mg is one-half the atomic mass of Ca and Mg because each has 2 mmol$_c$/mmol. Potassium and sodium are monovalent and have 1 mmol$_c$/mmol.

Divide the mass of each ion by the number of millimoles of charge per mole to determine the mass of 1 mmol$_c$. Next, divide the mass of each ion on the exchange complex of the sample by the mass per millimoles of charge. For calcium, 160 mg/20 mg/$mmol_c$ results in 8 $mmol_c$ of Ca per 100 g soil. The same calculation for each of the other ions results in 1 $mmol_c$ of Na, 3 $mmol_c$ of K, and 8 $mmol_c$ of Mg. The sum of millimoles of charge for the four ions is the cation exchange capacity of the sample, 20 $mmol_c$/100 g.

The same calculation can be reversed. If a soil has a cation exchange capacity of 32 $mmol_c$/100 g, and 30 percent of the CEC is $Ca^{2+}$, what is the mass of calcium on the exchange complex? Thirty percent of 32 $mmol_c$ is 9.6 $mmol_c$/100 g soil. One $mmol_c$ $Ca^{2+}$ has a mass of 20 mg. The total mass of calcium on the exchange complex is 9.6 $mmol_c$/100 g soil $\times$ 20 mg/$mmol_c$ = 192 mg/100 g soil.

## SUPPLEMENT 3–3   pH

Hydrogen ion concentration in the soil solution influences most solid–liquid interactions in soils. (The pH is considered in detail in Chapter 11.) Because of its importance, hydrogen ion concentration is one of the most frequently measured and reported soil properties. The pH is convenient for expressing the small concentrations and large concentration variation found in nature and is defined as follows:

$$pH = -\log[H^+]$$

where $[H^+]$ represents the hydrogen ion concentration in moles per liter. Each unit change in pH represents a 10-fold change in hydrogen ion concentration.

The **pH** of soils ranges from 4 to 10, but there can be extremes as low as 2 and as high as 12 (Figure 3–19). A pH of 7 is neutral because the concentrations of $H^+$ and $OH^-$ are equal at $10^{-7}$ mol/L. Soils with concentrations below pH 7 are acidic, and those with pH above 7 are basic. Those that are acidic have more

$H^+$ ions than $OH^-$ ions in solution, and the reverse applies to basic soils. The product of $H^+$ concentration times $OH^-$ concentration is always constant:

$$[H^+] \times [OH^-] = 10^{-14}$$

## SUPPLEMENT 3–4   Partial Pressure and Gas Solubility

At a constant temperature, the total pressure exerted by a mixture of gases in a definite volume is equal to the sum of the individual pressures that each gas would exert if it occupied the same total volume alone. The individual pressure of each gas is its *partial* pressure; the *total* pressure ($P_T$) of all soil gases is equal to the sum of the partial pressures of its constituent gases.

Henry's law relates a gas's partial pressure in the air to its concentration in a solution in contact with the gas mixture, at equilibrium. The solubility of $CO_2$ in a soil solution is given by

$$K_H = (H_2CO_3)/P_{CO_2} \text{ or } (H_2CO_3) = K_H \cdot P_{CO_2}$$

where $K_H$ is the Henry solubility constant of $CO_2$, $P_{CO_2}$ is the partial pressure of $CO_2$ gas, and $(H_2CO_3)$ is the concentration of carbonic acid formed when $CO_2$ dissolves in water. $K_H$ decreases markedly with increasing solution temperature.

## SUPPLEMENT 3–5   Measuring Bulk Density

A common method for measuring bulk density is to remove a 200- to 250-g clod (a sample, not necessarily a ped) from a soil horizon, determine its weight and volume at some moisture content, and divide the weight by the volume.

bulk density = clod oven-dry weight/clod volume

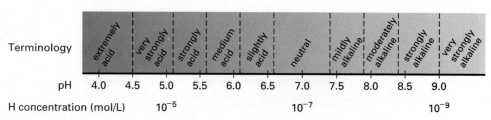

**FIGURE 3–19**   Hydrogen ion concentration, U.S. Department of Agriculture terminology for describing pH, and the "normal" range of soil pH.

Commonly, the oven-dry bulk density is used as a standard moisture state. Weight is determined simply by weighing the clod on a balance, and the moisture content is determined after the clod volume is determined. The clod volume cannot be calculated from measurements on the clod because most clods are irregular shapes. Rather, the clod volume is calculated according to the principle that the weight of fluid displaced by a body is proportional to the volume of the body. The difference between the clod weight in air and the clod weight when the clod is suspended in water is a measure of the clod volume. Another complication is that soils are porous and so must be coated with wax or some other waterproofing compound before they can be weighed in water. The clod weight thus must be corrected by subtracting the weight of the coating to accurately measure its bulk density.

To calculate the weight of a volume of soil, the bulk density is multiplied by the soil (or horizon) thickness and the soil area. For example, to calculate the weight of a 10-cm-thick A horizon in a 1-ha field that has a bulk density of $1.3 Mg/m^3$, use

$$\text{bulk density} \times \text{depth} \times \text{area} = 1.3\ Mg/m^3 \times 0.1\ m \times 10{,}000\ m^2 = 1{,}300\ Mg$$

Note that an Mg is a megagram ($10^6$ g).

## SUPPLEMENT 3–6  Organic Electrolytes

Electrolytes are strongly adsorbed on clay and humus surfaces by electrostatic (ion exchange) attraction. This attraction depends on the ions' charge and size and the soil's exchange capacity. There is usually a pH optimum for the adsorption of particular organic weak bases and anions of weak acids. (The effects are complicated because pH influences the soil's charges as well as the solutes' ionization, in opposing directions.)

Examples of weak organic electrolytes are benzoic acid (1) and 2,4-dichlorphenoxyacetic acid, or 2,4-D (2), which are mainly anionic at pH above 4 or 5; the triazines (3), which are mainly cationic at pH below 7 or 8; and picloram (4), which can be anionic or cationic.

Examples of strong organic electrolytes are paraquat (5), glyphosate (6), and hexadecyl trimethylammonium (HDTMA) (7).

These strong electrolytes are held by cation or anion exchange. The main effect of raising pH is to increase the soil's cation exchange capacity and reduce its anion exchange capacity. In addition, compounds such as glyphosate (6), with multiple amino, carboxyl, or phosphate groups, adsorb strongly and specifically by complexation of metals such as Al and Fe in the surfaces of minerals.

Hexadecyl trimethylammonium (7), a cation, adsorbs electrostatically onto negative clay surfaces. There its long hydrocarbon chains (akin to those of palm oil and soap) constitute a lipophilic coating into which nonpolar molecules can partition. Thus HDTMA addition can convert clays to "organo-clays." They may prove useful as linings or barriers, retaining nonpolar contaminants in wastewater ponds or hindering their passage through aquifers.

(7)

$$H-\overset{\overset{\displaystyle H}{|}}{\underset{\underset{\displaystyle H}{|}}{C}}-\overset{\overset{\displaystyle H}{|}}{\underset{\underset{\displaystyle H}{|}}{C}}-\overset{\overset{\displaystyle H}{|}}{\underset{\underset{\displaystyle H}{|}}{C}}-\overset{\overset{\displaystyle H}{|}}{\underset{\underset{\displaystyle H}{|}}{C}}-\overset{\overset{\displaystyle H}{|}}{\underset{\underset{\displaystyle H}{|}}{C}}-\overset{\overset{\displaystyle H}{|}}{\underset{\underset{\displaystyle H}{|}}{C}}-\overset{\overset{\displaystyle H}{|}}{\underset{\underset{\displaystyle H}{|}}{C}}-\overset{\overset{\displaystyle H}{|}}{\underset{\underset{\displaystyle H}{|}}{C}}-\overset{\overset{\displaystyle H}{|}}{\underset{\underset{\displaystyle H}{|}}{C}}-\overset{\overset{\displaystyle H}{|}}{\underset{\underset{\displaystyle H}{|}}{C}}-\overset{\overset{\displaystyle H}{|}}{\underset{\underset{\displaystyle H}{|}}{C}}-\overset{\overset{\displaystyle H}{|}}{\underset{\underset{\displaystyle H}{|}}{C}}-\overset{\overset{\displaystyle H}{|}}{\underset{\underset{\displaystyle H}{|}}{C}}-\overset{\overset{\displaystyle H}{|}}{\underset{\underset{\displaystyle H}{|}}{C}}-\overset{\overset{\displaystyle H}{|}}{\underset{\underset{\displaystyle H}{|}}{C}}-\overset{\overset{\displaystyle CH_3}{|}}{\underset{\underset{\displaystyle CH_3}{|}}{N}}-CH_3 \quad +$$

# 4

# SOIL CLIMATE

## OVERVIEW

*Our pedon has form and substance. Placed in the environment, it becomes part of the earth's reactive surface. Energy, water, and chemicals flow into, through, and out of the pedon. Energy in the form of solar radiation strikes the pedon's surface, where it may be absorbed or reflected. All soils normally gain heat by absorbing solar energy (shortwave radiation). Heat is then dissipated (spread, redistributed) by several processes: conduction from the hot surface to the subsurface and to the air, evaporation of water, and **reradiation** (longwave **infrared**) to the atmosphere and space. Daily heating and nightly cooling in field soils set up diurnal temperature fluctuations whose amplitude is greatest at*
the soil surface and progressively less with depth. The extremes of soil temperature depend on (1) how much energy comes in, (2) how rapidly heat is dissipated, and (3) how much heat the soil absorbs for a given rise in temperature (its heat capacity).*

*The control and management of soil temperature in the field are dominated by two variables: soil cover and soil water content. Soil cover moderates soil temperature extremes because it blocks incoming radiation by day and impedes outgoing radiation and heat transfer in moving air by night. Soil water moderates extremes of soil temperature because water adds to the soil's heat capacity, helps heat move through the soil, and cools the soil by evaporation.*

## 4.1   SUNSHINE: IRRADIATION AND HEATING

### 4.1.1   Radiant Energy

Electromagnetic radiation, behaving as both waves and fast particles (photons), transmits energy through space at or near the speed of light. The energy level of photons relates to their wavelength throughout the spectrum of different kinds of radiation (see Supplement 4–1 Figure 4–10).

**Radiation** is emitted by all objects, including soil, plants, and the sun. But the rate of emission and the wavelength depend on the emitter's temperature (Figure 4–1A). The hotter the emitter, the more photons it emits per unit time per unit area, and the shorter the photons' wavelength will be. The **emissivity** is a relative measure of an object's ability to emit radiant energy at a given temperature. Figures 4–1B and 4–1C contrast the rate and the wavelengths of emission from a hot surface—6,000 K, like the sun—and a cooler surface—

300 K, like the ground. Note that sunshine (Figure 4–1B) is mainly visible light and near-infrared radiation. In this context, these kinds of radiation are called shortwave radiation, in contrast with the longwave radiation in the far infrared that is emitted by the ground and its cover. This distinction is important. **Shortwave radiation** passes easily through the atmosphere, but **longwave radiation** does not; carbon dioxide and water vapor absorb much of it, preventing its escape and acting as a planetary "blanket" (the **greenhouse effect**). Similarly a cloudy sky decreases the net heat loss and lowers the risk of frost on a cold night.

When radiation encounters matter, several things may happen, depending on the kinds of radiation and matter: (1) radiation may be transmitted, as light is transmitted through transparent air and glass; (2) radiation may be scattered, as light is scattered in all directions by dust, fog, and clouds; (3) radiation may be reflected, and most surfaces reflect a great deal of the incident radiation; or (4) radiation may be absorbed.

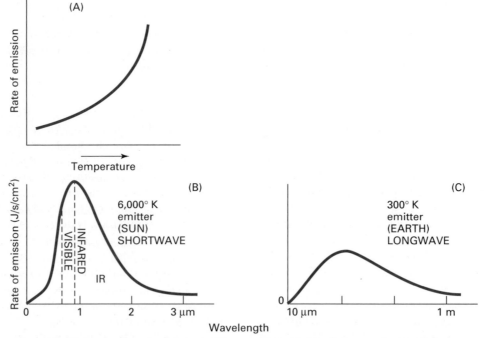

**FIGURE 4–1**   All things emit radiation. (A) The rate of emission and the energy of the emitted radiation increase with the temperature of the emitter. (B) The sun vigorously emits high-energy shortwave radiation (ultraviolet, visible light, and nearinfrared). (C) The cooler earth less vigorously emits lower-energy longwave radiation (far-infrared and radio).

## BOX 4–1
## Temperature Notation

Three scales are used to describe temperature: the Fahrenheit, Centigrade, and Kelvin (absolute) temperature scales. In the English (Fahrenheit) system, the freezing point of water is 32°F and the boiling point is 212°F. In the metric (centigrade) system, 0°C is defined as the temperature of freezing water and 100°C as the temperature of boiling water (at atmospheric pressure). In physics and some other branches of science, zero is defined as the temperature at which mechanical motion of molecules should stop. This is "absolute" zero on the Kelvin scale. Absolute zero is equal to −273°C and −459°F. The freezing point of water, 0°C, is defined as 273 K.

Transmitted, scattered, and reflected radiation remain radiation; only the direction is changed. But when radiant energy is absorbed, it ceases to be radiation; it is converted to electrical or chemical energy, or, most usually, sensible heat. **Sensible heat**—ordinary, everyday heat—can be sensed or felt and consists of an increase in the motion of molecules in materials. It exists only in materials and requires a material medium, or **conductor,** to be transmitted. Radiant energy, by contrast, needs no media; it is sensed by people and thermometers after they have absorbed it and converted it to heat.

In common usage, *heat* includes both sensible heat and certain kinds of radiation (e.g., we speak of heat lamps and the heat of the sun), but in this book, *heat* means sensible heat and excludes all forms of radiation.

### 4.1.2  Radiant Heating and Heat Dissipation

Soils and plants are heated by the absorption of solar radiation, and they in turn heat the air (Figure 4–2). The input and **dissipation** of energy at the ground govern the temperature regimes in the soil and in the plant canopy. In consequence, they strongly influence other processes: air movement near the ground, evaporation, transpiration, and the transfer of carbon dioxide and oxygen.

The intensity of heating—that is, the temperatures achieved—depends on the soil's heat capacity, the rate of absorption of radiation, and the rate of heat dissipation. The next three paragraphs discuss these in turn.

An object's **heat capacity** is the amount of heat (calories [cal] or joules [J]) that it must absorb or lose to produce a 1° change in its temperature. (Note the distinction between heat and temperature. If it is unclear, Figure 4–3 and its explanatory text may help.) The larger the heat capacity is, the more energy must be supplied to raise the object's temperature, and the more must be removed to cool it. For a material, it is common to express heat capacity on a unit mass basis (specific heat capacity in J/g or cal $g^{-1}$). For soil, a porous mixture, it is convenient to express heat capacity on a bulk volume basis. Various minerals, including soil minerals, have specific heat capacities of similar magnitude (Table 4–1). Soil's heat capacity on a bulk volume basis increases with increasing bulk density, because the mass and heat capacity of soil air are negligible. For a given bulk density, the soil's heat capacity increases with water content, because water has a large specific heat capacity. The larger a soil's heat capacity is, the less the soil changes in temperature when it gains or loses a certain amount of heat. Thus, temperatures tend to be least variable in soils that are dense and wet.

Radiation absorption at the ground surface depends on the incoming radiation rate and the reflectance of the surface. Nearly always, the radiation rate is more important. It decreases with decreasing sun angle and day length, at high latitudes, and in winter. Cloud, fog, haze, and dust intercept sunshine. So do plants, mulch, and other ground cover, which keep the soil cool while the cover itself either reflects the radiation or absorbs it and becomes hot. The reflectivity of the soil surface becomes important if the soil is bare, when light-colored soils absorb less sunshine and remain a few degrees cooler than dark soils do.

Heat dissipation from the ground surface depends on four dissipatory processes: reradiation,

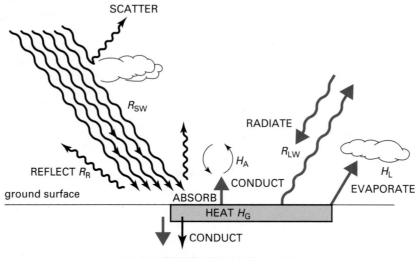

SCATTER

$R_{SW}$

RADIATE

$R_{LW}$

REFLECT $R_R$

$H_A$

CONDUCT

$H_L$

EVAPORATE

ground surface

ABSORB

HEAT $H_G$

CONDUCT

(A) *DAYTIME* ENERGY EXCHANGE

CONSERVATION EQUATION

$$R_{SW} - R_R = R_{ABSORBED} = H_G + H_A + H_L + R_{LW}$$

$R_{SW}$ = incoming shortwave radiation    $R_R$ = reflected SW radiation

$H_G$ = heat to ground    $H_A$ = heat to air    $H_L$ = latent heat

$R_{LW}$ = longwave radiation

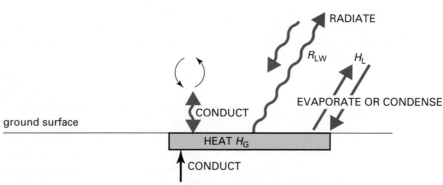

RADIATE

$R_{LW}$     $H_L$

EVAPORATE OR CONDENSE

ground surface

CONDUCT

HEAT $H_G$

CONDUCT

(B) *NIGHTTIME* ENERGY EXCHANGE

**FIGURE 4–2**  Heating and cooling processes at the ground surface.

evaporation, conduction to the air, and conduction to the subsurface soil (Figure 4–2A). The rate of heat dissipation is the sum of the rates of these processes. Heat capacity and conductivity are **thermal properties** of materials.

Loss of heat by reradiation accelerates sharply as the soil (or plant) temperature rises (Figure 4–1A), but even at low temperatures the rate is sig-

nificant. By night, with no incoming solar radiation, the heat loss by reradiation often exceeds the energy inputs so much that plant and soil surfaces become markedly cooler than the subsurface soil or even the air above. Frost and dew are signs of this condition.

**Evaporation** removes heat from the surrounding soil and air as energy (**latent heat** of vaporization) is consumed to break hydrogen bonds and accelerate

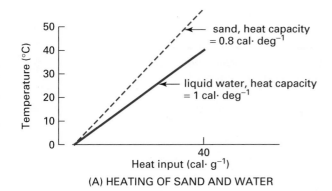

(A) HEATING OF SAND AND WATER

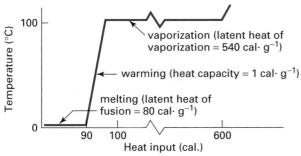

(B) 1 GRAM OF ICE HEATED SLOWLY AND VAPORIZED

**FIGURE 4–3** Graphical representation of heat capacity and latent heat. An object's heat capacity is the amount of heat needed to raise its temperature by 1°C. A material's latent heat of melting (or vaporization) is the amount of heat needed to melt (or vaporize) 1 unit mass of the material. (A) An object's temperature is easy to increase if its heat capacity is low: a comparison of dry sand and water. (B) The heat needed to warm water is small compared with the amount needed to vaporize it.

water molecules moving from the liquid to the gaseous state. Evaporation significantly cools plants and wet soil surfaces, but it is limited in soils with a dry surface layer.

Direct conduction of heat to the air, assisted by mixing when the air is turbulent, helps cool the soil and heat the air at the ground. This transfer goes fastest if the soil is hot, the air cold, and the wind strong. Finally, heat is conducted down into the soil when the surface soil becomes hotter than the subsurface, and up again whenever the temperature profile (or gradient) reverses. This is the normal daily sequence of events (Figure 4–4).

## 4.2 MOVEMENT OF HEAT THROUGH SOIL AND OTHER MEDIA

### 4.2.1 Conduction (Diffusive Transfer)

Conduction is the principal way in which sensible heat moves in soil. The conduction of heat requires a heat-conductive material, a medium. Heat flows more rapidly through some media than others. Thermal conductivity ($K$) is a measure of a material's ability to transmit heat. For example, metals are excellent heat conductors, hence the popularity of metal cookware. Air is a poor heat conductor, hence the common use of air trapped in clothing, blankets, and household insulation to hold heat. Note that the air must be trapped and held stationary to be a good **insulator** and do a good job of impeding heat flow. When air, water, or fluids in general are free to mix or flow, they transmit heat rapidly, by processes (turbulent transfer and mass flow, discussed later) that are important in the atmosphere but not in the soil, where fluids are nearly stationary.

Conductive flow also depends on the area of the conductor's cross section perpendicular to the direction of flow. The flow per unit cross-sectional area is the *flux*.

Finally, conductive flow depends on the driving force, the tendency for heat to flow. This driving force is the **gradient** of temperature. Heat tends to flow from the hotter to the cooler parts of a connected system (just as water tends to flow from wetter to drier parts of a soil). Heat flow is a process in which excited hot molecules share their excitement with cooler neighbors. It is this excitement—the rotational and vibrational energy of molecules—that flows. The process is slow because it depends on individual interactions among molecules more or less fixed in place. Net heat flow stops when it has eliminated temperature differences throughout the system—that is, when the system has reached thermal equilibrium. Thermal equilibrium is unusual in soils.

Fourier's heat flow equation, illustrated in Figure 4–5, relates flow rate ($Q$) to the three factors just mentioned—namely, the cross-sectional area ($A$), the thermal conductivity ($K$), and the temperature gradient ($dT/dx$):

$$\text{Heat flow} = Q = dq/dt = A \times K \times dT/dx \quad (1)$$

## TABLE 4–1

Thermal Properties of Materials

| Material | Density ($g \cdot cm^{-3}$) | Heat capacity ($cal \cdot cm^{-3} \cdot deg^{-1}$) | Heat conductivity ($mcal \cdot (s \cdot cm \cdot deg)^{-1}$) |
|---|---|---|---|
| **Solids** | | | |
| Quartz | 2.6 | 0.5 | 15–30 |
| Rocks, hard (e.g., basalt, slate) | 2.4–2.5 | 0.5 | 4–8 |
| Rocks, soft (e.g., chalk) | 1.8–2.0 | 0.3–0.4 | 1.5–3.0 |
| Sand, dry | 1.5 | 0.3 | 0.8–1.0 |
| Sand, 40 percent water | 1.9 | 0.7 | 4.5–5.5 |
| **Fluids (stationary)** | | | |
| Water | 1.0 | 1.0 | 1.4 |
| Air | 0.001 | 0.003 | 0.06 |

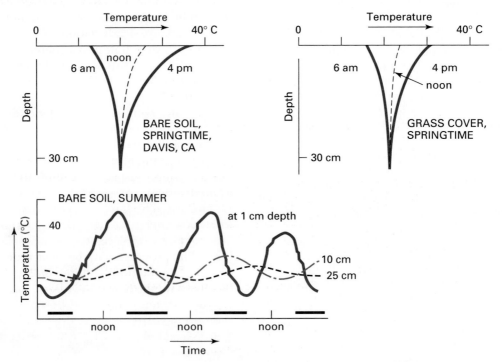

**FIGURE 4–4** Soil temperature profiles fluctuate during the day, especially in bare soil.

## SOIL AND ENVIRONMENT
### *Consequences of Exchanges between the Air and Ground Surface*

- In the long run, gas exchanges at the ground surface collectively affect the atmosphere's concentrations of carbon dioxide, nitrogen oxides, and methane—gases that originate from natural soil processes as well as human activities. They contribute to global warming by absorbing longwave radiation emanating from the earth's surface.
- Weather and climate depend more directly on energy exchanges at the ground surface. To some extent these effects can be manipulated to conserve water and reduce energy use for cooling and heating. Examples are common in greenhouse culture and other intensive horticultural land uses (Section 4.3.3), and urban summer climates can be cooled a few degrees by tree planting and reduction of exposed paving and masonry. Frequent speculations

that reforestation increases rainfall more than it increases evapotranspiration may be unfounded.
- Control of soilborne pests and weeds by means of solar heating is sometimes a feasible alternative to pesticide applications or fuel-intensive cultivation. The process, *solarization,* involves covering the soil with transparent plastic film for a few weeks of sunshine. Shortwave radiation passes through the plastic, heating the soil. Longwave radiation from the soil is trapped, maintaining the elevated soil temperature.
- Soil's low thermal conductivity makes it an effective insulating material for constructing energy-efficient buildings that stay cool in summer and warm in winter. Traditional examples include houses of adobe, mud brick, rammed earth, and sod, as well as underground houses, basements, and caves.

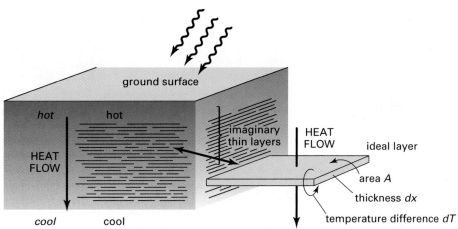

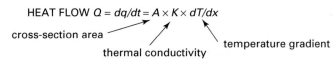

**FIGURE 4–5**   Heat conduction in relation to temperature gradient, area, and conductivity: Fourier's law.

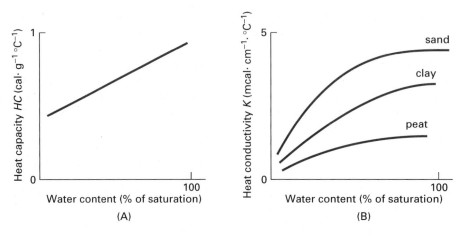

HC = quantity of heat required to change temperature by 1°C

K = quantity of heat (cal) transmitted per second through a layer 1 cm thick across an area 1 cm² when the temperature difference is 1°C (i.e., temperature gradient = 1°C/cm)

HEAT FLOW PATHS

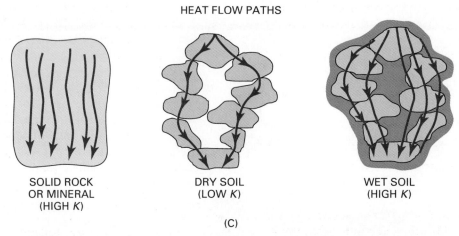

**FIGURE 4–6** Effects of water content on thermal properties of soil. (A) Heat capacity. (B) Heat conductivity. (C) Explanation of effect on conductivity: Water provides better pathways for heat flow than nonconductive air does.

Equation 1 and others derived from it enable the calculation of heat flow rates, temperature profiles, and temperature changes. They remind us that both conductivity and driving force control the heat flow. By combining this idea with the ideas of heat capacity and energy conservation (Section 4.1), you can explain most soil temperature regimes.

### 4.2.2 Heat Capacity, Conductivity, and Temperature Gradients

Heat capacity and **conductivity** in soils depend mostly on the soil's water content (Figure 4–6). The effect of water on heat capacity is simple: The heat capacity of the water is added to that of the solids.

But the effect of water on **heat conductivity** is less simple. Briefly, water, a good conductor, replaces air, a poor conductor. Silicate minerals lose their inherently good conductivity when divided into small particles that are largely separated by air, but water is almost as good a conductor as are the solid minerals themselves (compare the conductivity values for solid quartz, sand, air, and water in Table 4–1). Coating the solid particles with water brings them into better thermal connection.

The most distinct temperature gradients in soil are the vertical temperature **profiles,** which fluctuate and change direction daily in the field (Figure 4–4). At depths below 30 cm, temperatures fluctuate little, following the seasonal mean air temperature with a time lag of about a month. At the ground surface, by contrast, soil temperature can vary by 20 or 30°C daily. Typically, the vertical **temperature gradient** (temperature profile) is greatest when the surface is hottest or coldest—that is, in late afternoon and early morning. The heat flux is also greatest at these times. The temperature gradient becomes greatest in the top few centimeters of soil. The most rapid heat flux, however, may not occur near the surface, because the large temperature gradient will be offset by low thermal conductivity if the surface soil is dry, as it commonly is.

### 4.2.3   Turbulence and Heat Flow in Air and Water

In a solid medium, flows of heat, solutes, and gases are restricted to diffusion and conduction. But if the medium is a fluid in motion, such as air or soil solution, these flows speed up because they no longer depend solely on movement of individual molecules. Modes of rapid transfer in moving fluid media include mass flow and turbulent transfer. (Either or both of these are sometimes called **convection.**)

**Mass flow** is the transport of heat (or solutes) with a stream of fluid, with all the constituents moving together. The rate of mass transport ($Q$) is simply the product of the volume rate of flow of the medium ($V$) and the amount ($C$) of heat (or solute) per unit volume:

$$\text{transfer rate} = Q = V \times C \qquad (2)$$

Heat is carried by mass flow when hot water runs through a pipe. In soils, mass flow is seldom impor-

tant in moving heat, but it is always important in moving dissolved substances through the pedon or to the roots and through the conducting tissue of transpiring plants.

**Turbulent transfer** occurs when conductive flow, driven by a gradient of temperature or concentration, becomes accelerated by physical stirring or mixing of the medium (Figure 4–7). Moving "packets" of the turbulent fluid carry heat and solutes much faster than they could by individual molecular motions. Turbulent transfer resembles diffusion insofar as it depends on a driving gradient and does not require *net* movement of the medium. It resembles mass flow in speed. Water and air seldom move violently enough in soil to become turbulent, but in the atmosphere turbulence dominates the transfer of water, carbon dioxide, other gases, and sensible heat.

Turbulence in the atmosphere includes thermal or *convective* turbulence generated by the rise of air heated at the ground (Figure 4–7A) and the more generally significant turbulence of the wind (Figure 4–7B). Almost all winds are strong enough to produce eddies and gusts that move air not only laterally but also up and down. These vertical components of the wind flow accelerate severalfold the gas and heat flows between the ground and atmosphere.

Turbulent-transfer rates depend on the maintenance of a driving gradient, just as in diffusive conduction, following a rate equation such as Fourier's equation for diffusive heat flow:

$$\text{transfer rate} = Q = A \times Kt \times dT/dx \qquad (3)$$

As in the diffusion equation, $A$ is the cross-sectional area and $dT/dx$ is the temperature gradient. However, $Kt$, the turbulent-transfer coefficient, is a much bigger number than the diffusive coefficient ($K$), and its magnitude varies depending on the degree of turbulence.

## 4.3   PRINCIPLES IN ACTION: CONTROLLING SOIL CLIMATES

This section first summarizes the principles just described and then illustrates how they can be used to explain and predict temperature regimes and related changes in and near the ground surface.

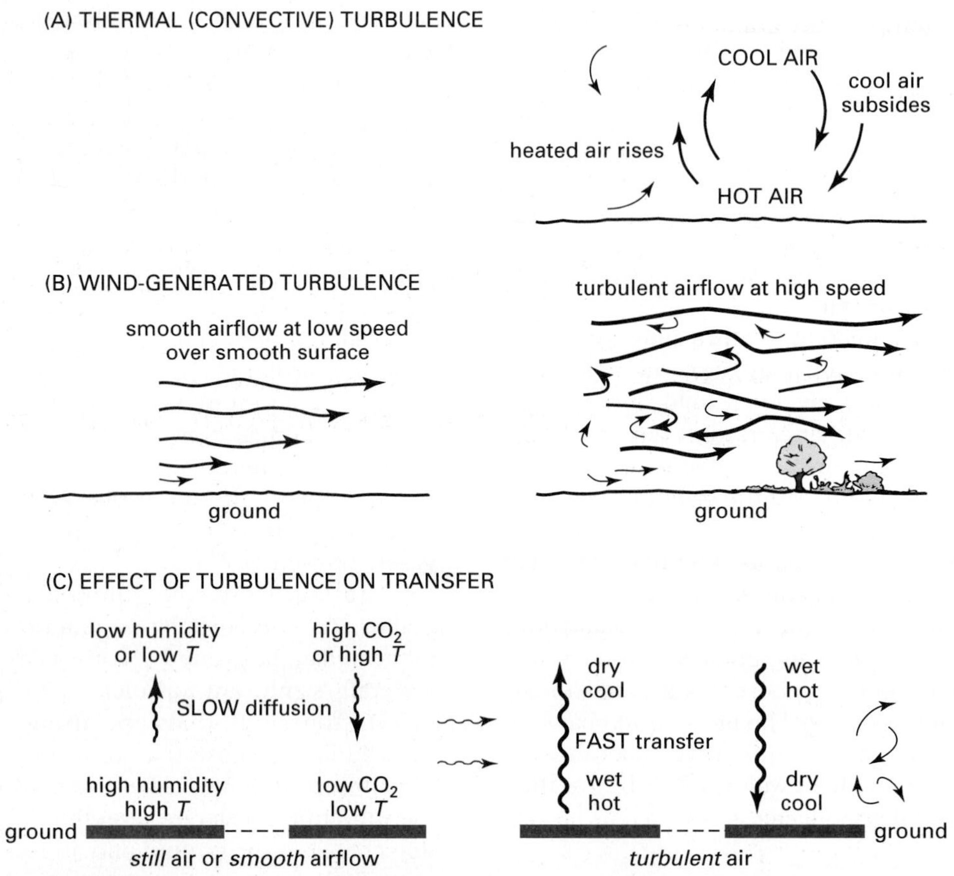

(A) THERMAL (CONVECTIVE) TURBULENCE

COOL AIR

cool air
subsides

heated air rises

HOT AIR

(B) WIND-GENERATED TURBULENCE

turbulent airflow at high speed

smooth airflow at low speed
over smooth surface

ground

ground

(C) EFFECT OF TURBULENCE ON TRANSFER

low humidity         high $CO_2$
or low $T$           or high $T$

SLOW diffusion

high humidity        low $CO_2$
high $T$             low $T$

ground

*still* air or *smooth* airflow

dry          wet
cool         hot

FAST transfer

wet          dry
hot          cool

ground

*turbulent* air

**FIGURE 4–7** Turbulence in the air near the ground: causes and consequences. (A) Convective turbulence caused by local heating of air at the
ground. (B) Turbulence in wind. (C) Turbulence accelerating vertical movements of heat and gases.

## 4.3.1 Summary of Principles

1. Incoming shortwave radiation ($R_{SW}$) that is not
reflected ($R_R$) is absorbed ($R_{ABS}$) and converted to heat. The heat is then distributed by
longwave radiation emission ($R_{LW}$), by heating
the ground (and plants) ($H_G$), by heating the
air ($H_A$), and by evaporating water ($H_L$). Energy is conserved; in the long run, the amount
absorbed equals the amount dissipated:

$$R_{ABS} = R_{SW} - R_R = R_{LW} + H_G + H_A + H_L \quad (4)$$

Thus, the ground gets hotter the more radiation it absorbs and the less it loses by radiation,
evaporation, and transfer to the air. At night,

energy loss normally exceeds energy absorption
(see Figure 4–2).

2. The soil's heat capacity is the amount of heat
that the unit volume (or mass) of soil must absorb per unit of temperature rise. The heat capacity determines how much the temperature
will change when heat is gained or lost, and it
increases as the soil's water content increases.

3. The soil's heat flow is usually conductive and
depends on area, thermal conductivity ($K$), and
temperature gradient ($dT/dx$), so that

$$\text{heat flow rate} = Q = A \times K \times dT/dx \quad (5)$$

The conductivity ($K$) in soil increases markedly with
soil water content. Flows of heat in the atmosphere

also depend on temperature gradients, but the transfer coefficients corresponding to $K$ vary, increasing up to a hundredfold as the air becomes more and more turbulent.

## 4.3.2   Seasonal and Geographic Variations in Soil Climate

Diurnal variations in soil temperature are superimposed on a seasonally shifting mean that is close to the temperature observed at 50 to 100 cm below the ground surface (Figure 4–4). The seasonal pattern reflects variations in sunshine reaching the ground. The sunshine received daily or weekly varies with day length, sun angle, and cloud or fog and tends to be highest where clear skies of the dry season coincide with long summer days—for example, in Mediterranean climates of lower latitudes in the temperate zone. In higher latitudes, less energy is received because of lower sun angles and more clouds. In the tropics, sun angles are always high, but cloud cover, moderate day length, and oceanic influence commonly prevent the development of extreme temperatures. Tropical soil climates are not characteristically hot (at high altitudes they are decidedly cool), but they are characteristically equable; that is, they lack the strong seasonal temperature variation characteristic of the nontropics.

These patterns of variation with season and latitude show the dominant effect of radiation input on soil temperature. But within a given zone and season, field soil temperatures depend mostly on water content and on cover. Wet soils, with high heat capacity and conductivity, tend to get less hot (and less cold) than do dry soils. The cover, normally plants and litter, shades the soil (Figure 4–8); that is, it intercepts some of the incoming radiation, heating the cover itself instead of the soil below. Any removal of cover (by clearing, fire, weeding, overgrazing, or tillage) increases the fluctuations of soil temperature. Another cause of local variation in radiation input is *aspect:* the compass direction of the slope of the land. Higher ground temperatures develop on south-facing slopes, which receive more radiation in the Northern Hemisphere, or on north-facing slopes in the Southern Hemisphere. This can be sufficient to produce different vegetation types—for example, trees and grass on a north slope but only grass on a south slope.

## 4.3.3   Management of Soil Temperature

All phases of plant growth are influenced by soil temperature (Table 4–2 shows some examples). For this reason, management of soil temperature is of interest. Soil temperature can be managed, the aim usually being to improve plant growth. Procedures vary in cost and practicality. In agriculture, the most common procedures are adjustments to fit the soil climate rather than alter it. For example, crops are selected to suit the local climate, planting is delayed until temperatures are right, and early-maturing crops are grown if the warm season is short. Actual modification of soil temperature is more expensive and less common, but the methods are interesting as illustrations of the principles we are discussing.

Raising the soil temperature often improves plant germination and growth. The roots of many agricultural species function poorly at soil temperatures below 15 or 20°C, and some species require even higher temperatures. Small but significant temperature increases can be achieved without the expense of direct artificial heating. For instance, cultivation to remove soil cover can improve germination and early growth because it allows the soil to receive more irradiation. Keeping the surface soil dry can also produce higher temperatures, at least some of the time and near the surface, by reducing both heat capacity and thermal conductivity. Mulching (covering the surface) can both minimize the heat loss and intercept incoming sunshine. The overall net effect is usually to reduce the temperature fluctuation and the average soil temperature. However, if the cover is transparent to solar radiation, it can lead to large increases in soil temperature. Transparent cover lets sunshine in but impedes loss of energy by longwave radiation, evaporation, and conduction to the air (Figure 4–8). Variations on this practice include full-scale greenhouses, smaller-scale hot frames for raising seedlings, and plastic film laid directly on the ground (Figures 4–8C and 4–9). In fact, leaving a clear plastic film in place for several weeks of hot clear weather can heat kill enough weed seeds, nematodes, and other pests to provide worthwhile improvement in subsequent crops.

Lowering the soil temperature is sometimes desirable. Roots grow or function poorly when soil temperatures exceed 40°C, or even 30°C in sensitive species. Such temperatures are commonly achieved, albeit briefly, in exposed soils. Several effective controls depend on curtailing the radiation reaching the

## (A) WIND SPEED PROFILES

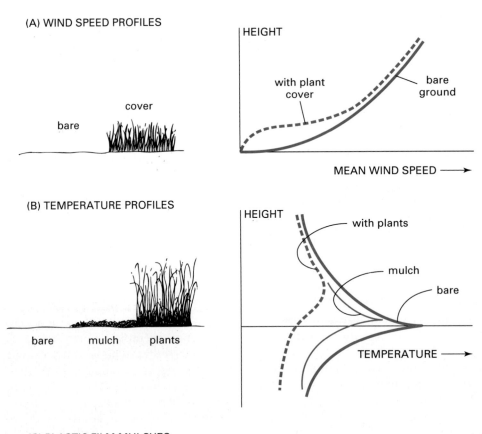

HEIGHT

with plant cover

bare ground

MEAN WIND SPEED ⟶

## (B) TEMPERATURE PROFILES

bare    mulch    plants

HEIGHT

with plants

mulch

bare

TEMPERATURE ⟶

## (C) PLASTIC FILM MULCHES

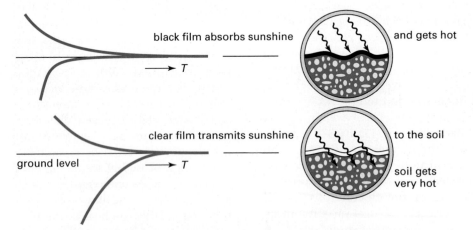

black film absorbs sunshine    and gets hot

⟶ T

clear film transmits sunshine    to the soil

ground level    ⟶ T    soil gets very hot

**FIGURE 4–8** Effects of ground covers. (A) Reduction of wind speed (and turbulence) by plant cover. (B) Moderation of soil temperature by mulch and vegetation. (C) Heating of soil under opaque and transparent plastic film.

**TABLE 4–2**

Soil Temperature and Plant Processes

| Plant process | Optimum temperature (°C) |
| --- | --- |
| Wheat seed germination | 4–10 |
| Bedding plants seed germination | 21–24 |
| Pea seed germination | 24 |
| Sorghum seed germination | >27 |
| Corn seed germination | 35 |
| Minimum for corn seed germination | 7 |
| Root growth of most plants initiated | 4–10 |
| Cool-season grass roots stop growing | 24 |
| Optimum potato (tuber) growth | 16–21 |
| Optimum apple tree growth | 18 |
| Optimum citrus tree growth | 25 |

**FIGURE 4–9**  Use of black plastic film to warm soil and control weeds in vegetable production. (Photo: USDA.)

soil: shading, mulching with nontransparent materials, establishing plant cover early, and maintaining the cover at critical times. Keeping the soil wet helps keep it cool by increasing heat capacity, heat conduction, and evaporation.

Excessive soil heating is a particular danger for plants grown in containers in greenhouses, for two reasons. First, there is no cool subsoil into which heat and roots can escape. Second, a container of soil receives radiation not only from above, as it would in a field, but also from the side. The containers, benches, and floor of a sunlit greenhouse, even in winter, can easily reach 40°C and are the main direct sources of heat for the greenhouse atmosphere. Cooling the air and improving airflow can make the greenhouse tolerable for people, but because of the low conductivity and heat capacity of air, air cooling by itself does not effectively control soil temperature. It needs to be supplemented. Operators can reduce the radiation input in several ways: whitewash or shade cloth on the roof, shading devices for the containers themselves, close spacing so that the plants themselves provide shade, white mulches on pot surfaces, and white or silver paint on benches and pots to reflect shortwave radiation. They can provide additional heat capacity with bigger containers or daytime watering. Or they can improve heat removal by standing the plant containers in cold water, which has sufficient heat capacity and conductivity to remove heat rapidly. This last practice is very effective, but it is messy and waterlogs containers with drain holes.

## 4.4  SUMMARY

Shortwave radiation from the sun is absorbed by the soil, and the soil temperature increases. Conduction, evaporation of water, and reradiation to the atmosphere redistribute and dissipate the heat. Soil surface temperature fluctuates greatly compared with subsoil temperatures. The heat capacity and conductivity of a soil depends mostly on the soil's water content. Moist soils conduct heat faster and have more heat capacity than do dry soils. It is possible (and sometimes expensive) to modify soil temperature. Blocking direct solar radiation reduces soil heating. Improved cover (such as plants, litter, and mulches) or increased soil water content tends to reduce the extremes of temperature fluctuation near the soil surface.

## QUESTIONS

1.  Explain the meaning of *radiation, conduction, turbulent transfer, heat content, heat capacity, temperature, heat of vaporization, shortwave/longwave,* and *thermal conductivity.*
2.  Design a greenhouse that would require minimal artificial heating and cooling.
3.  A mulch is likely to reduce the average temperature of the soil beneath it. But sometimes it is more important to keep the soil's minimum daily temperature from falling too far. How could a mulch help achieve this?
4.  The undersurface of plastic film spread on the ground becomes wet. List the processes of energy transfer and water movement that are involved. What can you do to stop water from collecting on the film?
5.  Why does a thermometer exposed to the sun give incorrect readings of the temperature of the surrounding air?
6.  On a frosty night, orchardists sometimes turn on irrigation sprinklers or large fans to try to prevent their trees from freezing. Explain these practices.
7.  Walking barefoot by the river on a summer afternoon may be more comfortable if you step on rocks rather than the sand. Why is a rock surface cooler than a sand surface under these conditions?
8.  Why does field soil temperature show little diurnal variation at a depth of 50 cm below the ground surface? Why does the seasonal temperature variation at that depth lag behind the seasonal mean air temperature?
9.  Plant nurseries must sometimes be placed in a climate zone either too hot or too cold for sensitive species that are to be cultured. For either case, draw up specifications for a site choice and for management of the plantings so as to minimize the climate's adverse effects.
10. List the conditions necessary for formation of dew. Explain the process.

## FURTHER READING

Hillel, D. W. 1982. *Introduction to soil physics.* New York: Academic Press.

Rose, C. W. 1966. *Agricultural physics.* Elmsford, N.Y.: Pergamon Press.

## SUPPLEMENT 4–1   Forms of Radiation

Forms of radiation are reviewed in Figure 4–10. The concept of wavelength and frequency are shown in the lower portion of the figure. The equation that links energy (E) to either the wavelength ($\lambda$) or frequency ($\nu$) is shown in the shaded box. Note that as the radiation wavelength decreases, the energy increases. Note also, the wide wavelength range in nature and how this range is utilized for communication, observation and science. On the right side of the figure, the molecular motion that coincides with each energy range is shown.

## SUPPLEMENT 4–2   Movement of Heat, Water, Gases, and Solutes

### General Similarities

Heat, water, gases, and solutes are obviously different, but the ways in which they move are similar and can be described by almost identical equations. Heat can move by conduction (diffusive transfer) and by the generally faster processes of mass flow and turbulent transfer, as discussed in Section 4.2. The carriage of solutes in water flowing through the soil is a mass flow process important to plant nutrition (Chapter 9), salt control (Chapter 11), and soil formation (Chapter 12).

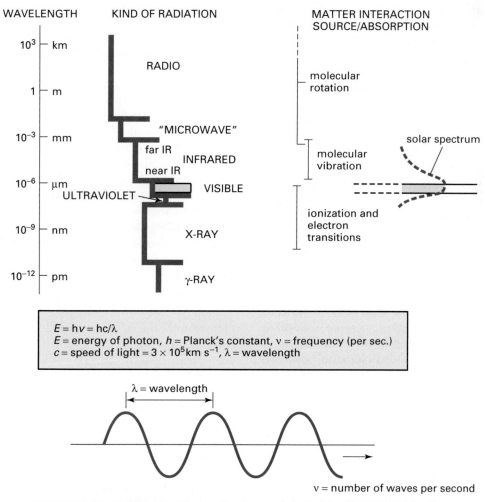

FIGURE 4–10  Forms of radiation. Radiant energy behaves as particles (photons) and waves and takes several forms. High-frequency short-wavelength photons are the most energetic.

The rate depends on the water flow rate and the concentration of the ions in the flowing water. Turbulent-transfer processes are seldom important in soil itself, because soils are not apt to become turbulent, but turbulence in the atmosphere plays a key role in speeding the movement of $CO_2$, $O_2$, and $H_2O$ vapor, as well as heat, through the plant canopy and between soil and atmosphere. The role of atmospheric turbulence in water loss is considered in Chapter 5.

## Conduction (Diffusive Transfer)

Diffusive movements all result from individual molecular-scale events, adding up to net movement because of an imposed gradient. Thus, heat conduction is the net effect of innumerable individual transfers of energy from hotter to cooler molecules. Water molecules tend to wander from wet surfaces and pores that hold them loosely to drier places that hold them more tightly. And molecules or ions tend to escape from regions where they are concentrated until they become evenly distributed.

In each of these cases, the net movement is driven by a gradient (a **driving force** or tendency to move; quantitatively, the rate of decline of temperature, wetness, or concentration per unit distance along the path of flow). For heat flow, the driving force is the temperature gradient. For water movement in soil, the driving force is the water potentialqgradient, arising from differences in soil water content (as well as height, pressure, and solute concentration). For diffusion of gases or solutes, the driving forces are concentration gradients, commonly induced by removals, reactions, additions, or releases of substances by localized chemical or biological processes.

Diffusive movements depend also on transfer coefficients, which reflect both the ease of movement of the heat, water, diffusing molecules, and so on and the constraints imposed by the medium through which they must pass. The overall relationship can be expressed by the following general equation:

flow rate (per unit cross section) = transfer coefficient × gradient

(Figure 4–5 illustrates the case for heat flow.)

Transfer coefficients for heat flow are called *thermal conductivities.* For water movement in soil they are called *hydraulic conductivities,* and for diffusion of gases and solutes they are called *diffusion coefficients.*

The soil water content strongly influences diffusive processes in soil by changing the transfer coefficients. Increasing the water content increases the hydraulic conductivity, the thermal conductivity, and the effective diffusion coefficients of solutes in the soil. This is because heat, solutes, and water itself move in water films and water-filled pores, so that movement becomes freer and more direct as the water content increases. By contrast, increasing wetness inhibits gas movement in soils because gases move almost entirely through the air-filled voids, which become constricted as the air is replaced by water.

## Diffusivity and Related Questions

Our single equation for diffusive flows gives an incomplete description. Some readers may wonder why it contains no capacity terms (heat capacity, water capacity) and what happens when the gradient changes in time and varies along the flow path. We need at least one more equation. Taking heat flow as our example, this other equation says that if we consider the temperature ($T$) of any of the "ideal layers" with cross section $A$, across which the heat is flowing (Figure 4–5), then the rate of change of $T$ with time ($dT/dt$) depends on the spatial r 11 of change of the local gradient ($dT/dx$) with distance ($x$) along the flow path:

$$\partial T / \partial t = D \cdot \partial^2 T / \partial x^2$$

The term $D$, the soil's thermal diffusivity, includes the thermal conductivity ($K$), bulk density ($\rho_b$), and specific heat capacity ($d$). This is saying, among other things, that the temperature shifts less readily if the soil has high density and heat capacity. Analogous forms of this *continuity equation* exist for soil water flow and chemical diffusion. Thanks to computers, they become useful to people who work in these fields.

# 5

# WATER IN THE SOIL–PLANT SYSTEM

## OVERVIEW

*Water is stored in pores and on inorganic and organic surfaces in our pedon. The pore size distribution, the amount of surface area, and the amount of water in the pedon control the rate of water movement. All soils store water, but the storage is never permanent, and the water is rarely stationary. Water is added intermittently by irrigation, snowmelt, and rain. As vapor, water escapes to the atmosphere, either directly (evaporation) or through plants (transpiration). As liquid, water flows out of the soil to the groundwater,*

*springs, and streams and through the soil to roots. Stored water is useless to plants unless it can move to their roots fast enough. The wetter the soil, the more readily the water can move.*

*Plants use hundreds of kilograms of water for each kilogram of plant product. Virtually all the water taken up by plants is transpired—that is, evaporated from leaves. The amount of plant growth in a season normally relates directly to seasonal transpiration.*

*Normally, water moves to leaves fast enough to compensate for the vapor loss from the leaves. A serious shortfall in the uptake rate leads to water stress, which inhibits the expansion of cells, thereby slowing the growth of leaves, stems, and roots and reducing the capacity for further growth.*

*Movement of water in soils, plants, and air—like the diffusion of heat and solutes—depends on a tendency to move, called a driving force or potential gradient. Water moves, if it can, from places where its potential is high to places where its potential is lower. Differences in water potential are made up of components called gravitational, pressure, solute, and matric potentials. This idea formalizes our common knowledge that water tends to move downward, to move under pressure, to move from dilute solutions to more concentrated solutions, and to move from wetter to drier surfaces. Even when very humid, the atmosphere has a much lower water potential than do soils or plant tissues; water in contact with air normally evaporates.*

*Rates of water movement depend not only on differences in potential but also on the ease of water movement, the hydraulic conductivity. In plants, water flows along pathways of different conductivity; healthy vascular tissue conducts freely, but transpiration is sometimes limited by the low conductivity across roots and is often limited by the low vapor conductivity of the leaf surface. In soil, hydraulic conductivity decreases sharply as the soil becomes drier. Without these hindrances to escape, water could not be stored in soil.*

This chapter describes water retention and movement in the continuum of soil, plant, and atmosphere. The next chapter deals with practical aspects of water use, irrigation, and drainage. Figure 5–1 illustrates several commonly applied terms.

## 5.1 WATER STORAGE IN SOIL

### 5.1.1 Water Retention and Capacity

Plants usually store little water. The water they take up is soon evaporated to the atmosphere; that is, it is **transpired.** This process is **transpiration.** Plants will wilt and perhaps die if their transpiration rate exceeds their uptake for long. They survive between rains because of water storage in the soil. But this storage itself suffers recurrent losses consisting of uptake and transpiration by plants, with some direct evaporation from the ground surface and **percolation** to the groundwater below.

Soils retain water because water molecules cohere to one another and adhere to the wettable surfaces of soil minerals and organic matter. The water partially occupies pores and is attached to solid sur-

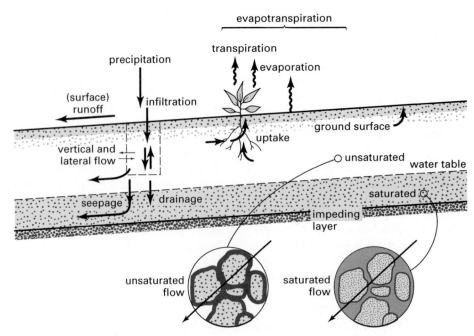

**FIGURE 5–1** Some terms applied to modes of flow in the soil–plant–atmosphere continuum.

faces. Likewise, it moves within the pores and surface water films. Not surprisingly, the water's storage capacity and freedom to move both depend on two soil properties:

1. The colloid content—clay and humus—which determines the total surface area on which water can be held
2. The pore properties—the amounts and proportions of pores of different size and their continuity (how well they are interconnected).

The soil's solid matrix retains water because of **adsorptive forces** that arise from water's properties of hydrogen bonding (the sharing of H among molecules) and dipolarity (charge separation within the molecule). These two properties cause water molecules to cohere to one another and to adhere to compatible surfaces, such as those of humus and soil minerals, that are charged and have many oxygen atoms to share H atoms with water. The same forces explain the general phenomenon of capillary rise, which is observed in the very wet soil immediately above a water table. Supplement 5–1 discusses capillarity.

In practical terms, soil water storage is maximized if

1. The soil is deep
2. The roots exploit a large volume of the soil
3. Each exploited horizon retains a large water content
4. A large fraction of the water is held loosely enough to move freely to roots

## 5.1.2  Water Content and Water Potential

*Volumetric water content* ($\theta$) is simply the volume of water (cm$^3$) in a volume of soil (or leaf or other material) (cm$^3$). Mass water content is the mass of water (g) in a mass of soil (g). This is frequently reported as grams of water per 100 g of soil or grams of water per kilogram (1000 g) of soil. Soil water content is commonly expressed as a fraction of the soil dry weight, or sometimes as a fraction of the bulk volume or of the pore space. Conventionally, soil water content is the mass of water that one can drive from a sample of the soil by drying it to steady weight in a 105°C oven. Box 5–1 outlines methods of measuring water content.

*Water potential* ($\psi$) is a measure of the relative energy level of water in a specimen of soil, plant, or air. At saturation, when all the soil pores are filled with water, soil water potential is normally zero ($\psi = 0$). As the soil dries, the water potential *decreases*, becoming increasingly negative (Box 5–2). Thus, when we say that the water potential in a soil is high, we usually mean the potential is less negative, closer to zero, than when the water potential is low. When the water potential in a soil is high, we also mean that the water is loosely held, highly available, ready to move somewhere else. Soil water potential can be estimated by measuring how much suction is needed to pull water from the soil (with a **tensiometer**) or how much pressure is needed to push water out of the soil (with a pressure membrane apparatus). It is easy to calculate water potentials in air from measurements of relative humidity and to calculate indirectly the water potentials of liquid, soil, or plant tissue samples by enclosing them in an air chamber and measuring the relative humidity after the sample and the air have equilibrated.

Water potential is the amount of work that must be done per unit quantity of pure water in order to transport, reversibly and isothermally, an infinitesimal quantity of the water from a specified source to a specified destination. The terms *potential, pressure* (or suction—negative pressure), and *head* are used when describing water movement in soil. These all describe "potential" but are not the same mathematically. Soil water potential is work/mass, with units of joules per kilogram. Pressure is work/volume, with units of pressure (pascals). Head is work/weight, with units of length (m or cm). *Head* is a convenient term because it only requires measurements of distance from a reference level to calculate differences in soil water potential.

We shall later discuss how water potentials in general depend on gravity, pressure, and solute concentration, but in soil they depend most strongly on water content. At low water content, water is held tightly in small pores and in thin films where all the water is close to the retentive surfaces. If water is added, water films will become thicker, larger pores will fill, and the added water will be held more loosely. This relationship between increasing water content and greater ease of removal—higher water potential—is crucial to water storage in soil. The relationship, an important soil characteristic, is expressed in the soil's water retention or water release curve (Box 5–2).

# BOX 5–1
# Ways to Measure Soil Water Content

(A) **Gravimetric** method (laboratory). Soil samples are weighed ($W_1$), oven dried at 105°C, and weighed again ($W_2$).

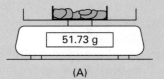

(A)

$$\text{(water removed)}/\text{(oven-dry weight)} = (W_1 - W_2)/W_2 = \text{gravimetric water content}$$

(B) **Neutron probe** (field). Probe inserted into access hole emits fast neutrons. In soil, neutrons are slowed and randomly scattered by collisions with water molecules. The water content of soil near the sensor relates to the count of slow neutrons scattered back to the sensor.

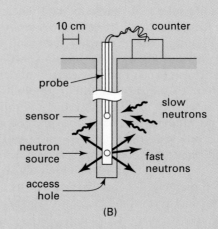

(B)

(C) Buried porous block (field). Buried plaster (gypsum) blocks absorb water, depending on the wetness of the surrounding soil, and water increases the block's electrical conductance. As the soil dries, water leaves the block and the conductance decreases.

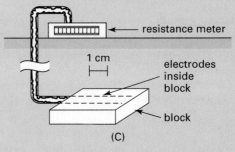

(C)

(D) Time domain reflectometry (TDR). The volumetric water content is measured indirectly by measuring the propagation velocity of an electromagnetic pulse traveling along a group of two or three parallel rods of length $L$ placed in the soil. The velocity is reduced as soil water content increases.

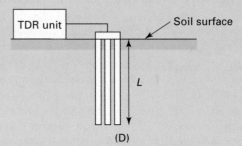

(D)

## BOX 5–2
## Water Retention Curves

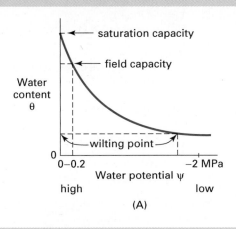

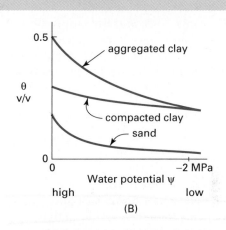

A soil's water retention curve relates the water content to the water potential (matric component). The curve also contains other information about the soil.

1. The soil's porosity (percentage of pore volume) corresponds to the volumetric water content when the soil is saturated, near zero matric potential. In this case, the well-aggregated clay has the greatest pore space.

2. The soil's pore size distribution is indicated by the slope and form of the curve. In the clay soils, for example, the abundance of adsorptive surface and fine pores that hold water tightly is indicated by high water contents held at low potentials, below −1 MPa.[a] Large pores that empty easily are indicated by sharp changes in water content at high potentials, near zero.

3. The soil's field capacity usually corresponds to the water content at about −0.01 to −0.03 MPa matric potential. It approximates the upper limit of useful storage in most soils, from which water in excess of field capacity drains too quickly for plants to use.

4. The wilting point water content usually corresponds to about −1.5 MPa matric potential. At potentials lower than this, water removal by plants goes too slowly to keep plants alive.

5. The available water-holding capacity approximates the amount of water held between the field capacity and the wilting point. Large field capacity need not mean large available capacity; the wilting percentage may also be large. This concept of available or useful water capacity has limited utility; well above the wilting point, water uptake can be slowed enough to restrict plant growth strongly. The degree of drought stress as water content declines in the range between the field capacity and the wilting point depends on the effects of plant and weather on water demand and the effects of plant and soil characteristics on rates of water movement to roots.

---

[a]A pascal (Pa) is a unit of pressure. (1 Pa = 1 newton of force per square meter.) Equivalent pressure is one way to express water potential. A megapascal is $10^6$ Pa.

The soil's water content and water potential always increase together (Box 5–2), but the numerical relationship between them differs from soil to soil, depending on the soil's structure and the kinds and amounts of clay and humus it contains. For instance, if a sand and a clay have the same water content, the water in the clay will be held more tightly—at a lower potential—than that in the sand. If a sand and a clay have equal water potentials, the sand will contain less water than the clay, because the clay has more colloid surface to hold water.

## 5.2   MODES OF WATER MOVEMENT

### 5.2.1   Vapor Flow and Liquid Flow

Water moves in plants, soils, and the atmosphere as both vapor and liquid. Vapor flow through air accounts entirely for the escape of water from plants and soils to the atmosphere during evapotranspiration. Vapor moves from zones of high water potential (i.e., high vapor concentration or humidity) toward zones of low potential. (Chapter 4 discusses diffusion, turbulent transfer, and mass flow.) Vapor flow to the atmosphere is quickest if the atmosphere is dry and windy (conditions producing a large diffusion gradient and greatest turbulent transfer).

In the soil, water vapor moves extremely slowly, for two reasons. First, the soil air is still, so that vapor moves only by the slow process of molecular diffusion. Second, water vapor normally has little tendency to move in soil because differences in vapor concentration are small; even at the wilting point the relative humidity in the soil air is near 98 percent, close to saturation. In soil, vapor movement is usually negligible relative to liquid movement.

Liquid water flows in soil pores and in water films coating particle surfaces. The actual process of movement is thought to be a slipping or sliding of water molecules over other water molecules (Figure 5–2). Flow rates depend on gradients in potential caused by differences in height, pressure, dissolved solutes, and soil wetness. Flow rates also depend on hindrances: friction between water and particle surfaces, as well as pore constrictions and other interruptions in the flow paths. These various hindrances affect the soil's hydraulic conductivity (permeability to water). Water flows faster in the larger pores than in the small pores, and in thicker water films than in thin films, because there is less friction and attraction to surfaces, and flow is interrupted least often in wet soil. For these reasons, the wetter the soil, the more freely the water can move.

### 5.2.2   Saturated and Unsaturated Liquid Flow in Soil

Saturated flow is flow through soil that is saturated (with the pores full or nearly full of water). Saturated flow in soils with large continuous pores can be rapid

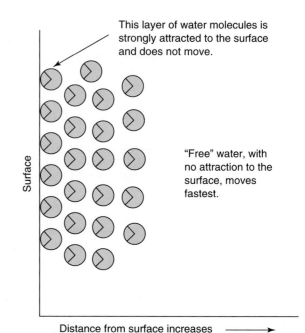

FIGURE 5–2   Water closest to a surface does not move. Water molecules farther from the surface are less strongly attracted to the surface and are free to move faster. Water in the center of a pore moves fastest.

if driven by large differences in gravity and pressure. Saturated flow occurs in aquifers (water-bearing sediments and rock layers), in flooded soils, and in lower horizons of soils with limited drainage. It is sometimes approximated in the upper horizons of ordinary soils as water infiltrates from irrigation or heavy rain.

Unsaturated soil is more common than saturated soil. With no saturation, the larger pores are emptied by gravity. With progressive drying, the remaining water becomes more tightly held and also subject to more friction and more interruptions of flow. The ease with which liquid flows through the soil declines sharply as the soil becomes increasingly unsaturated. On the other hand, gravity is now replaced by an additional force driving the water movement: the difference in water potential. Water moves from wetter soil, where it is held loosely and potential is high, toward drier soil, where it is held tightly in thin films and small pores and potential is low. In unsaturated soil with gradients of wetness, these matric effects, the potential gradient, often dwarf gravity and other driving forces.

**FIGURE 5–3**   In a stratified soil such as this, vertical unsaturated water flow is slow. Each layer has a different texture. (Photo: W. E. Wildman.)

## 5.2.3 Water Flow in Stratified Soils

A fine-textured horizon underlying a coarse-textured horizon retards saturated flow because the small pores do not transmit water as quickly as do large pores. When a fine-textured horizon overlies a coarse-textured one—for example, a clay loam over a sand or a loamy sand—water movement is also retarded (Figure 5–3), but for a different reason.

Under unsaturated conditions, the capillary and surface forces are stronger in the small pores and large surface areas of the finer texture. Until the pressure head is large enough to force the water out of the fine-textured horizon, the water will not flow. This can have several effects, some good and some bad. One good effect is that soils that you might expect to drain quickly and hold little water actually have quite a good water-holding capacity. On the other hand, stratification often impedes drainage. (Drainage of agricultural soils is considered in Chapter 6.)

Water flow need not be only vertically downward; it may be upward, downward, or horizontal. The upward flow of water by capillarity is important and not as rare as you might think. When a **water table** exists in a soil, there is a saturated zone and a zone above the saturated zone that is wet because of the upward movement of water by capillary flow. The upward movement of liquid water can be seen on the edge of an irrigation furrow (Figure 5–4).

**FIGURE 5–4**   Water from these irrigation furrows is moving upward into dry (light-colored) soil by means of capillarity. The dark bands are wet soil at the sides of each furrow. (Photo: W. W. Wallender.)

### 5.2.4   Principles of Water Movement

Soil scientists, plant scientists, and atmospheric scientists apply the same physical notions to describe water movement in soil, plants, the atmosphere, and the interfaces that separate them. We suppose that the water flow rate depends on two things: the **driving force (potential gradient)** and the ease of flow through the particular medium (**hydraulic conductivity**). The dependence of flow on potentials and conductivities is expressed mathematically in different ways. Some are more convenient for a particular application than others, but they all mean the same. Soil scientists, for instance, usually prefer Darcy's law:

$$Q = A \times K \times d\psi/dx \qquad (1)$$

where $Q$, the flow rate, is the amount of water passing per unit of time through cross-sectional area $A$, $K$ is the hydraulic conductivity, and $d\psi/dx$ is the water potential gradient—that is, the change of potential per unit distance along the direction of flow. Plant scientists commonly simplify—by lumping together the area, path length, and conductivity into one term—

conductance, or its reciprocal, **resistance** ($r$). This results in equations such as

$$Q = \Delta\psi/r \qquad (2)$$

where $\Delta\psi$ is the difference in potential along the flow path.

## 5.3   FORCES THAT MOVE WATER

### 5.3.1   Water Potential and Water Movement

The total **water potential** is the sum of gravity, pressure, matric, and solute components (Box 5–3). These components express the various familiar causes of water movement: gravity and pressure differences, the tendency of water to wet surfaces (suction or tension), and the tendency of water to make solutions more dilute (osmotic effect). In most real systems, some of these components dominate, and others vary so little that they can be neglected. For convenience, some people segregate or combine

## SOIL AND ENVIRONMENT
### Pollutant Transport in Soils

A significant environmental problem is determining how far and in what direction hazardous liquid and gaseous materials flow in soils. In general, the rules that constrain and describe liquid and vapor flow of water through soil also control the flow of other fluids and vapors.

When an underground tank used for storage of petroleum products such as gasoline leaks, both liquid gasoline and its volatile components move into the soil. The liquid and gas move at different speeds and react

differently with the solid soil particles than does water. Gasoline does not mix with water, so measuring where and how fast water is flowing does not always tell how fast and where the gasoline is flowing. Because gasoline is a complex of many organic compounds, some may be toxic to soil organisms and some may be suitable substrate (food) for soil organisms. This further complicates the ultimate fate of this and other pollutants as they flow through soils.

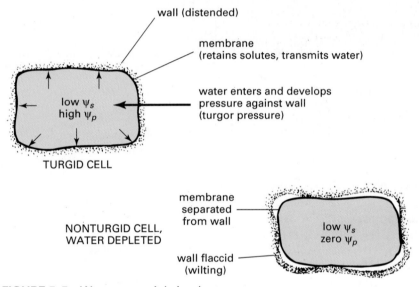

FIGURE 5–5 Water potentials in plants.

these components differently. For example, people dealing solely with soil water (see Hillel, 1982) sometimes combine matric and pressure potentials as one component, and hydraulic engineers often combine pressure and gravity potentials. In all cases, however, the differences in total potential dictate the movement.

In plants, water movement is dominated by solute or **osmotic potentials** caused by salts and organic solutes in cell water. These differences in solute concentrations are maintained by cell membranes.

Also, when the cell is "inflated" with water, it stretches its cell wall, developing a moderate pressure potential, or **turgor pressure** (Figure 5–5). Turgor pressure helps expand growing cells and stiffen plant tissue. Loss of turgor is wilting.

In unsaturated soil, water movement is dominated by **matric potential** differences arising from differences in water content. The terms *suction* and *tension* are common alternatives for *matric potential*, especially in older writings. A lowered matric potential is equivalent to higher suction or tension.

## BOX 5–3
## Principles of Water Movement: A Summary

### Basics

Water tends to flow from where its potential ($\psi$) is high to where its potential is low.

The flow rate depends on the potential gradient $\Delta\psi/\Delta x$, the cross-sectional area, $A$, and the conductivity, $K$.

This relationship can be expressed in different forms. One form is Darcy's law:

$$\text{Flow rate } Q = A \times K \times d\psi/dx$$

In another form, $A$, $K$, and $\Delta x$ are combined into one term—the resistance, $r$—and then flow rate $Q = \Delta\psi/r$.

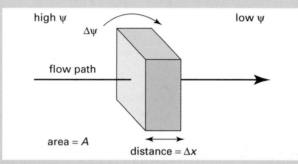

### Components of Water Potential

For practical purposes, we can say that differences in total water potential make water move and that the total potential consists of different components, thus

$$\psi = \text{total potential} = \psi_g + \psi_p + \psi_s + \psi_m$$

$\psi_g$ = gravity component (water flows down)

$\psi_p$ = pressure component (water flows so as to lose pressure)

$\psi_s$ = solute (osmotic) component (water moves from more dilute to more concentrated solutions)

$\psi_m$ = matric component (water moves from wetter to drier surfaces, and from larger to smaller pores)

### Summing the Components

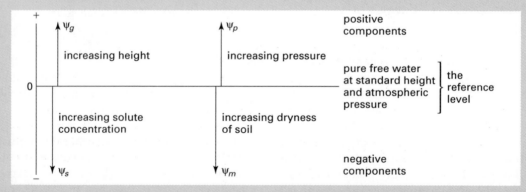

### Conductivity ($K$) (or Resistance [$r$]): Ease of Movement

Liquid flow in soil: $K$ decreases with drying and also depends on texture and structure.

Liquid flow in plants: $K$ is sometimes low in the flow path from soil across the root cortex to vascular tissue.

Vapor flow in soil and the leaf–air interface: $K$ depends on air movement; it is low if the air is still. Also, $K = 0$ through closed stomates.

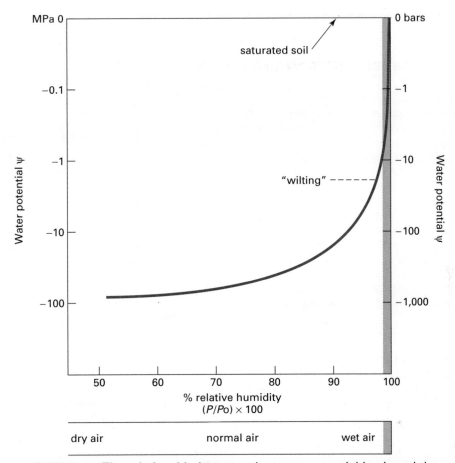

**FIGURE 5–6**   The relationship between the water potential in air and the relative humidity. Note that the potential rapidly drops below zero as the soil dries, reaching large negative values even at almost-saturated (98 percent) humidities. The shaded zone indicates the range of humidity in soils that supports live plants.

In saturated soil, water is moved mainly by gravity and pressure. The pressure increases with depth below the water table. The matric component of water potential is near zero (when the soil is saturated).

In air, the water potential depends entirely on the relative humidity ($P/P_o$), which is the ratio of the actual water vapor pressure ($P$) to the saturation vapor pressure ($P_o$). The saturation vapor pressure ($P_o$) is the pressure of water vapor in water-saturated air—that is, air in equilibrium with pure free water (Box 5–4). The saturation vapor pressure increases with air temperature, because hot air holds more water than cold air does. The water potential in air relates to the relative humidity as shown in Figure 5–6. Note the enormously low (negative) potentials of water even

in quite humid air. Soils and plants tend to dry out unless the air is nearly saturated. Only when the atmosphere is essentially water saturated will its water move *to* plants or soil, and that is what happens when dew and frost form. At all other times, the movement is the reverse: from soil and plants to the atmosphere.

### 5.3.2 Factors Affecting Water Potentials

**Water content**   Water potential decreases with decreasing water content not only in the air (Figure 5–6) but also in plants and soils. The plant can lose only 10 to 20 percent of its normal water content before its tissues will wilt or stop growing. (This damagingly low

## BOX 5–4
## Equilibrium between Liquid Water and Water Vapor

When liquid water comes to equilibrium with vapor in air, the water potential must be identical in both. If the water is pure and not closely bound to surfaces, the air becomes saturated, with relative humidity at 100 percent, and the total water potential is zero. If the water contains solutes or is attached to soil or plant surfaces, the solutes or surfaces hinder the escape of water into the air. Accordingly, the air does not become saturated, the relative humidity is less than 100 percent, and the total water potential has a negative value. Measuring the humidity at equilibrium is a good way to estimate the sum of matric and solute potential in soil and plant tissue. The biologically tolerable range of water potentials in plants and soils (from 0 down to perhaps −2 MPa) corresponds to a narrow range of atmo-

spheric humidities between 100 percent and about 98.5 percent saturation (Figure 5–6). In real soil and plants, the gas phase is rarely exactly in equilibrium with the liquid phase because vapor can move to and from the atmosphere. Nevertheless, the slowness of vapor diffusion usually ensures that the soil air and the intercellular gas of leaves are very humid—98 to 100 percent relative humidity—reflecting their proximity to liquid water with a potential high enough to support plants.

Equilibria between vapor and liquid water are shown in the figure below. Pure free water (A) (with total $\psi = 0$) supports a wetter atmosphere than does water that contains solutes (B) or is held by soil (C) or leaf cells (D). ($P$ = vapor pressure, $P_o$ = the saturation vapor pressure.)

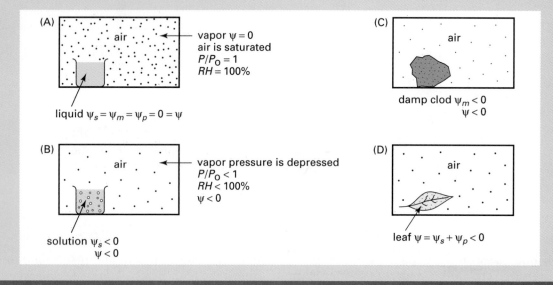

water content corresponds to a leaf water potential of −1.5 to −2.5 MPa, depending on the plant species and stage of growth.) A wider range of water content is acceptable in soil. Most soils hold much of their water at fairly high potentials (between −0.1 and −1.0 MPa) that allow ready uptake by plants. And even if the amounts of water are only a small percentage of the soil's weight or volume, they correspond to a large storage when summed throughout the root zone (Table 5–1).

In short, soils store large reserves of usable water, but plants store little reserve water and wilt rapidly if their supply is interrupted.

**Time changes**   In air, water potential or relative humidity depends partly on water vapor pressure ($P$) and partly on temperature. High temperature lowers the relative humidity because it raises the saturation vapor pressure, $P_o$. In most places and seasons, the relative humidity fluctuates daily, rising during the

## TABLE 5–1
Amounts of Water Available in Soil and Used by Plants[a]

| Water | In clay loam soil (1 m deep) | In sandy soil (1 m deep) |
|---|---|---|
| Available water | | |
| Percentage of soil bulk volume | 20% | 6% |
| Equivalent depth | 0.2 m | 0.06 m |
| Mass, Mg (metric tons)/ha of land surface | 2,000 | 600 |
| Water contained in vegetation | | |
| (20 tons live weight, 90% water) | | |
| Mass, Mg/ha of land surface | 18 | 18 |
| Monthly evapotranspiration from short grass cover mass, Mg/ha of land surface | | |
| Dry (inland Calif.), summer (July) | 2,000 | 2,000 |
| Dry (inland Calif.), spring (April) | 1,000 | 1,000 |
| Humid (n. coastal Calif.), summer (July) | 1,000 | 1,000 |
| Humid (n. coastal Calif.), spring (April) | 700 | 700 |

[a]Numbers are typical approximate round figures.

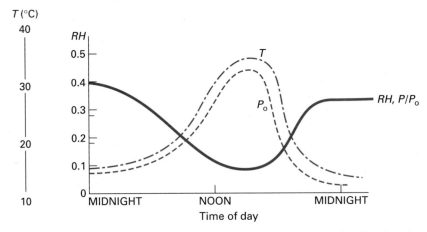

FIGURE 5–7   Daily trends in air temperature ($T$) and humidity. During the warm part of the day, relative humidity (RH) drops, mainly because warming raises the saturation vapor pressure.

night and falling as the air warms during the day (Figure 5–7).

In soils, the water content and water potential together increase (become less negative) when the soil receives rain or irrigation. They then decline rapidly as loosely held water drains out and then more slowly as the remaining water is removed by evaporation and transpiration (Figure 5–8). Usually the drying proceeds faster at the surface and progressively more slowly with depth. The average soil water potential fluctuates over the wetting cycle, rising sharply when the soil receives water and falling gradually between these times.

In plants, the leaf water potential usually fluctuates daily, as in the atmosphere, because the water uptake each day lags slightly behind the increase in transpiration rate (Figure 5–9). The leaf water potential normally declines during the afternoon hours. This decline is moderate if ample water is supplied, but it becomes more severe day by day as the soil dries and eventually will impair

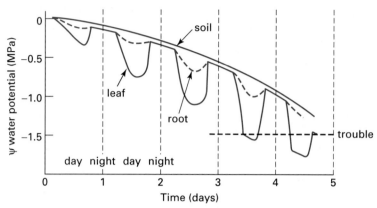

**FIGURE 5–8**   The drying cycle of a confined (potted) soil and a plant growing in it. Field plants can undergo similar, though usually less drastic, variations in water status.

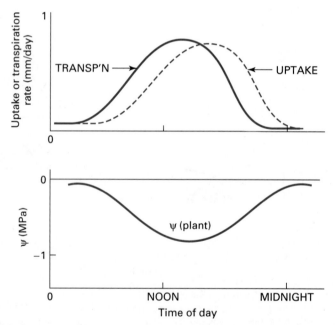

**FIGURE 5–9**   The daily development of temporary water stress in plants.

growth unless more water is added. Growth impairment may precede obvious wilting. Field measurements of leaf water potential are sometimes used as a sensitive indicator of the need for irrigation.

### 5.3.3   Potential Differences and Gradients

The overall gradient of potential differences that drives liquid water flow from soil to root to shoot and vapor flow from shoot to atmosphere is large except when the air is saturated. The steepest portions of the gradient occur at three junctions:

1. The leaf surface, especially if the air is still or the **stomates** are closed
2. The soil surface, especially if the surface layer is dry
3. The root (or root–soil interface), especially if the plant is under drought stress

## BOX 5–5
## Ways to Measure Water Potentials

(A) Tensiometer (measures matric potential of field soils). Water in sealed tube equilibrates with soil water by interchange through porous ceramic cup. Pressure drop in tensiometer numerically equals the matric potential. (See also Figure 3–11.)

(B) Pressure bomb (measures pressure plus solute potential of leaves and shoots in field or laboratory). Pressure needed to force a drop of fluid from a cut stem or petiole numerically equals potential in xylem cells.

(C) Pressure membrane apparatus (establishes matric potential of prepared soil samples in laboratory). Pressure needed to force water out of soil through porous plate or membrane numerically equals matric potential of remaining water in the soil sample. This is used to develop **water retention curves;** samples set to different potentials are removed for the gravimetric measurement of water content.

(D) **Psychrometer** (directly measures relative humidity of air, *RH*). Then $\psi = 2.3RT \log(RH)$. Indirectly estimates matric plus solute potential in solutions and in samples of tissue or soil with which the measured air has equilibrated (as in an enclosure or at depth in a pedon). The humidity of air controls the rate of evaporative cooling of the wet thermocouple, whose temperature registers as voltage on a voltmeter.

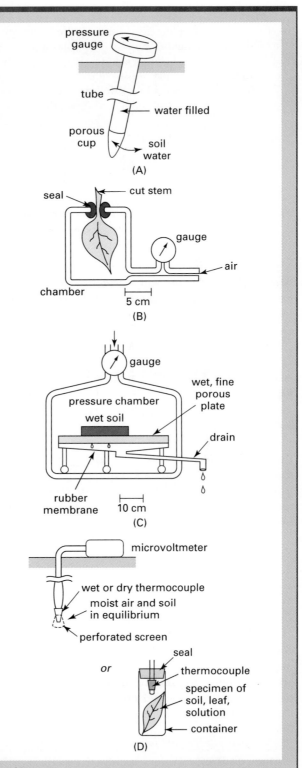

(Methods A through D do not include gravitational potential.)

*Continued*

(E) Test well (piezometer) (measures pressure plus gravity potential of saturated soil and aquifers in the field). The height that water rises above the bottom opening is the pressure head at the opening. The height of water in the well relative to a defined reference is the hydraulic head (pressure plus gravity) at the opening.

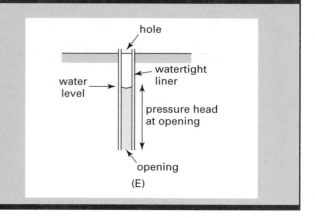

The gradients are steepest across these boundaries because they are zones where water movement is difficult and is limited to vapor diffusion through almost motionless air or to liquid movement in thin, badly connected films.

## 5.4 CONSTRAINTS TO WATER MOVEMENT

### 5.4.1 Conductivity in Soil

Pore properties have some influence on water conduction in soil, but soil water content has even more influence. We commonly distinguish water flow in saturated soil from flow in unsaturated soil as distinctly separate cases.

Saturated flow is important beneath water tables, in flooded soils, and as water infiltrates during flood irrigation or heavy rain. Essentially all the pores are filled with water, and water flows fastest with little hindrance through the largest interconnected pores. The hydraulic conductivity depends on the size and extent of these large pores. In unsaturated flow, by contrast, the largest pores are occupied by air. They obstruct the flow of water, because it has to pass around them. In unsaturated soil, water flows through the films of water coating the particles and through the intermediate-sized pores that remain filled with water.

**Water content** Water conductivity diminishes sharply as water content decreases, because more pores are empty and unavailable for water flow (Fig-

ures 5–10 and 5–11). There is more drag against surfaces when water flows in small pores and thin films, and the flow paths are less direct. The sharp decline in conductivity with decreasing soil water content has two consequences: (1) it limits drainage loss, and (2) it limits loss by evaporation and transpiration. Figure 5–12 summarizes this idea. Soil water in excess of field capacity is held so briefly that it is of little use for plant growth, unless the profile or substrata have some impeding layer to stop drainage. Thus field capacity is the upper limit of useful water-holding capacity. Further drying eventually lowers the conductivity so far that water can no longer get to roots fast enough to prevent severe stress in plants (the permanent wilting point).

Also important to water storage is the low conductivity of the surface layer when it dries. Within a day or so after rain or irrigation, the top few centimeters of soil become so dry that the upward movement of liquid water almost stops, despite the large difference in water potential between the soil below the dry layer and the atmosphere above. Thus a dry soil "mulch" develops naturally. For this reason, deliberate mulching practices do little for water conservation except to delay the drying of the surface layer.

### 5.4.2 Conductivity in Soil–Plant Systems

Once the soil surface has dried, transpiration becomes the main pathway of water loss. Soil water can be conserved by killing plants, which is one reason for weed control and for the practice of bare fallow that allows rainfall to accumulate in the soil between successive

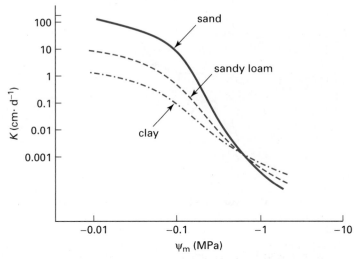

FIGURE 5–10   Effect of drying on hydraulic conductivity in soils.

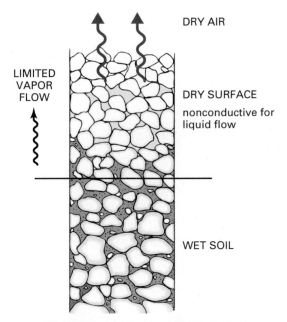

FIGURE 5–11   Low conductivity in dry surface soil retards the evaporative loss of soil water.

crops. Transpiration normally exceeds direct evaporation from soil because the soil surface is dry and unconductive, whereas the total leaf surface is large, receives sunshine energy to vaporize water, and is well supplied with water via the roots and conductive tissues.

Transpiration is sometimes limited by slow movement of water to the roots. This can result from sparsity of roots in the wet regions of the pedon, from drying of most of the pedon, or from temporary drying of the rhizosphere soil when transpiration is exceptionally fast. Occasionally, transpiration is limited when root diseases or soil disorders inhibit root development or when vascular diseases block xylem vessels. All these conditions, which impede water entry, can lead to the wilting and death of plants. But when the plant is well supplied with water, the principal resistance is at the leaf surface, where water movement is controlled by stomates (Figure 5–13). When the stomates close, transpiration almost stops. Stomates normally close in the dark and open in the light. In most plants they stay open in daytime unless the plant develops severe water stress (e.g., leaf water potentials below $-2$ MPa in corn).

## 5.5   WATER STRESS IN PLANTS

Drought damage results when the difference between water uptake and water loss causes desiccation of tissues sufficient to interfere with growth or reproduction. It seems that water stress primarily inhibits the expansive growth of shoot tissues. Cumulative day-by-day growth reduction is particularly damaging during early growth when the establishment of

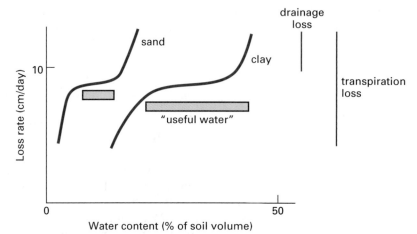

**FIGURE 5–12** Rates of loss of soil water in relation to water content in soils with different water capacities. Losses slow as the soil dries.

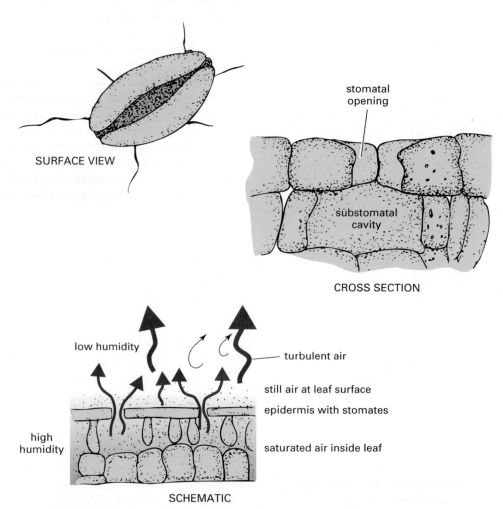

**FIGURE 5–13** Stomates and lack of air mixing retard the escape of water vapor at the leaf–air interface.

photosynthetic area is urgent. Less leaf area today means less growth tomorrow, and the effect is compounded over time. Desiccation can eventually reduce turgor to the point that cells collapse; this is wilting, a clear sign of severe water stress.

Less severe water stress can be diagnosed by measuring the relative leaf water content (relative to the water content at full turgor). But it is faster and more sensitive to measure the corresponding leaf water potential.

Although the primary effect of water stress is to slow expansive growth, this is not always the most important effect. Flower development or fruit set can be especially sensitive to drought at critical periods. Drought sensitivity at these late stages will be aggravated if previous vigorous growth has exhausted the soil's water supply.

Failure to grow or reproduce during drought does not mean failure to survive. For crops, growth and fruiting are critical, but drought survival is even more important to pasture, range, woodland, watershed, and wildland vegetation. Plants survive drought with the help of adaptations to conserve water and to get more water from the soil. The former include anatomical features for preventing water loss, such as sunken stomates, leaf hairs, narrow leaf form, ability to drop leaves, and ability to tolerate water loss. Some drought-tolerant plants reflect sunlight or turn their leaf edges to the sun. Root modifications may help the plant acquire water. Under moderate drought,

close root spacing helps by reducing the distance that water must move, and deep roots help exploit residual subsoil water.

## 5.6   EVAPOTRANSPIRATION

### 5.6.1   Factors Controlling Evaporation and Transpiration

Lack of water can limit transpiration, but the effect is minor unless the soil is so dry that the growth of plants is severely affected.

With ample water, transpiration increases with plant density or leaf area, up to the point that the plant canopy approaches complete cover (Figure 5–14). For a given climate and vegetation, the seasonal **evapotranspiration** relates fairly closely to the amount of growth. Large amounts of water—hundreds of kilograms per kilogram of plant biomass production—must be transpired.

With a given plant cover and ample water, the rates of evapotranspiration depend on the weather in the short term and on the climate in the longer term (see Chapter 4). Evapotranspiration is promoted by high wind, high temperature, and low humidity. Most important, evapotranspiration depends on the total (daily or seasonal) radiation received from the sun, the energy supply that converts liquid water to vapor.

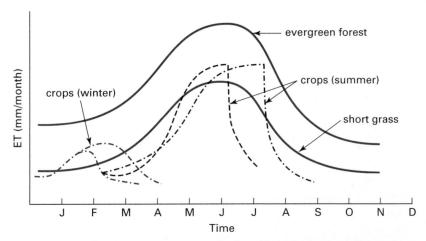

**FIGURE 5–14**   Seasonal evapotranspiration. Note the major effects of seasonal weather change and the additional effect of leaf area development in annual crops. Data of this kind (except for that of evergreen forest) originate from studies in lysimeters. (From Pruitt et al., 1972.)

## 5.6.2   Measurement of Water Use

Measuring or estimating evapotranspiration in the field is difficult but valuable for many purposes. The designers of irrigation systems need estimates of seasonal evapotranspiration to know how much water must be supplied or how much crop can be grown with the water available. Also, the design of water delivery systems and the scheduling of irrigations depend on short-term (daily or weekly) evapotranspiration as well as soil water storage. Parallel concerns exist in dryland management. Would supplemental irrigation be worthwhile? Can we extend the cropping season? Can we bring in plants with different seasonal water use? If the vegetation changes, what will happen to runoff water, groundwater, and salt? The answers all depend on evapotranspiration.

Evapotranspiration data have been obtained by three methods: water balance methods, calculation from meteorological data, and extrapolation from direct water loss measurements made under standard conditions.

The seasonal pattern of evapotranspiration for different crops can be related to the seasonal pattern for a standard ground cover (which might be measured directly). A favored standard ground cover is short grass, well watered and extensively surrounded by similar cover. Evapotranspiration from such a surface is called *potential evapotranspiration*. Once a crop's seasonal pattern is known, the crop's water use can be estimated from concurrent estimates of potential evapotranspiration.

The curves in Figure 5–14 illustrate the approach and some of the principles. Evapotranspiration varies seasonally, showing the strong influence of energy input; for this reason, a winter-grown crop such as barley uses less water than do summer crops. The curves for annual crops show the influence of leaf area and canopy development. At first, the water loss is restricted to evaporation from the soil, but then the developed canopy loses water rapidly until the crop matures and dies. Perennial tree crops show a seasonal pattern like that of grass, except for their greater overall evapotranspiration rate. In annuals, the length of the growing season strongly influences both total water use and crop yield. Finally, the seasonal total evapotranspiration depends on seasonal weather conditions and the plants' canopy structure and development pattern. For a particular vegetation and climate, most of the variables that influence evapotranspiration turn out to affect growth or yield similarly, with the result that seasonal evapotranspiration and plant growth are closely related.

## 5.7   SUMMARY

Water forms a dynamic continuum in soils, plants, and the atmosphere. The soil's store of water is sporadically renewed by infiltration from rain or irrigation and is more or less continually depleted by evapotranspiration. The useful storage capacity depends mainly on soil depth, clay content, humus content and structure, and the extent of the root system. As the soil dries, the remaining water moves more slowly and becomes less available.

Movement is the key issue. Water movement is easier to understand if you consider two questions: How much does the water tend to move (what is the driving force or potential gradient)? What are the constraints to movement (what is the conductivity or resistance)?

Water movement is driven by gradients in water potential, which integrate the factors that influence water's tendency to move. In saturated soil, water movement is driven by gravity and pressure differences; in unsaturated soil, these are commonly outweighed by matric potential differences arising from differences in wetness. Water enters plants from soil because the plant cell membranes keep the internal potential low by maintaining high solute concentrations. The massive vapor movement of water from soils and plants to the atmosphere is driven by potential differences between the high humidity within plants and soils and the usually lower humidity of the atmosphere.

Constraints to water movement—factors that influence resistance or conductance—vary throughout the system. In saturated soil, conductivity depends on soil structure; well-connected, coarse pores ensure high conductivity. In unsaturated soil, conductivity declines markedly as the soil dries. Resistance at the root depends on the soil water content and the health and extent of the root system. Within the plant, resistance is low if the vascular system remains healthy. The strong tendency for water to flow from leaves and soils to the atmosphere fortunately meets with a high resistance arising from the slowness of vapor diffusion in the still air trapped within these surfaces. Without that hindrance to the rapid depletion of soil water, life on land would be austere.

## SOIL AND ENVIRONMENT
### Water Movement and Environment

- Water limits plant growth in some season of the year nearly everywhere. The water intake, retention, and release principles outlined in this chapter are critical to agriculture and to the character of nearly all terrestrial ecosystems.

- Solid waste disposal in landfills is accepted more generally than oceanic disposal or incineration, because we hope that landfills will permanently contain mobile pollutants capable of moving out in seepage and runoff water. Suitable sites, isolated from sensitive stream and aquifer systems by distance, topography, and impermeable strata, are becoming more scarce and expensive to acquire. Even on a suitable site, landfills are lined and capped with layers of low hydraulic conductivity, such as compacted, clayey, noncracking soil materials, to restrict rainwater entry and effluent seepage.

- Wastewater disposal practices on land are meant to contain or destroy environmentally significant pollutants. Choices of site and soil are specific to each problem. For example, domestic sewage water, after primary treatment, might be disposed onto a "sewage farm" growing nonfood crops. The plants should absorb almost all the plant nutrients from the water, including nitrate, the main mobile contaminant generated. Other contaminants should be retained by the topsoil. Evapotranspiration removes much of the water; the rest percolates to the underlying groundwater body, which ideally is isolated from others and restricted to disposal use.

The soil needs to be level and rapidly permeable from the surface to a deep water table to accept copious irrigation without runoff or waterlogging.

- Unlike pollutants in sewage effluent, salts and sodium in high concentration kill plants and leach freely through soils. Salty, sodium-rich agricultural drainage may be delivered to evaporation ponds, where the concentrated salts accumulate (while someone thinks what to do next). Here the pond lining or substratum must prevent water flow to protect surrounding land from the accumulating brine.

- Sampling of soil or water below disposal sites is necessary for analytical monitoring of leakage and movement. Test wells yield samples only if they tap into saturated zones—that is, beneath a water table. However, water flow can also carry contaminants in unsaturated soil, from which water will not enter ordinary test holes. It becomes necessary to use less convenient methods such as soil sampling or to wait for deep test wells to detect the arrival of pollutants at the water table.

- Underground water in aquifers can suffer lateral intrusion of contaminated water. Control is a matter of maintaining potential gradients in the right direction to stop contaminated inflow. High hydraulic potential can be maintained in the water body to be protected by limiting outpumping or by pumping water into it. Or hydraulic potential in the contaminated body can be lowered by outpumping and disposal.

## QUESTIONS

1. Explain the meaning of the terms *water content, water potential, relative humidity, water saturation* (in soil and in air), *runoff, infiltration, seepage, transpiration, saturated and unsaturated flow, potential gradient,* and *hydraulic conductivity.*

2. Which soil properties most influence hydraulic conductivity in (a) saturated soil and (b) unsaturated soil?

3. How would you measure the water content and the water potential of a soil and of a leaf?

4. What are the major components of water potential? What factors most influence water potential in soil? In plants? In the atmosphere?

5. What are typical ranges of water content in soils and in plants?

6. In what ways can water be lost from soil? How would you measure the loss of water from (a) a planted container of soil and (b) a cropped field soil?

7. Water escapes from plants mainly because the stomates open to let carbon dioxide in. (a) Does transpiration offer any benefits to plants? (b) Why does a crop's total transpiration tend to correlate with the crop's biomass production?

8. What processes control water loss from a dry soil? How do they differ from the processes that control water loss from a wet soil? Contrast the role of plants in the two cases.

9. The following diagram shows the hydraulic head as a function of elevation for a soil. Assume that the reference level is a point below the soil surface.

   a. Draw $H_m$.
   b. At what elevation is hydraulic conductivity lowest?
   c. At what elevation is the soil saturated?
   d. Which way will water flow?

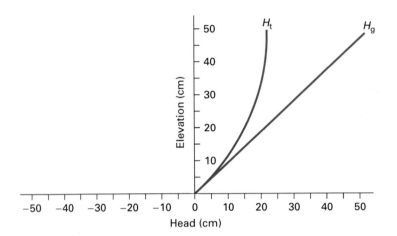

10. Can the aquifer illustrated here supply water to an alfalfa field that requires a volume of 235 m³ per day? Assume the pressure heads do not change. The aquifer has a cross-sectional area of 15,000 m² and a hydraulic conductivity of 0.75 m per day.

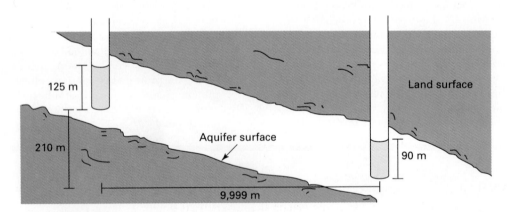

# FURTHER READING

Bradford, K. J., and T. C. Hsiao. 1982. Physiological responses to moderate water stress. In *Encyclopedia of plant physiology*, ed. O. L. Lange, P. S. Nobel, C. B. Osmond, and H. Ziegler. New York: Springer-Verlag.

Corey, A. T., and A. Klute. 1985. Application of the potential concept to soil water equilibrium and transport. *Soil Science Society of America Journal* 49:3–11. (This is a brief, advanced-level discussion that traces the history of the concept and argues that it is being incorrectly applied.)

Dirksen, C. 1999. *Soil physics measurements*. Reiskirchen, Germany: Catena Verlag.

Doorenbos, J., and W. O. Pruitt. 1975. Guidelines for predicting crop water requirements. *Irrigation and Drainage Paper 24*. Rome: Food and Agriculture Organization of the United Nations; New York: Unipub.

Hillel, D. W. 1982. *Introduction to soil physics*. New York: Academic Press.

Pruitt, W. O., F. J. Lourence, and S. von Oettingen. 1972. Water use by crops as affected by climate and soil factors. *California Agriculture,* October, Vol. 26, pp. 10–14.

Rose, C. W. 1966. *Agricultural physics.* Elmsford, N.Y.: Pergamon Press.

Taylor, H. M., W. R. Jordan, and T. R. Sinclair, eds. 1983. *Limitations to efficient water use in crop production.* Madison, Wis.: American Society of Agronomy.

Turner, N. C. 1997. Further progress in crop water relations. *Advances in Agronomy* 58:293–338.

Wilson, L. G., L. G. Everett, and S. J. Cullen. 1995. *Handbook of vadose zone characterization and monitoring.* Boca Raton, Fla.: CRC/Lewis.

## SUPPLEMENT 5–1   Capillarity

**Capillarity** is observed when water rises in narrow tubes (capillaries; see Figure 5–15); the effect becomes stronger the narrower the tube. Capillarity is most important in those soils with many fine, continuous pores. Water adsorption is the direct interaction of water molecules with wetted surfaces; the effect is strongest for the molecules closest to the surface. It is most important in colloid-rich soils with a large wettable surface area. Within a given soil, capillarity tends to become more significant at high water contents (near field capacity), and adsorption becomes more significant at lower water contents.

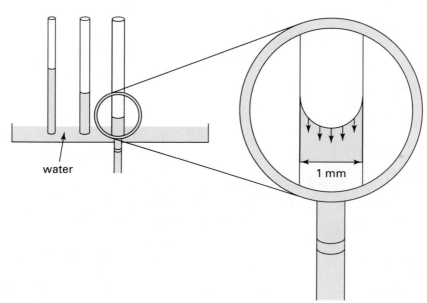

**FIGURE 5–15**   The attraction of water molecules to one another is stronger than that of air molecules to water molecules, causing the water molecules at a surface to act like a thin elastic sheet. When the end of a tube is placed into water, the water rises in the tube. The water molecules are attracted to the walls of the tube and adhere to them, thus forming a curved air–water interface, called a **meniscus.** The pressure under the meniscus is less than the atmospheric pressure over it. The water outside the tube is at atmospheric pressure. It flows into the tube, which forces the water in the tube upward until the reduced matric potential just below the meniscus is compensated for by the equivalent increase in gravity potential.

In moist soils, the pores act like thin tubes, and water is held by capillarity. Pores are not straight and smooth like the tubes in the figure, and so water does not reach the same height in pores as it would in tubes of the same diameter. A soil with many small pores holds more water by capillarity than does a soil with few large pores. In addition, because of capillarity, water rises higher from a water table in a clay soil than in a sandy soil.

# 6

# MANAGING SOIL WATER

## OVERVIEW

A soil temporarily stores water for plant use. The useful storage capacity depends on the soil's depth, texture, structure, and organic matter content and on the plants' root system. The amount currently stored depends on recent rainfall, irrigation, runoff, transpiration, evaporation, and deep percolation below the root zone. Water losses and plant growth increase with the wetness of the soil. The management of soil water seeks to optimize the use of soil water while it is available and to minimize nonproductive losses. In addition, irrigation management seeks to optimize the use of imported water as a supplement to rain.

Vegetation usually extracts soil water preferentially from the top of the pedon, establishing a drying profile. Growth becomes limited at some stage of drying that depends on soil water release properties, root distribution, stage of plant development, and the current weather's effects on water demand.

The entry of liquid water into soil—infiltration—is driven mainly by the attraction of water to soil surfaces and pores (matric potential). Infiltration proceeds by advance of a wetting front as the wetted zone extends into the dry soil. The rate of infiltration depends on soil hydraulic properties and always slows as wetting proceeds, unless it is limited by the rate of water application.

Irrigators usually try to recharge the root zone adequately and uniformly, without excessive evaporation, runoff, and deep percolation. Methods include flood, furrow, sprinkler, and drip irrigation, all of which can be efficient when used well under appropriate conditions.

Surface drainage minimizes undesirable surface accumulation of water, and subsurface drainage lowers water tables to improve aeration and salt leaching. Methods of subsurface drainage include the installation of trenches or tile drains (buried perforated pipes), set out and graded to collect and deliver water to a sump or sink. Pumping from

**108**

*the sump may be necessary. Because water escapes freely only from saturated soil, drains (such as springs, wells, and drain holes in pots) must be below the water table in order to operate. Drains draw down the water table. Uniform draw-down is not practical to achieve; compromises are imposed by costs and the soil's permeability.*

*Poor irrigation and drainage management can cause serious environmental problems. Perhaps the most impor-tant environmental problem is wasted water. Others include leaching of nutrients and amendments such as pesticides into groundwater and runoff of nutrients and amendments into surface water. Improper disposal of drainage water can lead to water quality degradation in the water receiving the drainage water.*

*There are different types of water use efficiency. In general, water efficiency is the ratio between the benefit gained and the water used, but the benefit can mean growth, yield, or value of product, and water used can mean transpiration, evapotranspiration, water applied, or total water supply. Economic efficiency—returns per unit of water available—would be maximal if all the water were used productively for growth of a crop that produced a high value of produce.*

This chapter describes practical aspects of soil water conservation, irrigation, and drainage. Although it can be read without reference to the physical princi-ples of water behavior given in Chapters 3 and 5, it il-lustrates most of those principles. The chapter deals mainly with field-scale processes that involve the pe-don as part of a system involving plants, atmosphere, drainages, and water supplies. The highlighted char-acteristic of the pedon itself is its constantly changing profile of water content, which together with water conductivities and root patterns controls the supply of water to plants.

## 6.1 WATER MANAGEMENT CONCEPTS

A soil is a leaky storage reservoir for water. In general, the addition of more water speeds up the growth of plants. It also speeds up the leaks: transpiration, evap-oration, percolation, and runoff (Figure 6–1). Soil water management minimizes avoidable losses and makes the best use of rainfall, stored water, and irri-gation water.

### 6.1.1 Nonirrigated Land

In rain-fed agriculture and other dryland systems, the input of water depends entirely on the weather, but cultivation, erosion control, and weed control prac-tices can reduce losses or enlarge intake and storage. Thus the cropping system is usually designed and the natural ecosystem evolved to exploit the soil water when it is available (Figure 6–2). Farmers select crops with maturation times and water consumption pat-terns to match the seasonal rain expectations and the amount of soil water storage. Farmers can also adjust planting times, planting densities, and fertilizer levels (Figure 6–2B). A short, wet season and limited stor-age capacity call for early-maturing crops, planted thinly and given less fertilizer. Excessive planting den-sity and fertilization are counterproductive when they lead to premature exhaustion of stored soil wa-ter while the crop is still drought sensitive. Note the importance of rainfall expectations: The reliability and timing of rainfall are as important as the quan-tity. If the rainfall is unreliable, large safety factors are needed to guard against disaster in bad seasons, which means that water will be left unused in good seasons. Supplementary irrigation (Figure 6–2B) is valuable in dryland crop and pasture systems. Just one small application of irrigation water can extend the growing season or save a crop in an unexpected dry spell.

Alternate-year cultivation, or fallowing, is a common (important) dryland water management practice. After harvest, the field is tilled to maximize water infiltration during the next rainy season. Fol-lowing the rainy season when the soil profile water-holding capacity is full, the field is again cultivated to kill weeds and to break the continuity of pores from subsoil to atmosphere. The rough dry surface soil makes an effective mulch and reduces evaporation. The absence of plants and their transpiration greatly reduces the use of water. The stored soil water plus the annual rain can be utilized for the next crop.

### 6.1.2 Irrigated Agriculture

Even with irrigation, rainfall is normally a valuable in-put, but the management challenge is to use irriga-tion water effectively and efficiently. Irrigation water management requires some compromise between wetness for good growth and frugality for economy.

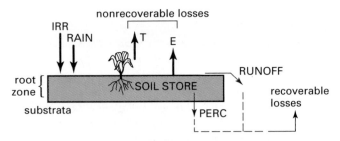

SOIL WATER STORAGE, INPUT, AND LOSSES
Over any time interval, change in amount of stored water
= IRR + RAIN − RUNOFF − PERC − T − E
where IRR = irrigation inputs, RAIN = precipitation,
RUNOFF = surface runoff, PERC = net movement from
root zone to deeper layers, T = transpiration,
E = direct evaporation.

(A)

DESIRABLE LOSS     INPUTS     UNDESIRABLE LOSSES

TRANSPIRATION                 LEAKS: E, RUNOFF, PERC

saturated

desirable range

STORED WATER

dry

(B)

**FIGURE 6–1**   Water storage. (A) Inputs, storage, and losses. (B) Soil regarded as a leaky tank. Lengths of arrows represent rates of loss at different soil water content.

The yield need not suffer from slightly suboptimal growth so long as water has been adequate during critical times such as early growth and establishment and, in some crops, early reproductive stages. Water must be applied in time to avoid the development of severe stress and in enough quantity each time to recharge the storage in the root zone.

Ideally, all the water loss would occur as transpiration, the one loss that is associated with crop growth. But other losses cannot be avoided. Most flood and furrow irrigation methods entail some runoff, by design, and salt control demands occasional percolation below the root zone to remove salts by leaching. Soil variation and imperfect water control during irrigation also cause runoff and seepage losses, unless the whole field is badly underirrigated. Much of the irrigator's art depends on judgment and experience.

## 6.2   WATER EXTRACTION AND INFILTRATION PATTERNS

### 6.2.1   Extraction by Plants

A potbound plant extracts water from its soil almost uniformly. Normally, however, water removal starts near the soil surface and progresses downward. Depletion of water from some of the soil does not greatly affect plant growth at first. The point at which water depletion begins to matter depends somewhat on the transpiration rate (i.e., on the weather and the amount of leaf) and on the pattern of root distribution. At the yield depression threshold, the point at which yield begins to suffer, profiles of water content might resemble those shown with advanced time in Figure 6–3. The **water profile** at incipient growth reduction varies among species, and it also varies with

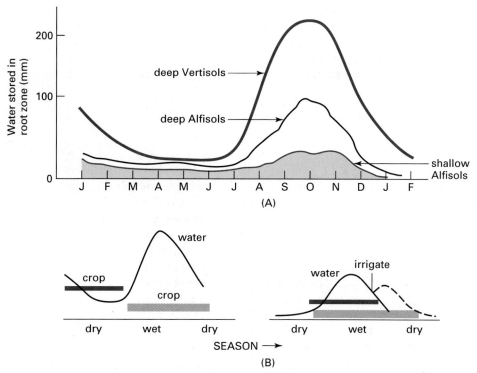

**FIGURE 6–2** Water storage in time. (A) Seasonal variation in a climate with distinct annual wet and dry seasons. Vertisols usually hold more water than Alfisols because they have more clay and it is dominantly smectite. (B) Improving use of rainwater (left diagram) by shifting planting date from early season (darker bar) to late season (lighter bar) or (right diagram) by supplemental irrigation and late-maturing crops (light bar).

time because root systems penetrate farther down as plants grow.

Clearly, the plants' current transpiration rates and root depth, as well as the soil moisture retention characteristics of each horizon (see Chapter 5), together determine how much water the soil can provide before growth suffers and irrigation or more rain becomes desirable. Nobody can define a single generally applicable value of soil water content or water potential at which water stress develops in the field. Farmers and smart gardeners do rather well by kicking the soil or shoving a shovel into it. These simple tests depend on sound physical principles: Hardness relates to dryness, and the experienced human mind can compute the need for irrigation from depth of drying, current weather, and a look at the plants. Large-scale irrigation can benefit from more quantitative guides to irrigation scheduling. In places where growing seasons have similar weather year to year, scheduling can be based on previous records of

the particular crop's water consumption and responses to irrigation, with allowances for the water storage capacities of individual soils. Scheduling can also be based on water consumption estimated by the methods described in Chapter 5, such as extrapolations from evaporimeter data or measurements of soil water. If water is available on short notice, daily monitoring of leaf water potential can help to determine when watering is needed.

## 6.2.2 Infiltration

**Infiltration** is the entry of liquid water into soil. It is driven mainly by the attraction of water to dry pores and surfaces, by the difference between the high matric potential of free water and the low matric potential of water held in pores and attached to surfaces in the soil. Gravitation helps if the infiltration is downward. A boundary between wet and dry soil called a

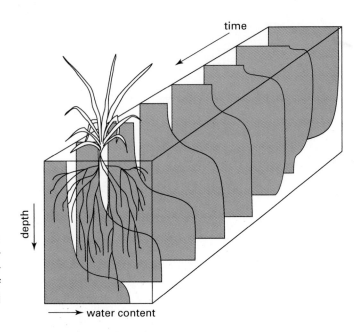

**FIGURE 6–3** Water extraction from field soil: Profiles of water content with increasing time after watering. Combined effects of vertical flow, evaporation, and transpiration.

**wetting front** normally develops and advances. You can see the wetting front if the dark color of the wet soil contrasts sufficiently with the lighter color of the soil that is still dry. Ahead of the wetting front, the soil remains at its original water content; behind the front, it is close to saturated (Figure 6–4). One consequence is that it is virtually impossible to wet a soil by infiltration to a water content much less than saturation. Another consequence is that the depth of water penetration indicates how well a rain or irrigation has recharged the soil water storage. The depth of water penetration is often easy to feel when you push a spade or rod into the soil. A tensiometer set at the required depth might do the same job. The depth to which a quantity of water wets the soil depends on the soil's water-holding capacity and its initial water content. A clay soil takes more water to wet to the same depth than does a sand because of the higher water-holding capacity of the clay. Conversely, less time is necessary to wet a sand to the same depth as a clay because of the larger pores in sand textures. It is easy to underestimate the amount of water and time needed to wet a soil to worthwhile depth. Sprinkling by hand relaxes a gardener, but a gardener with normal patience usually quits after a ritual wetting of the soil surface.

The infiltration rate is often limited by the rate at which water is applied. This is desirable during sprinkler irrigation and rain, so that water does not collect, or pond, on top of the soil. Ponding can in-crease runoff, erosion, and unevenness of watering. If water is applied faster than it infiltrates, the application rate can no longer limit the infiltration rate. This is common with most kinds of irrigation and with heavy or prolonged rain. Under these conditions, the rate of infiltration depends on the soil's physical properties and on time. The important physical properties are those that influence the soil's permeability (i.e., hydraulic conductivity) at high water contents. Soils that are highly permeable in the near-saturated state and whose structure does not collapse when they are wetted take in water rapidly, especially if they were initially dry or extensively and deeply cracked.

Infiltration slows as it proceeds (Figure 6–5). The driving force or average potential gradient declines when the water has to flow farther to extend the wetting front. Infiltration slows even more markedly if the wetting front meets an impermeable compacted, clayey, or cemented subsoil horizon or if clay swelling and dispersion block pores and close cracks during the watering. The slowing of infiltration with time has several consequences. The amount of water put into the soil is not simply proportional to the duration of watering or rain (unless the application rate is limiting infiltration, as in light rain or proper sprinkler irrigation). This must be taken into account in irrigation practice (see Section 6.3) and wastewater disposal (see Chapter 16). The fact that runoff increases during a rainstorm, as infiltration slows, helps explain why one long storm may contribute more than

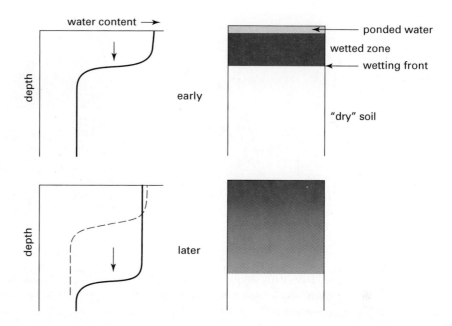

**FIGURE 6–4**   Infiltration.

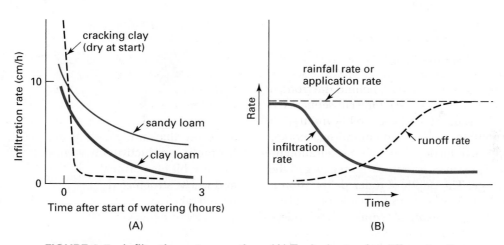

**FIGURE 6–5**   Infiltration rate over time. (A) Typical rates for different soils and slowing of infiltration as watering proceeds. (B) Dependence of runoff on infiltration.

many short storms to watershed yield, stream flooding, and soil erosion (see Chapter 15).

When water application stops, the wetting front can advance only by movement of the water already in the soil, lowering the water content of the wetted zone and making the water less free to move (the hydraulic conductivity drops precipitously with drying). Consequently, water movement within a day or two becomes practically negligible, when the wetted soil approaches its field capacity (Figure 6–6).

## 6.3   IRRIGATION METHODS

Water may be supplied by raising the water table so that water rises into the upper pedon. Such "subirrigation" is used in organic soils, where water moves up and sideways rapidly and control is easy. But surface application is much more common. Surface application methods include flood, furrow, sprinkler, and drip irrigation (Figures 6–7, 6–8, and 6–9). Ideally, any of these methods can wet the root zone uniformly

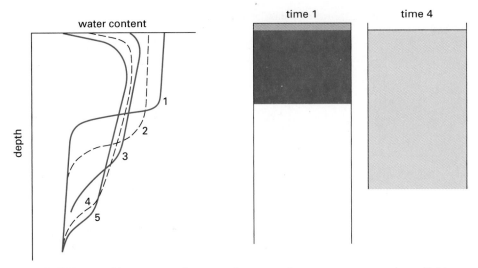

**FIGURE 6–6** Movement of water after watering stops: approach to field capacity. Curves 1 to 5 represent the water profile at successive times during 3 days after watering stopped.

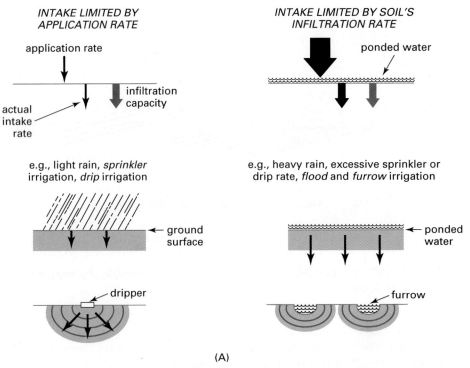

(A)

**FIGURE 6–7** Irrigation methods. (A) A classification.

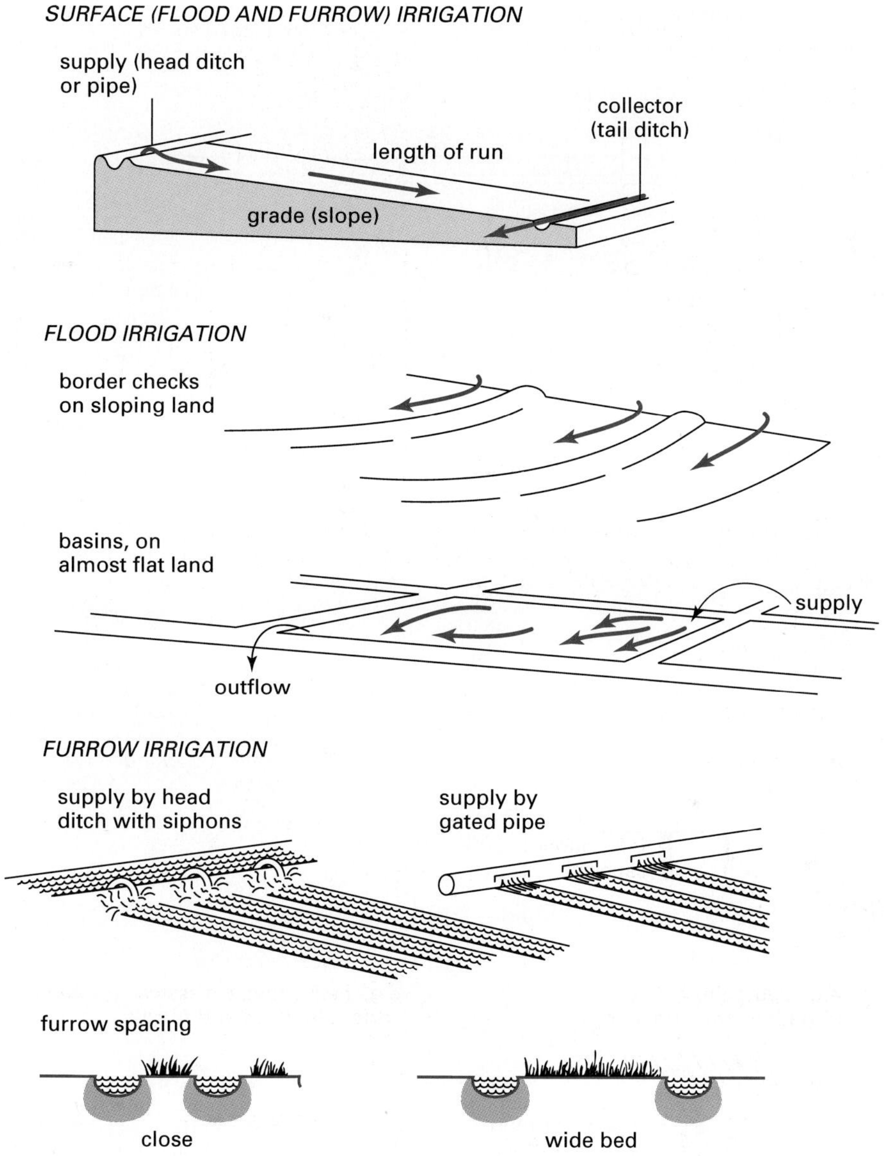

SURFACE (FLOOD AND FURROW) IRRIGATION

FLOOD IRRIGATION

FURROW IRRIGATION

(B)

**FIGURE 6–7  *Continued.*** Irrigation methods. (B) Schematic of some terms and techniques.

with little water loss. The choice depends partly on expense (such as increased costs of labor, pumping, and equipment for drip and sprinklers) and partly on slopes and soil permeabilities. So long as drip and sprinkler methods (Figure 6–7A) avoid ponding and runoff, they can succeed despite adverse or variable slope and soil conditions because intake into soil is controlled by the rate of water application. But flood and furrow methods (Figure 6–7B) require moderate uniform slope to achieve an even cover with ponded water and moderate uniform soil permeability for uniform intake.

## 6.3.1   Flood and Furrow Irrigation

In flood irrigation, water is applied fast enough to pond over the whole ground surface (Figure 6–7B).

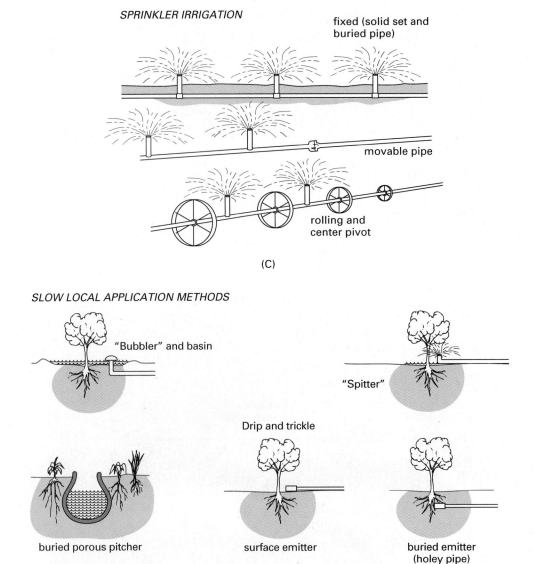

**FIGURE 6–7** *Continued.* Irrigation methods. (C, D) Schematic of some terms and techniques.

The ponded water is contained by banks, or levees, which follow the contour on sloping land or form a (usually) rectangular basin on flat land. Basins can be large, many hectares, if the land is flat enough and the soil uniform enough (Figures 6–8A and 6–9A). Large basins can be irrigated and farmed with little labor and the best use of large machinery. Bulldozers, graders, and land planes guided by laser beam can level large areas with extreme accuracy, and the high

cost can be compensated for by later economies. Sloping land requires small basins, border strips contained between contour levees, or furrow irrigation. Water is allowed to flow gently down the slope from inlets at the top, with some surplus tailwater escaping into a drainage ditch at the foot.

In furrow irrigation, water is run down parallel furrows (Figures 6–7B and 6–8B) typically about 35 cm deep. The soil between the furrows is not flooded; it

(A)

(B)

**FIGURE 6–8** Examples of irrigation. (A) Flood irrigation with contour levees in a field crop. (Note also evidence of uneven fertilizer application.) (B) Furrows, fed by siphons from supply ditch in foreground. (Photo A: William Wildman; photo B: J. Dahilig, U.S. Bureau of Reclamation.)

(C)

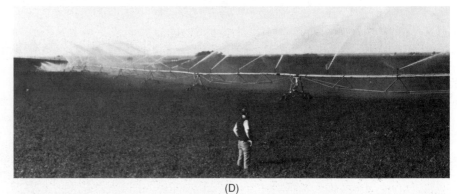

(D)

**FIGURE 6–8** *Continued.*
(C) Fixed sprinklers, with slight local ponding in low spot in left foreground. (D) Large moving sprinkler equipment. (E) Drip irrigation from emitters on the surface in a newly planted vineyard. (Photos C and D: J. Dahilig, U.S. Bureau of Reclamation; photo E: U.S. Department of Agriculture.)

(E)

(A)

FIGURE 6–9 Methods and design vary appropriately to suit conditions and facilities available. (A) Large (about 10 ha) basins on loam soils of moderate water-holding capacity and low infiltration rate, irrigated at 20- to 30-day intervals with a 100-hp electric pump (California alfalfa field). (B) Small (about 1 m²) basins on sand of low capacity and very high infiltration rate, irrigated twice a day by a person with two buckets (Senegal onion field). (Photo B: Harvey Liss.)

(B)

receives water by lateral unsaturated flow. The method suits row crops well and allows easier control of water on slopes than is possible with flooding methods. The **grade**—that is, the effective slope down which the water is guided—can be altered by changing the furrow direction in relation to the land slope (Figure 6–12). The overall average infiltration rate is slower than with flooding, so that the irrigation time needs to be perhaps 50 percent longer, because water is entering the soil through a smaller wetted area. The furrow spacing depends on the desirable row spacing of the crop and on the texture of the surface soil. It may be less than 1 m in sand and 2 m or more in clay. (Lateral water movement, in unsaturated soil, goes better if the soil is fine textured.)

In both flood and furrow irrigation, the application rate exceeds the rate of infiltration during most of the irrigation operation. Insufficient water will be supplied if the irrigation time is too short for the particular soil's infiltration characteristics (Figure 6–10).

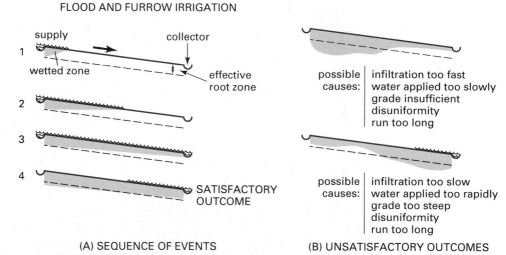

FLOOD AND FURROW IRRIGATION

**(A) SEQUENCE OF EVENTS**        **(B) UNSATISFACTORY OUTCOMES**

**FIGURE 6–10**  Factors to juggle in attempting uniform watering by flood and furrow methods.

Soils with extremely slow infiltration can require several days to irrigate, causing high labor costs, evaporation losses, and plant damage from waterlogging during the irrigation. Extremely fast infiltration makes uniform watering by flood or furrow almost impossible because most of the water infiltrates near the top of the run before any is delivered downhill. A partial solution is *surge irrigation,* in which the water supply is periodically turned on and off. The surging water erodes sediment from the furrow sides. When the sediment settles onto the ground surface, it slows the infiltration rate of subsequent surges.

Achieving uniform and adequate penetration of water over the whole field normally requires uniform soil properties (Figure 6–11), uniform slope, and a proper match among delivery rate, length of run, slope, and infiltration rate. Large areas of differing soil or slope are best irrigated separately (Figure 6–11). Smaller-scale variation is more difficult to manage. Grading to smooth the surface often aggravates soil variability, especially if it scalps the A horizon from high spots, completely exposing clayey, badly structured, less fertile subsoil. Steep slopes, high infiltration rates, or slow delivery can be compensated for by shortening the length of run (Figure 6–12). All these methods require reasonably uniform soil properties and slope. Sprinkler and drip irrigation methods are more forgiving in this respect.

The home gardener can assume that the garden soil is horizontally uniform throughout the garden without serious negative consequences. The farm, golf course, or wastewater disposal facility manager should not. Soil properties such as surface texture and structure, number and kind of horizons, and texture and structure of subsoil horizons vary spatially. This leads to much variation in infiltration rates (Figure 6–13). Figure 6–13 is a contour map in which the lines mark points of equal infiltration rate. The map was made from 1,280 measured values of infiltration rate in a field with one soil series. Note that the range of infiltration rate is from 4 to 12 mm hr$^{-1}$ and that this much variation was measured within sample distances of 1 m.

Irrigation based on the assumption of an average uniform infiltration rate causes inadequate refilling of soil water storage in portions of an irrigated area that have slower than average infiltration rates and extra water where infiltration rates are more rapid. The result is plant stress in some parts of the irrigated area and increased loss of plant nutrients such as nitrate in others. Leaching of salts is also not uniform, leading to highly variable salinity distribution (Figure 6–14). Conductivity of the saturation extracts from this field in Iran varies from <1 to >28 dS m$^{-1}$. Variation of this magnitude results in nonuniform plant growth and yield.

**FIGURE 6–11** Soil variability makes good water management difficult. The darker area in the upper left part of the vineyard is wetter because the soil has higher water-holding capacity than the rest of the field. Division to allow separate watering of the upper left area would allow better management of both portions of this field. (Photo: William E. Wildman.)

**FIGURE 6–12** Adjusting the grade and length of run in furrow irrigation.

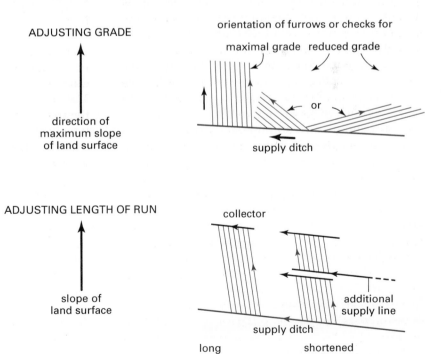

ADJUSTING GRADE

direction of
maximum slope
of land surface

orientation of furrows or checks for

maximal grade    reduced grade

or

supply ditch

ADJUSTING LENGTH OF RUN

slope of
land surface

collector

additional
supply line

supply ditch

long
run

shortened
run

121

**FIGURE 6–13** Contour map of 1,280 measured values of infiltration rate. The contour interval is 2 mm hr$^{-1}$. (From Vieira, S. R., D. R. Nielsen, and J. W. Biggar. 1981. Spatial variability of field-measured infiltration rate. *Soil Science Society of America Journal* 45:1040–1048.)

There is no easy solution to the problem of spatial variability. One part of the solution is to have a good soil map that details soil distribution on the landscape (see Chapter 13). Another is to monitor soil water content vertically and horizontally in the irrigated area and to irrigate when soil water content reaches preset minima. Automated systems that monitor soil water content with tensiometers and start-and-stop irrigation as needed are used on some large areas of irrigated farm acreage and on golf courses and in parks where the expense may be justified.

Irrigation is further complicated by changes in infiltration capacity as irrigation proceeds. The decrease in hydraulic gradient as a dry soil is wetted reduces the infiltration rate. In addition, changes in the surface soil structure reduce the infiltration rate. Aggregates at the soil surface slake as water entering the aggregates forces air out. The pressure of air trapped inside a weak aggregate causes the aggregate to fall apart. The disaggregated sand, silt, and clay particles block pores, reducing the water infiltration rate. In some soils, the disaggregated particles form a thin moist seal or dry crust (see Figure 15–9) on the soil surface, which reduces the infiltration rate by as much as 1,000 times.

## 6.3.2 Sprinkler and Drip Irrigation

In sprinkler and drip irrigation, water is applied more slowly than the soil can accept it, so that the application rate controls the rate of infiltration. Consequently, though uniform water application is important, the variability of soil and slope matters much less than with other methods. This flexibility and control is bought at additional cost: for equipment, interference with some kinds of cultivation operations, and energy to provide pressure. Sprinklers are used in many applications. Sprinkler equipment (Figures 6–7C, 6–8C, 6–8D) often must be movable for planting, tillage, and harvest. Mobile systems are feasible for smooth terrain and crops that allow clearance. Fixed sprinkler and drip methods (Figure 6–7D) suit permanent plantings such as vineyards, orchards, greenhouses, and gardens. Local application (drip) can be selective. It cuts water loss and helps control weeds because each emitter wets just the root zone of the adjacent plant and little of the ground surface. Drip can also supply dissolved fertilizers efficiently. But maintenance is difficult when dissolved Ca and Mg compounds deposit and clog the emitters. Underground drip systems place water and nutrients directly in the root zone and eliminate evaporation.

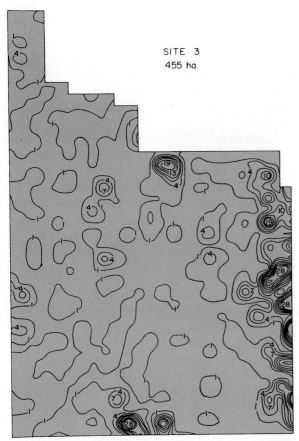

**FIGURE 6–14** Isosalinity contour map of a 455-ha field. Values are electrical conductivity in deciSiemens per meter. (From Hajrasuliha, S., N. Baniabbassi, J. Metthey, and D. R. Nielsen. 1980. Spatial variability of soil sampling for salinity studies in southwest Iran. *Irrigation Science* 1:197–208.)

## 6.4 DRAINAGE

Drainage is the escape of liquid water from the soil surface and subsurface. Surface drainage, so that water runs off without local ponding, is normally improved by grading and leveling for irrigation. For sprinkler or drip irrigation, surface drainage is not needed if the water is applied evenly and slowly enough. Subsurface drainage, to lower the water table, is often more problematic and expensive. Artificial subsurface drainage helps leaching and disposal of salts in irrigated agriculture. It reduces waterlogging and improves aeration of low-lying parts of irrigated areas and in the "reclamation" of marshlands (their conversion to other use).

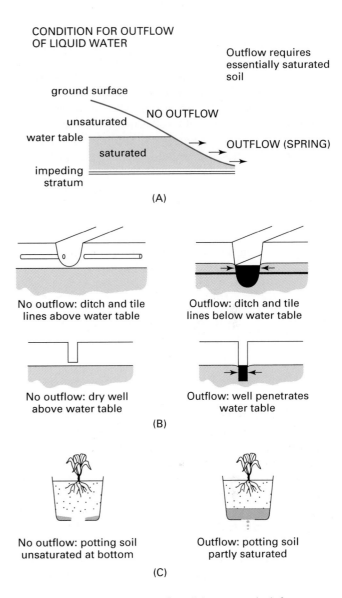

**FIGURE 6–15** Drainage. Conditions needed for escape of liquid water. The water must be able to lower its potential. Soil must be essentially saturated at the point of escape.

Soil loses water slowly by evaporation. The much faster liquid outflow requires certain conditions that always hold whether the outflow goes into a drain, a well, a spring, a seep, or the hole in the bottom of a flowerpot (Figure 6–15). Liquid water leaves the pedon only where the soil is essentially saturated. It does not flow out freely from unsaturated soil (you must apply pressure to make water escape from unsaturated soil in the pressure membrane apparatus—see

Chapter 5). Water moves only if the moving reduces its potential. As water moves out of the soil into the drain, drain hole, spring, seep, or well, its gravitational energy decreases only slightly. But to escape matric forces holding it in the soil, water must gain considerable energy unless the soil is saturated. Water flows out freely when the soil is nearly saturated (zero matric potential) or, better still, when the soil water is at a positive pressure, as below a water table.

For this reason, natural springs and seeps occur only where the water table meets the ground level (Figure 6–15A). Likewise, wells must penetrate below the water table or run dry (Figure 6–15B). The flowerpot stops draining through the hole as soon as the pressure potential at the bottom of the pot drops to

zero; that is, it stops draining when the bottom of the pot is still essentially waterlogged (Figure 6–15C). Consequently, potting soils should be coarse, with large pores that will drain easily and provide aeration. Field drains work only when they are below the water table, at positive pressure, and they stop removing water as soon as the water table drops to drain level.

Drains and wells lower the water table (Figures 6–16 and 6–17), but they cannot lower it uniformly. The **drawdown** is greatest near the points of water removal and less in between. Most of the flow to the drain or well occurs in the saturated zones, because saturated soil is much more conductive than unsaturated soil. For water in the saturated zone to flow to-

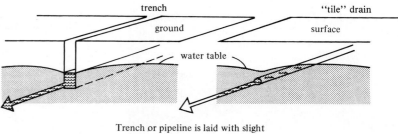

Trench or pipeline is laid with slight
slope to provide water flow

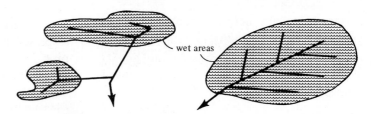

Patterns of tile or trench systems

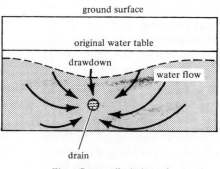

Water flow to tile drain, and
drawdown of water table

**FIGURE 6–16** Methods and water flow patterns in subsurface drainage.

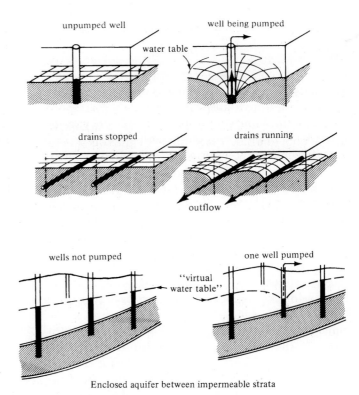

**FIGURE 6–17**  Water table draw-down by drains and wells.

ward the well or drain, the flow must be driven by a potential gradient (caused by gravity and pressure in this case). With a uniform, level water table, there are no differences in the sum of the pressure and gravity potentials. Water can flow laterally toward the drain only if the water table drops toward the drain, and the flow will slow as continued drainage lowers the slope of the water table. In practice this has several consequences. Heavy withdrawals of water from closely spaced wells tapping an underground water supply lower the overall water table, making it difficult to establish a respectable gradient and flow rate toward each well. Drainage cannot lower the water table evenly; it can lower only the average water table. To lower the water table adequately at its high points, between the drains, the drains must be spaced close enough together. The less permeable the soil is, the closer the drains must be, and the more costly. A soil with dense clay or hardpan subsoil can be exorbitantly expensive to drain and hence is unusable in badly drained or saline areas.

Drain lines are usually made from perforated plastic pipe or loosely fitting ceramic (**drain tile**) pipe

laid in a trench dug to the correct grade (bottom slope) and buried. A layer of gravel (the *envelope*) stops soil from blocking the holes or entering the pipe. In organic soils and some clay soils, *mole drains* are made by pulling a torpedo-shaped steel or iron "mole" through the ground, creating an unlined tunnel that will last for many years.

Water flows in drains under gravity—that is, downward. It must be disposed of into a **sink** where its presence is desirable or at least acceptable. If fortune smiles, such a sink exists downhill; otherwise pumping will be necessary.

## 6.5  WATER USE EFFICIENCY

Water is used most efficiently when the most advantage is gained from the least amount of water. Depending on how we define gain and water use, there are different measures of efficiency with different meanings. We shall look at some of these kinds of efficiency and possibilities for improving them.

### 6.5.1   Biomass–Transpiration Ratio

This efficiency measure is simply the amount of growth made per unit of water transpired. The ratio falls markedly in hot, dry climates or seasons, which promote vigorous transpiration without corresponding growth. Biomass–transpiration efficiency varies somewhat among crops, but by and large, little can probably be done to improve it. As implied in Chapter 5, most plant characteristics and management options exert similar effects on both growth and transpiration.

### 6.5.2   Biomass–Evapotranspiration Ratio

The growth per unit of evapotranspiration takes account of direct evaporative water use as well as transpiration. This kind of efficiency is improved by management to reduce direct evaporation and use more of the water productively for transpiration (Figure 6–18). Direct evaporation is less if the soil surface is kept dry, planting density is increased, or cover is established early. Some irrigation methods—especially drip irrigation—diminish evaporative loss because they keep much of the soil surface dry. Sprinkling in hot, windy weather aggravates evaporative loss.

### 6.5.3   Growth per Unit of Water Applied (or Available)

These measures of efficiency, based on water applied as irrigation or on water available from all sources, take into account losses through runoff and seepage as well as evaporation. Runoff and seepage can be minimized in irrigation by improving irrigation water management, by improving ground cover, or by practices such as grading and contouring.

It is neither possible nor always desirable to eliminate water losses completely. For example, occasional leaching is needed to control salt. And, although runoff or seepage water may be a loss to the individual farm, on a regional basis it may be recaptured for irrigation or find other uses (Figure 6–19). Sometimes, however, runoff and percolation from one site become a hazard elsewhere, by raising water tables and enhancing salt accumulation (Figure 6–20).

Transpiration tends to increase proportionately with growth, but direct evaporation, runoff, and seepage do not; in fact, they tend to decrease. Improved plant cover shades the soil, reducing evaporation, and plants on the surface may reduce runoff. Both runoff and seepage are decreased as a consequence of removing water from the soil through transpiration. Thus efficiency can be improved by double cropping, increasing planting density, or establishing a ground cover quickly, and in general it can be improved by practices that increase yield (such as soil fertility, control of pests and disease, and crops with more vigor or longer growing season).

### 6.5.4   Yield Efficiency and Economic Efficiency

For most crops, the yield of economically useful product matters more than biomass growth. Some crops should be more water efficient than others simply because they have a higher "harvest index"; that is, more of their total mass is usable product. Plant breeders often breed for this characteristic, which is valued for several reasons besides water use efficiency.

Finally, of course, there is economic efficiency: the net return of profits or other benefits per unit of expenditure on water. The obvious way to enhance this kind of efficiency, besides those already mentioned, is to get cheap water or grow valuable crops. High-value fruit and fresh-vegetable crops are prominent components of agriculture in regions with expensive water.

## 6.6   SUMMARY

The effectiveness and efficiency with which soil water can be managed depend greatly on climate, seasonal weather, and the plants' properties. The most influential soil properties are those that affect water storage, infiltration, and drainage. In dryland agriculture, effective water storage capacity may be the single most important soil property. Storage capacity in turn depends on depth, texture, organic matter, and freedom from constraints on root growth. In irrigated agriculture, or where rain falls frequently and dependably, storage capacity is less important. Variable infiltration rates, or extremely slow or fast

(A)

(B)

**FIGURE 6–18** Some examples of efficient water use. (A) Delivery by pipe or concrete-lined ditch reduces evaporation (pipe) and seepage (pipe and lined ditch) losses. (B) Intercropping can make better use of land and water. In this example, cowpea is being intercropped with fruit trees. Once the canopy expands, only shade-tolerant crops can be grown as intercrops. (C) Microsprinkler irrigation in a young orchard. The area wetted (dark area) is small and directed to the volume of soil where most of the roots will grow. In addition to decreasing water use, this method also helps to reduce unwanted plant growth by keeping the soil dry outside the root zone. (Photos A and B: J. Dahalig and W. W. Nell, U.S. Bureau of Reclamation; photo C: Dave Goldhamer.)

(C)

FIGURE 6–19 Water "lost" by runoff and percolation from one place may have a valuable use somewhere else. (Photo: A. G. d'Alessandro, U.S. Bureau of Reclamation.)

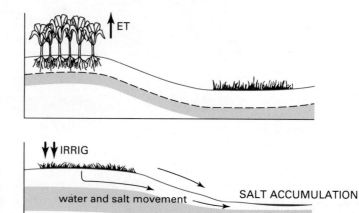

FIGURE 6–20 Water lost by runoff and percolation from one place may aggravate waterlogging or salinization somewhere else.

infiltration rates, interfere with irrigation, especially when water is applied by flood or furrow methods. Infiltration rates depend mainly on soil structure and structural stability in the topsoil. Subsoil impermeability due to clay accumulation or cemented horizons adds to the cost of drainage and salt control, sometimes to the point that irrigation agriculture cannot be sustained.

## QUESTIONS

1. What are the inputs and losses affecting soil water storage?
2. Describe the normal pattern of water removal from field soil by plants.
3. What soil properties and other factors affect infiltration rates?

# SOIL AND ENVIRONMENT
## *Land and Water Management*

- Soil properties governing water intake, storage, and release influence watershed management, stream flows, flood control, farm practices, natural vegetation, the erodibility of most classes of land, and appropriate land-use decisions, which are the first step in environmental protection. A smart society with choices would restrict agriculture to land that can best tolerate it, to preserve other lands for watershed, forest, wildlands, parks, and range. High-class agricultural land possesses a set of hydraulically significant properties: it is level and drained, with deep, permeable, uniform soils, moderate infiltration rates, no impeding horizons, and ample water storage capacity. These properties minimize soil erosion and several other hazards of agriculture. They also make for an austere landscape. More interesting uplands, deserts, and shorelands—with hills, views, shallow pedons, rocks, fascinating horizons, biomes, and variability—tend to be too fragile or inhospitable to use for sustained agriculture and too socially valuable as watershed, forest, wildlands, and the like.

- Water management affects soil erosion in many ways. In irrigation agriculture, risks of severe erosion are minor. Yet rapid overland flows arising from excessive grades, excessive application rates, or degradation of the soil surface structure can move sufficient soil on the field to interfere with operations, including irrigation itself.

   Dryland crop, range, park, and forest land uses often erode severely, with resulting loss of plant productivity as well as silting damage to downhill dams and waterways. Managers can reduce erosion by developing systems suited to the rainfall pattern, topography, and soils that will maintain protective plant cover and improve water infiltration. Control of grazing, fire, and traffic contributes to the preservation of plant cover and soil structure. Park and forest managers usually try to keep trails, camps, and vehicles off critical slopes and shallow, infertile soils.

- Water management can influence the leaching and erosion of contaminants from soil into water supplies. Surface runoff tends to contaminate surface waters with recently applied agrichemicals, sediment, organic matter, phosphate, and other substances that stay near the surface because of their limited mobility in soil. Subsurface drainage, having passed through the solum, carries leached substances that are mobile in soil, such as nitrate, selenium, boron, and most salts and alkalies. It delivers them to local groundwaters and eventually to lower lands and surface waters fed by springs and seepage. Such downhill accumulations can arise unexpectedly. For instance, tree felling in rangeland can reduce evapotranspiration and thereby increase percolation carrying subsoil salts to concentrate in croplands downhill.

   For watershed managers, maximal runoff would maximize the yields of water, but is unacceptable because it chokes dams too rapidly with sediment; in some cases partial conversion of woodland to grassland has successfully increased water yield without increasing erosion. In the urban context, runoff (or "stormwater disposal") has been seen as something to maximize. After all, what could erode from land covered by buildings, grass, and paving? The answers—garbage, oil, animal feces, and garden chemicals—become evident in streams, ponds, harbors, and beaches.

- Irrigation water is normally costly. Water prices that fully reflect costs, without direct or indirect subsidy, are often exorbitant for agricultural users, especially where urban demands are strong and sources are limited or distant. Besides the obvious costs of reservoirs, wells, delivery, and drainage disposal, there are usually social and environmental costs, beginning with competition effects on previously established farm communities. Water diverted for irrigation is not available for downstream functions such as groundwater recharge, salinity control, and the maintenance of wetlands and stream flows. Reduced flows promote eutrophication and streambed sedimentation. Improvement of water use efficiency in agriculture indirectly reduces some of these environmental costs. Low efficiency in the use of irrigation water is expected in warm, dry climates, because evaporation is rapid and local rainfall sparse. At least in principle, irrigation water can be used more efficiently to supplement rainfall in wetter regions.

4. What is "field capacity"?

5. How does soil water storage capacity influence (a) dryland cropping systems and (b) irrigation practice and design?

6. What soil properties influence a pedon's water storage properties?

7. The rate of infiltration usually decreases as infiltration proceeds. Give some reasons and consequences.

8. Figure 6–3 refers to a soil with no texture profile. Redraw it for a soil with a sandy A and a clay loam B horizon.

9. Why are so many different methods of irrigation used?

10. It is possible to drain a heavy clay soil—for example, a Vertisol—but it is seldom done. Why not?

11. Perfectly even application of irrigation water and perfectly even removal of subsoil water by drainage both are virtually impossible. Why?

12. During heavy rains, water in a stream comes mainly from surface runoff. The base flow that continues during dry periods comes mainly from groundwater. Compare the pollution load (kinds and quantity) in floodwater with the load in base flow.

## FURTHER READING

Hillel, D. W. 1982. *Introduction to soil physics.* New York: Academic Press.

Jury, W. A., and H. Flühler. 1992. Transport of chemicals through soil: Mechanisms, models, and field application. *Advances in Agronomy* 47:142–202.

Krantz, B. A., J. Kampen, and S. M. Virmani. 1979. *Soil and water conservation and utilization for increased food production in the semi-arid tropics.* Hyderabad, India: International Crops Research Institute for the Semi-Arid Tropics.

Schwab, G. O., R. K. Frevert, T. W. Edminister, and K. K. Barnes. 1981. *Soil and water conservation engineering.* New York: Wiley.

Taylor, H. M., W. R. Jordan, and T. R. Sinclair, eds. 1983. *Limitations to efficient water use in crop production.* Madison, Wis.: American Society of Agronomy.

Turner, N. C. 1997. Further progress in crop water relations. *Advances in Agronomy* 58:293–338.

Wilson, L. G., L. G. Everett, and S. J. Cullen. 1995. *Handbook of vadose zone characterization and monitoring.* Boca Raton, Fla.: CRC/Lewis.

# 7

# SOIL ORGANISMS

## OVERVIEW

*Soils are populated by plant roots, small animals, and many kinds of microorganisms: protozoa, algae, fungi, and bacteria. Each group is important to the soil's life, but most of the biomass and biological activity is provided by plant roots, fungi, and bacteria. A soil contains many habitats, each occupied by different species, and the habitats frequently change with plant growth and events of weather and management.*

*Soil microbes have general requirements for (1) space, within soil pores and on particles or root surfaces; (2) water of suitable salinity and pH, provided by the soil solution; (3) adequate nutrition for energy and essential constituents; and (4) satisfactory temperature. Space is usually ample; nearly always it is nutrition, water, or temperature that limits population growth. The microbes have evolved so many ways of adapting to different limitations that the total population in a soil is varied, versatile, and adaptable to changing conditions and food supplies.*

**131**

Soils are alive. Populations of roots, microbes, and animals are part of soil and carry out many of its essential processes: consuming and destroying organic matter, making humus, and recycling nutrients. Soil organisms vary in kind and size, from a tree root system extending for meters and weighing tons to a bacterium a micrometer long and weighing less than a picogram. Figure 7–1 indicates some of the diversity. The small creatures are the most numerous—in fact, so numerous that their total biomass can add up to many tons per hectare of land. Their favored habitats are the soil water and the surfaces of litter, plants, and minerals, especially in the soil's upper horizons.

## 7.1  SOIL COMMUNITIES

### 7.1.1  Kinds of Organisms in Soil

The larger organisms in soil include both plants and animals. Plant roots explore the soil, growing in cracks, spaces between aggregates, and other large voids. The plant feeds most other soil organisms with organic matter that originates from photosynthesis and is delivered to soil microbes in several ways. Microbes near the root feed on root excretions, and others invade roots. Still others are fed by plant litter that falls to the surface and sometimes forms a dis-

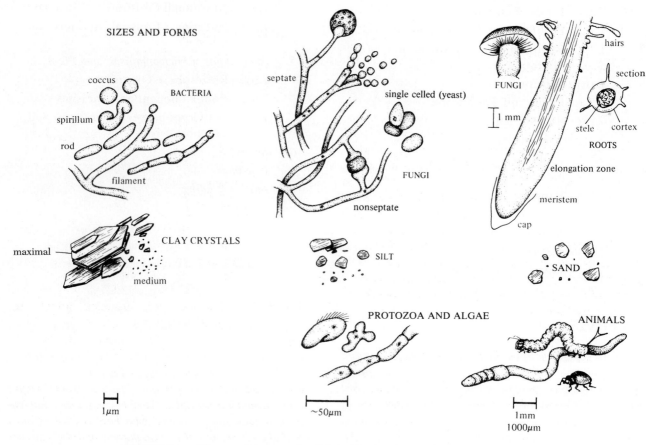

**FIGURE 7–1**  Sizes and forms of soil inhabitants, in relation to soil particle sizes.

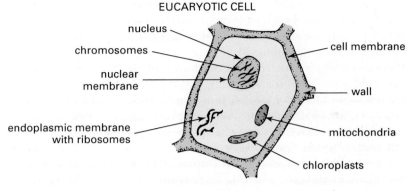

EUCARYOTIC CELL

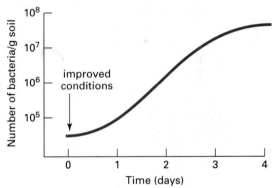

PROCARYOTIC CELL

**FIGURE 7–2** Eucaryotic and pro-caryotic cells. Procaryotic cells lack the distinct nucleus, containing chromosomes enclosed in a nuclear membrane, that distinguishes the cells of eucaryotes (algae, protozoa, fungi, plants, and animals). They also lack other organelles often found in eucaryotic cells.

tinct *litter (O) horizon.* Soil animals include burrowing mammals and reptiles, and, far more numerous, the smaller animals (e.g., insects, mites, and nematodes) that live in the litter, burrow in the soil, or infest plant roots.

The smallest organisms—microbes—include representatives of both types of fundamental cell structure: eucaryotic and procaryotic (Figure 7–2). The **eucaryotic** organisms include fungi, protozoa, and algae, as well as plants and animals. The **procaryotic** organisms are the Monera, or bacteria, which are the smallest soil organisms and physiologically the most diverse.

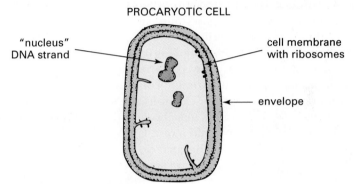

**FIGURE 7–3** Typically rapid bacterial growth response to improved conditions.

## 7.1.2  Proliferation, Spread, and Survival

Microbes proliferate rapidly. Under favorable conditions, many bacterial species can double in cell number in less than an hour. Typical soil bacteria under normal soil conditions divide less frequently, but they still can increase a hundredfold (about seven generations) within a day or two when dry soil is wetted, cold soil is warmed, or easily used organic food is added (Figure 7–3). This reproductive capacity enables soil microbes to respond rapidly to changes in conditions, not just multiplying but also replacing a population with organisms adapted to new conditions.

Most microbes spread readily, carried by larger organisms, water, and dust (airborne fine soil). Soil, difficult to sterilize, is impossible to keep sterile unless sealed from contamination.

Soil microbes are survivors. Conditions rarely favor growth and activity for long; much of the time the soil is too dry, cold, hot, or deficient in nutrients. The microbes' survival tactics vary. Dormancy is common. The cell ceases activity, perhaps desiccates (dries), and enters a resting state to await better times. Though mortality is high as cells enter and leave dormancy, enough remain viable to recolonize the soil when conditions improve. Spores specialized for long survival are formed by most fungi and many bacteria as their growth conditions deteriorate. (Some bacterial spores even survive boiling.) The filamentous habit of fungi and many bacteria might help these organisms survive unfavorable conditions that are temporary or localized. For example, the nutrient-rich surface horizon often gets hot or dry, while the soil below remains cool and moist. Unlike a single-celled microbe isolated in the surface, an extensive filamentous growth (mycelium) keeps functioning—with water supplied from below—until rain makes the topsoil usable once more. This may explain why fungi and **actinomycetes** (filamentous bacteria) tend to prevail in desert soils.

### 7.1.3 Requirements and Adaptations of Microbes

Organisms need favorable physical conditions and nutrition. The physical requirements include available space, usually on a surface (Figure 7–4). Space is probably adequate in nearly all conditions, because growth is usually limited by unsuitable temperature or inadequate water or nutrition.

The nutritional requirements include usable sources of energy and essential elements. Many organisms need certain biochemical compounds that they cannot make for themselves. Nutritional requirements of organisms in general are outlined in Tables 7–1, 7–2, and 7–3. Note the different kinds of energy-yielding processes that different organisms use (Table 7–1, p. 136).

**Heterotrophic** organisms get energy from reactions involving organic substrates such as **sugars,** from which energy is released by respiration or fermentation. In **respiration,** organic substrates are oxi-

dized with inorganic oxidizing agents, most commonly oxygen but sometimes nitrate or sulfate. In **fermentation,** which usually begins when oxygen is deficient, some of the organic substrate is itself used as the oxidizing agent, being reduced to a by-product such as an alcohol or organic acid. The heterotrophs include animals, fungi, and most bacteria. The plant root system functions heterotrophically, though the plant as a whole is autotrophic.

**Autotrophic** organisms do not need organic energy substrates; they derive energy from light (photosynthesis) or from inorganic oxidation reactions (Table 7–3). Most **assimilate** carbon as carbon dioxide from air or water. The autotrophs include green plants, nearly all algae, blue-green bacteria, and certain other bacteria.

Heterotrophs and autotrophs depend on each other, and the cycling of nutrient elements depends on both groups (Figure 7–5). The autotrophs are the prime synthesizers of organic material. The heterotrophs destroy it, releasing the nutrient elements for recycling. The soil ecosystem is the planet's busiest theater for nutrient cycling. The principal players are plants in the autotrophic role and fungi and bacteria in the heterotrophic role.

## 7.2 LARGE ORGANISMS

### 7.2.1 Plants and Their Roots

Plants produce the organic matter that feeds soil microbes and becomes changed into humus. This plant material, described chemically in Chapter 8, reaches the soil as litter—shoots, leaves, and twigs dropped onto the surface—or as living and dead roots. Do not underestimate the input from roots. In a typical cereal crop, roots constitute 30 to 50 percent of the total plant mass, up to 5 Mg/ha. In a forest, their mass could be several times larger.

Roots grow by cell division in the **meristem** just behind the root tip and by elongation of the cells in a zone a centimeter or so back from the tip (Figure 7–6). In the region of cell elongation, the root tissue begins to differentiate, developing a distinct **cortex** surrounding an inner **stele** that contains specialized conductive tissues, xylem and phloem. The **xylem** becomes the channel for movement of water and dissolved plant nutrients upward to the shoot. The **phloem** becomes the channel for delivery of sugar

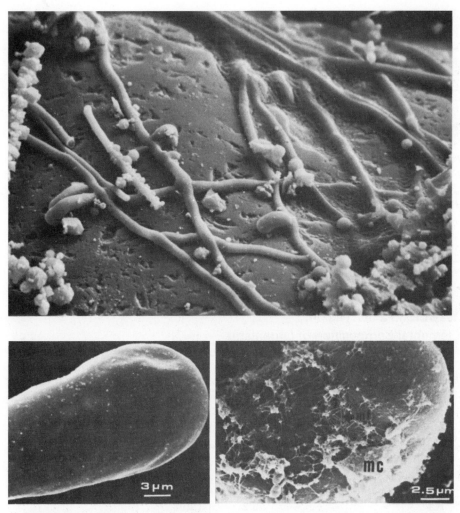

FIGURE 7–4 Most soil microbes live on surfaces. Top: Fungal hyphae on a sand grain. (Scanning electron micrograph of frozen specimen. Note ice crystals and water film. Reprinted with permission from *Soil Biology and biochemistry,* vol. 14, R. Campbell and R. Porter, Bristol University, 1982. Pergamon Press Ltd.) Bottom: Bacteria (*Rhizobium*) attached to a root hair. (Scanning electron micrograph. Reprinted with permission from *Applied and Environmental Microbiology, vol. 43,* G. Shantharam and P. Wong, American Society of Microbiology, New York.)

and other organic products to the root from the shoot. This same active region, where differentiation takes over from growth and elongation, is the most active in plant nutrient uptake. In many species, **root hairs** also develop here from **epidermal** cells.

The advancing root tip favors paths of least resistance between peds and through cracks and other large voids, but if the soil is moist and uncompacted, roots can penetrate peds, compressing the soil mate-rial as they advance and helping form new voids and aggregates. In either case, root growth in the soil is aided by the **root cap,** made from cells produced by the meristem and continually sloughed off as the root tip advances. The growth of roots contributes to the development of granular structure in topsoils. Granulation probably results from several processes: penetration, compression, addition of organic matter from root caps and exudates, production of gums

## TABLE 7–1
Kinds of Organisms in Soil

| Kind | Food | Energy source |
|---|---|---|
| **Large eucaryotes** | | |
| Animals | Solid organic | Oxidation of organics |
| Plants | $CO_2$ + ions | Light |
| **Small eucaryotes** | | |
| Protozoa | Solid and dissolved organic | Oxidation of organics |
| Algae | $CO_2$ + ions | Light |
| Fungi | Dissolved organic | Oxidation of organics |
| **Procaryotes (bacteria)** | | |
| Phototrophic | $CO_2$ + ions | Light |
| Lithotrophic[a] | $CO_2$ + ions | Oxidation of inorganics |
| Heterotrophic | Dissolved organic | Oxidation of organics |

[a]Chemoautotrophic.

## TABLE 7–2
Elemental Composition of Organisms

| Element | Bacteria[a] (% dry wt.) | Plants[a] (% dry wt.) | Main occurrences in tissue |
|---|---|---|---|
| C | 50 | 45 | All organic substances |
| O | 20 | 44 | Most organic substances |
| N | 12 | 1–2 | Proteins, nucleotides |
| H | 8 | 6 | All organic substances |
| P | 1 | 0.2 | Nucleotides, phospholipids, ion |
| S | 1 | 0.2 | Proteins |
| K | 1 | 1 | Ion |
| Na | 1 | 0.2 | Ion |
| Ca | 0.5 | 0.3 | Wall, chelates, other complexes |
| Mg | 0.5 | 0.3 | Chlorophyll, other complexes |
| Cl | 0.5 | 0.1 | Ion |
| Fe | 0.2 | 0.1 | Cytochrome, other complexes |

[a]Bacteria: *Rhizobium;* plant: midseason corn shoot (typical concentrations).

by microbes, and desiccation of the soil as the root absorbs water.

Roots influence microbial activities. The strongest influence is in the **rhizosphere,** the millimeter or so of soil nearest the root surface, where ample food from the plant stimulates fungi and bacteria to reach population densities 10 to 100 times higher than in the rest of the soil. The additional food includes dead root cap cells and root exudate.

**Root exudate** is a mixture of organic acids, sugars, and other soluble plant components that escape from roots. The root also hosts bacteria, fungi, and small animals that live in its cortex. Some of these are harmless, others are parasitic, and still others are symbiotic with the plant, helping it acquire nutrients (see Section 7.5). Overall, the root and rhizosphere environments support much of the soil's total microbial and biochemical activity.

**TABLE 7–3**

Oxidation Processes that Supply Energy to Microbes

| Reductant | Oxidant | Products | Organism/conditions |
|---|---|---|---|
| Sugars[a] | Oxygen | $CO_2$ | Most organisms/aerobic |
| Sugars[a] | $NO_3$ | $CO_2$, $N_2$, $N_2O$ . | Denitrifiers/low $O_2$ |
| Sugars[a] | Sulfate | $CO_2$, S | S reducers/low $O_2$ |
| Sugars[a] | Organic | $CO_2$, organic acids | Anaerobic fermenters/low $O_2$ |
| $NH_4^+$, $NO_2^-$ | Oxygen | $NO_2^-$, $NO_3^-$ | Nitrifiers/aerobic |
| S | Oxygen | $SO_4^{2-}$ | S oxidizers/aerobic |
| $Fe^{2+}$ | Oxygen | $Fe^{3+}$ | Iron bacteria/aerobic |

[a]And other organic compounds.

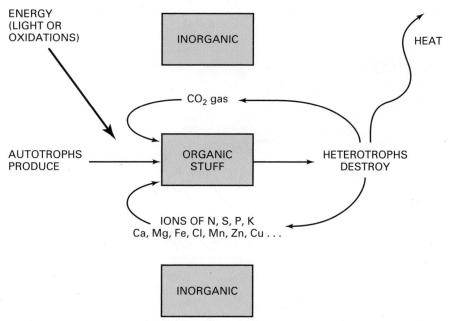

**FIGURE 7–5**  Interrelationships of autotrophic and heterotrophic organisms: nutrient cycling and energy flow.

The plant provides another important habitat for microbes, in the litter layer at the ground surface. In some soils this layer is negligible, but in others it is centimeters thick. It is a mixture of readily decayed leaf fragments and decay-resistant woody twigs, branches, and stems. This habitat is rich in organic food, but it exposes its inhabitants to more extreme temperatures and longer and more severe droughts than does the rest of the soil. It also develops a specialized population of microbes, enriched in fungi that excel at decomposing wood. Here, too, the soil animals chew, partly digest, and incorporate material into the mineral soil below so that it becomes more accessible to attack by microbes.

### 7.2.2  Soil Animals

Many kinds of animals live in soil (Figure 7–7). Some are visitors, and others are permanent residents.

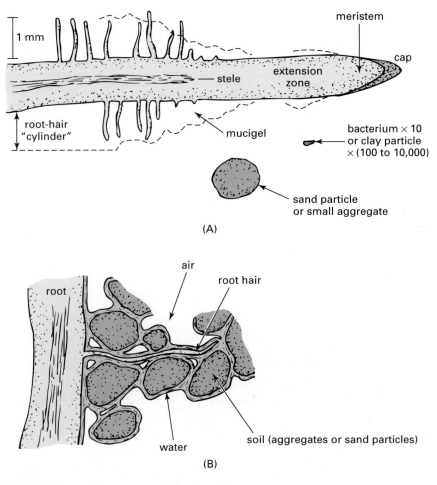

FIGURE 7–6    Plant root. (A) Basic structures. (B) Root hair–soil–water contacts.

Some must burrow, and others use existing channels. Some occupy soil air spaces, and others are aquatic, living in the soil's liquid phase. Some are large and conspicuous, and others are smaller, close to microscopic. Some eat living food—plants, other animals, fungi, bacteria—and others eat dead material. What they eat is ecologically significant; thus plant-eating insects and mollusks may add organic matter to the soil. Insects, arachnids, and worms that consume dung and plant litter mix it with soil and speed up its decay, and plant-parasitic nematodes and insects reduce the ecosystem's productivity.

Usually, the most numerous animals are the smallest: the nematodes, other helminths, and the arachnids (mites). Mollusks (slugs and snails) may be the most familiar soil animals. The soil arthropods include many species of mites, millipedes, springtails, insects, and their larvae. In soils with high organic matter and ideal moisture conditions, earthworms (annelids) may compose most of the animal biomass. They tend to die or emigrate when invited to improve bad soils.

The soil animals share several properties. All are heterotrophic, highly aerobic, and mobile. Though often smaller than their relatives who live above ground or in water, they are large compared with microbes. Only the smallest, the nematodes, protozoa, and rotifers, live aquatically in the soil solution (and even these are too big to remain mobile unless the soil is wet). The rest must live in large soil voids or make burrows. These factors restrict most soil animals to topsoil and litter, where oxygen, food,

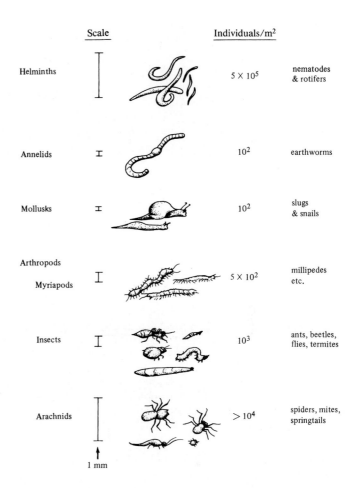

|  | Scale |  | Individuals/m² |  |
|---|---|---|---|---|
| Helminths | | | $5 \times 10^5$ | nematodes & rotifers |
| Annelids | | | $10^2$ | earthworms |
| Mollusks | | | $10^2$ | slugs & snails |
| Arthropods Myriapods | | | $5 \times 10^2$ | millipedes etc. |
| Insects | | | $10^3$ | ants, beetles, flies, termites |
| Arachnids | | | $> 10^4$ | spiders, mites, springtails |

1 mm

**FIGURE 7–7** Major classes of soil animals, indicating forms, relative sizes, and numbers (organisms per square meter of ground) expected in a fertile grassland soil. (From data of D. K. McE. Kevan in *Ecology of soilborne plant pathogens,* ed. K. F. Baker and W. C. Snyder.)

and structure are favorable. Animals are few in compacted or very wet soils.

## 7.3  SOIL MICROORGANISMS

### 7.3.1  Protozoa and Algae

Protozoa and algae (Figure 7–8) in soil are mostly single celled and smaller than their aquatic counterparts. The algae are capable of photosynthesis, like plants and blue-green bacteria, using chlorophyll and splitting water with the release of oxygen:

$$\text{light} + 2H_2O \xrightarrow{\text{chlorophyll}} \text{chemical energy} + \underset{\text{reductant}}{[4H]} + O_2$$

The protozoa in soil are nonphotosynthetic. The ability of some to move and to engulf objects makes them efficient consumers of bacterial cells and other fragments of organic matter. Some protozoa actively prey on live bacteria, whose populations they control. Protozoa are important in nutrient recycling because their excretions release nutrients that benefit other communities of microorganisms and plants. Both algae and protozoa are most active and abundant in wet soil; water films become too thin for them to move freely in dry soil. Protozoa often exist in a dormant state until the soil is sufficiently wet or sufficient organic matter is available to sustain their growth. The small size of protozoa gives them access to pores as small as 8 μm, where they are protected from predation by other organisms and by rapid desiccation.

### 7.3.2  Nematodes

More than 15,000 species of nematodes have been identified. These small, unsegmented roundworms,

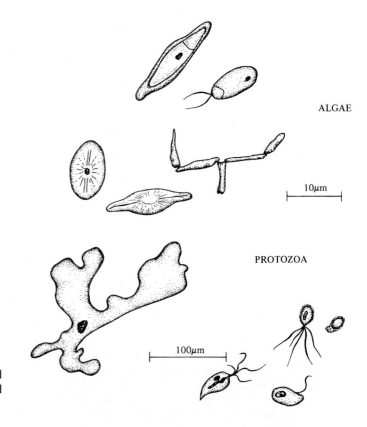

ALGAE

10μm

PROTOZOA

100μm

**FIGURE 7–8** Some forms and sizes of microscopic algae and protozoa occurring in soils.

0.5 to 2 mm long, are often present in large numbers in soils ($10^6$ to $10^7$ m$^{-2}$). Nematodes are an important part of the soil environment. The nonparasitic varieties are called "free-living" nematodes. In addition to their role as plant parasites, nematodes feed on soil bacteria, fungal mycelia, algae, and small invertebrates, and various predators such as mites feed on them. Nematode consumption of bacteria is an important step in releasing nitrogen that is immobilized by bacteria during decomposition of plant material.

Nematodes are best known for their role as plant parasites. Ectoparasites feed at root surfaces but do not enter roots. Endoparasites feed and live within root tissue. A very wide range of crops may be damaged by nematodes. Damage includes reduction in yield due to reduced plant vigor, wilting due to damage of root systems that reduces the plant's ability to obtain water, and nutrient deficiencies.

### 7.3.3 Fungi

Fungi are abundant and active in normal aerated soils. All are heterotrophic. One group, the yeasts,

are single celled and sometimes anaerobic, but most fungi, the molds, are strictly aerobic and produce filamentous growth (hyphae) that spreads through soil, litter, and (sometimes) plants. The total filamentous body, a mycelium, is essentially one large cell with many nuclei, although in the "higher" fungi it is divided by porous cross walls (septa). A single mycelium may extend many centimeters or even meters in the soil. Some species can organize part of their mycelium into conspicuous fruiting bodies (e.g., mushrooms, toadstools, and bracket fungi).

Fungal mycelium spreads over surfaces and helps bind mineral particles into aggregates. Many fungal species associate with plants, as neutral inhabitants of the root or rhizosphere, as pathogens, or as symbiotic partners (see Section 7.5). There are acid-tolerant species, wood-decomposing species, and drought-tolerant species. Not surprisingly, fungi as a group are prominent occupants of almost all microenvironments in aerobic soils. (Figure 7–9 shows some of the major kinds of fungi.) Fungi play a critical role in the decomposition of cellulose, hemicellulose, lignin, and chitin within soils and on soil surfaces (see Chapter 8).

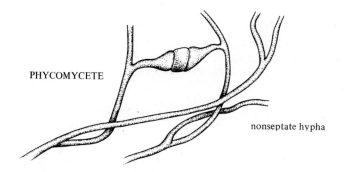

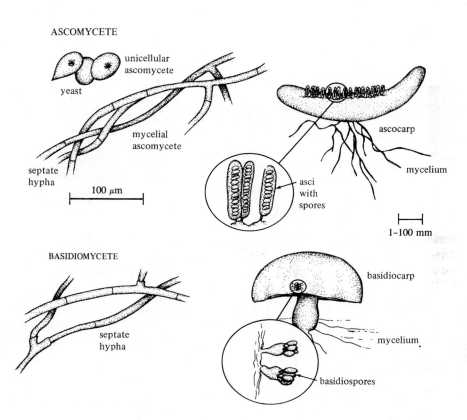

**FIGURE 7–9** Fungi: some forms and sizes occurring in soils and representing three major groups of fungi.

## 7.3.4 Bacteria (Monera)

Simple as bacteria might seem in form (Figures 7–1 and 7–2), they are extraordinarily diverse physiologically and biochemically, as discussed in Chapter 8. Bacteria of one kind or another colonize virtually every habitat within the soil. A fertile topsoil might contain $10^9$ living bacteria per gram of soil. Many bacteria reproduce rapidly—their cells divide every few hours under favorable conditions—and yet most soil bacteria also excel at surviving adverse conditions as endospores or dormant cells. Their small size enables them to disperse readily with dust particles. They are themselves within the upper part of the colloidal size range, of the same magnitude as coarse clay particles. Some bacteria can derive energy from light or from inorganic reactions. Some can use organic substances that most organisms find unassimilable or toxic. These substrates include macromolecules that compose the bulk of organic residues left by dead plants, animals, and bacteria (see Chapter 8, Figure 8–2). Toxic substrates include manufactured organic pollutants. Some bacteria can assimilate nitrogen from air (**nitrogen fixation**). Many bacteria tolerate or actually require the absence of oxygen (**anoxia**).

### 7.3.5  Viruses

Viruses are the smallest and most numerous ($10^{10}$ g$^{-1}$ soil) soil organisms (Figure 7–10). Most scientists do not consider viruses to be living organisms because they are unable to replicate without first invading a living host cell. Once produced by the host cell, a virus does not grow, as do other living organisms. Little is known about most viruses other than that they infect animals, plants, humans, fungi, and bacteria. They may be important in reducing the population of important soil bacteria such as *Rhizobium* that are responsible for symbiotic nitrogen fixation. Viruses, especially human pathogens, are frequently added to soil with the addition of sewage sludge (biosolids) to soils.

## 7.4  CONSTRAINTS AND ADAPTATIONS

### 7.4.1  Physical Constraints

Soil temperatures fluctuate daily and seasonally, particularly at the soil surface (see Chapter 4), where temperatures commonly range from subfreezing to 45°C. High temperatures, from 35°C upward, can progressively kill cells, depending on the heat tolerance of the species, though spore-forming bacteria are extremely tolerant. Temperatures high enough to cause temporary sterilization sometimes occur locally—for example, in compost heaps or during the burning of vegetation. Cold temperatures, approaching freezing, essentially stop microbial growth but without killing the organisms. Otherwise, heat and cold affect soil microbes less drastically. Yet, because certain organisms and processes are more sensitive than others, heat and cold alter the composition of the microbial population. For instance, heat-loving types proliferate at the expense of heat-sensitive ones in a hot climate or season (Figure 7–11). This ability of the population to adapt lessens the impact of adverse environment.

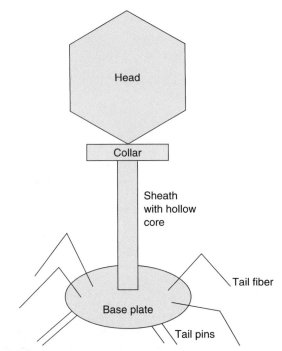

**FIGURE 7–10**   A soil virus showing the major parts. The head contains the DNA or RNA core.

Water conditions are seldom ideal for soil microbes. As soil dries, the water films become thinner, hindering the movement of both microbes and their nutrients. The myceliate growth habit of fungi and filamentous bacteria diminishes the impact by allowing the organism to move water and nutrients over large distances within itself. The greatest impact of drought is on those organisms that must swim to get food: the protozoa, nematodes, and rotifers. Most bacteria adapt to low water potentials and high salinity by maintaining high internal concentrations of solutes that help prevent desiccation. Even when they do desiccate, the cells' smallness, ruggedness, and simplicity allow some to survive. Dormant soil microbes can survive drought for many years in specialized forms—for example, encysted protozoa, fungal spores, and bacterial endospores. When water again becomes available, these survivors initiate a flush of microbial multiplication, fed by dead microbial cells.

Oxygen deficiency is most common in waterlogged soils—rice paddies, marshes and bogs, flooded areas—where the slow diffusion of oxygen through water-filled pore space fails to keep up with the oxygen demand of the roots and microbes. Se-

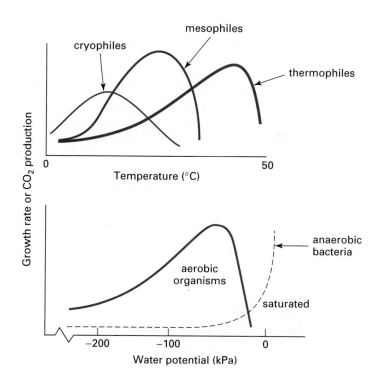

**FIGURE 7–11** Temperature and water requirements of differently adapted bacteria.

vere compaction also impairs aeration. Even well-aerated soils may contain anoxic microsites within peds, whose small internal voids remain filled with water, although the large voids between the peds are drained. As soil becomes less aerobic, the growth of aerobic organisms slows, and the anaerobic bacteria take over (see Chapter 8).

### 7.4.2   Acidity, Salinity, and Other Chemical Inhibitors

Soil acidity inhibits the growth of some soil organisms. As with plants, the adverse effect may involve shortage of calcium or toxicity of aluminum or manganese; strains and species vary in tolerance so that populations may adapt to local conditions. Most soil microbes probably tolerate soil salinity better than do plants. Toxic metals (copper, chromium, arsenic, mercury) and the extreme acidity associated with mining and smelting can almost sterilize soils, although there are instances of microbial adaptation to toxic pollution. Many soil chemical factors indirectly affect microbes when they reduce the growth of plants.

### 7.4.3   Nutritional Constraints: Nitrogen

Of the essential nutrient elements (Table 7–2), carbon and nitrogen are the most commonly deficient for soil microbes. The deficiency is rarely a simple lack of C or N, because most soils have plenty of carbon in organic matter and have nitrogen in organic matter and nitrogen gas. Rather, the deficiency problem lies in the inability of organisms to *use* these sources rapidly enough. For instance, most organisms can assimilate carbon in the form of simple sugars, but in soil, sugars are in short supply. Abundant sugar is tied up in large organic molecules, but only certain microbes can break down these large molecules and release sugar—and even for these organisms, the release is too slow for rapid growth. Likewise, the nitrogen supply often limits both soil microbes and plants because the N is tied up in large molecules and too little is in the available forms—ammonium and nitrate. (A related argument holds for other elemental nutrients.)

Adaptations to nitrogen deficiency include the ability to "scavenge" ammonium and nitrate ions released by organic matter breakdown. Most soil organisms, including plants, can absorb these ions rapidly from very low concentrations, and so the competition for available nitrogen becomes intense. Some organisms assimilate nitrogen from organic sources, such as humic compounds, that other organisms cannot use; certain procaryotes can assimilate atmospheric nitrogen.

**Biological nitrogen fixation** is the assimilation of dinitrogen, $N_2$, the gaseous form that composes 80 percent of the atmosphere. In nitrogen fixers, dinitrogen is reduced to ammonia ($NH_3$), which is in turn quickly consumed in the synthesis of organic nitrogen compounds. Nitrogen fixation has been identified in bacteria of several genera that live in soils, water, the roots or leaves of many plants, and the guts of various animals. All evidently possess a similar nitrogen-fixing enzyme system.

Nitrogen fixation requires much energy. Different nitrogen fixers get energy from different sources: organic food, photosynthesis by the bacterium, photosynthesis by a host plant, and so forth. The enzyme nitrogenase, which catalyzes the reaction, is similar in all the organisms from which it has been isolated. It consists of two unusual proteins. Both contain Fe and S; one contains molybdenum as well, and deficiency of Mo can inhibit nitrogen fixation in pristine soils and lakes. Free oxygen destroys nitrogenase, and different organisms have developed different ways to protect the enzyme from oxygen. Like other bacteria, nitrogen-fixing organisms can use ammonium and nitrate and usually prefer these to dinitrogen. Thus mineral nitrogen (ammonium and nitrate) inhibits nitrogen fixation. This response prevents energy from being wasted on unnecessary fixation. Where mineral nitrogen is limiting, rather than energy, nitrogen fixation gives an obvious advantage to the fixing organisms. When they eventually die and decay, the nitrogen becomes available to the rest of the ecosystem.

Biological $N_2$ fixation is important in most ecosystems and is actively managed in many agricultural and pastoral systems as a substitute for nitrogen fertilizers (see Chapters 8 and 10). In most soils, fixation rates are small: a few kilograms per hectare per year. But higher rates are possible, especially when the nitrogen-fixing bacteria operate in symbiosis with plants, as in the root nodules of leguminous plants (see Section 7.5). Symbiotic associations are especially effective because (1) the plant's photosynthesis provides ample energy; (2) the plant assimilates ammonium that would otherwise inhibit fixation; and

(3) the root nodule (or corresponding structure) provides suitable conditions, including protection from free oxygen.

### 7.4.4  Nutritional Constraints: Energy and Carbon

Table 7–3 lists reactions and processes that yield energy to organisms. The reactions are oxidations. (Supplement 8–2 explains oxidation and reduction.) Because most soil organisms, like humans, are heterotrophic, they need assimilable organic compounds to serve as a source of carbon and to provide energy through oxidation. The large organic molecules that compose almost all the organic matter of soil and plants cannot be used until broken down into simple small molecules. This breakdown involves specialized organisms, and its slowness limits the food supply to the bulk of the microbial population. Autotrophs escape this particular limitation, but in soils their special energy sources are usually even more scarce than the organic materials needed by the heterotrophs.

There is direct evidence that energy and carbon limit microbial action in soil: the addition of decomposable organic matter causes a microbial population explosion (Figure 7–3) and an outpouring of carbon dioxide from respiration or fermentation. The importance of energy is reflected in the variety of adaptations that microbes have evolved to deal with different energy-limited conditions. The most important energy-winning adaptations are autotrophy, opportunism, antibiotic production, specialization (to use unattractive substrates), predation, parasitism, and symbiosis. (These adaptations are discussed in Supplement 7–3.)

## 7.5  ASSOCIATIONS OF MICROORGANISMS WITH PLANTS

### 7.5.1  Rhizosphere Organisms

The rhizosphere is the cylindrical space extending a millimeter or so from the root surface. Root products stimulate bacteria and fungi, especially fast-growing heterotrophs, so that microbial populations are normally at least 10 times more dense in the rhizosphere than in the rest of the soil. The rhizosphere's boundaries are inexactly defined. The outer boundary is diffuse because the root's influence extends more or less indefinitely, diminishing with distance. The inner boundary may be taken to include the cell surfaces of the root epidermis or even the outer cortex, to which many bacteria and fungal hyphae adhere (some people distinguish this as a separate region, the *rhizoplane*). Rhizosphere microbes ordinarily account for much of the soil's heterotrophic activity, decomposing and assimilating dead plant cells and exudates. Dead plant cells come from the root cap, from disruption of the cortex by emerging lateral roots, or from damage by parasites and pathogens. Exudates are soluble organic metabolites—for example, sugars and citric, malic, and oxalic acids—as well as cell wall polysaccharides and pectic substances (see Supplement 8–1 for descriptions of these substances).

Respiration by the microbes produces carbon dioxide in addition to that produced by the root and tends to acidify the rhizosphere of alkaline soils. At least some of the rhizosphere bacteria are known to produce gummy coatings that help them stick to root cells (Figure 7–4) and glue the bacteria and soil particles together (Figure 7–4). In this way they might contribute to the development of granular soil structure in the root zone. Under the right conditions, significant rates of nitrogen fixation can be achieved, for example, by *Azospirillum* and related bacteria in the rhizospheres of tropical grasses and some cereal crops. There is evidence that rhizosphere organisms can help mobilize plant nutrients such as iron, copper, and zinc by producing soluble complexing agents. Others suppress root pathogens.

Some rhizosphere fungi and bacteria take their association with the plant a stage further, by invading the root cortex and becoming pathogens or symbionts.

## 7.5.2  Root Pathogens and Parasites

Soilborne bacteria and fungi cause several important diseases of roots or underground stems. Examples are damping-off, root rot, and wilt diseases caused by species of *Fusarium, Verticillium, Phytophthora, Pythium, Rhizoctonia, Alternaria, Armillaria,* and other fungi. There are soilborne root-rotting bacteria, and the crown gall or plant tumor disease agents (*Agrobacterium* spp.). About half the species of nematodes in soils are plant parasitic, invading roots of susceptible

plants, producing galls, and sometimes spreading to the vascular system and the shoot.

Some of these diseases are aggravated by wet soil conditions that favor the parasite's spread. Control measures include the choice or breeding of resistant crop plants and the reduction of the pathogen population through soil fumigation, control of soil water, rotation with nonsusceptible crops, or adjustment of soil conditions to hinder the pathogen or to favor microbes antagonistic to it.

## 7.5.3  Symbioses between Plants and Soil Microorganisms

The best-known symbiotic associations are of two general kinds: (1) nitrogen-fixing associations with bacteria, which supply the plant host with nitrogen from the air, and (2) mycorrhizae (or mycorrhizas), involving fungi, which help the plant get nutrients such as phosphate and zinc from the soil.

**Nitrogen-fixing symbioses**    At least three groups of nitrogen-fixing bacteria form symbioses with plants:

1. *Blue-green bacteria* colonize specialized cavities in the shoot system of certain cycads and ferns. Because they are photosynthetic, these bacteria probably need no organic food from their hosts, but the plant does provide an enclosed aquatic environment complete with inorganic nutrients. An important example is the association of a blue-green bacterium, *Anabaena,* with *Azolla,* a tiny floating fern cultivated throughout Southeast Asia to provide nitrogen for rice fields.

2. *Frankia* spp., a group of actinomycetes, infect roots of shrubs and trees of many genera, forming nitrogen-fixing root nodules. Widely distributed hosts include desert and brushland shrubs—for example, *Ceanothus* (deerbrush) and *Purshia* (bitterbrush)—and tree species—for example, *Myrica* (myrtle), *Alnus* (alder), and *Casuarina.* In some respects this symbiosis resembles the *Rhizobium*–legume symbiosis.

3. *Rhizobium* and *Bradyrhizobium* spp. (rhizobia) are bacteria that infect roots of plants of the legume family, forming nitrogen-fixing root nodules (Figures 7–12 and 7–13). Legumes

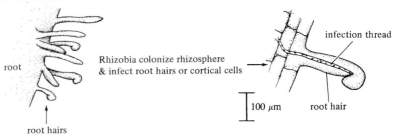

1. INFECTION

Rhizobia colonize rhizosphere
& infect root hairs or cortical cells

root

root hairs

infection thread

100 μm      root hair

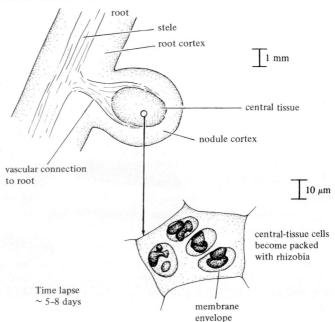

2. NODULES DEVELOP FROM INFECTED CORTEX

root

stele

root cortex

1 mm

central tissue

nodule cortex

vascular connection
to root

10 μm

central-tissue cells
become packed
with rhizobia

membrane
envelope

**FIGURE 7–12** Legume–*Rhizobium* symbiosis: typical stages in development.

Time lapse
~ 5–8 days

include many economically important species: alfalfa, clover, bean, pea, soybean, vetch, lentil, chickpea, mung bean, cowpea, peanut, lupine, lotus, and other food, forage, and pasture plants, as well as ornamentals, firewood, and fine timber species, drug species, and weeds. The legume–*Rhizobium* symbiosis is important in nature, and it has been exploited for centuries in agriculture to get nitrogen directly into valuable high-protein foods and indirectly into nonlegume crops via the soil through residue incorporation, animal feed, and crop rotation.

Rhizobia are rod-shaped, motile, heterotrophic bacteria. They normally grow aerobically in soil but require low oxygen levels if they are to fix nitrogen. There are countless strains; a few thousand are maintained in culture collections. Strains are distinguished by growth characteristics, immunological reactions, DNA typing, and host specificity—that is, the ability to infect and nodulate different legume species or groups of species. Figure 7–12 describes the infection of the plant and establishment of the symbiosis. Fixation rates up to 500 to 600 kg/ha yr$^{-1}$ have been estimated. Rates of 50 kg/ha yr$^{-1}$ are common, but various constraints (see Chapter 8) frequently slow fixation or stop it. (Further explanation of this symbiosis and the kinds of rhizobia appear in Supplement 7–1.)

**Mycorrhizae**   Mycorrhizae (Figure 7–14) are mutualistic, symbiotic associations between plant roots and soil fungi. (See Supplement 7–3 for more details of nutritional adaptations.) The relationship is benefi-

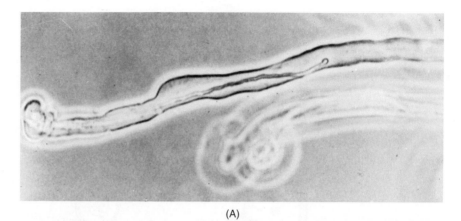

(A)

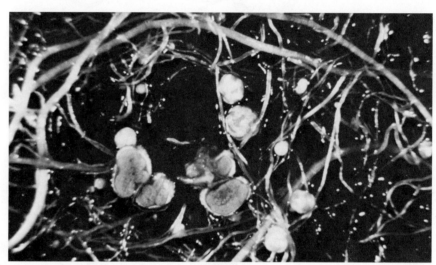

(B)

**FIGURE 7–13**  Legume–*Rhizobium* symbiosis. Top: Infection of an alfalfa root hair by *Rhizobium.* Bottom: Root nodules on bean, *Phaseolus vulgaris.* Largest nodules in center have been cut to expose a cross section.

cial to both organisms because the plant provides the fungus with carbon compounds and the fungus provides the plant with inorganic nutrients. Not all fungi–root associations are mycorrhizal. To be considered mycorrhizal, fungi must be intimately associated with plant roots. Certain plant nutrients move with difficulty in the soil and become deficient for this reason (see Chapter 9) unless the plant can extend its root system fast enough. An efficient way of extending the root system is to develop mycorrhizae. A mycorrhizal fungus infects the root, getting organic food from the plant but also extending its hyphae into the soil, where they absorb nutrient ions and deliver them to the plant root. To explore the soil, hyphae need to produce little biomass because they are thinner than roots. They may also have special mechanisms for extracting nutrients, but this point is still in doubt.

The several kinds of mycorrhizae involve different groups of host plants and fungi. The two that seem most important are the *ectomycorrhizae,* which develop on many evergreen trees and shrubs, and the *endomycorrhizae* (formerly called vesicular–arbuscular [VA] mycorrhizae), which develop on most deciduous trees and most annual crops and other herbaceous species (see Figure 7–15 and Supplement 7–2). Mycorrhizae occur on most plant species and in most soils and are generally beneficial. Plants occasionally benefit from the inoculation of soil or seedlings with cultures of mycorrhizal fungi (spores, infested soil, or fragments of mycelium or infected root), as long as pests and pathogens are not transferred with the inoculum. Mycorrhizal inoculation is practiced in nurseries that raise orchard or forest-tree species in sterilized soil mixes. Unsterilized soils commonly have sufficient natural populations of effective mycorrhizal fungi.

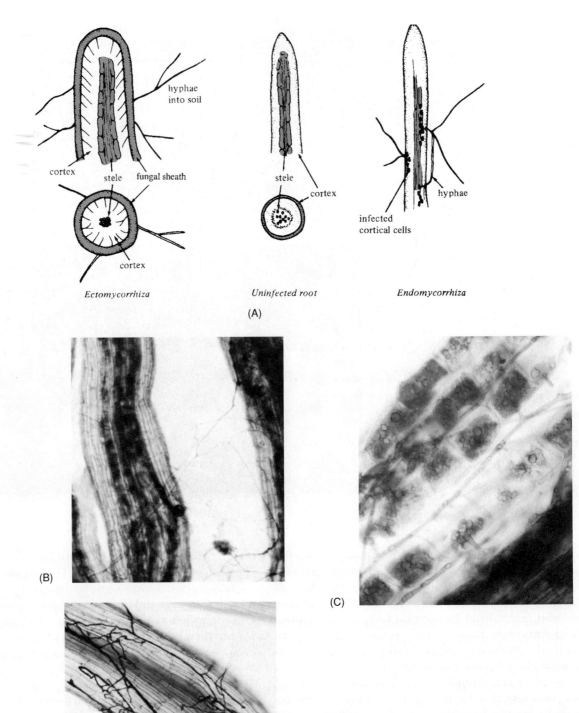

FIGURE 7–14 Mycorrhizae. (A) Major types of mycorrhizae. (B, C) *Endomycorrhizae* not only penetrate the root cortex but also enter cells of the plant root cortex, forming arbuscules. (D) Both types of mycorrhizae extend hyphae into the soil. The roots represented here are typically 1 to 2 mm in diameter. (Micrographs: Linda Geiser.)

root hairs                    hyphae

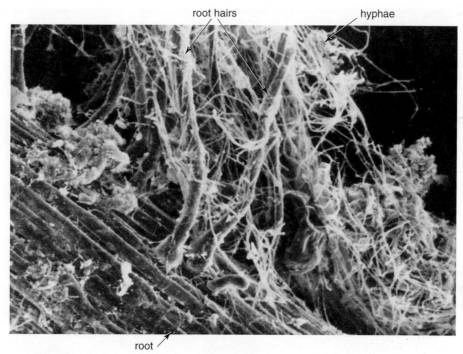

root

**FIGURE 7–15** Mycorrhizal association between root hairs and the fungal hyphae are illustrated in this photo. (Photo courtesy of C. S. Bledsoe; photograph taken by Kelly Leslie, University of Washington.)

## 7.6 SUMMARY

Soil populations are large and diverse and rapidly adapt to changing conditions. Each group plays vital roles in several soil processes, to be described in the next chapter. Plants play a major role, supplying organic matter and providing the important rhizosphere habitat. Fungi and bacteria are the major decomposers of organic matter and recyclers of nutrient elements.

Soil microbes are frequently constrained by unfavorable temperature, drought, or shortage of nitrogen, usable carbon compounds, and energy. Many different adaptations help them deal with these constraints. Some of the most important adaptations lead to parasitic or symbiotic association with plants.

## QUESTIONS

1. Describe the major groups of autotrophic and heterotrophic soil organisms—bacteria, fungi, algae, and protozoa. What are their habitats, sizes, and sources of nutrients and energy?

2. Explain the meaning of *rhizosphere, symbiosis, root nodule,* and *mycorrhiza.*
3. Why do so many different kinds of organisms inhabit soil?
4. In which respects can it be said that plants support the existence of all other soil organisms? Are there exceptions?
5. Soil microbes adapt to food shortages by producing antibiotics or becoming opportunist, specialized, predatory, parasitic, or symbiotic. Think of cases in which one or another of these adaptations might have a practical outcome.
6. Soil algae, protozoa, and animals are usually restricted to the uppermost soil layers and are usually smaller than their aquatic or aboveground counterparts. Why is this? Why is it not true for soil fungi and bacteria?
7. Why are few bacteria found in groundwater?
8. Speculate about the outcome of the following scenarios (the next three chapters may make further speculation fruitful):
   a. Animals are exterminated in (1) a cultivated crop soil and (2) a forest soil.

## SOIL AND ENVIRONMENT
### *Environmental Bugs*

- Soil organisms are the crunchers, buriers, and decomposers of organic wastes, from dead plants to synthetic chemicals. In the process, they recycle the carbon dioxide and other nutrients needed for life on earth. Without them, topsoils would lack the marvelous porosity that keeps them ventilated, and the land would be deep in litter.
- Most microbes stick to soil; thus underground waters tend to be microbe free, unlike surface waters contaminated by microbes carried in with eroded soil and litter.
- Populations within soils are complex and adaptive to changing conditions. Many people have tried unsuccessfully to improve soil microflora by introducing (inoculating) laboratory-bred bacteria and fungi with desired properties but poor adaptation to field conditions. The most effective modifications of soil microflora have come from alteration of the microbial environment—that is, the soil and plant community. Thus nitrifiers and most heterotrophs increase in response to ammonium addition; consumers of petroleum or chlorinated hydrocarbons increase in response to spills of those materials.

Enhanced microbial nitrogen fixation also requires environmental changes, starting with depletion of the inorganic forms of N that microbes and plants normally use. Desirable fixation by rhizobia in soil depends mostly on growth of legumes and correction of the usual P deficiency by appropriate fertilization, just as undesirable fixation by blue-green bacteria in water depends on sunlight and the elimination of the normal P deficiency by pollution with phosphated detergents, fecal material, or runoff from recently phosphated fields.

---

   b.  Heterotrophic bacteria are exterminated in (1) an aerobic soil and (2) a flooded soil.
   c.  Autotrophic soil bacteria are exterminated.
   d.  Fungi are exterminated in (1) an agricultural soil and (2) a forest soil.
9. Why do only some plants form symbioses with nitrogen-fixing bacteria, whereas almost all plants form mycorrhizal symbioses?

## FURTHER READING

Angle, J. S. 2000. Viruses. In *Handbook of soil science,* ed. M. E. Sumner. Boca Raton, Fla.: CRC Press.

Baker, K. F., and W. C. Snyder, eds. 1965. *Ecology of soilborne plant pathogens.* Berkeley and Los Angeles: University of California Press.

Klironomos, J. 2000. Mycorrhizae. In *Handbook of soil science,* ed. M. E. Sumner. Boca Raton, Fla.: CRC Press.

Madsen, E. L. 1995. Impacts of agricultural practices on subsurface microbial ecology. *Advances in Agronomy* 54:1–65.

Paul, E. A., and F. E. Clark. 1996. *Soil microbiology and biochemistry.* New York: Academic Press.

Richards, B. N. 1987. *Microbiology of terrestrial ecosystems.* London: Longman Group.

Russell, E. W. 1975. *Soil conditions and plant growth.* 10th ed. London: Longman Group.

Stanier, J. A., S. C. Adelberg, and J. Ingraham. 1993. *The microbial world.* Englewood Cliffs, N.J.: Prentice-Hall.

Tate, R. L. 1995. *Soil microbiology.* New York: Wiley.

## SUPPLEMENT 7–1
### *Rhizobium*–Legume Symbioses

The legume root nodule bacteria (rhizobia) comprise two genera: *Rhizobium* and *Bradyrhizobium.* Within each genus are hundreds of known strains, differing in performance with different hosts and other characteristics and designated by letter and number codes.

Symbiosis begins with the colonization of the rhizosphere of young roots by rhizobia that can infect the particular host. Then some of the rhizobia infect either root hairs, depending on the species, or cortical cells exposed by emergence of lateral roots. The plant provides a sheathed *infection thread* (Figure

7–12), which penetrates to the inner cortex and stimulates cortical cells to divide and produce the nodule structure. The developing nodule differentiates vascular tissue, connected to the plant's vascular system, through which will flow water, nutrient ions, the sugars to supply energy, and the N-containing products (amines or ureides). The nodule also develops a cortex, which helps control the inward diffusion of oxygen, and central tissue whose closely packed cells receive bacteria from the infection thread and become the actual site of nitrogen fixation. Each infected plant cell contains rhizobia enclosed in plant-produced membrane envelopes, with a solution of hemoglobin similar to the hemoglobin of blood. The hemoglobin makes healthy legume nodules pink, and its function is probably to combine with free oxygen (which would inactivate nitrogenase), while allowing fast diffusion of combined oxygen to support fast respiration (needed for fixation). In combination, the plant's photosynthesis, the rhizobial nitrogenase and respiration, the oxygen-control system, and the vascular plumbing can form a very effective nitrogen-fixing system.

## SUPPLEMENT 7–2   Mycorrhizae

*Ectomycorrhizae* (Figure 7–14A) are developed by basidiomycete and ascomycete fungi, which envelop the ends of young roots and establish a mycelial network among the cortical cells without penetrating the cells (hence *ecto-*). The fungal hyphae extend into the soil, from which they evidently help the plant get ammonium (an especially important form of available N in forest soils), phosphate, and zinc. Typical basidiomycete fruiting bodies—mushrooms—frequently develop from the mycelium in the soil.

*Endomycorrhizae* (Figure 7–14B) are developed by phycomycete fungi and are the most common type of mycorrhizae. The fungal spores germinate in the rhizosphere, perhaps stimulated by exudates and the plant's removal of inhibitory ions. The hyphae penetrate the surface of young roots, establishing mycelium (and poorly understood vesicles) among the cortical cells. Some hyphae also enter cortical cells, forming *arbuscules*—many-branched, tree-shaped mycelial structures that appear to fill the plant cell. In fact, the cell's membranes and nucleus remain intact and recover in time as the fungus withdraws the arbuscule and forms new ones in newer sections of the root. The arbuscule may function by pro-

viding a large active surface for the transfer of sugars to the fungus and nutrient ions to the plant. The extensive hyphae outside in the soil effectively absorb phosphate and zinc and move them rapidly through the mycelium to the plant. The nonseptate form of the phycomycete hyphae might facilitate this movement. Formation of arbuscular mycorrhizae (AM) is often suppressed if the plant receives a high level of phosphate.

## SUPPLEMENT 7–3   Nutritional Adaptations

### Autotrophy

Sunlight supplies energy directly to photoautotrophic soil microbes, which include several kinds of photosynthetic bacteria (blue-green bacteria are the most abundant in soil) and eucaryotic algae (diatoms are the most common in soil). But their dependence on sunlight restricts these microbes to the surface layers of flooded or very wet soils. Other autotrophs—the chemoautotrophs—are bacteria that derive energy from inorganic oxidation reactions that they catalyze. Most important are bacteria that promote the oxidation of ammonia (e.g., *Nitrobacter, Nitrosomonas,* and *Nitrosococcus*) or the oxidation of sulfur or sulfide (e.g., *Thiobacillus*). These organisms affect soil fertility by accelerating the transformations and cycling of N and S (see Chapter 8). Though important, they rarely become as numerous as the heterotrophs. Their energy sources—ammonium or reduced sulfur—are seldom abundant, except where oxygen is too deficient for these oxidizing bacteria to operate.

### Opportunism

The addition of organic material usually occasions a flurry of microbial activity (Figure 7–3), which then subsides as the readily assimilated compounds are consumed. During the period of intense competitive activity, the advantage goes to those species that can grow rapidly, as long as they are also able to stay alive and ready during intervals of starvation. Fungi and bacteria with these properties are prominent rhizosphere organisms and colonizers of dead or dying plants and animals, dung, and plowed-in crop residues.

## Inhibitor Production

Bacteria and fungi sometimes release organic substances that inhibit the growth of competitors. These inhibitors include bacteriocins, specifically directed at close relatives of their producer, and more general inhibitors called *antibiotics*. Some antibiotics have become valuable drugs for treatment of infections—for example, penicillin (from fungi of the genus *Penicillium*), streptomycin, actinomycin, aureomycin, tetracycline, and others (from actinomycete species).

## Specialization

An organism can avoid competition if it can grow under especially difficult conditions or use a food source that others cannot. Examples include the wood-rotting fungi and bacteria. Organisms that rot wood must produce and release extracellular enzymes that break down the macromolecules of cellulose and lignin (see Chapter 8) into smaller molecules that can be assimilated. The extracellular enzymes cost the organism energy but give it access to ample food with little competition from common fast-growing species. Many examples of the evolution of specialized populations come from pollution studies. Soils commonly lack organisms capable of using exotic substances such as petroleum, manufactured plastics, herbicides, and pesticides. However, given time, soils polluted with such substances frequently develop microbial populations that can decompose the pollutant. Variants that can use (and destroy) the pollutant either evolve in the soil or arrive as colonists and then increase in numbers. The ability to live anaerobically may also be regarded as a special adaptation to a limited energy supply; anaerobes can get energy from organic matter without competition from the abundant and vigorous aerobes.

## Parasitism and Predation

Much of the soil's population feeds by parasitizing or preying on other organisms. Examples include the many species of soil fungi and bacteria that cause disease in nematodes, insects, other animals, and plants. Protozoa consume smaller organisms such as bacteria. Bacteria may fall prey to other bacteria; a bizarre example is the small, fast-moving predatory *Bdellovibrio*, which inject digestive enzymes into the larger bacteria that they attack. Many soil animals consume other animals or fungi, and some ants and termites even cultivate fungi to eat.

## Symbiosis

**Symbiosis** includes different types of beneficial and detrimental biological interactions that can occur in soils. Among the beneficial interactions are *mutualism,* in which both partners benefit; *neutralism,* in which neither partner benefits; *commensalism,* in which one partner benefits while the other is unaffected by the interaction; and *synergism,* in which the combined activity exceeds the sum of individual actions. Negative interactions include *parasitism, predation, competition,* and *amensalism* (one species is inhibited by toxin from another).

An important synergistic symbiosis is the **lichen** symbiosis among fungi, algae, and blue-green bacteria. Lichens colonize rocks, tree trunks, and exposed soil surfaces; the algae and blue-green bacteria photosynthesize and fix nitrogen, and the fungus provides water and mineral nutrients. Forest lichens are especially sensitive to poisoning by particulate airborne metals; they have become valuable indicators of the spread of atmospheric pollution from smelters. Bacteria and protozoa that live in termites' guts assist with nitrogen fixation and cellulose digestion in return for a supply of wood chips.

# 8

# MICROBIAL PROCESSES

## OVERVIEW

*Soil organisms collectively decompose organic matter, returning its elements to the mineral state utilized by plants. Decomposition has three phases: Animals chew up raw material and mix it with the soil, special fungi and bacteria release enzymes that break large molecules into small molecules, and the general microbial population assimilates and metabolizes these soluble breakdown products. Metabolism, which energizes heterotrophic microbes, proceeds via respira-* *tion if oxygen is available or fermentation if oxygen is deficient. Fermentation is the slower process. During decomposition, solid organic by-products accumulate as humus, which itself decomposes slowly. The rates of breakdown of both fresh and humified organic matter depend on the composition and physical state of the organic material and on temperature, water supply, nutrient supply, and oxygen supply. Organic matter decay is the largest microbial process.*

*Ammonium, sulfur, and other inorganic substances also are oxidized in soil. These oxidations by aerobic autotrophic*

**153**

*bacteria are critical steps in the cycling of nutrients and the development of soil acidity.*

*When oxygen becomes deficient, anaerobic bacteria institute a new set of conditions and processes. Nitrate is reduced to gases and transferred to the atmosphere, manganese and iron oxides are reduced and dissolved, sulfate is reduced and either precipitated or volatilized, pH is changed toward neutrality, and secondary changes follow. Reductive processes dominate rice paddy and marsh soils, and some are also important in aerated soils.*

*Individual processes interact, often in recurrent cycles. The nitrogen and carbon cycles integrate the microbiological and chemical transformations and the physical movements of these elements in the soil, water, and atmospheric systems. The N and C cycles illustrate the usefulness of these cyclic schemes for explaining and relating complex events.*

Microbial processes are chemical changes that microbes mediate. *Mediate* here implies accelerating, assisting, moderating, or actually doing something. Our view of these processes is doubtless not shared by the soil microbes. Each soil microbial process is the sum of countless individual processes of consumption, metabolism, growth, excretion, death, and decay by organisms driven to get energy and nutrients for propagation.

The soil biology and ecology incompletely sketched in Chapter 7 illuminate these processes and their behavior, but the illumination is far from ample. Fortunately for soil science, a "black-box" view of the processes is useful, much as we can use a complex instrument, although its works mystify us.

# 8.1 DECOMPOSITION OF ORGANIC MATTER

## 8.1.1 General Process of Organic Matter Decay

**Starting materials; substrates**   Soil microbes decompose exotic substrates such as petroleum, rubber, plastics, processed natural fibers, celluloid, American food, DDT, 2-4D, BHC, PCB, cyanamide, dung, dead animals, gum wrappers, and old shoes. Microbes that destroy or detoxicate these substances rescue the human environment daily. But most of the decomposer activity is directed toward food sources (**substrates**) produced by plants and the soil microbes themselves.

Plant materials (Figure 8–1) are the primary and most massive substrate. They differ in composition and decomposability depending on the plant species, age, and part—roots, woody stems, leaves, or fruit. Tables 7–2, 8–1, and 8–2 and Supplement 8–1 describe plant materials and their elements and compounds. A live plant is 90 to 95 percent water; the rest is nearly all C, H, O, and N in the form of organic compounds. A small proportion of these organic materials consists of soluble, monomolecular "simple" substances: sugars, organic acids, **amino acids, lipids,** and **nucleotides.** A much larger proportion consists of the less soluble structural **macromolecules** produced by **polymerization** (joining together) of smaller molecular units. (Supplement 8–1 briefly describes the main groups of biological substances. Consulting it as you read this and the next three chapters may enlarge your understanding or recollection of the terms and help you relate the subject matter to general biochemistry.)

The soluble cell components of small molecular size include organic acids, amino acids, lipids, nucleotides, sugars, sugar phosphates, amino sugars, and sugar acids (**uronides**). They are readily absorbed and metabolized by most organisms and do not last long in soil. Far more abundant are the macromolecular (polymeric) compounds. Their large molecules are made by joining together (polymerizing) smaller molecules of sugar, amino acids, or other **monomer** units. They include the following.

**Peptides** and **proteins** are short chains of a few amino acid units and long chains of scores or hundreds of amino acid units, respectively. The proteins of plants and microbes are important as enzymes and components of cell membranes. They account for most of the N and S in the plant. They are readily broken down by plant and microbial enzymes.

**Nucleic acids,** or nucleotide **polymers,** comprise thousands of nucleotide units, encode the organism's genetic information, and regulate its transcription for the synthesis of enzymes. Nucleic acids and nucleotides contain much of the P in cells.

**Polysaccharides** (**carbohydrate** polymers and their relatives) form the bulk of organic matter naturally added to soils. Most abundant is *cellulose,* a linear polymer of the simple sugar glucose. The long molecules of cellulose align side by side and cross-bond together into microscopic fibers that give strength and resilience to plant cell walls (Figure 8–2). Cellulose is

(A)

Fruiting body

Mycelia

(B)

(C)

**FIGURE 8–1** Plants are the chief source of soil organic matter. (A) Cereal crop residue. Cutting for better contact with the soil increases the decay rate. (B) Decomposition in progress: colonization by fungi (×20). (C) Use of residues for building material (in this case), fuel, animal feed, or fiber competes with use for soil amendment. (Photos A and B: Francis Broadbent; photo C: Harvey Liss.)

## TABLE 8–1

Major Compounds in Plant Tissues: Approximate Percentage of Dry Weight

| Type of compound | Young leaves | Old leaves | Stems |
|---|---|---|---|
| **Polymers** | | | |
| Polysaccharides | | | |
| Cellulose | 30 | 50 | 50 |
| Other | 25 | 30 | 25 |
| Proteins | 25 | 5 | 1 |
| Nucleotides | 5 | 2 | — |
| Lipids | 5 | 2 | 1 |
| Lignin | 2 | 10 | 20 |
| **Simple molecules** | | | |
| Sugars, acids, etc. | 8 | 3 | 1 |

## TABLE 8–2

Nitrogen Concentrations and C/N Ratios in Organic Materials

| | %N | C/N |
|---|---|---|
| **Plant tissues: trees** | | |
| Deciduous leaf (poplar, maple) | 0.5–1.0 | 45–70 |
| Evergreen leaf (pine) | 0.5–0.8 | 60–70 |
| Tree roots (fine) | 0.5–1.0 | 60–90 |
| Bark (oak) | 0.5 | 80 |
| Wood (oak) | 0.1–0.3 | 130–400 |
| **Plant tissues: herbaceous** | | |
| Clover or bean leaf | 2.5–5.0 | 8–16 |
| Grass shoots (young) | 3–4 | 12–15 |
| Grass shoots (mature, yellow) | 0.5–2.0 | 20–80 |
| Alfalfa shoot, 10 days old | 5.2 | 9 |
| Alfalfa shoot, 40 days old | 2.8 | 16 |
| Alfalfa root, 10 days old | 3.6 | 12 |
| Alfalfa root, 40 days old | 2.7 | 17 |
| **Animals and microbes** | | |
| Insects, mammals | 6–12 | 5–10 |
| Fungi (grown on leaf) | 3–4 | 11–16 |
| Fungi (grown on wood) | 1–2 | 20–45 |
| Bacteria | 4–12 | 5–14 |
| Dung (horse, cow, chicken, etc.) | 1–2.5 | 25–60 |

Source: Data from Swift, Heal, and Anderson, 1979; Tisdale, Nelson, and Beaton, 1985.

familiar in almost pure form as cotton, linen, and fine-quality papers. Other sugar polymers, involving different sugars and more cross-branching, are abundant. For example, in plants, *hemicelluloses* bond together the cellulose microfibers in cell walls (Figure 8–2), and *starch* is an important storage substance. The outer layers of plant cell walls and the gelatinous coating on roots contain pectic substances—**polyuronides** (polymers of sugar acids)—which are sticky, hold water, and possess a pH-dependent surface charge because of the

## (A) POLYMERS (MACROMOLECULES)

monomer sugar (e.g., glucose)    shorthand glucose monomer

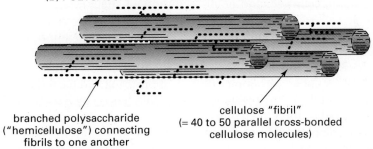

polymer polysaccharide (e.g., cellulose)

## (B) POLYSACCHARIDE COMPLEX IN PLANT CELL WALL

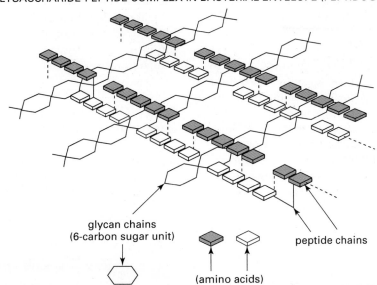

branched polysaccharide ("hemicellulose") connecting fibrils to one another

cellulose "fibril" (= 40 to 50 parallel cross-bonded cellulose molecules)

## (C) POLYSACCHARIDE-PEPTIDE COMPLEX IN BACTERIAL ENVELOPE (PEPTIDOGLYCAN)

glycan chains (6-carbon sugar unit)

peptide chains

(amino acids)

**FIGURE 8–2**  Monomers and polymers. (A) Structure of the simple sugar glucose (monomer) and its polymer cellulose, a macromolecular polysaccharide. (B) Polysaccharides combined in a plant cell wall. (C) Polysaccharide combined with peptides in a bacterial cell envelope.

dissociation of their carboxyl groups. (This gives roots a cation exchange capacity.) Fungal cell walls and insect exoskeletons contain the amino–sugar polymer *chitin.*

*Lignin* compounds, polymers based on phenolic compounds, are abundant in wood and other plant tissue with thickened cell walls. Lignin is incompletely characterized. It is resistant to decay and helps preserve the carbohydrates associated with it. Resistant ligninlike compounds accumulate in humus.

Combinations of polymers occur extensively. In plant cell walls, cellulose is combined with other polysaccharides (Figure 8–2B) and often with polyuronides and lignin as well. Other examples are cell membranes (protein–lipid combinations) and the bacterial envelope's peptidoglycan layer (Figure 8–2C). In woody tissues, lignin, cellulose, and hemi-celluloses are combined.

Biological substances decay at different rates in the soil. Generally, cellulose decomposes more slowly, and lignin much more slowly, than the other macro-molecules of plants. Simple soluble (monomeric) compounds disappear most rapidly. Decay resistance has little to do with a compound's inherent chemical stability. (Wood can burn as well as starch or protein.) Rather, a substance resists decay in soil if microbes cannot absorb it and if common microbes do not release enzymes that break it into absorbable units. **Extracellular enzymes** that hydrolyze cellulose and lignin are specialized products of certain fungi and bacteria. Because extracellular enzymes are energy costly, decomposers of lignin and cellulose need to grow without competition in close contact with their inhospitable substrate. Also, tissues high in lignin and cellulose provide too little nitrogen for decay microbes. Low nitrogen content correlates with high cellulose and lignin content, for two reasons. First, lignin and carbohydrates contain no N. Second, as tissues age or become woody, their protein breaks down, and the amino acids move to more active leaves and growing points. In general, plants are lower in N than are bacteria and fungi (Table 8–1), and slow decay caused by high lignin, high cellulose, and low nitrogen is particularly common in woody, stemmy, or old plant tissue.

The decomposability of material depends on physical state as well as composition. Large pieces expose less surface to attack and decay especially slowly if they are also too hard, tightly structured, dry, or wet for the easy entry of microbes, oxygen, and nutrients.

The rates of decay also depend on environmental factors (see Section 8.1.5).

**Stages and process of decomposition** Decomposition is a sequence of processes that converts complex organic materials into simpler organic compounds, $CO_2$, $NH_4^+$, and other inorganic compounds. A small amount of the decomposition products is the complex and stable material humus. Dying and dead plant material is rapidly colonized by fast-growing bacteria and fungi fed by soluble, easily degraded cell constituents. The destruction of resistant components takes time and a concerted attack, often led by the animals of the litter layer and topsoil, who chew what they can into small pieces, partly digest them with the help of gut microbes, and defecate most of their intake, putting it into closer contact with the soil. Tissue high in cellulose and chitin decays following colonization by specialized bacteria, and massive wood requires the efforts of specialized fungi or insects and their symbiotic gut populations of protozoa and bacteria. To the eye, the material becomes less recognizably plant material and more humic—black and amorphous—as tissue structures disappear and microbes and altered lignin accumulate. The dominant processes become the slow decay of humus and the recycling of microbial biomass. All the organisms at all stages give out energy and carbon, about half their intake, as heat and carbon dioxide. Thus the mass of residual organic matter steadily diminishes, and the constituent elements eventually return to inorganic states.

## 8.1.2 Carbon Dioxide Production and Oxygen Consumption

The main end product of organic matter decay, carbon dioxide ($CO_2$), is produced either from respiration, under aerobic conditions,

$$\text{carbohydrate} + O_2 \rightarrow CO_2 + H_2O + \text{energy} \quad (1)$$

or from fermentation, under anaerobic conditions

$$\text{carbohydrate} \rightarrow CO_2 + \text{acid or alcohol} + \text{energy} \quad (2)$$

The rate of carbon dioxide release is a useful measure of the decomposition rate and the overall microbial activity in the soil. Carbon dioxide production in the soil raises the $CO_2$ concentrations in the soil air to exceed the 0.03 percent found in the general atmosphere by 10 to 1,000 times, depending on the rate of production and the rate of diffusion out of the soil (see Section 3.1 and Supplement 4–2). Diffusion is hindered by soil compaction and wetness, both of which reduce the air space through which the gas can freely diffuse. Elevation of the $CO_2$ concentration in the soil air can lower the soil pH, by formation and dissociation of carbonic acid:

$$CO_2 + H_2O \rightarrow H_2CO_3 \tag{3}$$

$$H_2CO_3 \rightarrow H^+ + HCO_3^- \tag{4}$$

Carbonic acid is too weak to dissociate much at pH below 5, but it has a significant acidifying effect in alkaline and neutral soils, where it can help release plant nutrient ions and promote mineral weathering.

In aerated soil, microbes consume $O_2$ about as fast as they release $CO_2$. Consumption of soil $O_2$ is countered by diffusion from the overlying $O_2$-rich atmosphere (20 percent $O_2$). A moderate lowering of oxygen concentration has little direct effect on soil microbes and roots, but in warm soil that is compacted or wet, the $O_2$ concentration may fall substantially below 5 percent of the soil air, and roots of sensitive species become deficient. The direct effects of $O_2$ shortage may be less important than the indirect effects, such as the encouragement of root disease organisms and the bacterial production of inhibitors such as ethylene. Waterlogged soils develop extreme anoxia, with spectacular biological and chemical effects (see Section 8.3).

### 8.1.3   Humus Formation

Our knowledge about humus composition is changing rapidly as new methods of analysis are developed. Typically, strong bases such as sodium hydroxide (NaOH) are used to separate humus from the inorganic fraction. The extract is then treated to separate the chemical constituents into various fractions. Artifacts, new chemical compounds produced by the extraction and fractionation treatments, lead to erroneous ideas about the chemical structures and components of humus. To avoid this problem, instrumental methods of analysis including modern spectroscopic techniques are applied to samples without chemical extraction.

Humus is a mixture of colloidal organic decay products that accumulate because they decay slowly.

Several changes occur during the humification of fresh organic matter. First, large anatomical structures break up, and smaller colloidal molecules accumulate. Bonding of the humic colloidal molecules with surfaces of clay particles promotes stable aggregates, and the large colloidal humus surface adds to the soil's capacity to hold water.

**Phenolic** macromolecules accumulate and become the main components (Figure 8–3). Phenolic groups may be residual from lignin or formed by decay bacteria. Their toxicity to most microbes helps protect them from microbial decay, and their partial oxidation (to form quinones) gives humus its dark brown color.

With oxidation, the proportion of **carboxyl (COOH) groups** increases (Figure 8–4). Both the carboxyl groups and the phenolic OH groups are weakly acidic. They release or take up $H^+$, depending on the pH of the solution with which they are in contact. These reactions give humus three related properties: a capacity to **buffer** pH, a pH-dependent surface charge, and an ability to chelate certain cations. These properties are important because they help the soil maintain a steady pH, hold nutrient metals, and immobilize potentially toxic metals. Together, carboxyl and phenolics buffer the pH over a wide range (releasing $H^+$ as the soil tends toward alkalinity and accepting $H^+$ as the soil tends toward acidity). The carboxyl groups release or accept $H^+$ as the pH fluctuates between 4 and 6.5. The phenolic groups, more weakly acidic, release and accept $H^+$ as the pH fluctuates between 7 and 9. If the pH is raised, the humus releases $H^+$ and becomes more negative (its cation exchange capacity [CEC] increases). If the pH is lowered, the humus gains $H^+$ and becomes less negative (its CEC decreases). These surface reactions of humus give almost all topsoils a strongly pH-dependent CEC. Furthermore, the closely spaced negative groups on humus molecules chelate Cu, Zn, Co, Ni, and Mn. The metals are thus immobilized, because the humus to which they are bound is immobile (by contrast, soluble chelates are mobile). Immobilization on humus can cause deficiencies of these metals or prevent toxicities.

Both N and S accumulate in humus colloids (Figure 8–4). Humus is the main reserve of these plant nutrients in soil. Humus contains some N as peptides, but 50 to 70 percent exists in organic ring structures (Figure 8–3), which are difficult to break down. Some of the S in humus is in amino acids; most

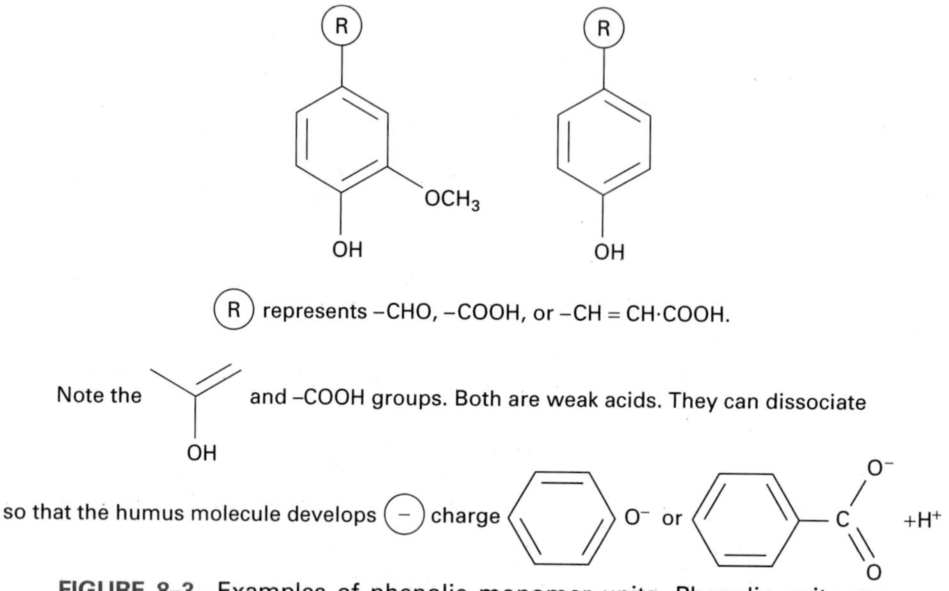

(R) represents –CHO, –COOH, or –CH = CH·COOH.

Note the [structure] and –COOH groups. Both are weak acids. They can dissociate

so that the humus molecule develops (−) charge [structure] O⁻ or [structure] +H⁺

**FIGURE 8–3** Examples of phenolic monomer units. Phenolic units are major components of macromolecules in lignin and humus.

is in other forms. During humification, the organic matter's ratio of carbon to nitrogen (C/N) decreases, from high values characteristic of plant material (between 25 and 100) to low values like those of bacteria (about 10) (Figure 8–4). The C/S ratio also decreases. The release of N and S from humus as available inorganic ions depends mainly on the slow process of humus decay.

### 8.1.4 Release and Immobilization of N and Other Nutrients

The decay of organic matter can release nitrogen, sulfur, and phosphate as free ions. It releases N in the form of ammonium, $NH_4^+$, or it can immobilize N as the decay microbes assimilate ammonium and nitrate. The net effect—mobilization or immobilization—depends on how well the decomposing substrate material itself provides N to the decomposer microbes. If it provides plenty of N, the microbes will release the surplus (Figure 8–5). But if it provides insufficient N, the microbes will absorb ammonium and nitrate from the soil, and the resulting shortage of available N can limit plant growth, microbial growth, and the decay process itself. The nitrogen deficiency is usually temporary because decay, though

slowed, continues to dissipate organic carbon while N is retained; eventually energy and carbon again become limiting, and a nitrogen surplus becomes available for release (Figure 8–5). Net release or net immobilization can be predicted from the organic substrate's C/N ratio or percentage of N (the two correspond closely). If the ratio C/N is below 20 or the percentage is above 2.5, N will be released. If C/N is much above 20, N is likely to be immobilized until decomposition and respiration lower the C/N ratio.

Values of the percentage of N and the C/N ratio are given in Table 8–2. Note the low C/N ratio of bacteria and the high C/N of plants, especially old or woody plants with abundant structural carbohydrates and lignin. The addition of organic materials high in N improves the N supply to plants; these materials are effective N fertilizers. The addition of organic materials low in N (high C/N) induces N deficiency unless remedial steps are taken. One remedy is to supply the missing N as fertilizer. Another is to leave the organic material on the surface (as a mulch), where separation from the soil's water and soluble N retards decay. Another remedy is to delay planting until the soil microbes lower the average C/N ratio enough to release the N again.

Many examples of these phenomena are important to agriculture and horticulture. Sawdust, rice

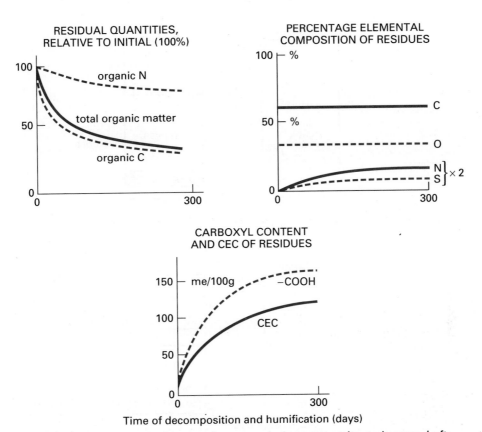

RESIDUAL QUANTITIES,
RELATIVE TO INITIAL (100%)

PERCENTAGE ELEMENTAL
COMPOSITION OF RESIDUES

CARBOXYL CONTENT
AND CEC OF RESIDUES

Time of decomposition and humification (days)

**FIGURE 8–4**   Changes in organic matter during conversion to humus. Left: Loss of N and even greater loss of C (mainly as $CO_2$). Right: Percentage enrichment in N and S. Center: Increase in carboxylic groups and consequent increase in cation exchange capacity.

hulls, and wood chips mixed with soil usually cause severe N deficiency in crops, although the same materials are useful surface mulches. Gardeners often mix "peat moss" or other humic materials with soil ingredients to make a potting soil that holds water well, drains well, and does not compact (see Chapter 16). This practice seldom results in nitrogen immobilization, because the peat moss, like soil humus, decays slowly and has a fairly low C/N ratio. Crop residues last better as protective mulches and immobilize N less severely if they are kept loose and dry on the surface in minimal contact with the soil's water and nutrients. Bulky crop residues can be incorporated into the soil for disposal without air pollution from burning, but time must be allowed for the material to decay and for the N fertility to recover before the next planting.

### 8.1.5  Decay Rates and Controlling Factors

The rate at which organic matter decays in soil depends on several things: the properties of the organic matter, the amount of organic matter, the stage of the decay process, and environmental factors.

**Properties of the organic matter**   Organic matter is a complex material consisting of a wide variety of compounds, each with a decay rate. In addition, the size of organic particles varies greatly. The quality of the organic matter—that is, the composition of the constituent parts and their physical size—will contribute to determining the overall decay rate. Decay will be retarded if the material is in large pieces, dense, or nonporous; if the material is high in lignin and cellulose; or if it is low in N or some

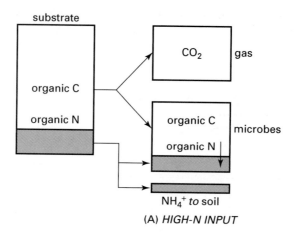

**FIGURE 8–5** Mobilization or immobilization of N during decay depends on C/N (or % N) in the organic residue: a simplified view. (A) Mobilization of surplus N as $NH_4^+$ during decay of N-rich organic matter. (B) Immobilization of soil $NH_4^+$ and $NO_3^-$ to meet N demands of microbes during early stages of decay of N-deficient organic matter.

other nutrient. Soil animals alleviate these constraints: Chewing and burying make the material more accessible to microbes, water, and mineral ions. Likewise, farmers can accelerate the decay of bulky crop residues by chopping and spreading them, cultivating them into the soil, or turning livestock onto the field.

**Amount of organic matter and stage of decay**
The rate of decay is roughly proportional to the amount of organic matter present. If this proportionality were exact, the absolute decay rate would decline, but the relative rate would be constant; that is, the amount decayed per unit time would be a constant fraction of the amount of organic matter still present. But in fact this simple relationship holds only for substrates consisting of a single substance or for real-life systems averaged over intervals of years. For a single addition of a complex material, the *relative* decay rates also decline as decay proceeds because the composition of the residue changes. The most easily assimilated substances decay first, leaving resistant plant constituents and humified material to accumulate progressively. Thus there is a marked decline in both the relative decay rate and the amount of material available for decay. High relative decay rates approaching 50 percent per week are common in fresh plant material, but for humus a typical rate would be only 3 percent per year.

**pH, salinity, and mineral nutrients** Decay is inhibited by extremes of soil pH (below 4.5 and above

9), very high salinity, and deficiencies of mineral nutrients (most commonly N).

**Water**   Decay, like plant growth, goes fastest at water potentials in the range of −10 to −50 kPa and slows progressively as the soil becomes drier (Figure 8–6). Activity of bacteria is limited by low water content before that of fungi because of decreased mobility. Fungi have been shown to survive water potentials as low as −10 MPa. One result is that organic matter persists in litter and mulch layers, which are dry most of the time. (Another result, also familiar, is that desiccated foods do not rot.) At very high water potentials (approaching saturation, −10 to 0 kPa), decay slows markedly because of the shortage of oxygen.

**Temperature**   Although the members of the soil population have diverse temperature optima, overall soil respiration ($CO_2$ production and $O_2$ consumption) declines steadily as the soil temperature falls from 20 to 5°C (Figure 8–6). Laboratory incubation studies have shown that decay rates increase about 2× for each 10°C increase in temperature between 10 and 40°C. For this reason, microbial activity varies seasonally, and in cold weather, crop residues can be slow to decompose and organic manures slow to release nutrients. Likewise, soils in cool climates commonly develop high organic matter contents.

**Oxygen**   Aerobic decomposition rates slow when oxygen concentration in the soil atmosphere reaches approximately 10 percent. This threshold is less important than the rate at which oxygen is replenished as it is used in microbial respiration. If oxygen is not replenished by diffusion or mass flow at the rate it is consumed by microorganisms, decomposition slows. Anoxia slows the decay of organic matter in general and of lignin and humus in particular. Bogs, swamps, and other waterlogged sites are normally high in humus and may contain long-preserved woody plant remains and wooden artifacts.

**Accessibility**   Microorganisms must be in contact with organic matter to utilize it. Organic matter located in small pores or inside aggregates may be protected from decomposition because the spaces are too small for bacteria to enter. Organic matter may also be protected from decomposition by adsorption on clay mineral surfaces. Enzymes needed for organic matter decay may be adsorbed on clay minerals

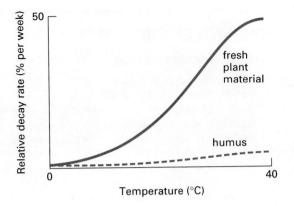

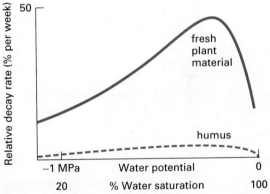

**FIGURE 8–6**   Effects of temperature and water availability on the rates of decay of both fresh plant material and humus. Note the slow decay of humus.

and become inactive. (See the Soil and Environment box on carbon sequestration.)

### 8.1.6   Managing Soil Organic Matter

The management of soil organic matter has one or more of the following purposes:

1. Disposing of organic materials, such as urban wastes and crop residues
2. Building or maintaining the level of humus to improve soil structure, water retention, N and S supply, cation retention, and pH buffering
3. Directly alleviating the effects of clay and poor structure on water infiltration and aeration
4. Providing a mulch to protect the soil surface from crusting, erosion, or overheating

# SOIL AND ENVIRONMENT
## *Carbon Sequestration*

- Global climate change is the slow warming and cooling of the earth's atmosphere. The slow increase in the average temperature of the earth's atmosphere, followed by cooling, is a recurring phenomenon recorded in cores of ocean mud and in soils and sediments on land. Since the start of the industrial revolution, humans have been converting vast amounts of carbon from petroleum and coal to $CO_2$. The $CO_2$ concentration of the earth's atmosphere thus increased from 280 ppm by volume in 1850 to 365 ppm by volume in 1996. Most scientists agree that the increase in $CO_2$ has produced a corresponding increase of about 0.5°C in the earth's temperature in the past 100 years. This human-induced increase in temperature could have serious consequences including worldwide rise of sea level, changing weather patterns, and major changes in ecosystems.

- The $CO_2$ and other gases such as water vapor, carbon monoxide (CO), methane ($CH_4$), nitrous oxide ($N_2O$), nitrogen oxide (NO), nitrogen dioxide ($NO_2$), and ozone ($O_3$) allow shortwave solar radiation to enter the atmosphere to heat the earth but do not allow the longwave radiation to escape. The atmospheric capacity to retain heat is a natural process that has made the earth suitable for life as we know it. What concerns people is the human-induced potential for rapid global climate change and subsequent (unknown) effects.

- One method to combat the rising temperature is to reduce the $CO_2$ emitted by vehicles and electric power–generating plants. As the scientific search continues for alternative fuel sources, other techniques for slowing the release of $CO_2$ to the atmosphere have been considered. One of these is carbon sequestration by soil.

- Carbon is sequestered (protected) in soils as humus. Cultivation increases the rate of humus decomposition and increases the release of $CO_2$ to the atmosphere. Cultivation also increases the potential for soil erosion and loss of humus. It is estimated that reduced cultivation has the potential to sequester 5 billion metric tons of carbon over the next 50 years. This is about 10 percent of the total soil organic carbon in the lower 48 states. Erosion prevention, conservation tillage, residue management, improved cropping systems, land restoration, conversion of land from cultivation to natural ecosystems, and water management have potential to sequester more carbon per year than is emitted from agricultural activities in the United States. The percentage that each may contribute to carbon sequestration is shown on the pie chart.

- The successful sequestration of carbon depends on our knowledge of soils and the behavior of soil organic matter, as well as a willingness to implement creative ways to manage soils.

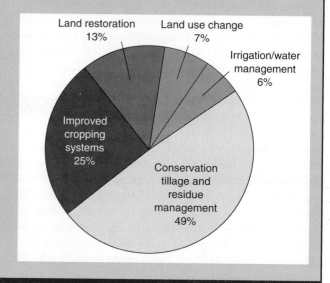

These purposes are not all compatible. Rapid decay, for instance, is desirable for disposal, and a degree of decay is necessary for making humus. By contrast, infiltration will improve most if the added organic matter is coarsely fibrous; the benefit diminishes when the fibers decay. Effective mulching materials need add nothing to the soil's humus or chemical fertility and must resist decay if they are to last.

Organic matter levels can be increased by reducing losses or increasing inputs. Organic matter losses can be increased, for instance, by

1. Promoting soil erosion (see Chapter 15)
2. Selecting the kind of organic matter added (small particle size, low lignin and cellulose content, and high N content)
3. Optimizing environmental conditions for decay microbes (incorporating material into the soil; providing warm moist conditions; adding deficient nutrients such as N; adding lime to correct soil acidity; enhancing aeration by drainage)

Organic matter input can be modified by adding organic wastes, manures, and composts and by retaining organic matter grown on the site. On-site production includes **green manures**—crops grown solely for incorporation into the soil. Green manure crops can be economically worthwhile if grown cheaply, between seasons, without interfering with other operations. Normally the most important inputs are the roots and shoot residues of ordinary commercial crops, pastures, forest, or other vegetation. Management choices for conservation of this resource include (1) removal (burning or harvesting) versus conservation (incorporation or mulching) of the aboveground residues and (2) general management practices that increase plant biomass production and often also increase yield (lengthened production season, optimal plant density, and the

improvement of varieties, pest control, disease control, water management, and plant nutrition).

## 8.2 OXIDATION OF AMMONIUM, SULFUR, IRON, AND MANGANESE

### 8.2.1 Processes

Supplement 8–2 explains oxidation and reduction reactions. The oxidations of ammonium and sulfur in soil yield energy for the chemoautotrophic bacteria that mediate these reactions (Box 8–1). Microbes are involved in manganese oxidation in soil; sterilization or even temporary drying leads to a rise in levels of reduced Mn ($Mn^{2+}$) in solution. Iron-oxidizing bacteria are prevalent in pools and streams, but in soils the oxidation of iron is mostly nonbiological. Box 8–1 lists these reactions, as a shorthand guide to the most important requirements and consequences of these processes.

### 8.2.2 Requirements and Consequences

Oxidations occur in soil only if it is aerated. The oxidations of ammonium, sulfur, and manganese also require the appropriate microbes. The need for these

---

## BOX 8–1
## Important Inorganic Oxidation Reactions in Soils

Organic matter is oxidized in soil by the respiratory metabolism of heterotrophic organisms. In addition, certain important inorganic oxidation reactions mobilize nutrient ions and affect soil pH. Each of these reactions has $O_2$ as the electron acceptor:

*Nitrification* oxidizes ammonium first to nitrite and then to nitrate:

$$2NH_4^+ + 3O_2 \rightarrow 2NO_2^- + 4H^+ + 2H_2O$$

$$2NO_2^- + O_2 \rightarrow 2NO_3^-$$

The overall reaction is

$$NH_4^+ + 2O_2 \rightarrow NO_3^- + 2H^+ + H_2O$$

*Sulfur oxidation* produces sulfuric acid from sulfur or sulfides:

$$2H_2S + O_2 \rightarrow 2S + H_2O$$

$$2S + 3O_2 + H_2O \rightarrow 2SO_2^- + 4H^+$$

*Iron oxidation* converts ferrous ion to the less soluble ferric oxides:

$$2Fe^{2+} + 1/2O_2 + 3H_2O \rightarrow 2FeO(OH) + 4H^+$$

*Manganese oxidation* converts manganous ion to the less soluble manganic oxides:

$$2Mn^{2+} + O_2 + 2H_2O \rightarrow 2MnO_2 + 4H^+$$

microbes to multiply sometimes causes oxidation to lag for some days following addition of the substrate: ammonium, sulfur, or manganese ion. The substrate itself must be supplied by addition, previous reduction reactions in the soil, or migration from another site in the soil where the conditions are more reducing.

Oxidation mobilizes N and S in soil. Sulfate and nitrate, the oxidized products, move freely in soil because they tend neither to adsorb on soil colloids nor to form insoluble compounds. By contrast, ammonium is held by cation exchange; sulfur is an insoluble solid (yellow brimstone); and sulfide either precipitates with copper, zinc, and iron or vaporizes from the soil as hydrogen sulfide. Mobilization enhances leaching, and thus most of the N and S in rivers and groundwater is nitrate and sulfate. Mobilization also enhances **availability** to plants. Thus elemental S is a useful, slow-acting S fertilizer, because sulfur bacteria oxidize it to sulfate.

In contrast with N and S, iron and manganese are immobilized by oxidation, because their oxides are only slightly soluble. For this reason, strongly oxidizing, high-pH conditions can induce deficiencies of both Fe and Mn in plants, and reduction under wet, acidic conditions can induce Mn toxicity. Movement of Mn and Fe in their reduced forms and subsequent precipitation when oxidized help form gravel-sized concretions and hardpan layers (see Chapter 12).

All these oxidations produce hydrogen ions and lower the soil pH. The oxidation of ammonia from synthetic or organic N fertilizers is an important cause of long-term soil acidification in agriculture (see Chapter 11). Oxidation of Fe and S in mine spoils, drained wetlands, and marine sediments can produce acid in enormous amounts, damaging concrete, plants, and aquatic life nearby. Sometimes soil acidification is desirable. This process usually calls for large quantities—tons per hectare—of sulfuric acid or materials that produce acid when oxidized. Toxicity and expense preclude using large quantities of ammonium, manganous salts, or sulfides of copper and zinc. But elemental sulfur and iron sulfide are safe and cheap, and ferrous sulfate is used horticulturally because it acidifies quickly and provides soluble iron as well.

Acidity inhibits and alkalinity promotes the oxidation reactions. The sulfur- and iron-oxidizing bacteria have remarkable acid tolerance, yet manganese and ammonium oxidation are significantly slowed at pH much below 5. These pH effects can be exploited

in practice. For example, ammonium sulfate fertilizer concentrated in a band, rather than spread or mixed through the soil, provides a slow release of N for vegetable crops. The production of mobile nitrate near the band is quickly self-inhibited by the acidity produced. The effects of soil pH on Mn and Fe are frequently important. Raising the pH with **lime** aggravates iron deficiency, but liming is useful for controlling Mn toxicity, a condition that often arises in badly aerated acidic soils or in nurseries where sterilization promotes Mn reduction.

## 8.3 ANOXIC PROCESSES

### 8.3.1 Depletion of Oxygen: Reducing Conditions

Oxygen becomes depleted if microbes and roots consume it faster than it is replaced from the atmosphere. Consumption is accelerated by warmth, moisture, readily decomposable organic matter, and anything else that promotes respiration. Replacement of oxygen is hindered when large interconnected pores are absent or blocked by water. Oxygen is consumed as the final oxidant (electron acceptor) for the respiratory oxidations that energize the organisms. If oxygen becomes deficient, other substances will be reduced instead. They are more difficult to reduce than oxygen is; that is, their reduction requires more intensely reducing conditions. Reduction intensity can be measured and expressed as *redox potential* (oxidation-reduction potential). Free oxygen holds the redox potential above $+300$ mV, but in waterlogged soils the potential may drop as low as $-400$ mV, allowing the anaerobic microbes directly or indirectly to reduce nitrate, manganese oxide, sulfate, and iron oxide (Box 8–2).

The products of these reactions and fermentation include gases that accumulate in the soil air and eventually escape to the atmosphere. These gases include dinitrogen ($N_2$), nitrous oxide ($N_2O$), hydrogen sulfide ($H_2S$), carbon dioxide ($CO_2$), carbon monoxide (CO), and methane ($CH_4$). All except $N_2$ affect the atmosphere, either by absorbing reradiation and thus enhancing warming ($CO_2$, CO, $N_2O$, $CH_4$) or by increasing the acidity of precipitation ($N_2O$, $H_2S$, $CO_2$).

Other products accumulate in the soil—for example, manganous and ferrous ions, precipitated sul-

## BOX 8–2
## Important Reduction Reactions in Soils

The substances on the left of these equations can be reduced, as microbial activity creates reducing conditions when the supply of $O_2$ is impeded. The reactions are listed in order: The first require less intensely reducing conditions than those at the end, other things being equal. This generalization can be disturbed by differences in reactant concentrations.

1. Reduction of manganic oxides:

$$MnO_2 + 4H^+ + 2e^- \rightarrow Mn^{2+} + 2H_2O$$

2. Reduction of ferric hydroxide:

$$Fe(OH)_3 + 3H^+ + e^- \rightarrow Fe^{2+} + 3H_2O$$

3. Denitrification:

$$NO_3^- + nH^+ + ne^- \rightarrow N_2, N_2O + H_2O$$

4. Sulfate reduction:

$$SO_4^{2-} + 8H^+ + 6e^- \rightarrow S + 4H_2O$$

5. Methane production:

$$CH_3OH + 2H^+ + 2e^- \rightarrow CH_4 + H_2O$$

6. Hydrogen production:

$$2H^+ + 2e^- \rightarrow H_2$$

fides, humic material, organic acids, and alcohols. Some of the reductive processes are done by fairly specific groups of organisms; others are more general. Fermentation processes are common to anaerobic microbes in general, as well as to many aerobes. Sulfate is reduced by particular anaerobic bacteria such as *Desulfovibrio*. The accumulation of FeS and other sulfides during formation of wetlands and tidelands can later cause intense acidification if these soils are drained and aerated (see Section 8.2.2).

The process of **denitrification**—the reduction of nitrate to gaseous $N_2$ and $N_2O$—is done by a variety of denitrifying bacteria (and sometimes by nitrifiers during the oxidation of $NH_4^+$). Denitrification frequently causes about 30 percent of the total loss of N from agricultural soils and is favored by warm, wet conditions with plenty of decomposable organic matter and nitrate, conditions desirable for crop growth.

Some substances are reduced more easily than others are. For instance, nitrate reduces more easily than sulfate, so that nitrate is denitrified even without waterlogging, whereas it takes complete waterlogging and ample organic matter to generate hydrogen sulfide from sulfate.

### 8.3.2  Flooded Soils

Many soils flood occasionally and briefly, but regular long-term flooding is characteristic of tide flats, marshes, bogs, alluvial basins, lily ponds, taro fields and other water gardens, and the world's single most important crop, wetland rice.

Flooding quickly, within a half day in warm weather, drops the average concentration of free oxygen to near zero, and the redox potential begins dropping from 300 mV toward −200 mV. Yet some parts of the soil remain aerobic, even in permanently flooded marshlands. Convectional mixing transfers oxygen through the water standing above the soil. Rice and many marsh plants can transfer oxygen rapidly through spongy tissues in their stems and roots. Consequently, the topmost centimeter or so of soil remains aerated, and so does much of the rhizosphere (Figure 8–7). In these aerated zones, nitrate and sulfate remain in their oxidized forms, and oxidized iron may color the soil red. In the rest of the soil, reductive processes dominate, and the soil becomes dark colored or in some cases pale blue-green, the color of ferrous salts. Here the anaerobic microbes take over. Restricted to fermentation as their energy source, they decompose organic matter only partially so that organic acids, alcohols, carbon dioxide, methane, and hydrogen accumulate. The soil atmosphere (the gas that has not been displaced by water) becomes a mixture of $CO_2$, $N_2$, $CH_4$, and $H_2$.

Mineralized N remains reduced as ammonium, which becomes a major exchangeable cation. Ammonium is more mobile than in aerobic soil, because its high concentration in solution and the high soil water

(A)

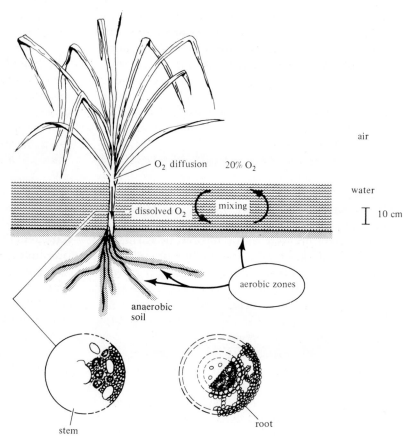

air

$O_2$ diffusion    20% $O_2$

water

dissolved $O_2$    mixing

10 cm

aerobic zones

anaerobic soil

stem

root

**FIGURE 8–7** (A) Flooded agriculture: an example. (Taro patch, island of Kauai.) (B) Aerobic and anaerobic zones in a flooded soil, with adapted plants that transmit oxygen to roots through stems (e.g., rice).

(B)

content allow it to diffuse rapidly. Ammonium salts are the favored nitrogen fertilizers for rice. If nitrate is added, most will be lost through denitrification.

Manganous and ferrous ions both increase in concentration in the soil solution and become important exchangeable cations, displacing ammonium and other cations into solution and enhancing their availability to plants. As iron oxide is reduced and dissolved, the phosphate adsorbed on it is released into solution and made more available to plants.

The pH usually moves toward neutrality when soils are flooded. If the pH was initially high, it is lowered by carbonic acid from the trapped $CO_2$. But if the pH was initially low, it is raised by denitrification, manganese reduction, iron reduction, and sulfate reduction, all of which consume $H^+$.

Flooding helps correct several common soil problems: acidity, P deficiency, cation deficiencies, water stress, and weed competition. It also allows N release from organic matter at C/N ratios that would be too high in aerobic soil and encourages $N_2$ fixation by blue-green and anaerobic bacteria. Most crops cannot exploit these benefits because they cannot maintain an aerobic rhizosphere or are sensitive to anaerobic microbiological products.

# 8.4   CARBON AND NITROGEN CYCLES

## 8.4.1   Carbon Cycling in the Soil–Plant System

The carbon cycle is a sequence of movements, reactions, and transformations of carbon-containing substances in air, soil, water, and organisms. It couples with the nitrogen cycle (see Section 8.4.2) and other cycles of nutrient elements (Chapter 10). To represent one of these cycles legibly, we have to omit detail and select an emphasis. Figure 8–8A summarizes the soil–plant part of the planetary C cycle and shows a linkage to the aquatic C cycle. The arrow width represents the relative amount of material moving, and the box size represents the relative size of each carbon pool.

The scale and pool sizes are large. The ocean is estimated to contain about 40,000 petagrams (Pg) of carbon—that is, $40,000 \times 10^{15}$ g. The upper 100 cm of the world's soils contains an estimated 1,000 to 1,500 Pg of organic C, perhaps more than in the vegetation above ground, and there is another 800–900 Pg as soil carbonate (see Batjes, 1996). The atmosphere contains about 750 Pg, mainly as $CO_2$.

Photosynthetic plants and other autotrophic organisms fix $CO_2$ from the atmosphere into organic forms. Food chains of heterotrophic scavengers, predators, and parasites, mostly microbes, convert and ultimately destroy this organic material. All the organisms return gaseous carbon to the atmosphere, some through fermentation, most through respiration (Figure 8–8A, 8–8C, and 8–8D). The mainstream of this return is generated by decomposer microbes in soils and oceans.

The balance of fixation versus release of $CO_2$ affects the atmosphere's $CO_2$ concentration, currently rising and contributing to the enhanced greenhouse effect thought to be warming the planet's climate. Carbon dioxide is the main atmospheric component that absorbs shortwave radiation coming from the ground surface, converting it to heat (Chapter 4). Other greenhouse gases connected with C cycling and soil microbes include CO, $CH_4$, and $N_2O$ (see Soil and Environment box).

Only a small fraction of the C is in rapid flux. In the soil–plant system, standing wood, litter, and humus decompose slowly. Mobilizing these pools by deforestation and burning is probably enhancing the greenhouse effect, though these pools are small compared with the planet's vast accumulation of coal and petroleum—"fossilized humus"—now being mobilized by mining, drilling, and burning.

Large amounts of C are also immobilized as solid calcium carbonate in soils, rocks, and the ocean. Carbon dioxide in water hydrates to carbonic acid ($H_2CO_3$), and under the conditions of alkalinity (pH 8.3) and high $Ca^{2+}$ concentration found in oceans or alkaline soils, it precipitates as calcium carbonate (Figure 8–8B). This chemical process, exploited by creatures who make shells and skeletons of calcite, has evidently been scrubbing $CO_2$ from the atmosphere for billions of years and building strata of calcareous sedimentary rocks that form another vast dump of carbon immobilized in the earth's crust. The ocean will absorb $CO_2$ from the air but will do so to a lesser degree if the water warms (Figure 8–8B).

Within the soil, Figure 8–8C and Figure 8–8D hint at some of the detail missing from Figure 8–8A, in the form of (1) food chains involving different groups of organisms; (2) separate pools of structural,

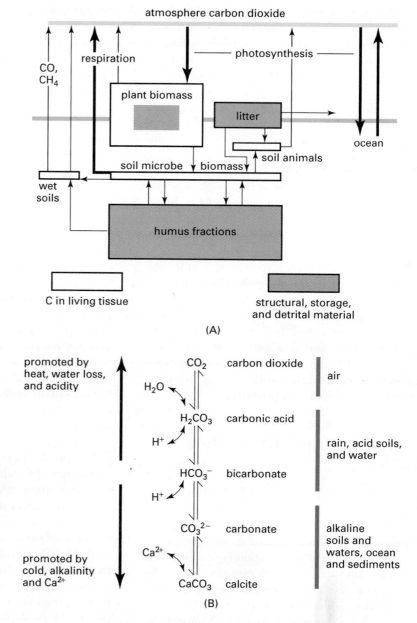

**FIGURE 8–8** Aspects of the carbon cycle. (A) Overview emphasizing soil and plant and identifying the largest pools of biomass C and less labile structural, stored, and detrital C. (B) Chemical reaction sequence linking carbon dioxide, water, and the carbonate minerals (calcite, dolomite, magnesite) of alkaline soils, the ocean, and sedimentary rocks.

storage, and metabolite components within organisms; and (3) detrital organic matter pools of different *lability* (speed of decay). Attempts to define such compartments are useful in research (especially in computer modeling). An especially useful measurement is the microbial biomass C (or N), the biochemically "live" fraction of the soil's total organic C (or N), whose variations help explain rates of heterotrophic activity.

## 8.4.2 Nitrogen Cycling in the Soil–Plant System

We have already described the processes involved in N cycling (Sections 8.1 to 8.3). Now we shall connect them in a cyclic scheme, represented from different points of view in Figures 8–9, 8–10, and 8–11. Figure 8–9 emphasizes the movement of N among parts of an ecosystem and its surrounding systems;

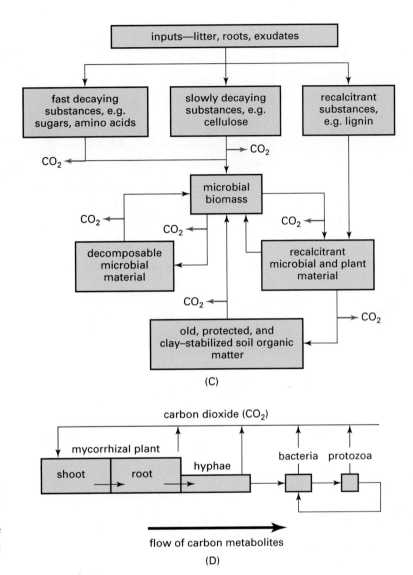

FIGURE 8–8 *(Continued)* (C) Partial expanded view of belowground carbon cycling, recycling, and dissipation as $CO_2$. (D) One of many food chains within the soil–plant system.

Figure 8–10 emphasizes chemical processes; and Figure 8–11 emphasizes "pool sizes," relative amounts in different parts of the ecosystem.

Central to N cycling is the release and assimilation of ammonium. Ammonium is released as organic material decays (ammonification or nitrogen **mineralization**). Ammonium is taken up and assimilated again as plants and microbes grow and synthesize new organic nitrogen compounds. In most soils this simple cycle is complicated by nitrification and nitrate assimilation. Autotrophic soil microbes oxidize ammonium to nitrate (nitrification). Then plants and microbes take up the nitrate and reduce it to make organic nitrogen compounds. This assimilatory reduction produces amino N, used by the organism; do not confuse it with denitrification, in which nitrate is reduced to $N_2$ or $N_2O$ gases. The assimilatory reduction of nitrate uses energy (just as nitrification yields energy), but despite this energy cost, plants usually absorb more nitrate than ammonium in aerated soils, because nitrate is abundant and mobile.

Besides cycling N within itself, the soil–plant system also loses and gains N, as part of a larger cycle involving the atmosphere, groundwater, rivers, lakes,

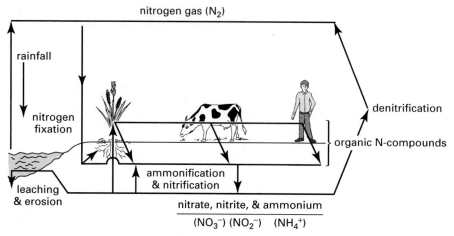

nitrogen gas (N₂)

FIGURE 8–9    A simple representation of a nitrogen cycle.

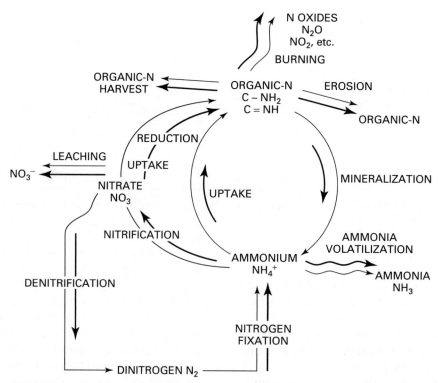

FIGURE 8–10    The nitrogen cycle, emphasizing major processes.

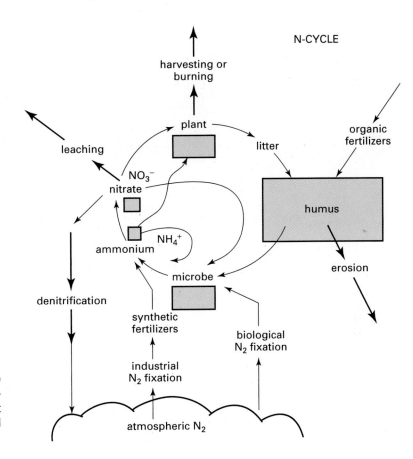

N-CYCLE

**FIGURE 8–11**   The nitrogen cycle in an agricultural ecosystem, emphasizing pools of N in different compartments of the system and losses and gains of N.

and oceans. The main losses from the soil–plant system are (Figure 8–11) as follows:

1. Erosion: removal of organic N in litter and humus
2. Harvesting: removal of organic N in biomass
3. Leaching: removal of nitrate and some ammonium
4. Denitrification: conversion of nitrate to gases
5. Burning of plants: conversion of organic N to gases

These processes vary in importance, but together they can remove hundreds of kilograms of N per hectare per year. The N cycle on land is leaky.

The aquatic N cycle also leaks; denitrification in ponds and oceans passes $N_2$ and N oxides to the atmosphere, the ultimate nitrogen sink. The atmosphere is also the largest source for the N we find in soils, plants, microbes, and water. The atmosphere's

$N_2$, 80 percent of its mass, is an enormous pool of N ($4 \times 10^{18}$ kg, enough to provide about $10 \times 10^7$ kg/ha of the planet's surface if it all were fixed).

From the atmosphere, N is recycled by nitrogen fixation reactions. The most important is the reduction of $N_2$ to form ammonia, either biologically in certain bacteria (see Chapter 7) or industrially in ammonia synthesis for fertilizer manufacture (see Chapter 10). Fertilizers and biological nitrogen fixation are the significant inputs that offset losses of N from agriculture, horticulture, and forestry. (Oxidation of $N_2$ to $HNO_3$ by lightning is an additional input, minor in most localities.)

### 8.4.3   Nitrogen Cycling and Soil Fertility

The biggest pools of N in the soil–plant system (other than the atmospheric pool) are organic N in plants, humus, and litter (Figure 8–11). Their relative pro-

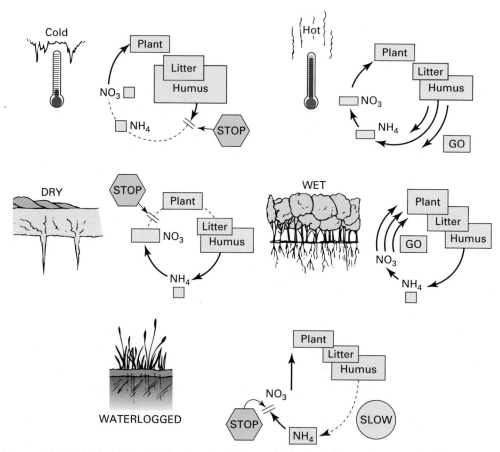

**FIGURE 8–12** Effects of some soil-forming and management factors on the N cycle and pool sizes.

portions vary; forests, for instance, hold most of their N in plant biomass and litter, whereas grasslands and croplands usually hold most of their N in humus. In contrast with organic N, ammonium and nitrate rarely compose more than a small percentage of the total (Figure 8–11).

Pool sizes and process rates vary with the season and local conditions. Because each process reacts differently to any change, the quantitative behavior of the whole system is difficult to predict, but common experience and observation allow some generalizations. For instance, the amounts of N in the major pools—litter, humus, biomass, and solution—can be related to the soil-forming factors. Some of these factors are discussed in connection with soil development in Chapter 12, and others are shown in Figure 8–12.

Management influences are important because nitrogen is the most commonly deficient plant nutri-

ent in managed ecosystems. The rate at which a soil naturally supplies N depends on the N mineralization rate, which in turn depends mainly on the amount of organic nitrogen and favorable conditions of temperature, water, and aeration. Common sense dictates that temperature, water, and aeration will be managed, if at all, for the benefit of plants, which luckily suits decay microbes well enough. The remaining management variable is the level of soil organic nitrogen. This level can be raised by adding organic matter taken from somewhere else, by growing green manure crops to dig into the soil, or by retaining residues of crops grown on the land itself. The use of residues, nature's way, is often economic and desirable for several reasons (see Section 8.1.5). In the long term, soil organic matter and soil N depend most strongly on plant growth, the source of soil organic matter. This generalization leads to the simple

## SOIL AND ENVIRONMENT
### The Work of Microbes

- Microbial decomposition is the biggest waste disposal process on this planet. It recycles plant nutrients back to the inorganic forms in soil, water, and air. Unlike burning, the main alternative, microbial decay releases more methane but less nitrogen oxides and carbon monoxide to pollute the atmosphere, and no sulfur dioxide or smoke. The atmosphere's carbon dioxide level, so important to global climate, depends on the balance between photosynthetic assimilation of carbon dioxide and its release by burning and decomposition. Symbiotic nitrogen fixation and mycorrhizae reduce agriculture's need for nitrogen and phosphate fertilizers, as well as associated resource consumption and pollution risks.

- Neither nitrogen fixation nor organic farming eliminate the main hazards associated with the use of nitrogen fertilizer. Whether nitrogen originates from synthetic fertilizers, organic fertilizers, biological nitrogen fixation, or soil organic matter, it eventually becomes biologically available in soils as ammonium. Ammonium oxidation to nitrate, coupled with leaching of excess nitrate not absorbed by plants, promotes soil acidification and is the main source of nitrate pollution in groundwater. Both risks can be minimized in systems that avoid or retrieve surplus soil nitrate, regardless of the source of nitrogen.

- Soil microbes decay, destroy, or detoxicate most organic herbicides, insecticides, fungicides, and other materials added to soil. Too rapid a decay can limit the material's usefulness; too slow a decay can cause the material to persist in soil or move to groundwater. Once in groundwater, persistence is almost guaranteed by lack of microbial activity.

- Wetlands present special environmental problems. Their flatness and proximity to water make them targets for agricultural or urban development, with a variety of disastrous results. Breeding grounds for birds and fish disappear. Sediments and waterways rearrange in unforeseen ways. And wetland soils, formed in water, can punish those who remove the water. Drainage means aeration and microbial oxidation, which in organic soils means subsidence (10 cm a year in parts of California's San Joaquin Delta). Also, in marine sediments that contain sulfides, oxidation means release of sulfuric acid, death to estuarine ecosystems and concrete.

rule that inherent N fertility, like soil organic matter level, is improved most effectively by residue conservation combined with practices that sustain high biomass production.

Even with high soil organic matter, mineralization rarely goes fast enough to meet the demands of vigorously growing crops. Farmers therefore supplement the soil's ammonium or nitrate, either directly by adding inorganic N fertilizers or indirectly by adding organic materials or promoting biological nitrogen fixation. The reduction of losses is also beneficial in the long term, because it saves expensive N fertilizer and reduces the contamination of air and water. Chapters 9 and 10 deal with fertilization and fertility maintenance in general, but denitrification and biological nitrogen fixation are two microbiological aspects we consider here.

Denitrification loss can be eliminated by keeping the soil dry and infertile, but this is rarely recommended agricultural practice. Otherwise, little can be done in the field to control denitrification, except to avoid adding nitrate to wet, warm soils.

Enhancing biological nitrogen fixation in agriculture usually means growing legumes as rotational crops, rotational pastures, green manures, **intercrops,** or range plants. If a legume is planted in soil lacking sufficient numbers of the right *Rhizobium,* the seed can be *inoculated* with commercially available **inoculant** cultures. These usually contain live bacteria of the specified strains in a carrier of moist peat or some other humuslike material. The carrier keeps the cells alive and facilitates spreading them with the seed. The seed and inoculant are mixed shortly before planting. Few symbioses fix more N than the host plant needs. Fixation will stop if the soil provides ample N or the host grows poorly. Thus vigorous fixation requires N-deficient soil and vigorous growth of effectively nodulated plants.

# 8.5 BIODEGRADATION OF CHEMICALS IN SOIL

Biological degradation reduces the persistence and spread of organic chemicals in their everyday extensive use or misuse as pesticides, herbicides, fuels, and industrial reagents. It also forms the basis of the bioremediation of land spectacularly polluted by major spills or dumping. Biodegradation does not proceed in isolation; it is coupled with sorption (Section 3.4) and diffusive movement (Supplement 4–2) of the chemicals.

## 8.5.1 Biodegradation, Sorption, and Diffusion

The rate of biodegradation of an organic chemical in a soil depends on (1) its rate of metabolism by bacteria (or sometimes fungi), (2) its rate of diffusion from sorption sites to decomposer microbes (Figure 8–13A), and (3) its affinity for the soil's solid surfaces (Figure 8–13B).

The rate of metabolism depends on the biomass and activity of microbes capable of decomposing the particular chemical. This population is likely to be large and active if the conditions are warm and moist and if the chemical itself supports growth of the microbes. It is likely to be inactive if the chemical poisons them. In numerous cases, contamination has induced the growth of adapted or selected populations able to tolerate and decompose the particular contaminant. This can take time.

Diffusion of chemicals in soil occurs almost entirely in the soil solution and depends partly on properties of the chemical itself (Supplement 4–2 and Section 9.3). The diffusion of ions and highly polar molecules is hindered by attraction to the soil's charged surfaces. Charge has little effect on the many organics with no charge and little or no polarity; the main difference in this group (mostly alkanes and their close relatives) is that larger molecules go more slowly. Solutes of all kinds diffuse faster as soil temperature and water content increase, and diffusion is always slowed by increasing the length of the diffusion path (Figure 8–13A). Substances on sorption sites inside aggregates must travel via tortuous paths to reach colonies of decomposer microbes. For this reason, reduction of aggregate size by cultivation or crushing can accelerate biodegradation.

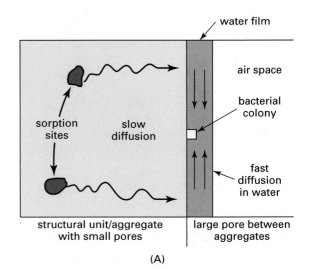

(A)

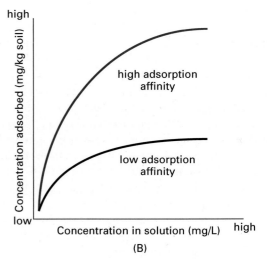

(B)

**FIGURE 8–13** Diffusion and sorption can limit rates of microbial degradation of chemicals. (A) The tortuosity of diffusion pathways through fine pores slows the transfer of chemicals from sorption sites to bacterial colonies. Movement is faster in the larger pores between aggregates, at least when the soil is wet. (B) Movement to degradation sites also depends on adsorption. Degradation is slowed if the chemical has a high adsorption affinity or the soil has high adsorption capacity, as in the upper curve of these sorption isotherms.

Strong adsorption affinity reduces a chemical's decomposition rate; glyphosate, an extreme example, may be virtually inactivated by surface complexation with hydrous iron oxide (Section 3.4 and Supplement 3–5). But among the nonionic, weakly polar substances (alkanes and the like), biodegradation is little affected by affinity differences.

There are parallel cases of soil biological processes coupled with sorption and transfer processes. These include most other microbial processes, such as the oxidation and reduction reactions of Sections 8.2 and 8.3 and the uptake of plant nutrients from soil (Sections 9.1 through 9.4).

### 8.5.2   Bioremediation

Bioremediation is the use of microorganisms to convert organic contaminants to less harmful products. It employs biological oxidation and assimilation that convert hazardous and nonhazardous contaminants to the nonhazardous products, $CO_2$ and water. Bacteria are most commonly used, but fungi can be used. Most soil bacteria can degrade the kinds of organic contaminants found in soils. The most difficult problems for successful bioremediation are the relatively slow rates of contaminant conversion and the large area needed for treatment. Half-lives (the time it takes to transform 50 percent of a contaminant) of common hydrocarbons vary from 6 to over 230 days. In general, lighter, low-viscosity, soluble, and simple compounds are more quickly biodegradable than heavier, less soluble, more strongly adsorbed, and complex materials. Contaminants toxic to microorganisms greatly slow biodegradation. Petroleum hydrocarbons are among the more common soil contaminants and are generally biodegradable, as are the nonchlorinated solvents such as alcohols, ketones, ethers, and carboxylic acids. Lightly chlorinated compounds, such as dichlorobenzene, and chlorinated phenols are biodegradable under aerobic conditions. More highly chlorinated compounds are less easily degraded under aerobic conditions but may be biodegraded under anaerobic conditions. Bioremediation has been used in situ (at the site where soil is polluted) or off site (on soil that has been removed from the site of contamination). It has also been used to remediate groundwater and surface water.

For bioremediation to be successful, microorganisms that can utilize the pollutant as a substrate must be present or must be introduced into the soil. An A horizon with adequate C, O, and nutrients may contain $10^7$ to $10^9$ microorganisms per gram of soil. Of these, only 0.1 to 1 percent are petroleum degraders. According to Bossert and Bartha (1984), the most common two genera able to degrade petroleum are *Pseudomonas* and *Arthrobacter*. Upon exposure to petroleum, the number of petroleum-degrading bacteria increases. In addition, the soil must provide other essentials for the growth of the microorganisms, including appropriate temperature, water content, and chemical conditions such as pH. Frequently, subsoils are contaminated and nitrogen and phosphorus must be added to the soil to promote microbial growth. Soil pH in the range of 6 to 8 appears to be most favorable for bioremediation. Bioremediation takes place over a much wider range of pH, but at slower rates. With petroleum products, biodegradation often releases protons with a subsequent lowering of pH. Lime ($CaCO_3$) is often added to the soil to raise the pH to the most favorable range.

The most common electron acceptor for microbial growth is oxygen. Successful bioremediation sites must be well oxygenated. In general, about 1.5 kg (3.1 lb) of oxygen is required to degrade 0.4 kg (1 lb) of hydrocarbon. In a well-aerated soil, the oxygen content is approximately 20 percent, or 200,000 ppm. In saturated soil, there is only about 8 ppm oxygen. In well-aerated soils, permeability of the soil to air is the main parameter controlling oxygen content and vapor movement in soils. As oxygen is consumed by microorganisms, it must be replaced by oxygen flow to the decomposition site. Permeability to gas is a function of particle size distribution, pore size distribution, soil moisture content, and bulk density.

Microorganisms require moisture to assimilate the contaminants most efficiently. Irrigation is sometimes used to maintain soil moisture at between 70 and 80 percent of field capacity. Too much water slows biodegradation by reducing the oxygen supply; insufficient water slows biodegradation by limiting microbial growth. If the soil or underlying geologic material is part of an aquifer, the groundwater flow rate, nutrient content, and oxygenation must be known for remediation to be successful.

Most bioremediation uses the natural soil microorganism population. In some cases, the soil may be sterilized by the contaminant, or the contaminant may be well below the zone of major microbial population. In such cases, microorganisms can be introduced into the soil. There has been little experience with introducing microorganisms to soil. To be effective, the organisms must be transported to or injected into the zone of contamination. Permeable soils allow for rapid movement of microorganisms introduced into subsurface contaminants by leaching. Rapid permeability generally ensures well-oxygenated soils and rapid microbial growth. Microorganisms must also attach themselves to the solid phase, proliferate, and remain active degraders.

Nitrogen is typically added as urea or ammonium nitrate. Reduced nitrogen can also be added in the vapor phase as anhydrous ammonia gas ($NH_3$). This practice has been used in agricultural production since the 1940s. Ammonia is toxic to soil microorganisms at concentrations above about 300 ppm, so it must be added carefully.

Organic compounds are biodegraded by two mechanisms: use of the contaminant as a primary substrate and cometabolism. In the first mechanism, the organism consumes the organic compound directly. **Cometabolism** is the transformation of an organic compound by enzymes or cofactors produced by organisms for other purposes. The compound is decomposed because it happens to be in the vicinity of the microorganism, not because the organism is targeting it for consumption. The organism may not obtain a direct benefit from the transformation. This process is of great interest because most common halogenated aliphatic hydrocarbons (a large group of pollutants in soils and groundwaters) can be biodegraded by cometabolism. The soil ecological conditions for cometabolism are the same as for primary substrate consumption. An appropriate substrate must be present for the microorganisms, and an active population of microorganisms with the appropriate enzymes or cofactors must be present.

Either process may occur aerobically or anaerobically, but the decomposition products will be different under aerobic and anaerobic conditions. Aerobic transformations generally are oxidations. Anaerobic processes generally are reductions. The specific reaction and transformation product depends on the substrate, the microorganism, and the soil conditions.

Treatment may be in situ or on excavated material. Biological treatment to remove petroleum from soils is sometimes most effective if the soils are excavated, stockpiled, and treated in batches. This requires excavation and mixing equipment that can oxygenate the soil during the treatment period. Such treatment allows for the addition of needed nutrients and water and monitoring of pH, pollutant constituents, and degradation products as remediation proceeds.

In situ biological treatment includes procedures in which a solution containing nutrients and hydrogen peroxide (to provide oxygen) is used to saturate the contaminated soil. As the treatment proceeds, contaminated water is removed, cleaned, and recirculated through the soil or is discharged and fresh water is added. Another method is to add oxygen and nitrogen gas to the soil. This has the advantage of maintaining an aerobic, unsaturated condition in the treatment volume and a higher oxygen level than can be obtained with hydrogen peroxide.

## 8.6  SUMMARY

Major microbial soil processes are concentrated in the soil's upper horizons and are moderated by diverse groups of fungi and bacteria. Most of these creatures collectively oxidize an enormous array of natural and manufactured organic materials, mineralizing them to simple carbon gases and dissolved ions. Humus is the major by-product. In addition, important minority groups of bacteria bring about oxidative and reductive transformations of N, S, and other elements important in soil chemistry and soil interactions with plants, water, and the atmosphere.

Many factors can limit one or another of these processes at some stage. The unavailability or difficulty of assimilation of organic substances strongly limits the breakdown of humus and resistant-added materials (such as woody residues and nonbiodegradable pollutants). The lack of N in assimilable forms such as amino acids, ammonium, or nitrate often retards the decay of plant residues as well as the processes of N mineralization, nitrification, nitrogen assimilation, and denitrification. Insufficiency of oxygen slows oxidative processes such as organic decay and the production of nitrate and sulfate in flooded or wet compacted soils. Excess oxygen limits reductive processes such as nitrogen fixation, denitrification, and sulfate reduction. All these processes are frequently hampered by unfavorable temperatures or the onset of drought.

## QUESTIONS

1. Explain the meaning and significance of *oxidation, reduction, anoxia, decomposition, humification, nitrification, denitrification, nitrogen fixation, ammonification,* and *bioremediation* in soils.
2. Contrast the most important effects of plants and heterotrophic microbes on the levels and kinds of organic matter in a soil–plant ecosystem.
3. Design two management systems for a piece of land: (a) a system to produce the maximum amount of organic litter on the ground surface as mulch, and (b) a system to minimize litter. You may neither burn material nor bring in organic matter produced somewhere else, and you may grow only one kind of crop or plant cover.
4. In what ways would Question 3 be an easier problem if you were free to grow any kind of crop or plant cover?
5. Why do you suppose the percentages of N and organic C in a soil vary together, so much so that either can be used as an index of organic matter content?
6. Why does humus contain macromolecules with properties and composition similar to those of lignin? In what important ways does humus differ from lignin?
7. Is it possible to build up a soil's organic matter and at the same time make it less fertile?
8. Why do microbes oxidize ammonium and sulfide? What would happen to soil fertility if they did not?
9. Why do microbes reduce nitrate to $N_2$ (denitrification)? This process causes loss of N from soils to the atmosphere. How can it be minimized? In what ways can it be beneficial?
10. Flooding a soil has beneficial effects on fertility for plants that can tolerate the flooding. What are some of these beneficial effects?
11. If you increase the rate of one step in the N cycle, what is likely to happen to the other steps?

## FURTHER READING

Bartlett, R. J., and B. R. James. 1993. Redox chemistry of soils. *Advances in Agronomy* 50:152–208.

Batjes, N. H. 1996. Total carbon and nitrogen in the soils of the world. *European Journal of Soil Science* 47:151–163.

Bossert, I., and R. Bartha. 1984. The fate of petroleum in soil ecosystems. In *Petroleum microbiology,* ed. R. M. Atlas. New York: Macmillan.

Kostecki, P. T., and E. J. Calabrese. 1990. *Petroleum contaminated soils.* Vol. 3. Chelsea, Mich.: Lewis Publishers.

Lovley, D. R. 1995. Microbial reduction of iron, manganese, and other metals. *Advances in Agronomy* 54:175–232.

MacCarthy, P., C. E. Clapp, R. L. Malcolm, and P. R. Bloom, eds. 1990. *Humic substances in soil and crop sciences: Selected readings.* Madison, Wis.: Soil Science Society of America/American Society of Agronomy.

Norris, R. D., et al. 1994. *Handbook of bioremediation.* Robert S. Kerr Environmental Research Laboratory. Boca Raton, Fla.: CRC Press.

Paul, E. A., and F. E. Clark. 1996. *Soil microbiology and biochemistry.* New York: Academic Press.

Ponnamperuma, F. W. 1972. Chemistry of submerged soils. *Advances in Agronomy* 24:29–96.

Richards, B. N. 1987. *Microbiology of terrestrial ecosystems.* London: Longman Group.

Scow, K. M., and C. R. Johnson. 1997. Effect of sorption on biodegradation of soil pollutants. *Advances in Agronomy* 58:1–56.

Swift, M. J., O. W. Heal, and J. M. Anderson. 1979. *Decomposition in terrestrial ecosystems.* Berkeley and Los Angeles: University of California Press.

Tate, R. L. 1995. *Soil microbiology.* New York: Wiley.

Tisdale, S. L., W. L. Nelson, and J. D. Beaton. 1985. *Soil fertility and fertilizers.* 4th ed. New York: Macmillan.

# SUPPLEMENT 8–1   Biological Compounds

## CARBOHYDRATES AND RELATED COMPOUNDS

### SIMPLE SACCHARIDES

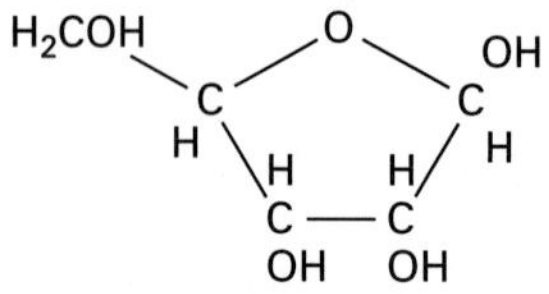

a pentose (ribose)

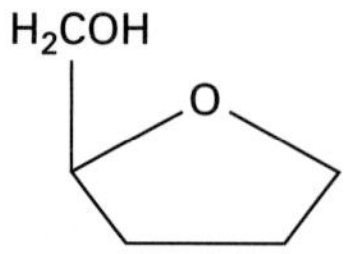

pentose

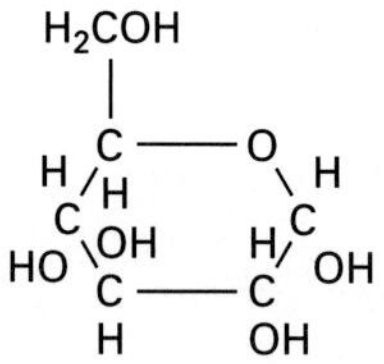

a hexose (glucose)

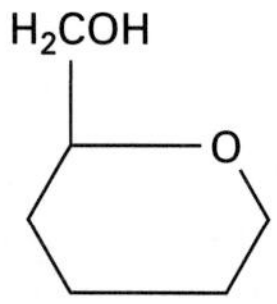

a hexose

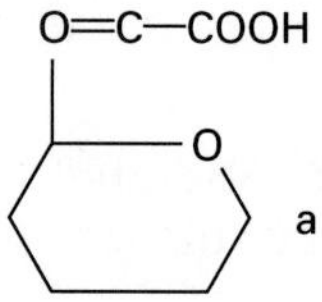

a (hex) uronic acid

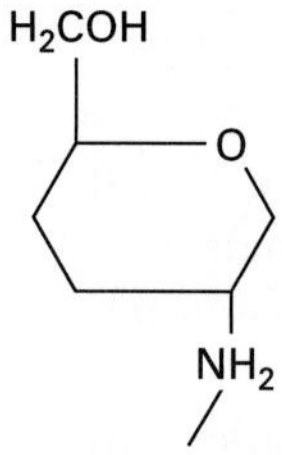

a hexosamine

CARBOHYDRATES AND RELATED COMPOUNDS

POLYSACCHARIDES

OH
O
OH

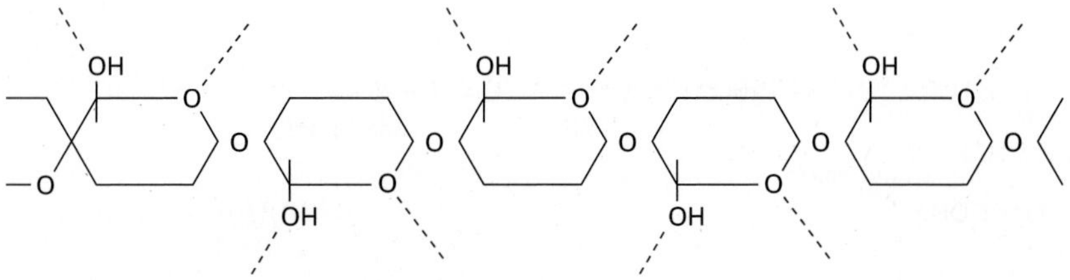

UNBRANCHED POLYSACCHARIDE CHAIN (cell wall), e.g., CELLULOSE (glucose units;
--- indicates H-bonds holding parallel chains together)

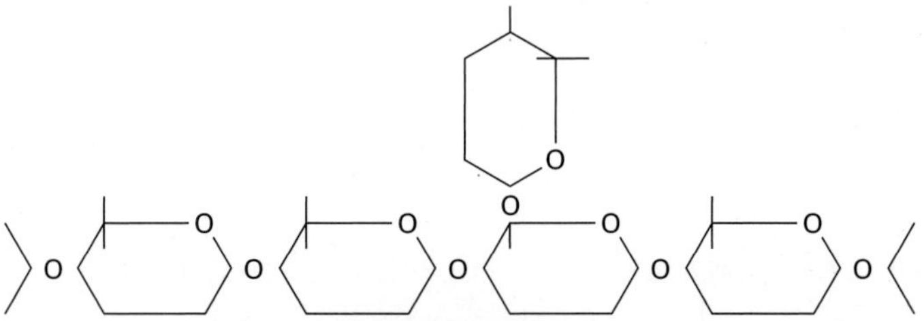

BRANCHED CHAIN
e.g., STARCH (storage), XYLOGLUCAN (cell wall), ARABINOGALACTAN (cell wall)

COOH
CO CH3
COOH
COOH

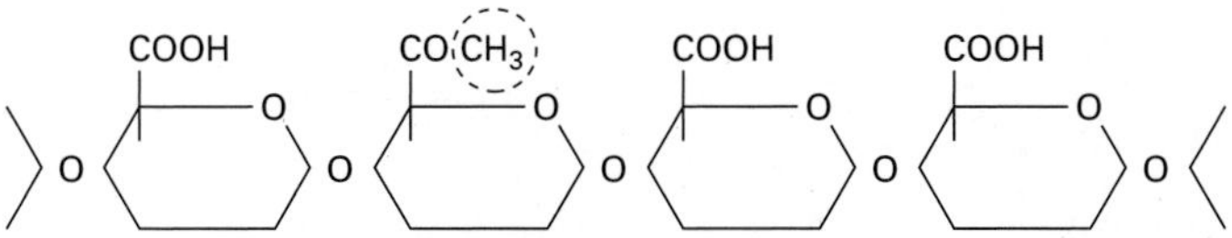

POLYURONIDE (partly methoxylated–$CH_3$), e.g., PECTIN (cell wall)

NH
O=C—CH3
NH
O=C—CH3

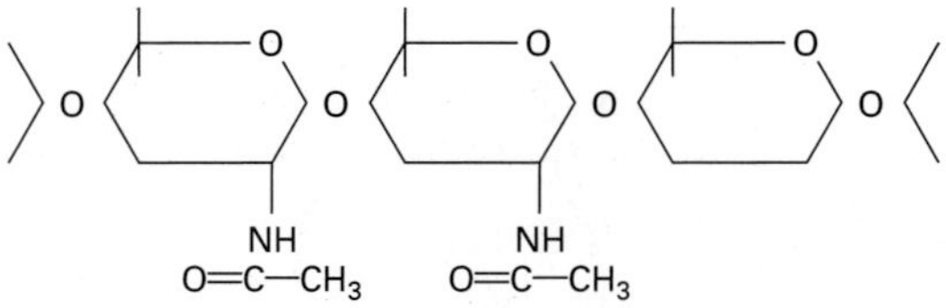

AMINO–SUGAR POLYMER, e.g., CHITIN (insects and fungi)

LIPIDS

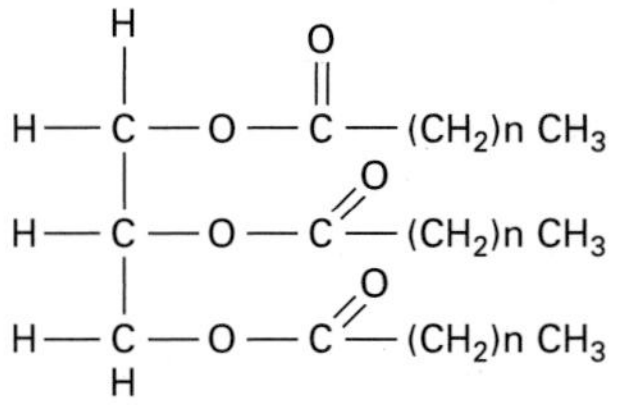

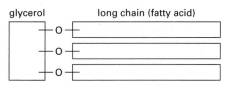

glycerol
long chain (fatty acid)

PHOSPHOLIPIDS

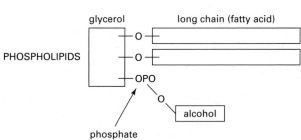

glycerol
long chain (fatty acid)
OPO
alcohol
phosphate

## ACIDS

### "ORGANIC" ACIDS

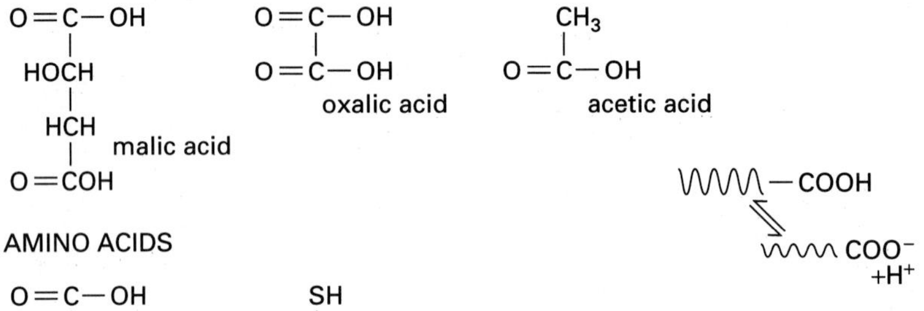

O=C—OH
|
HOCH
|
HCH
|
O=COH     malic acid

O=C—OH
|
O=C—OH
oxalic acid

CH₃
|
O=C—OH
acetic acid

### α AMINO ACIDS

O=C—OH
|
HCH
|
C—NH₂
|
O=C—OH
aspartic acid

SH
|
HCH
|
C—NH₂
|
O=C—OH
cysteine

H—C—NH₂
|
O=C—OH
glycine

WWW—COOH
WWW COO⁻
+H⁺

COOH
NH₂

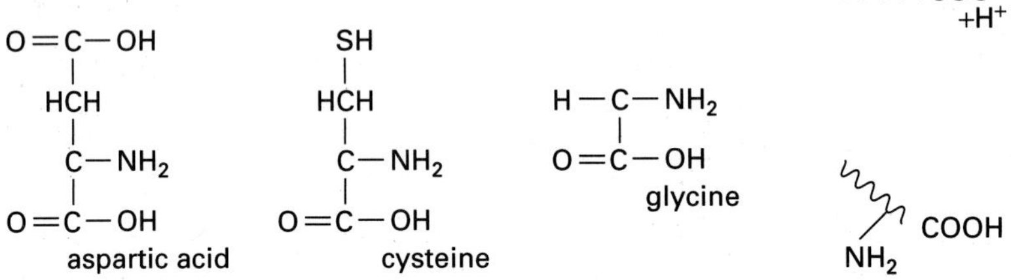

### AMINO ACID POLYMER  –  PEPTIDES  +  POLYPEPTIDES (PROTEINS)

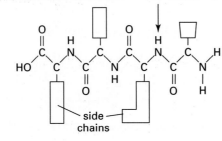

side chains

### NUCLEOTIDES

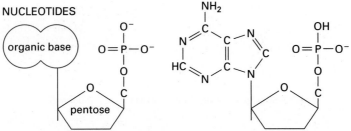

generalized nucleotide, e.g., adenosine monophosphate (AMP)

### POLYNUCLEOTIDES: NUCLEIC ACIDS

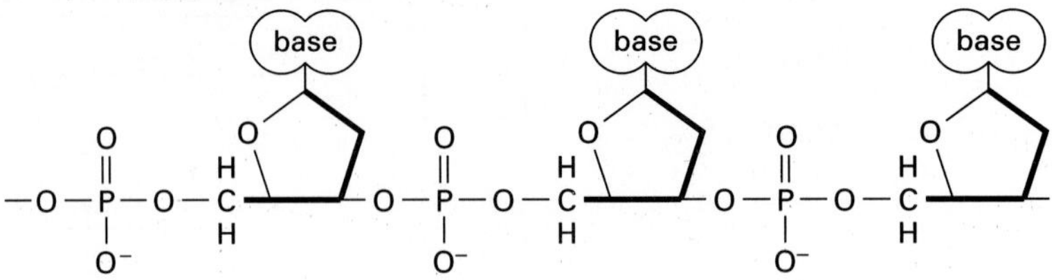

PHENOLICS

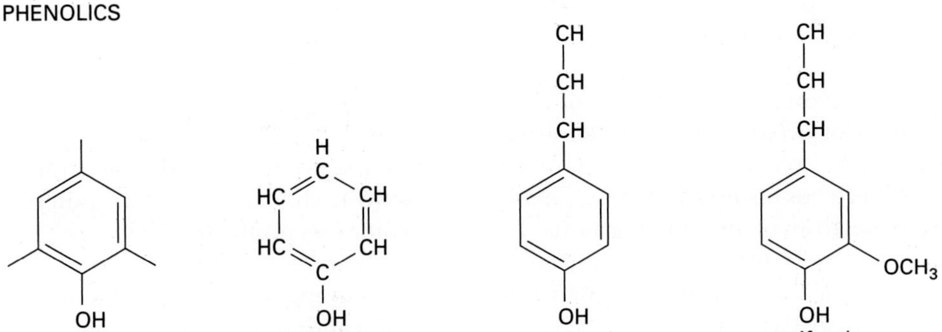

phenol

coumarol

coniferol

PHENOLIC POLYMER: POSSIBLE LIGNIN STRUCTURE

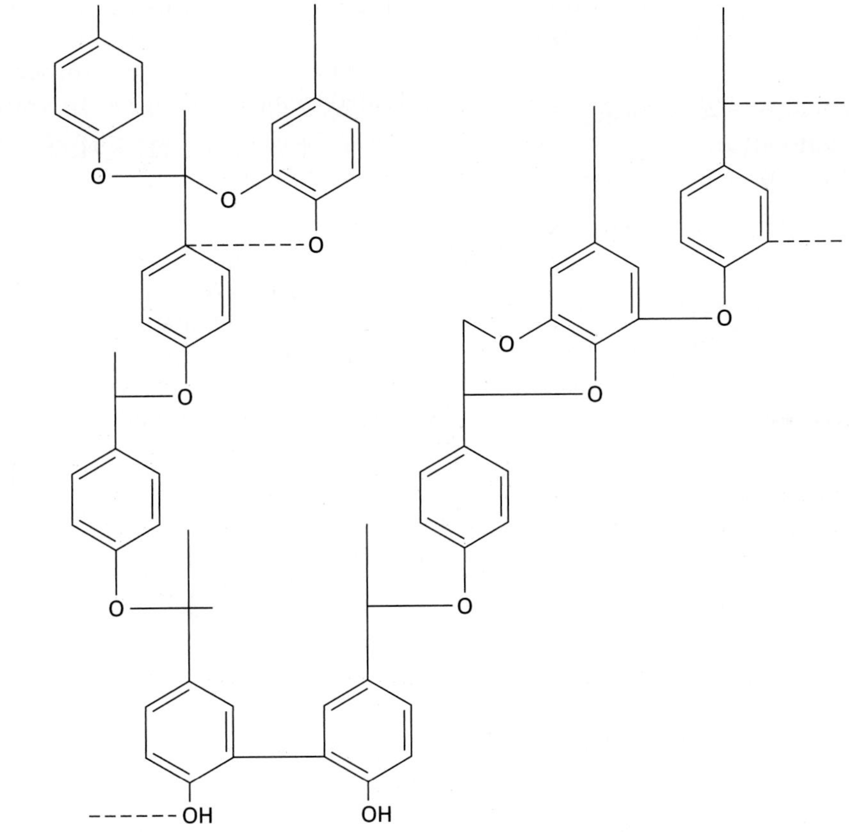

## SUPPLEMENT 8–2    Oxidation and Reduction

Many reactions can be thought of as transfers of electrons, from one atom or group of atoms to another. Such reactions are called oxidation-reduction, or *redox*, reactions. The electrons involved are loosely held, outer shell, valence electrons on the reacting atoms. In every redox reaction, an electron donor (reductant) "gives" electrons to an electron acceptor (oxidant) as the donor and acceptor react. In so doing, the donor becomes oxidized, and the acceptor becomes reduced. Oxidation is a loss of electrons, and reduction is a gain of electrons. Oxygen ($O_2$) is the most abundant oxidizing agent in natural processes on earth.

We can write generalized redox reactions like this:

$A$ (oxidized) + $D$ (reduced) = $A$ (reduced) + $D$ (oxidized)

acceptor + donor = acceptor + $e^-$ + donor − $e^-$

Or we can write a pair of half-reaction statements like this:

$D$ red. = $D$ oxid. + $e^-$
$A$ oxid. + $e^-$ = $A$ red.

The half-reactions must be coupled to go; one supplies the electrons, and the other uses them.

In some cases the transfer of electrons is easy to visualize. An example is the reduction of ferric ion by reduced tin ion:

$$2Fe^{3+} + Sn^{2+} \rightarrow 2Fe^{2+} + Sn^{4+}$$

Here the tin ion loses electrons (and becomes more positive), while the ferric ions gain the electrons (and become less positive).

In most cases the transfer of electrons is less obvious. For example, the production of water from hydrogen and oxygen can be regarded as the reduction of oxygen and the oxidation of hydrogen. Electrons are "transferred" from H to O in the sense that they become shared. (Indeed the oxygen atom, being attractive to loose electrons, receives most of their time and attention.) Oxygen's two empty orbits become occupied by electrons from hydrogen, somewhat as indicated by the following electron-sharing diagram:

$$O + 2H = H_2O$$

In soil, microbial metabolism is the main source of reactive electrons, reducing power. When oxygen runs short, electrons are accepted by more reluctant oxidizing agents. We then refer to *reducing soil conditions*. These other oxidants include nitrate, manganese and iron oxides, and sulfate (see Box 8–2).

# 9

# MINERAL NUTRIENTS IN SOILS AND PLANTS

## OVERVIEW

*A plant must contain at least the 16 elements currently known to be essential: C, O, H, N, K, Mg, Ca, P, S, Fe, Cl, Mn, Zn, B, Cu, and Mo. All but C, O, and H are "mineral nutrients," normally absorbed from soil as ions or other simple dissolved forms. Absorption through the plant cell membrane is controlled and selective. Cells may accumulate an element to a higher concentration than in the outside medium, by metabolically driven ion pumping or by the continuous assimilation of ions into organic plant constituents. Within a plant, nutrient elements are transported through the xylem from root to shoot. Nutrients are also remobilized into the phloem for transport from old parts of a plant to growing tissues.*

*Nutrients must be released into solution from the soil's solid components for adequate uptake by plants. The several release processes—microbial decay, ion exchange, dissolution, and desorption—vary in importance depending on the nutrient element and the soil.*

*Movement of a nutrient to the root can limit uptake when the dissolved concentration is low, so that movement depends mainly on diffusion. To acquire immobile nutrients fast enough, root systems must actively explore the soil by extending rapidly, branching densely in nutrient-rich zones, and developing root hairs or mycorrhizae. Root growth, form, and function depend on genetic, physical, chemical, and microbiological factors. Root extension is frequently inhibited by dryness, high soil density, oxygen shortage, Al toxicity, Ca deficiency, and disease.*

*All these plant and soil processes and factors influence nutrient* **availability**—*that is, the adequacy with which the soil supplies the plant's need for the nutrient. Because availability is complex and incompletely understood, plant response remains the only truly reliable basis for diagnosing deficiencies. The same processes and factors influence the toxicities of nutrients and other elements.*

The next three chapters describe soil as a growth medium providing nutrients to plants and as part of a system in which the plant and soil influence each other. This chapter begins by describing functions of mineral nutrients in plants and then the chain of processes that release nutrients in the soil and move them into and through the plant. Here we emphasize individual small-scale chemical, physical, and biological processes and their relationship to "soil fertility."

Chapter 10 will then take a larger, field-scale view of soil fertility management, and Chapter 11 will describe the special problems of acidic soils and salty soils.

## 9.1 NUTRIENT ELEMENTS IN PLANTS

### 9.1.1 Elemental Composition of Plants

Water is the principal constituent of plants, about 90 percent of their fresh, live weight. The rest—dry matter—consists mainly of carbon, oxygen, and hydrogen. The C and O derive from atmospheric $CO_2$, and the H derives from water during photosynthesis. A small but indispensable remainder consists of elements absorbed from the soil as inorganic ions and simple molecules (Table 9–1).

Essential plant nutrients are those elements necessary for the plant to complete its life cycle. In general, each **element** has at least one function in which substitution by other elements is not possible. Given the essential elements in suitable form, the plant can synthesize all the other nutrients that its cells need. The essentiality of nutrients can be demonstrated clearly in simple growth media, such as

## TABLE 9–1
Concentrations of Essential Mineral Elements in Leaves

|  | Sugarcane | Rice | Corn | Soybean |
| --- | --- | --- | --- | --- |
| **Macronutrients (%)** | | | | |
| Nitrogen (N) | 1.5 | 2.5 | 3.0 | 4.0 |
| Potassium (K) | 2.2 | 1.0 | 1.9 | 1.7 |
| Phosphorus (P) | 0.05 | 0.1 | 0.25 | 0.26 |
| Magnesium (Mg) | 0.1 | 0.1 | 0.25 | 0.25 |
| Sulfur (S) | 0.1 | 0.1 | 0.2 | 0.25 |
| Calcium (Ca) | 0.1 | 0.15 | 0.15 | 0.15 |
| **Micronutrients (ppm)** | | | | |
| Iron (Fe) | 10 | 70 | 15 | 50 |
| Chlorine (Cl) | — | — | 20 | 25 |
| Manganese (Mn) | 20 | 20 | 15 | 20 |
| Zinc (Zn) | 10 | 10 | 15 | 20 |
| Boron (B) | 1 | 3 | 10 | 20 |
| Copper (Cu) | 2 | 5 | 5 | 10 |
| Silicon (Si) | — | 5 | — | — |
| Molybdenum (Mo) | <0.1 | <0.1 | 0.1 | 0.7 |

Data, expressed as percentage (%) or parts per million (ppm), represent the lowest concentrations found in recently matured leaves of healthy plants. Higher levels are commonly found.

**FIGURE 9–1** A plant (*Leucaena leuco-cephala*) growing in solution culture. Healthy plants need only light, $O_2$, $CO_2$, and other nutrients supplied as ions dissolved in aerated water.

aerated solutions of nutrient salts called *hydroponic cultures* or *solution cultures* (Figure 9–1). For research on micronutrient essentiality, it is necessary to purify the nutrient salts, the water, and sometimes the air supply.

Sixteen elements have been shown to be generally essential to plants:

> C, O, H, N, K, Ca, Mg, P, S (the **macronutrients**)
> Fe, Mn, Zn, Cu, Mo, Cl, B (the micronutrients or **trace elements**)

In addition, at least some plant species need Na and Si. Legumes dependent on symbiotic nitrogen fixation need cobalt (see Chapter 7), and there is evidence of a need for nickel. Research is likely to extend the list of essential nutrient elements. Deficiencies of the essential elements frequently limit plant growth or quality. Even when the plants themselves do not

suffer from a deficiency, they sometimes provide animals with a diet deficient in certain elements—for example, Mo, Se, Co, Cu, P, Na, Ca, or I.

Plants frequently contain unnecessarily high and sometimes toxic concentrations of an essential or nonessential element. This characteristic has been exploited to remediate contaminated soils. Plants with the ability to accumulate one or more toxic elements are grown, harvested, dried, incinerated, and the ash disposed of as hazardous waste. This process, *phytoremediation*, is slow, requiring many generations of plants to be successful, but it is less expensive than engineering solutions. Most commonly toxic to plants are Al and Mn (in acidic soils), Na and Cl (in saline soils), boron, and copper. Elements such as copper, lead, arsenic, cadmium, molybdenum, and selenium are sometimes present at concentrations that injure foraging animals, though the plant itself is little affected. Toxicities will be discussed in Chapters 11 and 15.

### 9.1.2   Forms and Functions of Nutrients in Plants

With few exceptions, every cell of a plant seems to need each essential element. Nutrient ions are not just absorbed but are assimilated, becoming components of numerous cell constituents with different functions. Table 9–2 lists some of these constituents and functions (and Supplement 9–1 gives an outline description).

### 9.1.3   Uptake and Movement of Nutrients within Plants

Leaves and other green tissues absorb most of the plant's C and O as $CO_2$. Leaves can also absorb nutrient ions: **Foliar uptake** is important to aquatic plants and when nutrients are sprayed on foliage in horticulture and agriculture. Ordinarily, however, roots absorb the plant's nutrient ions and load most of them into the **vascular** system (**conductive tissue**) for transport to the shoot.

A root takes up ions and water most actively in the region beginning about 10 mm behind the tip and extending back 50 to 100 mm. Beyond that, the root usually absorbs less actively and develops an *endodermis* separating the root's outer region (cortex) from its conductive tissue (stele) (Figure 9–2).

## TABLE 9–2

Some Forms and Functions of Elements in Plants

| Element | Forms and functions |
|---|---|
| C, H, O | All plant organic components |
| N and S | Amino acids—constituents of proteins |
|  | Proteins—enzymes, storage compounds, and membrane components |
| N and P | Nucleotides—energy transfer (e.g., ATP), electron transfer (e.g., NADP), genetic information (DNA and RNA) |
| P | Phospholipids—membranes |
|  | Inorganic phosphate—synthesis of ATP |
| K | K ion—enzyme activator, osmotic regulator |
| Ca | Complexed as calmodulin—regulator of many cell processes |
|  | Attached to cell membranes—stabilizer |
| Mg | Complexed as chlorophyll—photosynthesis |
|  | Complexed with ATP—energy transfers |
| Fe | Complexed as cytochromes—electron transfers |
| Mo | Component of enzymes—$N_2$ fixation and nitrate reduction |
| Ca, Mg, Mn, Cu, Zn | Associated with enzymes—activators |

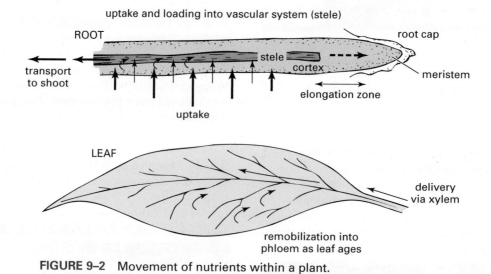

**FIGURE 9–2**   Movement of nutrients within a plant.

At the level of the single cell, the absorption process is similar in all plant tissues. Ions diffuse to and through the cell wall but then encounter the outer cell membrane, a barrier to the free movement of most ions. Uptake through the membrane discriminates among different ions, but the rate of an ion's uptake also depends on the ion's concentration in the cell's surroundings (Figure 9–3). A nutrient element often becomes more concentrated in plant cells than in the soil solution. The energy needed to accumulate the nutrient elements to high concentrations comes from the respiratory breakdown of products of photosynthesis. For this reason, nutrient uptake, like respiration, is sensitive to cold and lack of

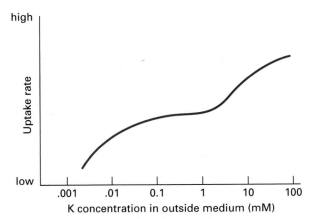

**FIGURE 9–3** Rates of nutrient uptake by roots as influenced by nutrient concentration in the growth medium.

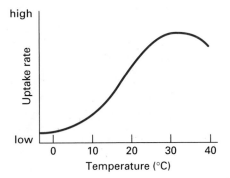

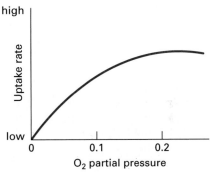

**FIGURE 9–4** Rates of nutrient uptake as affected by temperature and oxygen supply.

oxygen (Figure 9–4A and 9–4B). (Supplement 9–1 gives more details on ion uptake.)

The uptake of ions must be electrically balanced. In general, cells and plants that absorb more cations than anions maintain their electrical balance by exporting $H^+$, and cells that absorb more anions than cations export $OH^-$ or $HCO_3^-$. Because of these uneven acid and base flows, ion uptake by plants changes the external pH, especially in the rhizosphere (Figure 9–5).

Plants also transport (translocate) nutrients within themselves. Nutrients move from cell to cell across the root cortex and into the xylem, which carries them by mass flow to the shoot (Figure 9–6). The nutrient load dissolved in the xylem fluid consists mostly of ions but also includes some organic materials made in the roots (e.g., amino acids). Upon their delivery to growing shoot tissues, the nutrients are reabsorbed from the xylem fluid and assimilated.

As leaves and roots age, some of their nutrients are set free and retranslocated to the young growing leaves, roots, fruits, or storage organs. Retranslocation occurs through the phloem, the same vascular tissue that carries organic nutrients from the leaves to the roots. Not all nutrient elements retranslocate with equal ease. Calcium and most micronutrients do not move readily. Nitrogen and sulfur are readily remobilized in the form of amino acids when proteins break down in old or stressed tissue, and other nutrients readily remobilized are Mg, K, and P. Symptoms of nutrient deficiency will appear first in older leaves if the deficient nutrient is remobilized but will appear

in growing points and young leaves if the deficient nutrient is not remobilized (see Chapter 10).

These plant processes—uptake, translocation, assimilation, and remobilization—are only part of the plant's overall process of nutrient acquisition. Equally important and often limiting to the plant's welfare are two sets of soil processes: nutrient release and nutrient movement to the root.

## 9.2 RETENTION AND RELEASE OF NUTRIENTS IN SOIL

### 9.2.1 Soil Solution and Solid Nutrient Reserves

Nutrients occur in the soil solution mostly as ions, although organic complexes are significant in some cases (Fe, Zn, Cu), and uncharged molecules are important in others (Si, B). Figure 3–9 suggests likely concentrations. Uptake by plants and leaching by rain or irrigation would deplete the ions in solution unless they are replenished from reserves contained

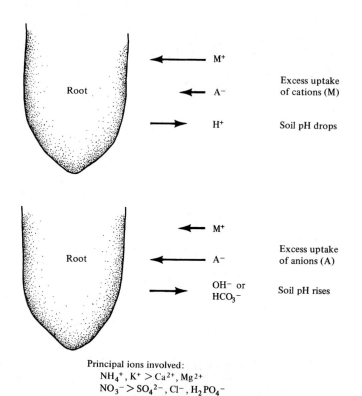

FIGURE 9–5   Charge balance and effects on soil pH during nutrient uptake. Note the consequences of the form of N absorbed. Plants using nitrate raise the pH around their roots, plants using $N_2$ often lower it, and plants using ammonium lower it strongly.

Excess uptake of cations (M)

Soil pH drops

Excess uptake of anions (A)

Soil pH rises

Principal ions involved:
$NH_4^+$, $K^+$ > $Ca^{2+}$, $Mg^{2+}$
$NO_3^-$ > $SO_4^{2-}$, $Cl^-$, $H_2PO_4^-$

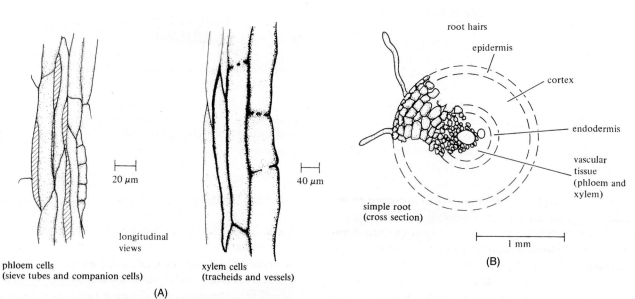

FIGURE 9–6   Vascular conductive tissues in plants. (A) Xylem carries water and dissolved mineral nutrients from roots to leaves. Phloem carries organic and remobilized mineral nutrients from leaves to other parts of the plant. (B) Cross section of root.

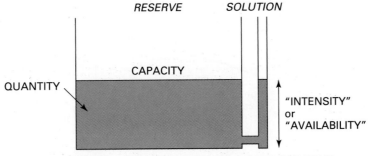

RESERVE          SOLUTION

CAPACITY

QUANTITY

"INTENSITY"
or
"AVAILABILITY"

(A) HYDRAULIC "MODEL" (TANK AND GAUGE) FOR SOLID
RESERVE/SOIL-SOLUTION NUTRIENT RELATIONSHIP.

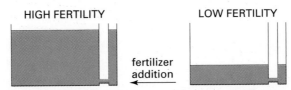

HIGH FERTILITY          LOW FERTILITY

fertilizer
addition

(B) ANALOGY FOR CHANGES IN FERTILITY (DEPLETION
OR FERTILIZATION). NOTE THAT *RESERVES* MUST
CHANGE AS WELL AS SOIL SOLUTION.

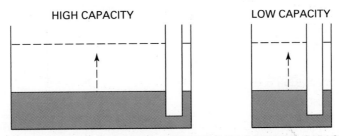

HIGH CAPACITY          LOW CAPACITY

(C) ANALOGY FOR DIFFERENCES IN RESERVE OR SORPTION *CAPACITY.*
SOILS WITH HIGH CAPACITY NEED MORE FERTILIZER TO RAISE FERTILITY
AND RESIST DEPLETION BETTER. THEY ARE "BETTER BUFFERED."

**FIGURE 9–7**  Concepts of nutrient reserve and availability in soil. (A, B, and C) A hydraulic model. The tank's capacity represents the soil's capacity to hold a nutrient, the fluid volume represents the quantity of nutrient in the soil, and the depth of the fluid represents the nutrient's availability ("intensity").

in minerals and organic matter and attached to colloids, as pointed out in Chapters 3 and 8. Some processes of release are rapid, with the solution and the solid phase remaining close to equilibrium. Cation exchange is an example. Other release processes are slower, and the rates of release are critical. An example is the release of N and S by the decay of soil organic matter. Both the concentrations in solution and the rates of release depend on the amounts and the kinds of solid-phase nutrient, soil water content, temperature, aeration, microbial activities, and properties of the soil solution.

Fertilizers are materials that dissolve or decompose to release useful amounts of the ionic forms of some nutrient, but seldom do all the released ions remain in solution. Most become *fixed,* or immobilized, on the soil's solid surfaces. The fertilizer thus adds to the soil's solid nutrient reserves as well as to the immediately available dissolved pool of nutrients.

The relationships between solid and dissolved nutrient forms are generalized in Figure 9–7, with specifics for individual elements in Tables 9–3 and 9–4. (Supplement 9–2 adds to this information.)

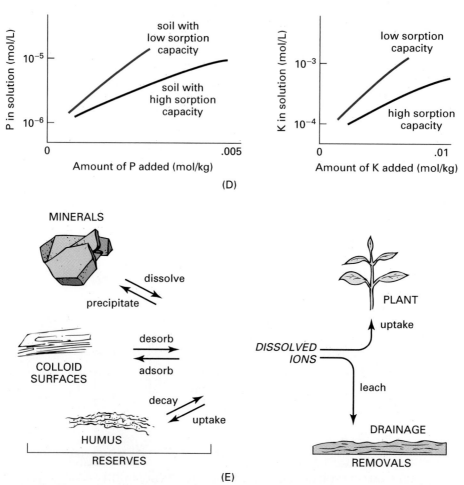

FIGURE 9–7 *Continued.* (D) Examples of "quantity/intensity" plots for some nutrients in individual soils. The nutrient concentration in solution (availability or intensity) relates to the amount added (change in quantity), depending on the soil's capacity to absorb the nutrient. (E) Major reserve forms of soil nutrients. Processes of release and immobilization.

## 9.2.2 Release Processes

Organic matter decay is most important to the release (mineralization) of N and S (as well as P in some soils). Concentrations of dissolved nitrate and sulfate fluctuate with season, cropping, and weather. Their release is fastest when there is abundant, readily decomposable organic matter rich in the nutrient elements and when the soil is moist, warm, and well aerated (see Chapter 8).

Cations of all kinds are held in the soil solution near the negatively charged surfaces of clay and humus. Unlike the ions held by more specific kinds of adsorption, exchangeable cations remain hydrated and exchange readily with one another. Cation exchange is rapid. The equilibrium level of an ion in solution depends on the amount present, the cation exchange capacity and selectivity, and the kinds and levels of competing cations. Cation exchange underlies the special problems of acidic and saline soils (see Chapter 11) and provides a reserve of Ca, Mg, and K.

The dissolution of minerals, a weathering process, also releases the nutrients K, Ca, Mg, and other metals from the common "weatherable" (i.e., not quartz) silicate minerals: feldspars, ferromagne-

**TABLE 9–3**

Major Reserve Forms and the Ions They Release

| Solid reserve forms | Principal nutrients involved (common forms in solution) |
|---|---|
| **Adsorbed on colloid surfaces** | |
| Held electrostatically (ion exchange) | $K^+$, $NH_4^+$, $Ca^{2+}$, $Mg^{2+}$, $Na^+$ |
| Surface complexes on humus | $Zn^{2+}$, $Mn^{2+}$, $Cu^{2+}$, $Fe^{3+}$, and soluble complexes |
| Chemically bound on oxide clays | $H_2PO_4^-$, $HPO_4^{2-}$, $MoO_4^{2-}$, $H_3BO_3$ |
| **Incorporated into organic matter** | |
| Released by decay | $NH_4^+$, $NO_3^-$, $SO_4^{2-}$, $H_2PO_4^-$, $HPO_4^{2-}$ |
| **Incorporated into minerals** | |
| Released by dissolution (weathering) | $K^+$, $Na^+$, $Ca^{2+}$, $Mg^{2+}$, other nutrients in small amounts |

**TABLE 9–4**

Release of Nutrient Reserves in Soils

| Solid forms (reserve) | Release processes | Solute forms |
|---|---|---|
| **Nitrogen** | | |
| Organic matter | Microbial decay | $NH_4^+$ |
| Interlayer $NH_4^+$ in mica and vermiculite | Mineral weathering | $NH_4^+$ |
| Exchangeable ammonium | Cation exchange | $NH_4^+$ |
| **Sulfur** | | |
| Organic matter | Microbial decay | $S^{2-}$, $SO_4^{2-}$ |
| Sulfides (e.g., FeS) and sulfur in anoxic soils | Microbial oxidation | $SO_4^{2-}$ |
| Gypsum (calcium sulfate) | Mineral dissolution | $SO_4^{2-}$ |
| **Phosphorus** | | |
| Organic matter | Microbial decay | $H_2PO_4^-$ and $HPO_4^{2-}$ |
| Attached to Fe and Al oxides and Ca carbonate | Ligand exchange and mineral dissolution | |
| **Potassium, calcium, magnesium** | | |
| Silicate minerals, (e.g., feldspars, micas, and 2:1 clays) | Mineral weathering (dissolution) | $K^+$ $Ca^{2+}$ |
| Exchangeable cations | Cation exchange | $Mg^{2+}$ |
| Carbonates and sulfates | Mineral dissolution | $Ca^{2+}$ |
| **Iron, manganese, zinc, copper** | | |
| Precipitated as hydroxyoxides | Mineral dissolution | Cations and soluble chelates |
| Adsorbed on Fe, Al, and Mn oxides | Desorption | |
| Bound on humus (chelated) | Dissociation | |
| Exchangeable | Cation exchange | |
| **Boron, molybdenum** | | |
| Adsorbed on Fe and Al oxides and other clay minerals | Desorption | $H_3BO_3$ $MoO_4^{2-}$ |

sian silicates, and micas (see Chapter 2). This source of K, Ca, and Mg is most significant in soils that are still rich in weatherable minerals, if the mineral particles are small and the soil is wet, warm, and somewhat acidic. In most soils, the dissolution of hydroxyoxides of the clay fraction supplies most of the soluble Fe, Mn, and Al. Some soils also have solid calcium sulfate or metal sulfides, which release S.

Specific adsorption and desorption processes control the levels of dissolved P, Mo, B, Cu, and Zn in most soils. These processes resemble precipitation or dissolution from surfaces. They include the anion sorption process that retains phosphate on Al and Fe oxyhydroxides (see Chapter 3) and holds other oxygen-containing anions: molybdate, arsenate, borate, and sometimes sulfate. The concentration of the ions in solution increases with increasing amount on the solid phase: The more there is, the more loosely it is held. Ions of some kinds are held more tightly than others. For example, phosphate is so tightly held that exceedingly little is in solution even when large amounts are on the solid surfaces. This is especially characteristic of soils rich in colloidal Fe and Al oxyhydroxide minerals, such as the Oxisols and Ultisols of warm, wet climates, and soils with poorly crystallized minerals formed from **volcanic ash.**

Surface **chelation** on humus can bind Fe, Cu, Zn, and Mn tightly and specifically. These polyvalent metal cations bind to closely spaced negative groups on the humus; that is, they form a **chelate,** or are chelated. Like other binding processes, chelation is reversible; ions escape whenever the removal of free ions or the microbial destruction of the chelating agent disturbs the equilibrium. Chelation by humus occasionally causes a deficiency of Cu or Zn in alkaline soils high in humus. Chelation by humus immobilizes cations because the humus is itself immobile. In contrast, chelation with small, soluble, organic molecules can increase the mobility of cationic nutrients.

**pH relationships**   The retention and release of nutrients in soils show several general trends with soil pH. These generalizations have some value as a guide to possible problems in diagnosis, but they are full of exceptions. One reason is that many soil properties change as the soil pH changes, and much depends on how the pH is changed. In particular, the natural variation in pH due to soil-forming processes has different consequences than does pH variation produced artificially by liming or acid application. (This issue is discussed further in Supplement 9–3 and Chapter 11.)

## 9.3   MOVEMENT OF IONS TO ROOTS

### 9.3.1   Significance and Processes

The plant's mineral nutrition depends, as we have seen, on the physiological processes of uptake and translocation and on the chemical processes of nutrient release from the soil's solids. No less important are the intervening physical processes that move nutrients to the roots, especially as the same processes are also involved in leaching and in the movement of nutrients into the soil from fertilizer particles, weathering minerals, or decaying plant residues. Slow movement to roots is the chief reason that plants are often deficient in phosphate and other so-called immobile nutrients.

Nutrient ions, like other solutes, move in soil by a combination of diffusion and mass flow, processes discussed in Chapters 3 and 4. Mass flow occurs when moving soil water carries solutes with it. Mass flow is the principal mode of leaching, and it also helps carry nutrients to the roots of transpiring plants. Diffusion is the movement of individual ions or molecules, even without water movement, when differences in ion concentration develop in the soil solution. Such differences in concentration commonly result from uptake at the root surface (Figure 9–8).

### 9.3.2   Nutrient Transport Rates

The rate of mass flow of a solute is proportional to the volume rate of water flow and the concentration of the particular solute in the soil water. Mass flow is most effective when the plant is transpiring rapidly and ions are present in the soil solution at high concentration relative to the plant's uptake rate. Normally, mass flow alone is sufficient to meet the plant's needs for certain nutrients (e.g., Ca and Mg). Sometimes it even supplies an ion faster than the root absorbs it, so that the ion accumulates around the root (Figure 9–8A). In saline soils this accumulation at the root doubtless aggravates the effects of salts on plants. In alkaline soils high in Ca, the accumulation of $Ca^{2+}$ and $HCO_3^-$ can lead to the precipitation of

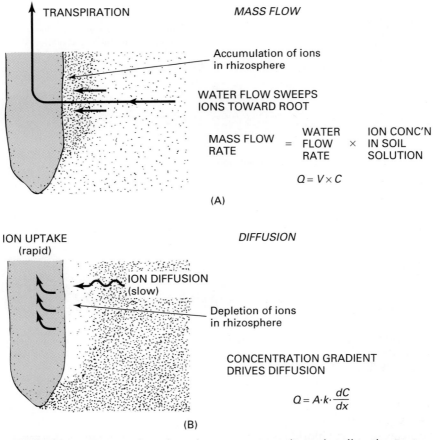

**FIGURE 9–8** Two modes of nutrient movement through soil to the roots: (A) mass flow and (B) diffusion. In nature, both modes operate together.

$CaCO_3$, encasing the root in lime. More commonly, mass flow falls short of uptake, so that the concentration in the rhizosphere drops (Figure 9–8B). Now, with a lower concentration near the root than farther out, the ion diffuses inward toward the root. The diffusion complements the mass flow, so that they go together, superimposed.

The rate of diffusion of a solute depends on the magnitude of the **concentration gradient** (driving force) and on the ion's **diffusion coefficient** ($k$) (ease of movement) in the soil (Figure 9–8B). (Chapter 4 contains a general description of diffusion.) High availability in the soil and vigorous uptake by the plant together establish a large concentration gradient to drive diffusion. All solutes will diffuse less freely if the soil is cold and dry and if they react with the soil's solid surfaces. Phosphate, zinc, and copper, being strongly sorbed, diffuse more slowly through

the soil solution than do nitrate, chloride, calcium, and potassium.

### 9.3.3 Nutrient Mobility in Soil

Nutrient mobility in soil denotes the ease with which the nutrient can move to supply the plant's demand. It is an important aspect of availability; in fact, if the total quantity in the soil is sufficient, mobility and availability become almost the same thing. Note that mobility here depends partly on the magnitude of demand for the nutrient. Thus, a nutrient can show mobile behavior because much of it is free, dissolved in the soil solution (the normal case for nitrate, sulfate, calcium, potassium, and magnesium). Or it can show mobile behavior, even though little is in solution, simply because the plant takes up little (the normal case

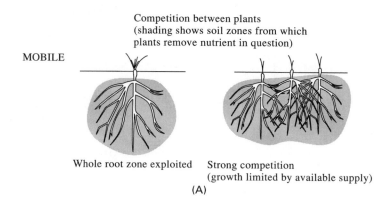

MOBILE

Competition between plants
(shading shows soil zones from which
plants remove nutrient in question)

Whole root zone exploited          Strong competition
(growth limited by available supply)

(A)

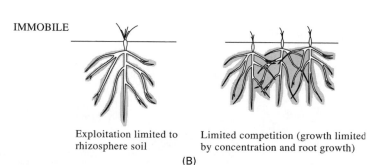

IMMOBILE

Exploitation limited to          Limited competition (growth limited
rhizosphere soil                 by concentration and root growth)

(B)

**FIGURE 9–9** Plant competition is greater for mobile nutrients (A) than for immobile nutrients (B).

for boric acid, manganese, sodium, chloride, or toxic ions).

According to one mobility concept expounded by R. H. Bray, there are both mobile and immobile nutrients. Mobility influences plant competition for nutrients. A crop can deplete a mobile nutrient from the whole root zone (Figure 9–9A), so that plants grown together compete strongly for the supply, and the amount required as fertilizer increases in proportion to the crop yield. By contrast, plants deplete an immobile nutrient only from the immediate vicinity of individual roots, so that there is little competition among plants, and the amount required as fertilizer depends less on the crop yield than on the crop's root growth and the soil's capacity to immobilize the nutrient (Figure 9–9B).

Viewed in relation to transport processes, an ion shows mobile behavior if mass flow alone can supply the plant's needs and immobile behavior if rhizosphere depletion makes diffusion an important supply mechanism. From this viewpoint, every nutrient ion species becomes immobile when it is deficient, and most nutrients are mobile when they are adequately supplied. Some nutrients (e.g., P, Fe, Zn, and Mo) show immobile behavior even under normal, nondeficient conditions. The amounts of phosphate and Fe needed by plants are similar to the amounts of magnesium and sulfur needed. Yet soil solutions normally contain 100 to 1,000 times less phosphate than Mg or sulfate, and even less Fe (see Chapter 3, Figure 3–8). Very little Fe might get to plant roots without mobilization by soluble complexing agents produced by plants and some microbes.

The plant's demand for immobile nutrients is unlikely to be satisfied unless the plant shortens the travel distance by extending its root system.

## 9.4  ROOT GROWTH AND DEVELOPMENT

### 9.4.1  Characteristics and Functions of Root Systems

Root systems differ in total length, depth of penetration, lateral spread, and **root density** (amount of

ROOT SYSTEMS

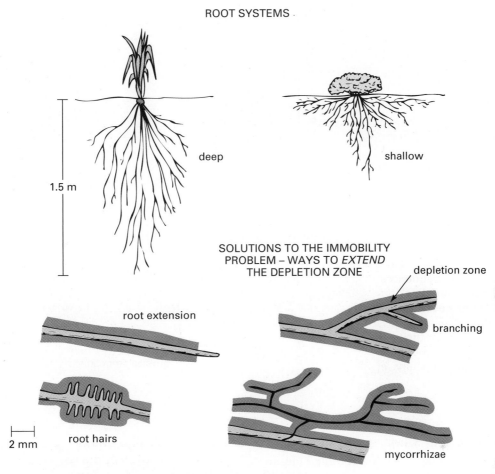

deep

shallow

1.5 m

SOLUTIONS TO THE IMMOBILITY
PROBLEM – WAYS TO *EXTEND*
THE DEPLETION ZONE

depletion zone

root extension

branching

root hairs

2 mm

mycorrhizae

**FIGURE 9–10** Forms of roots and root systems.

roots per unit soil volume) at each depth through the pedon. Roots also differ in form: thickness, degree of branching, and development of root hairs and mycorrhizae (Figure 9–10). Finally, roots differ in growth rate and persistence. All these characteristics influence root functions. Soil properties, plant, and climate determine **root distribution** in the soil.

Roots absorb immobile nutrients, mobile nutrients, and water. They translocate these materials to the shoot. Roots also anchor the plant, penetrate and spread through the soil, and sometimes double as storage or propagative organs. Their production and maintenance must not make excessive energy demands on the plant. These functions and requirements demand different characteristics, and compromises are necessary.

To get immobile nutrients efficiently, roots must rapidly extend the rhizosphere **depletion zone** with minimal cost of energy and material. Thus length and slenderness are desirable, but if the roots are too thin they cannot penetrate the soil (Figure 9–11). As the roots extend, they require more vascular tissue to deliver the nutrients to the shoot. *Multiple branching*, the formation of laterals, enables new roots in fresh soil to join the existing conductive tissue. Root hairs widen the depletion zone, and mycorrhizae efficiently extend it. Most immobile nutrients are more available in surface horizons because the surface horizons contain the most densely spaced, fine, actively growing roots, with ample root hairs or mycorrhizae.

But roots should also penetrate deep to rescue mobile nutrients and tap subsoil water—tasks for

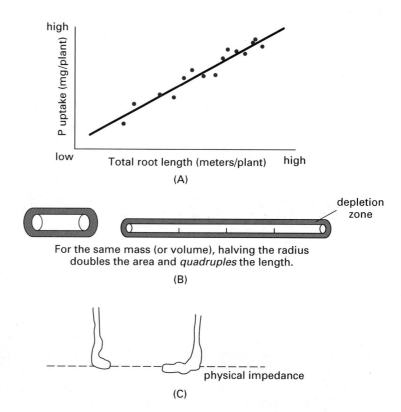

**FIGURE 9–11** Root length and root thickness: some consequences. (A) Acquisition of "immobile" nutrients by a plant relates closely to total root length. (B) Slenderness gives the most length for the least cost to the plant, but (C) too much slenderness causes difficulty in penetrating hard soil.

which fineness, density, hairiness, or mycorrhizae provide no great advantage. Shallow rooting helps in acquiring immobile nutrients but is useless during drought.

Most root systems are compromises among the different requirements. Normally they include both shallow, fine-branched roots and deep, coarse roots. Nearly all plant species form mycorrhizae. Those that do not (e.g., members of the crucifer family) usually develop many root hairs and fine lateral roots.

### 9.4.2   Factors Affecting Root Development

Root development depends on the plant **genotype** and on soil physical, chemical, and microbiological factors.

**Plant genotype**   Some attributes of individual roots and root systems are characteristics of the species (see Figure 9–10). Within species, varieties sometimes differ in their degree of root development and form.

**Soil physical conditions**   Root growth in soil is strongly influenced by soil structure, texture, and water content, because all three affect the ease with which the roots can extend. If the soil has granular structure or sandy texture, the roots will grow through the large pores, pushing aside sand grains or structural granules. But if the soil has a clay or clay loam texture and less favorable structure, the roots will be confined to cracks unless they can penetrate structural units, and so the soil's penetration resistance becomes important. Soil penetrability depends on water content and bulk density (Figure 9–12). Dense soils with few large pores are difficult to penetrate, especially when they become dry and hard. Increasing the water content softens the soil and helps penetration, except near saturation, when poor aeration begins to inhibit root growth.

**Soil chemical and biochemical factors**   Oxygen deficiency, Ca deficiency, and Al excess directly stunt root growth. Drought and most nutrient deficiencies limit root growth indirectly, by slowing the growth of the shoot that supplies energy to the roots. In soil,

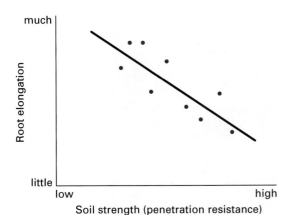

**FIGURE 9–12**   Soil physical effects on root growth. Both compaction and dryness hinder elongation because they make the soil harder to penetrate.

roots are also inhibited by a variety of poorly understood organic inhibitors. Some plants release substances that inhibit the growth of roots of other species. Microbial root toxins may be important, especially under limited aeration; ethylene in excess may be one of these toxins. Most important, root disease organisms and parasites are widespread among the soil bacteria, fungi, and animals.

# 9.5   NUTRIENT AVAILABILITY AND DEFICIENCY

The readiness with which a soil supplies a nutrient to plants depends on how much of the nutrient the soil contains, other soil conditions, the type of plant, and the environment. It is good to recognize and confront this complexity. There is little point in attempting simple strict quantitative definitions of nutrient availability.

The quantity of solid reserves of any nutrient is one aspect of availability, and the release rate is another aspect if the nutrient is released slowly (e.g., N or S from humus). A third aspect, the concentration maintained in solution, indicates the nutrient's freedom to move in the soil and is said to reflect the "intensity" aspect of availability. The quantity of readily mobilized reserves (such as exchangeable cations, adsorbed P, or readily decomposable organic matter) relates positively to intensity measures and release

rates. The retention curves for phosphate shown in Figure 9–7D are examples of quantity–intensity relationships.

Increasing the amount of nutrient in the soil increases the concentration in solution, the speed of replenishment of the nutrient in solution, and the nutrient's mobility in the soil. In short, it increases the availability. This is the idea behind fertilization. A fertilizer is a material—natural or synthetic, inorganic or organic—that provides useful quantities of a plant nutrient in forms that can become soluble in soil. Most of the soluble nutrient that a fertilizer releases becomes immobilized in soil (adsorbed, precipitated, incorporated into humus), building up reserves and thus improving the nutrient's availability.

Availability depends not only on the amount and nature of reserve nutrient but also on other factors that vary from soil to soil. The following are three examples:

1. The rate of N mineralization (release of ammonium and nitrate from organic matter) depends not only on the quantity and kind of organic matter but also on the soil's water content and temperature.
2. The concentration of soluble phosphate depends not only on the amount of adsorbed P but also on the soil's P-sorption capacity. The more capacity the soil has, the more tightly it will hold a given quantity of P, and the more P must be added or removed to cause a significant change in the P supply to plants. A soil with a high capacity is better "buffered" with respect to P than is a soil with less capacity.
3. The availability of K increases with increasing exchangeable K, but it also depends on the cation exchange capacity and on other sources of K. If the cation exchange capacity is large, a given quantity of exchangeable K will support a lower and less variable concentration of K in solution. But if the soil has other useful sources of K, such as interlayer K in 2:1 clay lattices or feldspar or mica in the silt fraction, then the exchangeable K will diminish in importance.

Nutrient deficiency is not some arbitrary, analytically measured, low level of a nutrient. Rather, nutrient deficiency is the condition in which a plant is getting a nutrient so slowly as to limit growth or impair a significant attribute of quality. How can you tell

(A)

(B)

**FIGURE 9–13**  Plant responses to nutrient additions. (A) These four field plots show the effects of different levels of phosphate added to chrysanthemums growing on the island of Maui, Hawaii. Plants growing on the plot in the upper left have the least P, and those on the upper right have the most. Note the differences in height and leaf area in the two plots. (B) Alfalfa response to sulfur addition in a farmer's test strip. The dark band starting on the left edge of the photo and extending to the bottom right is a strip of alfalfa that is healthier because of the addition of S. (Photos: W. E. Martin.)

whether a nutrient is deficient? Test whether the plant improves when you increase its supply of the nutrient. This idea should not be overlooked just because it is obvious.

How does one test for nutrient response? First, establish test plots or containers, add the nutrient in question to some of them, and leave others untreated as controls. Then grow the relevant test plant, and measure its growth, yield, or whatever feature is important: height, wood production, juiciness, protein content, sugar content, vitamin content, oil content, drug content or potency, size or color of flowers or fruit, stem length, weight gain of grazing animals, or whatever. The soil is deficient in the nutrient if the nutrient addition causes a significant improvement, in comparison with the control plants. Figure 9–13 indicates some possible responses, and Chapter 10 develops these ideas.

# SOIL AND ENVIRONMENT
## Nutrients, Fertilizers, and Sustainable Land Use

Harvesting of agricultural and forest products entails the removal of plant nutrients. This removal is desirable because it is the source of human nutrition. Unless the soil's pool of each nutrient is maintained affordably, no agriculture, horticulture, or forestry is sustainable. Inputs from mineral weathering, irrigation water, and sediment deposition normally supply most of the plant nutrients sufficiently. But nitrogen and, in the long run, potassium and phosphate almost always have to be added (plus sulfur, calcium, and alkalies in areas of high rainfall). Such additions, as fertilizers and soil amendments, have environmental costs to be recognized sooner or later.

Fertilizer manufacture and use consume petroleum or gas for collection, preparation, and delivery. Fertilizer represents 30 to 50 percent of the agricultural energy expenditure in most North American and European countries. For perspective, this is 1 to 2 percent of these societies' total use of coal, gas, and oil.

Plant nutrients are transferred, not created or destroyed. Fertilizer use depletes the fertility of the land that is the source of organic fertilizers or the geological deposits that are the source of most other fertilizers. (Many countries or regions lack useful high-quality sources of P or K.) Plant nutrients discarded in landfills, dumps, and sewage may become pollutants.

Catastrophic loss of plant nutrients can irreversibly change a sensitive ecosystem. For example, frequent cutting or burning in forests and woodlands on shallow, low-nutrient soils allows so much nutrient loss by erosion and leaching during rainstorms that trees fail to regenerate. Likewise, grassland and semidesert soils exposed by drought, fire, or overgrazing can lose their fertile topsoil to wind erosion, becoming incapable of supporting plant cover.

### Food Chains

Plant nutrients and toxins are important components of food chains that begin with plants and include animals—wild, domestic, and human. Even in the early twenty-first century, human populations suffer deficiencies of fluoride or iodide, and extensive protein malnutrition disorders relate to shortages of nitrogen and/or sulfur. Herbivorous animals, with less varied diets than humans, suffer a wider range of soil-related deficiencies and toxicities, including deficiencies of phosphate, calcium, sodium, cobalt, copper, and selenium and toxicities of nitrate, cadmium, arsenic, molybdenum, copper, and selenium.

Land disposal of cumulative nondegrading toxins (lead, cadmium, mercury, manganese, nickel, chromium, arsenate, and so on) is especially hazardous. Nevertheless, the spraying of lead arsenate as an insecticide on apple trees continued well into the twentieth century, even rendering some soils unproductive. Phosphate sources loaded with cadmium, uranium, or fluoride are no longer used to make fertilizer in most countries. Less hesitancy seems to apply to urban sewage sludges, though these may contain toxic metals derived from industries and plumbing. The metals are further concentrated when sludge is composted to make organic soil amendments.

# 9.6 SUMMARY

Sixteen elements—C, O, H, N, K, Mg, Ca, P, S, Fe, Cl, Mn, Zn, B, Cu, and Mo—are currently known to be essential to plants. Essential means the plant cannot complete its life cycle if the element is missing. Nutrients are not simply absorbed by plants, they are assimilated, used, and transformed into essential plant components. Plants have different strategies to obtain nutrients from soils that retain nutrients in many ways. Nutrient uptake depends on roots and nutrient availability.

Availability, viewed chemically, is a soil's ability to maintain high concentrations of a nutrient in solution. But viewed biologically, availability is less simple. The plant's nutrition depends not only on chemical availability but also on nutrient movement and plant characteristics. From this wider point of view, nutrient availability is complex. Chemical soil tests can help assess it in some cases but fail in others. Someday we might understand the overall process well enough to assemble the chemical, physical, and biological information into predictive computer simulations. Meanwhile, there is only one truly dependable criterion of nutrient availability: plant response.

## QUESTIONS

1. Describe the main forms and functions of N, C, P, S, and K in plants. Which one occurs and functions in the plant mainly as the free ion?

2. Why must plants contain large amounts of N? Why must they contain almost 10 times more C than N?

3. Aging root tissue, more than a few centimeters behind the root tip, loses its capacity to absorb nutrients. Would the plant be better off if the old roots could still actively absorb? Or would events in the soil make this a useless property?

4. A cell's outer membrane does not freely transmit ions. Why is this property necessary?

5. How does nutrient uptake by roots influence the pH in the soil next to the root? What is the difference in this respect between a plant assimilating ammonium and a plant assimilating nitrate? What effects might the pH change exert on plant nutrition, soil chemistry, and soil microbes?

6. What is a fertilizer? Think of 10 supermarket items that could be useful fertilizers (for valuable plants).

7. Name the dominant forms of N, K, P, and S in soil solutions and in soil solid reserves.

8. Increasing the level of soil organic matter may immobilize some nutrients in soil. Which ones? How?

9. In most soils, deep rooting is more beneficial to a plant's N nutrition than to its P nutrition. Why is this so? Think of an exception.

10. Both drought and cold sometimes aggravate P deficiency, but they tend to make the effects of N deficiency less severe. Explain this difference.

11. Define the following terms: *nutrient essentiality, nutrient deficiency, nutrient mobility, nutrient availability, soil fertility,* and *soil productivity.*

12. The availability of a nutrient usually increases as more is added to the soil, but availability also depends on other factors besides the total amount. Think of some factors that you would expect to influence the availability of a nutrient other than N, K, or P.

## FURTHER READING

Barber, S. A. 1995. *Soil nutrient bioavailability: A mechanistic approach.* New York: Wiley.

Marschner, H. 1995. *Mineral nutrition of higher plants.* London: Academic Press.

McBride, M. P. 1994. *Environmental chemistry of soils.* New York: Oxford University Press.

Mengel, K., and E. A. Kirkby. 1987. *Principles of plant nutrition.* 3d ed. Bern, Switzerland: International Potash Institute.

Nye, P. H., and P. B. Tinker. 1977. *Solute movement in the soil-root system.* Berkeley and Los Angeles: University of California Press.

O'Toole, J. C., and W. L. Bland. 1987. Genotypic variation in crop plant root systems. *Advances in Agronomy* 41:91–145.

Russell, R. S. 1977. *Plant root systems.* Maidenhead, England: McGraw-Hill.

Tan, K. H. 1993. *Principles of soil chemistry.* New York: Dekker.

Tisdale, S. L., W. L. Nelson, and J. D. Beaton. 1993. *Soil fertility and fertilizers.* 5th ed. New York: Macmillan.

## SUPPLEMENT 9–1    Nutrient Elements in Plants

### Roles of Nutrient Elements

The macronutrients C, O, and H occur in all major organic constituents of plants: saccharides, proteins, lipids, and nucleotides. (If these chemical terms are unfamiliar, Supplement 8–1 might help.) Nearly all the N and S in plants is contained in proteins. Most proteins in plants function as **enzymes,** catalysts that direct and control the cell's chemical reactions. Proteins are large molecules of changeable shape. They catalyze specific reactions because they can change their configuration so as to alter substrate molecules. Common cations—Ca, Mg, and K—balance negative charges on protein molecules much as they do on soil colloids. The active, or prosthetic, groups of an enzyme often contain atoms of S, Fe, Zn, Cu, Mn, or Mo.

Complexes of nutrient metals take part in many biochemical reactions. Thus, **chlorophylls,** the green pigments that trap light energy to drive photosynthesis in plants, are Mg complexes; **cytochromes,** which transfer electrons during photosynthesis and respiration, are Fe complexes; and **adenosine triphosphate (ATP),** which helps transfer energy during biochemical processes in general, functions in the form of a Mg complex.

Phosphate, along with C, H, O, and N, is part of all the nucleotides. Nucleotides include ATP as well as deoxyribonucleic acid (DNA), which encodes genetic information, and ribonucleic acid (RNA), which transcribes genetic information as enzymes are being made.

Several nutrients are essential to the structure of the membranes that surround the cell and enclose its vacuole, nucleus, mitochondria, and chloroplasts. All these membranes contain lipids and phospholipids (C, H, O, and P) as well as protein (N, S), and they are stabilized by the presence of Ca.

Finally, nutrients occur in the cell to some extent as free ions and simple metal complexes that are readily moved and may have important functions themselves. For instance, K and Cl help control the water potential in cells, and the synthesis of ATP depends on an abundance of phosphate ions.

## Ion Uptake

Uptake through the membrane may require "carriers" (transfer enzymes or ion pumps) associated with the protein molecules embedded in the membrane. Some of these carriers are specific, or selective; they preferentially absorb certain kinds of ions and exclude others. The rate of uptake typically increases with the concentration of the ion in the external solution, as shown in Figure 9–3. Here there is a high-affinity mechanism that becomes fully saturated (reaches its maximum rate) at low concentrations. Such mechanisms are specific; they discriminate between ions as closely related as Na and K, for instance. They ensure absorption of an ion even when it is scarce and similar ions are abundant. Less specific mechanisms absorb ions at higher concentrations (Figure 9–3); these mechanisms allow similar ions to compete. Nonspecific uptake mechanisms that work well at high concentrations are valuable under saline or drought conditions, when the plant needs a rapid intake of solutes to keep internal water potentials low so that water intake can continue.

Plants can accumulate nutrient elements. *Accumulation* means the concentration of the nutrient element becomes higher inside the plant cell than in the soil solution outside. The cell membrane stops the ions from diffusing out. The work (energy expenditure) of accumulation against a concentration gradient takes either of two general forms (Figure 9–14A, and 9–14B): (1) incorporation of the element into new forms inside the cell, thereby lowering the internal concentration of free ions, and (2) chemical "pumping" of the ions, with metabolic energy applied to a chemical carrier system in the membrane.

Once absorbed by cells, nutrient ions of some kinds are immobilized by complex formation, adsorption, or assimilation into complex cell substances. These processes keep the ionic form of the nutrient at low levels in the cell. By contrast, other nutrient elements remain free as ions at high concentration in the cell. For example, potassium ion concentrations are typically several times higher in cells than in soil solutions, and phosphate ion concentrations may be one thousand times higher.

## SUPPLEMENT 9–2 Nutrient Reserves and Release in Soil

This supplement to Section 9.2 expands on the information in Tables 9–3 and 9–4. The text subdivides the subject according to processes. This supplement subdivides it into groups of nutrients that behave alike.

### Nitrogen and Sulfur

Soil organic N and S is the main solid reserve, and decay by microbes is the main release process (discussed in Chapter 8). The release rate depends on the following:

- Temperature and soil water; both factors are difficult to manage, but the optimal conditions resemble those for plant growth
- The amount of organic matter, its N or S content, and its decomposability

Humus, with about 5 percent N and 0.5 percent S, decays slowly, at an average relative rate of about 3 percent per year. Other kinds of organic matter, such as plant residues, decay faster, but some of them are deficient in

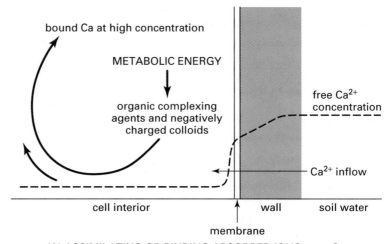

FIGURE 9–14 Accumulation of nutrient elements by cells: two ways of applying energy to develop high internal concentrations of the element. (A) Assimilation or binding of absorbed ions (example: calcium). (B) Ion pumping at the cell membrane (example: potassium).

(A) ASSIMILATING OR BINDING ABSORBED IONS, e.g., Ca

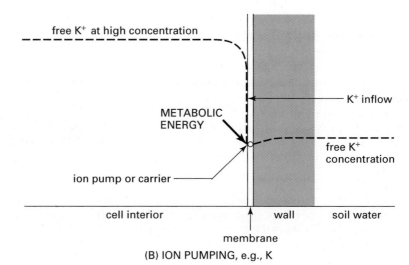

(B) ION PUMPING, e.g., K

N or S. The decay of N- or S-deficient organic matter temporarily immobilizes the nutrients because the microbes now absorb ammonium, nitrate, and sulfate to satisfy their own requirements.

Exchangeable ammonium is an important N source in soils recently fertilized with ammonium, in flooded soils, and in acidic forest or grassland soils. In other cases, ammonium oxidizes rapidly to nitrate, and exchangeable ammonium is at low or fluctuating levels.

Other reserves of N and S are important only in special cases:

1. Adsorbed sulfate in highly weathered soils whose reactive oxide minerals may retain sulfate much as they retain phosphate, though more weakly

2. Ammonium between the lattice layers of expanding 2:1 clay minerals of weakly weathered soils rich in these clays
3. Sulfide precipitated with iron and zinc in intermittently flooded soils
4. Calcium sulfate in soils so arid that this salt does not leach out

## Phosphorus

Inorganic adsorbed or surface-precipitated phosphate is the main reserve source in most soils. In alkaline calcic soils, the phosphate is associated with calcium carbonate or apatite-like precipitates, and its solubility will increase if the pH drops toward neutrality. In neutral or acidic soils, the phosphate is associated with Fe and Al hydroxyoxides and the edges of other clay minerals, mainly coordinated with the Fe and Al in place of hydroxyl or water groups (anion sorption or ligand exchange); this sorption process occurs over a wide pH range. Adsorption can become irreversible if the phosphate on the surface becomes occluded—that is, covered by further precipitation of the mineral. Otherwise, these adsorption processes are reversible, but even so, they hold down the phosphate in the soil solution to very low concentrations (about 1 $\mu$mol/L). Virtually all the P added as fertilizer quickly becomes adsorbed. (Also, many kinds of precipitates of P with Ca, Al, K, and Fe appear temporarily during reactions between fertilizer granules and adjacent soil.)

Organic phosphate is the main P reserve in Histosols, forest soil A horizons, and other soils rich in humus and low in minerals with the capacity to adsorb phosphate. Release depends on microbial decay, as for N and S. Some of these soils have significant levels of organic P in the soil solution.

## Potassium, Calcium, and Magnesium

Exchangeable cations are the main reserve of Ca and Mg in most soils and of K in many soils. Most of the K added as fertilizer appears almost immediately as exchangeable $K^+$. As plants remove nutrient ions, the resulting changes in the soil solution are partly counteracted by cation exchange reactions (see Chapter 3) involving the major cations ($Ca^{2+}$, $Mg^{2+}$, $K^+$, $Na^+$). The net effect is that exchange releases into solution those ions being taken up in the largest quantity. (*Total* exchangeable Ca, Mg, K, and Na is not reduced unless the plant is acidifying the soil, thereby reducing the cation exchange capacity and/or causing the accumulation of exchangeable Al.)

Other sources of K, Ca, and Mg are significant in soils that have not been highly weathered:

1. K held between the crystal layers of vermiculitic clays. In soils rich in these clay minerals, much of the K added as fertilizer becomes fixed by entrapment within the clay crystals, as these expand when wet and then dry and collapse. The reverse process is slow.
2. Ca in calcium carbonate and calcium sulfate. The calcium salts are abundant in arid, usually alkaline, soils.
3. K, Ca, and Mg in weatherable primary minerals of the silt fraction. Weatherable primary minerals are common in neutral or alkaline soils de-

veloped with little weathering from alluvial sediments and rocks high in feldspars and carbonates.

## Iron, Manganese, Zinc, and Copper

Iron and manganese oxides and hydroxyoxides are important reserves of Fe and Mn, as well as Zn, Cu, Co, and other metals that coprecipitate along with the Fe and Mn oxides and also attach to their surfaces. Release involves the desorption and dissolution of the metal cations. It is favored by acidity and the presence of soluble chelating agents that combine with the cations. The specialized Fe-complexing agents released by some microbes and plants at the onset of Fe deficiency are called *siderophores*. Reducing conditions also contribute to the dissolution of the oxides; Fe oxides are partly reduced in flooded soils, and Mn oxides are reduced even without flooding if organic matter supply and temperature favor microbial activity.

Humus strongly binds these micronutrient metals to its negatively charged sites (surface chelation), providing another major reserve in topsoils. In alkaline soils high in organic matter, this binding can cause deficiencies of Zn or Cu. As with the oxides, the release of the metals is favored by acidity and by agents that form readily soluble chelates.

Not surprisingly, Mn toxicity occurs in wet, warm, acidic soils high in organic matter, and deficiencies of Fe, Mn, and Zn are common in naturally alkaline soils and in acidic soils that have been excessively limed (see Chapter 11).

## Boron and Molybdenum

Sorption on hydroxyoxides and other clay colloids holds borate and molybdate. Raising the soil pH alleviates Mo deficiency and aggravates B deficiency.

## SUPPLEMENT 9–3   Nutrient Availability and Soil pH

In nature, soils are acidified by processes that also result in severe leaching (see Chapter 11). Because of the leaching, they are more likely than are neutral soils to be deficient in K, Ca, Mg, or S. If the acidic soil has little clay or humus to retain nutrients, leaching may even have produced a deficiency of such tightly held nutrients as Zn, Cu, or B. Often, acidic soils are also highly weathered soils, rich in the reactive oxide clays that immobilize P and Mo, making deficiencies of these nutrients both likely to exist and difficult to correct. If low fertility has meant little vegetation, then little humus will have accumulated to provide reserves of N and S. None of these difficulties of *naturally* acidic soils needs to appear when soil is artificially acidified. However, Al toxicity and Mn toxicity, caused by the acid dissolution of Al and Mn oxides, are likely consequences of both natural and artificial acidification (see Chapter 11).

Adding lime to an acidic soil has little immediate effect on the humus levels and clay types in the soil. It raises the pH, the cation exchange capacity, and the exchangeable Ca, and it lowers the soluble Al and Mn, correcting these toxicities if they exist. The effects of liming on P availability are complex and permit no useful generalization. Liming acidic soils lowers the availability of P in some situations and appears to improve it in others. Naturally neutral soils seem somewhat less likely to be P deficient than do soils that are naturally acidic or alkaline.

In naturally alkaline soils, S, B, Mo, and the major cations are rarely deficient. Many alkaline soils are in arid climates. If so, they may have naturally high levels of nitrate in solution but little reserve N because of poor plant growth and the resulting lack of humus. They are also likely to be salty and too high in Na (see Chapter 11). These are problems of naturally alkaline soils not likely to be induced by artificial alkalinization (overliming). However, in both overlimed and naturally alkaline soils, Fe, Zn, and Mn may become immobilized by precipitation, chelation with humus, or association with oxide clay minerals.

(A)

(B)

(C)

(D)

Nutritional symptoms in plants. ***General chlorosis*** (yellowing). (A) Sulfur deficiency on an unfertilized test strip in alfalfa. (B) Sulfur deficiency in tomato, with a normal plant in background. ***Interveinal chlorosis***. (C) Zinc deficiency in tomato, with normal leaves in background. (D) Zinc or Fe deficiency in maize (corn).

**Purpling.** (E) Phosphorus deficiency in tomato, with normal plant in background. (F) Phosphorus deficiency in maize. **Local necrosis** (dead patches). (G) Potassium deficiency in clover, with normal leaf on left. (H) Excess ammonium nitrate application to bean. Other symptoms include stunting and dieback; see black-and-white photographs in text.

# 10

# MANAGING PLANT NUTRIENTS

## OVERVIEW

The most commonly deficient nutrients are nitrogen (N) and phosphorus (P), followed by sulfur (S), potassium (K), and zinc (Zn). The quantities needed depend, in the short term, on fixation or immobilization and, in the long term, on the quantities of nutrient harvested and lost during nutrient cycling. Usually, harvesting removes more N and K than other nutrients, particularly if the harvest includes much leaf. Fertilizer amounts, though substantial, are typically a tenth the amount of amendments needed to treat acidity and sodium-salinity conditions.

Most fertilizer elements are mined from nonrenewable sources, but N is obtained from the atmosphere. Energy is a large component of fertilizer cost, especially for the industrial fixation of atmospheric N. For any purpose the appropriate fertilizer is the one that supplies plants with the needed (i.e., the deficient) nutrients cheaply, conveniently, and with acceptable side effects. A diagnosis of fertilizer requirements answers two questions: "What is needed?" and "How much?" Soil tests, plant analyses, symptoms, and experience can help, but sound diagnosis is based on plant response trials. Nutrient response typically diminishes as sufficiency is approached. The best level of application, economically and *environmentally, is less than full sufficiency. Fertilizers are applied by broadcast spreading, dissolution in irrigation water, "drilling" and injecting into soil, or spraying onto foliage. Placement and timing can affect efficiency of use.*

*Nutrient efficiency can also be improved by nutrient conservation—that is, management to reduce erosion, leaching, escape of gases, and undesirable harvest. Crop response and nutrient use efficiency commonly increase when general culture conditions are improved.*

Nutrient deficiencies are major constraints to commercial plant production, along with drought, disease, pests, and politics. They are among the more readily managed constraints. In intensive agriculture, the use of fertilizers is responsible for much of the productivity and is one of the biggest inputs of energy, materials, and money.

The manufacture and use of fertilizers do not create nutrients; they merely recycle them or move them from one place to another. For instance, animal manures, green manures, and composts contain nutrients that plants took from soil, and most inorganic fertilizers contain nutrients mined from some enriched source. Essentially, we are extracting nonrenewable resources. Even where the resource is vast—such as the atmosphere—its processing, transport, and application still consume energy. Inefficient fertilization wastes resources and can pollute other ecosystems with excessive nutrients.

Nitrogen is the nutrient added as fertilizer most extensively and in the greatest amount (Table 10–1). Nitrogen deficiency is normal; few soils can supply

enough for sustained cropping without supplementation. Phosphorus deficiency occurs on perhaps 70 percent of agricultural soils and perhaps most other soils. Deficiencies of K, S, and Zn also are common. Deficiencies of Fe, B, Mo, Mg, Cu, and Mn are less common, and deficiencies of Cl, Co, and Na are rare. Calcium deficiency is a special case. Simple Ca deficiency is unusual, but shortage of Ca aggravates problems associated with soil acidity, sodicity, and salinity (see Chapter 11).

The quantities of material required for fertilization are smaller than for the correction of acidity and salinity problems (see Chapter 11); nevertheless they are substantial. The amounts required are largest for N and K and vary depending on the crop, the portion of the plant harvested, and the manner of harvest (Table 10–2).

## 10.1 FERTILIZERS

### 10.1.1 Common Fertilizer Materials

Tables 10–3 and 10–4 list some fertilizer materials and their sources. Manufactured (synthetic) fertilizers have become popular during the past hundred years, but natural organic and inorganic materials are still used. (We use the terms *organic* and *inorganic* in a restricted chemical sense: Organic compounds have C–C bonds, but inorganic compounds do not.)

**Nitrogen**   The N in soils, plants, and us comes from the fixation of atmospheric $N_2$. Worldwide, biological nitrogen fixation (discussed in Chapter 7) is estimated

## TABLE 10–1
Quantities of Nutrients Removed by Crops

| Crop | Yield Mg/ha[a] | Amounts of nutrients (kg/ha per crop) | | | | | | |
|---|---|---|---|---|---|---|---|---|
| | | N | P | K | Ca | Mg | S | Micronutrients |
| Alfalfa hay | 20 | 500 | 45 | 350 | 250 | 50 | 50 | 2–7 |
| Maize grain | 15 | 200 | 40 | 40 | 20 | 30 | 20 | 0.2–0.4 |
| Rice grain | 6 | 80 | 15 | 15 | 5 | 7 | 8 | 0.1 |
| Wheat grain | 6 | 120 | 30 | 30 | 25 | 15 | 15 | 0.2 |
| Tomatoes, fresh | 20 | 150 | 25 | 200 | 10 | 15 | 20 | 0.2–0.4 |
| Cotton | 2 | 55 | 13 | 17 | 3 | 5 | 6 | — |

[a]1 Mg = 1,000 kg = 1 metric ton. The yields quoted are high yields, not average yields.

**TABLE 10–2**

Typical Rates of Nutrient Application in Fertilizers for Crops in California

| Crop type | Rates applied (kg/ha per year) | | |
|---|---|---|---|
| | N | P | K |
| Vegetables | 150 | 35 | 40 |
| Fruit crops | 120 | 7 | 15 |
| Field crops | 100 | 7 | 4 |
| Alfalfa | 0 | 30 | 15 |
| Other hay crops | 30 | 15 | 2 |
| Pasture | 7 | 2 | 0 |

**TABLE 10–3**

Composition of Inorganic Fertilizer Materials

| Material | % N | % P | % K | % S |
|---|---|---|---|---|
| Ammonium nitrate $NH_4NO_3$ | 35 | 0 | 0 | 0 |
| Monoammonium phosphate $NH_4H_2PO_4$ | 12 | 27 | 0 | 0 |
| Diammonium phosphate $(NH_4)_2HPO_4$ | 21 | 23 | 0 | 0 |
| Ammonium sulfate $(NH_4)_2SO_4$ | 21 | 0 | 0 | 24 |
| Anhydrous ammonia $NH_3$ | 82 | 0 | 0 | 0 |
| Aqua ammonia $NH_4OH$ | 40 | 0 | 0 | 0 |
| Calcium nitrate $Ca(NO_3)_2$ | 17 | 0 | 0 | 0 |
| Urea $CO(NH_2)_2$ | 47 | 0 | 0 | 0 |
| Rock phosphate various | 0 | 9–12 | 0 | 0 |
| Superphosphate (single) mixture $CaSO_4$, $Ca(H_2PO_4)_2$ | 0 | 17 | 0 | 9 |
| Superphosphate (triple) impure $Ca(H_2PO_4)_2$ | 0 | 20–26 | 0 | 0 |
| Potassium sulfate $K_2SO_4$ | 0 | 0 | 45 | 18 |
| Potassium chloride $KCl$ | 0 | 0 | 52 | 0 |
| Gypsum (calcium sulfate) $CaSO_4 \cdot 2H_2O$ | 0 | 0 | 0 | 18 |
| Sulfur $S$ | 0 | 0 | 0 | 100 |

## TABLE 10–4

Composition of Organic Fertilizer Materials
(Concentrations, Dry Weight Basis)

| Material | % N | % P | % K | % S |
|---|---|---|---|---|
| **Manures** | | | | |
| Chicken, fresh | 5.1 | 2.0 | 1.8 | 0.7 |
| Chicken, partly composted | 2.9 | 3.3 | 3.3 | 0.7 |
| Cattle, fed alfalfa | 2.5 | 0.4 | 0.7 | 0.4 |
| Cattle, fed alfalfa and rice hulls | 2.0 | 0.3 | 3.9 | 0.5 |
| Sheep | 2.3 | 0.4 | 0.8 | 0.3 |
| Horse | 1.8 | 0.5 | 1.2 | 0.2 |
| **Plant materials** | | | | |
| Young alfalfa or vetch | 3.5 | 0.3 | 1.4 | 0.3 |
| Old alfalfa (flowering) | 2.0 | 0.2 | 1.2 | 0.2 |
| Grain legume residues | 1.5 | 0.15 | 1.2 | 0.15 |
| Cereal straw, old grass, etc. | 0.6 | 0.1 | 1.0 | 0.1 |
| Sawdust, rice hulls, etc. | <0.5 | 0 | <1 | 0 |

to supply about half the N, the rest coming from industrial fixation for fertilizer manufacture. In the most common industrial fixation process (the Haber process and its modifications), $N_2$ is reacted with $H_2$ at high temperature and pressure in the presence of metal catalysts, to form ammonia, $NH_3$. Hydrogen, the energy-expensive component, is derived from natural gas, petroleum, or coal. The resulting ammonia, a gas at ordinary pressure and temperature, can be transported in liquid state when compressed and refrigerated. This form is used as a fertilizer (anhydrous ammonia, 82 percent N). A form that can be used with simpler equipment is *aqua ammonia*—that is, ammonia dissolved in water (like the familiar household ammonia). Both anhydrous and aqua ammonia are injected into the soil or irrigation water; spreading them on the ground surface would allow too much gas to escape.

Ammonia is also used for making other N fertilizers (Figure 10–1A). Ammonia can be oxidized to nitric acid, which is reacted with ammonia or lime to produce ammonium nitrate or calcium nitrate. Ammonia is reacted with carbon and steam to form urea, $CO(NH_2)_2$, a soluble, white, granular solid that decomposes quickly when added to soil, forming ammonium carbonate. Ammonia reacted with sulfuric acid, nitric acid, or phosphoric acid produces the white, soluble, crystalline solids ammonium sulfate, ammonium nitrate, and ammonium phosphate.

These fertilizers are used more widely than is ammonia itself because of their easier handling.

Alternative sources of N are biological $N_2$ fixation (see Chapters 7 and 8) and organic fertilizers such as compost, manure, biosolids, and green manure.

**Phosphorus** The main industrial source of P is phosphate rock. Most deposits of phosphate rock were lime-rich sedimentary rocks that became enriched in P under shallow marine conditions. They contain apatite, a group of minerals with the composition $Ca_5(PO_4)_3(OH, F, Cl)$. (The proportions of OH, F, and Cl vary.) The fertilizer value of most phosphate rocks is reduced by the presence of clays, calcium carbonate, and silica, which interfere with processing, and by minor toxic constituents that are extremely costly to remove (e.g., uranium and cadmium). Removal of clay, carbonate, and silica from phosphate rock by sieving and flotation is **beneficiation.** The world has enormous supplies of low-grade phosphate rock, but many countries lack high-grade deposits.

Rock phosphate, finely ground, is used directly in soils acidic enough to dissolve apatite. More commonly, the rock is treated with acid during processing, to improve P solubility (Figure 10–1B). Sulfuric acid treatment produces ordinary (or single) **superphosphate,** an impure mixture of calcium phosphate

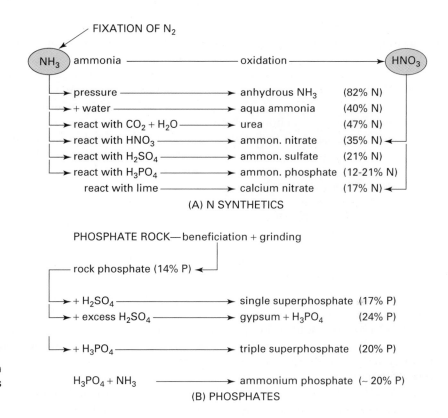

FIXATION OF $N_2$

NH$_3$ ammonia ———————— oxidation ————————→ HNO$_3$

→ pressure ————————→ anhydrous $NH_3$ (82% N)
→ + water ————————→ aqua ammonia (40% N)
→ react with $CO_2 + H_2O$ ————→ urea (47% N)
→ react with $HNO_3$ ————————→ ammon. nitrate (35% N)
→ react with $H_2SO_4$ ————————→ ammon. sulfate (21% N)
→ react with $H_3PO_4$ ————————→ ammon. phosphate (12-21% N)
react with lime ————————→ calcium nitrate (17% N)

(A) N SYNTHETICS

PHOSPHATE ROCK—beneficiation + grinding

rock phosphate (14% P)

→ + $H_2SO_4$ ————————→ single superphosphate (17% P)
→ + excess $H_2SO_4$ ————————→ gypsum + $H_3PO_4$ (24% P)

→ + $H_3PO_4$ ————————→ triple superphosphate (20% P)

$H_3PO_4 + NH_3$ ————————→ ammonium phosphate (~ 20% P)

(B) PHOSPHATES

**FIGURE 10–1** Synthetic nitrogen and phosphorus fertilizers: sources and manufacture.

and sulfate. Also, phosphoric acid is produced by separation from acidified rock phosphate. The phosphoric acid treatment of phosphate rock produces concentrated (or triple) superphosphate, and the reaction of phosphoric acid with ammonia produces ammonium phosphate.

**Potassium** Potassium is mined from deep sedimentary deposits or from salt-lake deposits of KCl and $K_2SO_4$. Potassium salts are separated from the Na salts usually present by procedures based on their different solubilities.

**Sulfur** Sulfur sources are several. The most abundant is gypsum—calcium sulfate—a major constituent of extensive sedimentary rocks. Gypsum is soluble enough to be a good S and Ca fertilizer with no treatment except grinding. Elemental sulfur (S, yellow brimstone), another source, is extracted from deposits associated with petroleum. The element is negligibly soluble; its use as a fertilizer depends on oxidation to sulfate by soil microbes. Superphos-

phate made with sulfuric acid is a widely used sulfur fertilizer.

**Organic fertilizers** **Organic fertilizers** are diverse, but all ultimately derive from plants. Some consist of unprocessed plant material (such as tree leaves, grass clippings, crop residues, and green manure crops grown specifically to be used as fertilizer). Others have been processed, by industry (cannery wastes), by animals (manure, litter, blood, bone, offal, sewage sludge), or by microbes (lees from fermentation, composts). Organic materials are frequently composted—piled to allow partial microbial decomposition (see Chapter 7). Composting reduces bulk, eliminates some unpleasant odors, and kills weed seeds and disease organisms if the composting process generates high enough temperatures. Composting may improve the fertilizer value by lowering the ratio of C to N, S, and P. However, it commonly allows gaseous losses of N and the leaching of other nutrients. Accordingly, long-term experimental comparisons have shown that crops benefit less from composted plant residues than from the addition of the same residues directly to the soil, fresh, without composting.

## 10.1.2   Comparisons and Choices

A farmer or gardener must choose which fertilizer to use. A good choice has these properties:

1. It contains the needed nutrients.
2. It releases them at the right time.
3. It can be obtained at the right price.
4. It is convenient to use.
5. Its side effects are acceptable.

**Nutrient content**   Fertilizer formulations vary (see Supplement 10–1). The simple fertilizers most often used in field agriculture supply only one or two plant nutrients, but there are more complex formulations. "Complete" fertilizers sold to home gardeners normally supply N, P, and K, and sometimes S, Ca, Mg, and micronutrients are included, along with an increase in price. Organic fertilizers are likely to contain all the essential plant nutrient elements. However, few soils are deficient in more than one or two plant nutrients. Complete fertilizers eliminate the need to diagnose which nutrient is deficient. That is an overwhelming advantage for a gardener but not for a farmer, whose large fertilizer bills and environmental responsibility justify the cost of diagnosis.

**Release rates**   Most inorganic fertilizers release nutrient ions almost immediately, but some dissolve slowly in the soil (e.g., rock phosphate, elemental S), and others are granulated, **pelleted,** or coated during manufacture to slow down the release rate. Sulfur-coated urea is an expensive example: Urea, though highly soluble, escapes from the fertilizer granule into the soil solution only as the coating of elemental S dissolves.

Most organic fertilizers release nutrients slowly as they decay. Decay is especially slow if the soil is cold or the material is coarse, woody, fibrous, or low in nutrients. Slow release is desirable, if it reduces nutrient losses through leaching, volatilization, and immobilization, but the release must coincide with the plants' nutrient demand.

**Availability and cost**   Farmers usually choose the locally available fertilizer that most cheaply provides

the nutrient they need to add. The percentage of active ingredient is the most significant basis on which to compare fertilizer prices. Competitively priced fertilizers have similar costs per unit of N, S, or whatever nutrient is wanted. Even then, however, the high-analysis fertilizer (the one with the high percentage of active ingredient) has the advantage that less bulk needs to be transported and spread. This economy in energy and labor costs has caused a trend toward more use of high-analysis fertilizers, especially when they must be transported long distances or by expensive means.

Organic fertilizers generally contain lower nutrient concentrations than do inorganic fertilizers (Table 10–4). Some are so low, even in N, that they have doubtful fertilizer utility or even immobilize soil N (see Chapter 8). Because of their low nutrient concentration and bulk, organic fertilizers are used close to where they are produced.

**Convenience and ease of use**   Concentrated, high-analysis fertilizers usually are the easiest and cheapest to apply. Other considerations include physical state, solubility, stability of the material, and suitability to the user's application equipment. For example, if injection equipment is not available, one might choose urea or ammonium nitrate rather than ammonia, even if ammonia is less expensive. Likewise, a farmer equipped to apply fertilizer in irrigation water would prefer completely soluble ammonium phosphate over poorly soluble superphosphate.

**Side effects**   Not all the effects of fertilizers are beneficial, and some indirect costs are not always fully apparent. We have already mentioned the **eutrophication** of streams, groundwater, and nonfarmland—pollution with unwanted fertility. Fertilizers can also damage plants and soil microbes, through toxicities of a nutrient ion or a secondary ingredient. Soluble salts such as KCl and $K_2SO_4$ can damage emerging seedlings. Organic materials in excess can impede root growth and make soils physically unmanageable.

Soil acidification (see Chapter 11) results from inputs of N as ammonium or as materials that release ammonium—that is, almost all N inputs: biological nitrogen fixation, most organic fertilizers, and inorganic N fertilizers except nitrates. Acidification is sometimes harmless or beneficial, but detrimental acidification (e.g., to pH below 5) requires correction that must be counted as one of the indirect costs of improving N fertility.

Fertilization can also cause new deficiencies to appear. Bigger crops need more nutrients, and a change to higher-analysis fertilizers may bring to light a deficiency that had previously been corrected by secondary fertilizer ingredients. For example, an S deficiency may appear following a switch from single superphosphate to concentrated superphosphate.

Costs and side effects are minimized, and efficiency improved, by accurate diagnosis to guide the proper selection of fertilizers and fertilizer rates, by good timing and methods of application, and by nutrient conservation.

## 10.2   DETERMINING FERTILIZER NEEDS: DEFICIENCY DIAGNOSIS

### 10.2.1   Kinds of Diagnoses

Two questions arise: "Which nutrients are deficient?" and "How much of what fertilizer is required?" The surest answers come from tests of plant response to nutrient addition under relevant conditions: *field trials* at representative sites for field crops, row crops, orchard, or forest, and *container trials* under realistic conditions for nursery and greenhouse plants. Field tests to measure plant response are time-consuming and expensive, but they can be complemented with easier, faster tests. All these complementary methods are based on data from representative plant-response tests and include

- Guesswork guided by experience
- Interpretation of visual symptoms
- Analysis of soil samples
- Analysis of plant samples

### 10.2.2   Nutrient Response Experiments with Plants

Figure 10–2A shows the results of a simple comparison between untreated and N-fertilized plants in field plots in three soils. The fertilizer improved growth in field A; therefore, that soil was deficient in N. The fertilizer had a negligible effect in the other fields; therefore, they were not N deficient. In field B, the soil probably supplied ample N. In field C, the soil may have supplied ample N, but growth was limited by something else, perhaps drought, pests, disease, or some other nutrient deficiency.

Fertilizer trials are useful when the growth conditions are good enough for the deficiencies and responses to be expressed. If the soil had multiple nutrient deficiencies, plant responses to nutrient additions would be inhibited if just one of the deficiencies were left uncorrected. A solution is to give every plot a "basal" fertilizer containing the nutrients not under test, in case one of them is deficient enough to interfere. Another solution is the **subtractive trial,** or missing element trial, in which a control treatment containing all nutrients is compared with similar treatments from which single nutrients have been left out, or subtracted. In Figure 10–2B, for example, the yield is less from the minus-P and minus-N treatments than from the complete control treatment, indicating that the soil is deficient in both P and N. Subtractive trials have been useful in surveying deficiencies of infertile areas for which no information existed previously. Another approach is the **factorial trial,** which tests combinations of individual nutrients (Figure 10–2C). Factorial experiments are seldom used to find out *what* is deficient, however, because they become cumbersome if more than a few nutrients are to be tested. Rather, their main value is to identify interactions.

**Rate trials**   How much of a nutrient should be added? Fertilizer rate trials compare plant responses to a nutrient applied as a fertilizer at different rates (amounts per unit area). The data can then be plotted as **response curves** (Figure 10–3). Ideally, growth, yield, or quality improves with increasing nutrient addition until the deficiency is eliminated. The response diminishes as the point of sufficiency is approached. Beyond this point, more fertilizer has no obvious benefit, or it damages the plant (Figure 10–3A). The responses will be small if the field is not deficient or if growth is suppressed by other constraints (Figure 10–3B). Response curves and fertilizer requirements often depend on the plant characteristic that is observed and measured; growth, yield, and quality attributes need not respond alike (Figure 10–3C).

It is seldom economical to apply enough fertilizer to maximize yield, because fertilizer application is seldom free, and the plant response diminishes as

SAMPLE LAYOUT

| No N | +N | |
|------|------|--|
| +N | No N | |

→
replications

SAMPLE RESULTS

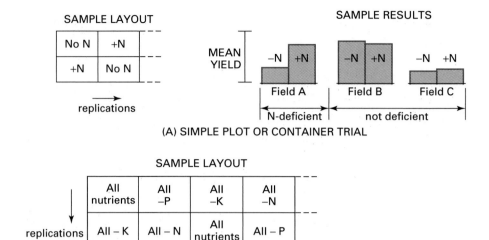

(A) SIMPLE PLOT OR CONTAINER TRIAL

SAMPLE LAYOUT

↓
replications

| All nutrients | All −P | All −K | All −N |
|---------------|--------|--------|--------|
| All − K | All − N | All nutrients | All − P |

SAMPLE RESULTS

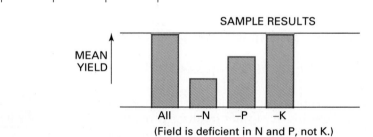

(Field is deficient in N and P, not K.)

(B) SUBTRACTIVE TRIAL

SAMPLE LAYOUT

| No treatment | +N | +P | +K |
|--------------|-----|-----|-----|
| +NP | +NK | +PK | +NPK |

↓
replicated

SAMPLE RESULTS

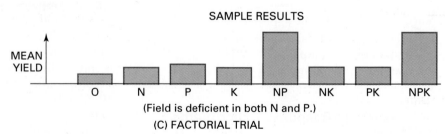

(Field is deficient in both N and P.)

(C) FACTORIAL TRIAL

FIGURE 10–2 Determining which fertilizers are needed. Plant response trials, showing example plot layouts and plant responses. (A) A simple test, showing a need for N. (B) A subtractive trial, useful when deficiencies may be multiple, showing a need for both N and P. (C) A factorial trial, used mainly for defining interactions, showing a response to N and P together but not separately.

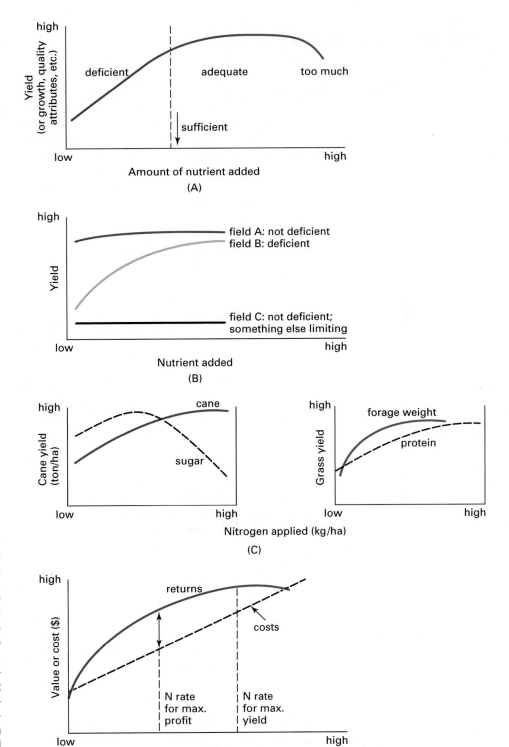

**FIGURE 10–3** Determining how much fertilizer is needed: fertilizer rate trials and response curves. (A) Generalized response curve for a nutrient applied at a wide range of rates to a deficient soil. (B) Large and small responses. (C) Different responses expressed by different plant characteristics. (D) Economic interpretation of response data, showing a graphical analysis to find the most profitable fertilization rate.

maximum yield is approached. Usually the additional return ceases to justify the cost of additional fertilizer at a point well below the maximum yield (Figure 10–3D). How close to the maximum yield it is desirable to go depends on the value of the crop and the cost of fertilization. Fertilizers are used most heavily when they are cheap and the crops are valuable.

**Nutrient interactions**   When the effect of one factor depends on the level of another, the two factors are said to **interact.** Figure 10–4 shows some interactions involving nutrients. In general, more benefit is obtained from one input if other factors are improved. The better the plant can grow, the more it will respond to fertilization with a limiting nutrient. This general rule is another reason that the most

productive agricultures tend to use the most fertilizer. It also implies that improvement of crop yield improves the efficiency of fertilizer use (the useful return per unit of fertilizer applied).

**Changes in fertilizer requirements**   Fertilizer requirements change as soil fertility adjusts to new sets of soil-forming factors, whenever there are major changes in land use, management, or productivity. For example, during the 20 years or so after forest or grassland is turned into cropland, the soil organic matter declines and the fertilizer requirements for N and S increase. As another example, P-deficient soils require much phosphate fertilizer to satisfy their sorption capacity when first cropped, but small maintenance applications are sufficient thereafter.

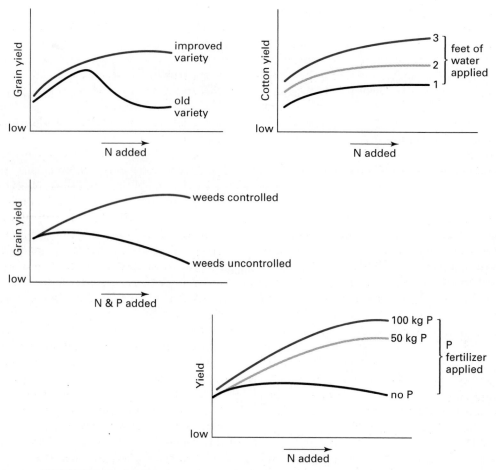

**FIGURE 10–4**   Interactions: some examples of interactions between fertilizer applications and other factors.

## 10.2.3 Supplemental Diagnostic Procedures

**Plant symptoms of nutrient disorder** Visible signs of nutrient disorders are often ambiguous but helpful to skilled observers who know the local plant nutrition problems. The main signs include patterns of chlorosis, purpling, local necrosis, and stunting (Figure 10–5 and Color Plates A–H). Developing symptoms show up in either old or young parts of the plant, depending on the ease with which the plant can move the particular deficient nutrient. The symptoms appear in old leaves first if the deficient nutrient is one that can retranslocate readily from old leaves to young growing tissue (e.g., K, Mg, N, and S). But if the nutrient is not mobile in the plant (e.g., Fe, Mn, Zn, or B), then the symptoms first appear in young leaves and growing points.

*Chlorosis (yellowing)* Chlorosis is a generalized yellowing of old leaves, or all leaves, and may indicate deficiencies of nitrogen or sulfur, though it may also indicate old age, waterlogging, or lack of light. Chlorosis due to N deficiency is a familiar sight in pastures and lawns, where the yellowness is highlighted by the deep green of patches that received extra N from gardeners, livestock, or the neighbor's dog. Chlorosis of part of the mature leaves is sometimes a sign of P or Mg deficiency.

Interveinal chlorosis (yellowing between veins, with the veins forming a green pattern) is a symptom of a deficiency of Fe, Mn, or sometimes Zn or Cu. Of this group, Fe and Zn are the most commonly deficient. Some viral diseases produce interveinal chlorosis, and some ornamental plants show it even when healthy. Interpretation as an Fe, Mn, or Zn deficiency is more certain if the soil is alkaline and the plant is known to be susceptible to the deficiency.

*Purpling (accumulation of anthocyanin pigments)* Purpling, an overall dark green color with a purple, blue, or red tint, is the common sign of P deficiency. It can also occur in Al toxicity and drought. Some plant species and varieties do not show this coloration when P is deficient but become yellow instead.

*Local necrosis (death of tissue)* "Firing" or **necrosis** of patches, spots, or margins on leaves is a sign of K deficiency, salt damage, or sometimes Mo or chlorine deficiency. It can also result from drought or frost.

*Stunting* Overall stunting (reduced growth) is a sign of all nutrient disorders. It is the hardest to detect unless normal plants are nearby for comparison. Acute stunting of growing points produces distinct growth patterns characteristic of certain deficiencies. Zinc deficiency produces "little leaf" (in which the youngest leaves do not grow as large as normal full-grown leaves) and "rosetting" (in which the stem fails to elongate at the tip, so that the terminal leaves become bunched). Boron deficiency commonly kills growing points.

**Diagnostic plant analyses (tissue tests)** Usually as the supply of a nutrient increases, its concentration in the plant tissues increases, as shown in Figure 10–6. In the deficient range the extra nutrient produces more growth, so there is only a moderate increase in the plant's concentration of the nutrient. When the concentration exceeds the *critical concentration,* at which the plant is no longer deficient, growth no longer increases to balance the greater uptake, and the concentration rises sharply. If the critical value or critical range has been established, plant analysis can be used as a diagnostic guide (Figure 10–7).

Critical concentrations depend on several factors. They vary with the nutrient and the plant species, part, age, and maturity. Nutrient concentrations are higher in leaves than in woody stems or petioles, and, accordingly, the critical values for leaf analyses generally will also be higher. Old leaves have a lower concentration of many nutrients than young leaves do, and lower critical values as well.

For all these reasons, tissue for tests must be sampled by standardized methods, and critical levels must be established for the particular sampling procedure, nutrient, and plant species. Establishment of critical values—test calibration—nearly always depends on the analysis of plant samples from fertilizer rate trials. The tissue tests can be viewed as a way to extend information from field trials.

**Tissue tests** are most often used for diagnosing deficiencies (and some toxicities) in perennials such as fruit trees, forest trees, or alfalfa. They are less often used for annuals, because the results tend to arrive too late for corrective action. High-value crops may justify periodic plant testing, which can indicate

**FIGURE 10–5** Plant symptoms of nutritional disorders. (See also Color Plates A–H.) (A) Bright spots on these alfalfa leaves are dead (necrotic) spots where cells have died because of potassium deficiency. Necrotic spots and edges are common symptoms of nutrient deficiency and disease. (B) Rosetting due to stunting of young leaves and shortening of internodes at growing point. In this photo of alfalfa leaves, zinc is deficient. (C) The onion plants on the left have sufficient zinc. Those on the right are deficient. Note the smaller plants and the crooked tops. (D) On the left is a normal almond branch, and on the right the upper portion of the branch is dead and has no leaves. This twig dieback is caused by boron deficiency. (Photo A: R. Meyer; photos B and D: S. Pettygrove; photo C: U.S. Department of Agriculture.)

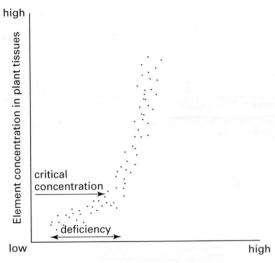

**FIGURE 10–6** Element concentration in the plant increases slowly as growth increases in response to added nutrient. At a critical concentration, growth does not increase to balance the additional nutrient uptake, and concentration in the plant tissues increases rapidly.

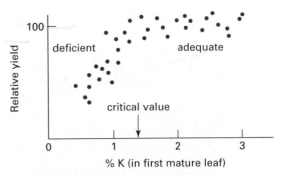

**FIGURE 10–7** An example of calibration and interpretation of diagnostic plant analyses. Yield data from field trials are plotted against analytical values from the same plots.

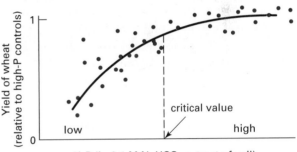

**FIGURE 10–8** An example of calibration and interpretation of soil test values. As for calibration of plant analyses, data from field fertilizer trials are used to establish which ranges of analytical values correspond to deficiency and sufficiency.

fertility trends during the season in an annual crop and long-term trends in orchards and plantations.

**Diagnostic soil analyses (soil tests)** The availability of a nutrient in soil is difficult to define, let alone measure (see Chapter 9), and yet diagnostic tests must be fast, cheap, and therefore simple. Nevertheless, some simple chemical tests on soil can help diagnose deficiency because they correlate with deficiencies or toxicities as measured by plant response.

**Soil tests** have been calibrated and used for many nutrients and crop species in many states and countries. They are most generally successful for deficiencies of phosphate, potassium, and zinc and for acidity and salinity conditions (see Chapter 11). All use some convenient, arbitrary extraction procedure to remove part of the soil's total supply of the element in a dissolved form convenient for analysis. For example, phosphate might be extracted with 0.1 M $NaHCO_3$, with $H_2SO_4$, or with dilute 0.1 M $CaCl_2$. Each gives a different amount of phosphate. Which procedure correlates best with plant response depends on the properties of the plants and soils. Not surprisingly, therefore, different regions have their

favored procedures. Like other test procedures, soil tests are calibrated against field fertilizer trials (Figure 10–8).

Few simple soil tests have been satisfactory for nitrogen or sulfur, whose availability depends on rates of microbial processes. (Measurements of rate of mineral N release when soil is kept moist and aerated in the laboratory have had some success but are slow and costly for routine use.)

Soil variation makes it difficult, and sometimes futile, to obtain samples representing the field's "average" fertility. Thus sampling, rather than analytical procedure or calibration, may cause the most error and uncertainty in practice.

**Soil sampling**    How to take good samples depends on the purpose of the analyses and the pattern of soil variation. Soils in a field vary vertically and horizontally, and the variation is often increased by fertilization, leveling, and irrigation. Large sections of the field that seem greatly different should be sampled and analyzed separately. Small variations within these areas can then be averaged by mixing together several subsamples. Thus, one first subdivides the field into homogeneous units in which past treatment, crop appearance, topography, and soil appearance are uniform. Then from each of these units one takes at least 10 small subsamples and mixes them to produce a convenient-sized sample. The process is repeated to provide replicates so that variance can be measured. Shallow sampling (top 10 cm) is often suitable for pH, P, and K in pasture and shallow-rooted crops. Deeper sampling, perhaps to root depth, is better for testing mobile constituents (such as nitrate, sulfate, and salt) and for deep-rooted plants (such as alfalfa and tree crops). Samples are usually air-dried to prevent changes during storage, but freezing or freeze-drying is better if the soil is to be analyzed for nitrate, manganese, or other constituents that change during ordinary drying.

## 10.3    FERTILIZER EFFICIENCY AND NUTRIENT CONSERVATION

### 10.3.1    Methods and Timing of Fertilizer Application

Ideally, application methods (Figure 10–9) and timing (Figure 10–10) maximize the uptake of fertilizer nutrients by the crop, minimize nutrient losses, and reduce the impact of nutrient immobilization in the soil. Actual methods are compromises between these considerations and others, such as cost and suitability to existing equipment and cultivation practices. However, all good methods apply the right rate uniformly or, better, adjust it to match the variations in fertility (see Soil and Environment box). Overfertilizing some parts of the field while underfertilizing the rest wastes fertilizer and reduces yield.

**Broadcast application**    Fertilizer can be spread onto the ground surface, either by hand or by machines ranging from hand-operated spreaders to aircraft. Aircraft do especially well over rough, steep, wet, or inaccessible terrain and where ground traffic would excessively damage plants or soil. Broadcast application is common, fast, convenient, and usually the only method cheap enough to use on range, pasture, or forest. It is unsuitable for volatile fertilizers such as ammonia.

**Application in irrigation water**    Soluble fertilizers can be injected as a concentrated solution at a controlled rate into irrigation water. This method is cheap and can be used frequently and safely during crop growth. Fertigation is the application of nutrients through an irrigation system.

**Drilling and injection**    Placement below ground surface often improves fertilizer efficiency, especially when the fertilizer is banded at a close but safe distance from young plants (Figure 10–9E and 10–9F). On a garden scale, the fertilizer might be laid along a hoed furrow and covered. On a farm scale, a fertilizer drill drops or injects fertilizer behind a tool (shank, knife, or tine) that opens a furrow as the machine advances. The soil is then swept back to cover the fertilizer. Field and row crops are commonly planted with a combined seed and fertilizer drill. A good drill operator can also make midseason applications of fertilizer (side-dressing) to precisely planted row crops, with little damage to plants. Ammonia is mostly applied by tractor-mounted drills with injectors.

**Foliar spray**    Nutrients in solution can be sprayed onto foliage by hand, spray rig, or aircraft. Foliar spray gives quick response, accurate timing, and no immobilization in the soil, but only limited amounts can be applied, because the leaf surfaces hold little water before the surplus drips off, and concentrated solutions damage the plants. Micronutrients—for example, Fe and Zn—are commonly applied by foliar spray.

Timing fertilization to coincide with maximum need is desirable. A crop usually needs the most nutrients during maximum growth and nutrient uptake (Figure 10–10). With tree crops, timing is complicated by storage in stems and roots and by delay in the penetration of applied nutrients to active roots

(A)

(B)

**FIGURE 10–9** Methods of fertilizer application. (A, B) Broadcasting with wheeled spreaders. Manure is being spread in (B). (C) Loading fertilizer for aerial broadcasting.

(C)

(D)

(E)

(F)

FIGURE 10–9 *Continued.* (D) Aerial broadcasting of superphosphate on hilly rangeland. (E, F) Drill equipment for band application of fertilizer near seed during planting.

(G)

(H)

**FIGURE 10–9** *Continued.* (G) Equipment for injecting liquid fertilizer or compressed gas into soil. (H) Filters and injection equipment for applying liquid fertilizer in irrigation water. (Photos A, C, E, F, G: Stuart Pettygrove; photo B: U.S. Department of Agriculture; photo D: Charles Raguse; photo H: Roland Meyer.)

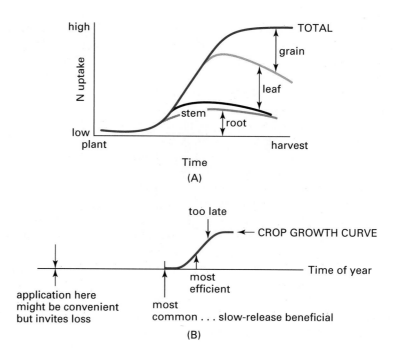

**FIGURE 10–10** (A) Nutrient up-take during crop growth, showing time of greatest need, and (B) optimum timing of fertilizer application.

deep in the soil. Ideally, nutrients are applied during or just before maximum demand, by a method that gives quick response (drill placement or foliar spray) or minimizes damage to the crop (aerial application or application with irrigation water). An alternative is to apply slow-release or timed-release fertilizers that release nutrients at rates that match crop demand. Elemental S, sulfur-coated urea, rock phosphate, and most organic manures dissolve or break down slowly. Ammonium salts, though readily soluble, produce a similar effect when banded; acidity and concentrated ammonium keep roots from invading the band and retard the bacterial production of mobile nitrate.

## 10.3.2 Nutrient Cycling and the Control of Losses

Each nutrient undergoes a cyclic set of transformations and movements. (Chapter 8 describes N cycling.) Figure 10–13 shows the cycles of the major fertilizer elements: N, S, K, and P. The diagrams in the figure summarize much information about soil fertility. In each case, the element moves among several "compartments" or "pools" representing different forms or states of the element in the ecosystem. The

pool sizes vary from one soil and production system to another; those in the diagrams represent a typical agricultural ecosystem such as a wheat crop growing on a reasonably fertile soil. A forest would have more biomass, litter, and humus, and a sandy desert soil would have less humus and less inorganic reserve K and P.

The cycles are leaky. The cycling of nutrients involves losses from the ecosystem, more so with greater fertility. The high levels of mobilized nutrients that promote uptake by plants also promote loss. Good fertility management limits the losses while accelerating the nutrient gain by desirable plants. Nutrients are lost through four main mechanisms: soil erosion, removal by crops, volatilization of gases, and leaching.

Soil erosion removes topsoil, enriched in nutrients by natural soil formation processes and fertilization. The loss of P and N is the most serious, because they are often deficient and accumulate markedly near the ground surface. Erosion losses can be minimized by appropriate land use and erosion control (see Chapter 15). An especially rapid "erosion" of nutrients occurs when snowmelt or rain carries off fertilizer recently applied to bare, frozen, or snow-covered ground. The lost phosphate can be a serious pollutant of natural water bodies, because these are often phosphate deficient and the additional phosphate permits rapid, undesirable growth of algae.

Removal of nutrients with crops is unavoidable, because the nutrients are essential constituents of the product. Harvested in the greatest quantity are N and K (Table 10–1), especially in leafy products such as green vegetables and fodder crops. Loss of N and K is lower if leaf material is left in the field and only grain or fruit is harvested. Even so, N losses are substantial in grain, especially grains rich in protein (compare wheat, maize, and rice in Table 10–1). Losses can be reduced by conservation and return of crop residues. Previous composting (piling for partial microbial decay) reduces bulk and helps make residues manageable, but it is laborious and allows loss of nutrients, especially N. Grazing animals quickly reduce and recycle standing residues. Their urine and dung return most of the N, K, and other nutrients from the portion they eat, and the rest they trample, hastening decay by soil microbes.

Gaseous N and S are lost to the atmosphere when plant material is burned; when ammonia escapes from soil, plants, or fertilizer; and when microbes reduce nitrate and sulfate to gaseous forms. Fire, be it a natural event or agricultural practice, conserves most plant nutrient elements as carbonates, oxides, and phosphates in the ash, but much of the N and S escape as the gaseous oxides ($N_2O$, $NO_2$, $SO_2$) eventually to fall somewhere else as acid components of rain. This loss is eliminated if burning can be replaced by aerobic microbial decay. Ammonia ($NH_3$) volatilizes from alkaline or neutral solutions and decaying plants and microbes. Avoidable losses occur when ammonium fertilizers are mixed with lime or applied to soils that are dry or alkaline or when ammonia is injected at insufficient depth. Microbial volatilization of N and S results from the reduction of nitrate to $N_2$ and $N_2O$ (denitrification) or the reduction of sulfate to $H_2S$ (see Chapter 8). Denitrification is favored by the combination of wet, warm conditions, with ample nitrate and decomposable organic matter, but this combination is desirable in agricultural practice.

Leaching can cause losses of all soluble nutrient forms, especially potassium, nitrate, and sulfate. Phosphate leaching is insignificant except in extraordinarily sandy soils. Leaching losses can be reduced by not fertilizing in excess or at the wrong time such as during wet seasons when crops are not growing. Otherwise, little useful can be done directly to reduce leaching of nutrients. Where it would be most readily feasible—in arid, irrigated agriculture—nutrient loss is of less concern than the absolute need for periodic leaching to prevent salt accumulation (see Chapter 11). Claims that one can reduce leaching by increasing the soil's exchange capacity are not overly practical. Regardless of the exchange capacity, fertile soils have dissolved nutrients available for both plants and leaching. One can, however, reduce leaching losses in low-capacity soils by using fertilizer in smaller and more frequent applications. The point is to avoid excesses that the plant cannot use.

Leaching, erosion, volatilization, and cropping losses can be reduced by keeping the soil infertile and cutting down on product removal. Obviously, these practices are inconsistent with most agriculture, but they are not inconsistent with other important land uses such as watersheds and natural areas.

### 10.3.3   Improving Nutrient Efficiency of Plants

Less fertilizer will be required if efficiency can be improved by selecting plants or manipulating their environment. Some of these methods are proven practices, but others are not.

**Nutrient-efficient plants**   Some productive plants contain only low levels of a nutrient. This property is *nutrient utilization efficiency.* It is not restricted to wild plant species. For example, rice and corn produce more grain per kilogram of N used than does wheat (Table 10–1). High utilization efficiency can mean that the product is less nutritious, and indeed, neither corn nor rice has as much protein as wheat does. Similarly, some forage species produce well with exceptionally low levels of N, P, Ca, Cu, Co, Cl, Na, or Mg, but animals grazing these plants on deficient soils may develop mineral deficiencies. These considerations should not prevent the use of low-nutrient plants on poor soil if there are simple ways to correct the deficiencies in the animals—for example, by mineral supplements for livestock or a varied diet for humans.

Plants also differ in their ability to get nutrients from deficient soil. Plant nutritionists have discussed deliberate selection and breeding for superior nutrient acquisition, but so far with limited

# SOIL AND ENVIRONMENT
## *Farming by Soil*

In most agricultural systems, soil or plant tests are used to diagnose the average fertilizer needs across an entire field. This results in application of too much fertilizer to some parts of the field and too little to other parts because most agricultural fields contain several soils with different natural fertility. Computer technology, geographic information systems, and global positioning satellites are combined to allow for fertilizer addition to fields according to the kinds of soil in the field, reducing excess fertilizer additions and potential pollution from runoff and leaching. This is precision agriculture, or farming by soil. These new technologies allow fertilizers to be added to fields on a point-by-point basis. The necessary elements include a yield map for the field, an ultradetailed soil map based on many soil samples and analyses, and a detailed knowledge of where in the field each point is located and where the tractor is located in the field when fertilizer is applied.

The first step in precision agriculture is collecting yield information on a point-by-point basis within a field. Yield data are collected with a yield monitor mounted on the harvester as a crop is harvested (Figure 10–11). The location of the harvester is recorded by computer using a global positioning system in the tractor or on the harvester. The dark crosses in Figure 10–11 are points in the field where yield samples were taken. The yield data are compared with soil data collected in the field and remote sensing data collected from aerial photographs (Figure 10–12). In Figure 10–12, the darker pattern indicates less clay content in the soil and better soil drainage. When there is a good correlation between soil property distribution and yield, as there is between the two figures, then the yield map can be used to plan soil management strategies. In this case, drainage of the upper end of the field could be improved. A similar map (not shown) was derived for soil phosphate levels and weed competition, which were shown to be major causes of wheat yield variability (see Plant et al., 1999). These results can be used to plan how much fertilizer to add to the field in each location and where to apply herbicide to reduce weed competition.

The tractor operator can apply the appropriate amount of fertilizer or herbicide for each location based on the yield and soil data. The result is more efficient fertilizer and herbicide application, less potential pollution, higher yields or better crop quality, and less wasted fertilizer. These benefits come at an increased initial cost for data collection, mapping, and equipment.

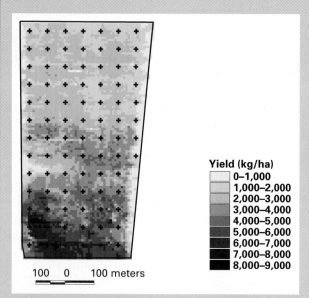

**Yield (kg/ha)**
- 0–1,000
- 1,000–2,000
- 2,000–3,000
- 3,000–4,000
- 4,000–5,000
- 5,000–6,000
- 6,000–7,000
- 7,000–8,000
- 8,000–9,000

100   0   100 meters

**FIGURE 10–11** This yield map was collected as part of a precision agriculture study. The darker the pattern, the higher the wheat yield in this irrigated field in Yolo County, California. (From Plant et al., 1999. Trans. ASAE 42(5):1187–1202).

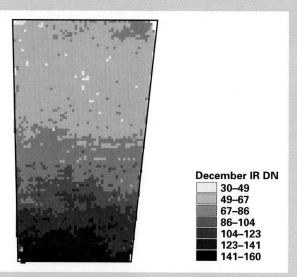

**December IR DN**
- 30–49
- 49–67
- 67–86
- 86–104
- 104–123
- 123–141
- 141–160

**FIGURE 10–12** Infrared band photos and soil texture analyses yielded this map of soil drainage capacity. The darker the pattern, the better the soil drainage. (From Plant et al., 1999. Trans. ASAE 42(5):1187–1202).

practical outcome. Similarly, mycorrhizae are clearly important to plant nutrition, but there may be little scope for improving their performance.

**Nitrogen fixation**   Legumes and other plants with effective N-fixing root nodules can acquire adequate N and be independent of N fertilization. This is ancient agricultural technology that in recent years has become better understood and a little improved (see Chapters 7 and 8). Legumes provide some nitrogen to the soil unless the fixed N is efficiently harvested. Sometimes legumes yield valuable high-protein products with little consumption of soil and fertilizer N. Pasture, range, forest, and natural ecosystems depend on biological N fixation, which can be enhanced by introducing and fostering N-fixing herbs, shrubs, trees, and their microbial partners.

**Better crops**   Finally, as most farmers know, healthier crops use nutrients more efficiently. The most general prescription for efficient use of a fertilizer is twofold: (1) use only the right fertilizer at the right rate and (2) alleviate other constraints on productivity by improving crop varieties, soil management, water management, and the control of weeds, pests, and diseases.

## 10.4  SUMMARY

Careful management of plant nutrients is an important aspect of environmental protection and land stewardship, as well as efficient sustainable agriculture, forestry, range management, and horticulture. These concerns are not inconsistent with one another. For instance, control of the pollution of groundwater with nitrate and of surface water with phosphate and organic N—major environmental concerns—is improved if fertilizer application rates are optimized (not excessive), erosion is minimized, and water is used efficiently. Tools for improving nutrient efficiencies include recycling, control of avoidable losses, and accurate quantitative diagnosis to assess fertilizer needs.

## QUESTIONS

1. Explain the meaning and significance of *fertilizer, organic, synthetic, high analysis, slow release, eutrophication, subtractive trial, factorial trial, rate trial, response curve, interaction, tissue test, soil test, test calibration, sampling procedure, broadcast application, foliar application, nutrient cycling, leaching, volatilization, fixation,* and *nutrient use efficiency.*

2. Why does the harvesting of crops remove more N and K than other nutrients from the soil?

3. If you continuously grow corn that removes 200 kg/ha of N each year from the soil, you may have to supply 400 kg of N as fertilizer. What happens to the extra N? Why are applied P and K likely to be recovered by the crop more efficiently than N is in the long run?

4. Although most vegetable crops remove less N, P, and K than do alfalfa or corn, they receive more N as fertilizer. Why?

5. Tables 10–1 and 10–2 list nutrient removals by crops and typical fertilizer additions. Allowing for the atypical high yields on which the removals are based in Table 10–1, it is still clear that fertilization does not entirely replace the removal of all nutrients by crops. Is this a stable situation? What other sources could supply nutrients besides fertilizers?

6. In what forms are N and P removed from soil by (a) leaching and (b) erosion? Why are erosion losses probably more important for P than for other nutrients?

7. Why is it uneconomical to fertilize to the point of sufficiency at which no further crop response occurs? Think of an exception to this generalization.

8. Design a test to determine the best rate of zinc fertilizer to apply to azaleas grown in a zinc-deficient soil. What symptoms would you look for? What rates of Zn would you add? How many times would the trial need to be repeated (replicated) to give credible results? What response would you measure? What would the response curve look like? How would you interpret it? If you had no zinc but had sulfuric acid and lime, which one might help cure your zinc deficiency?

## FURTHER READING

California Fertilizer Association. 1975. *Western fertilizer handbook.* Danville, Ill.: Interstate Printers and Publishers.

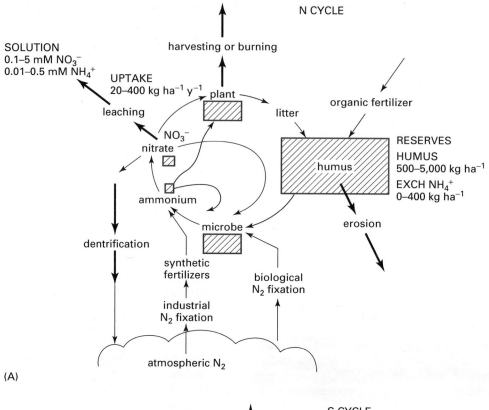

(A)

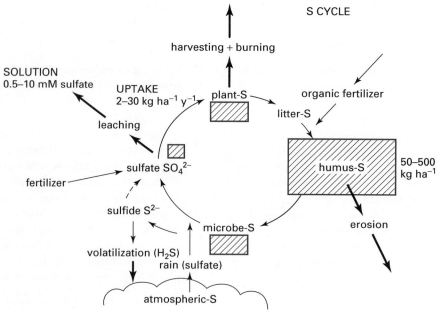

(B)

**FIGURE 10–13**  Major nutrient cycles, summarizing opportunities for losing, conserving, and supplementing soil–plant reserves. (A) Nitrogen. (B) Sulfur. The relative size of each nutrient pool is illustrated by a cross-hatched box. Arrow thickness is proportional to transfer amounts.

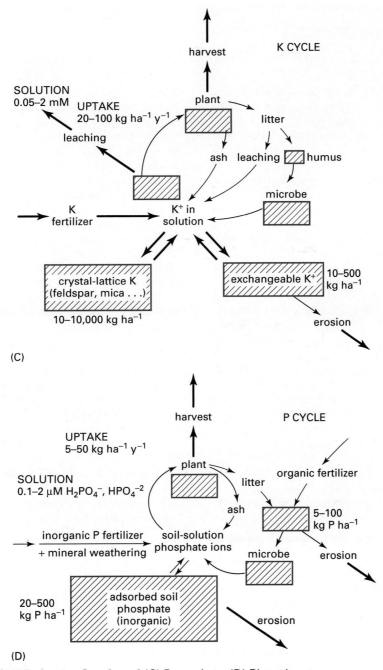

**FIGURE 10–13** *Continued.* (C) Potassium. (D) Phosphorus.

# SOIL AND ENVIRONMENT
## *Nutrient Efficiency, Energy Conservation, and Soil Conservation*

Maximizing the efficient use of plant nutrients helps to conserve energy and nutrient resources and to reduce eutrophication and groundwater pollution. These combined advantages can result from precisely calculated fertilizer rates; better nutrient recovery by plants; reduced nutrient losses from leaching, erosion, and volatilization; and increased exploitation of biological nitrogen fixation. Maintaining adequate fertility is essential for establishing and maintaining plant cover for erosion control. Likewise, control of the erosional loss of P, K, and organic N is part of maintaining soil fertility and reducing stream pollution.

### Eutrophication and Groundwater Pollution

Nitrate is the one plant nutrient form that commonly and significantly pollutes groundwater, because of its mobility and its abundance in fertile soils. In surface waters, as on most soils, photosynthesizing organisms are growth limited by at least one mineral nutrient and grow more when its supply increases. Nutrient enrichment (eutrophication) of surface waters can result from pollution with sewage, stormwater, livestock, or runoff and drainage from fertile soils. Eutrophication encourages aquatic weeds and may enhance growth of algae and blue-green bacteria enough to turn water green and opaque. Bacteria decomposing masses of dead algae and plants can depress oxygen levels enough to be lethal to aquatic animals.

Phosphate is the commonest growth-limiting nutrient in surface waters because it is generally immobilized by tight sorption on colloidal minerals, in both soils and aquatic sediments. Sorption prevents significant P movement from soils except as solid forms carried by runoff—that is, organic residues from plants or animals, unreacted fertilizer, and phosphated soil mobilized by erosion. Phosphate inputs to streams from land are therefore limited, diffuse, and manageable by controlling erosion and the access of livestock to streams. The much more concentrated inputs of soluble phosphate in sewage effluents have been reduced in many communities by curtailing use of phosphate in washing preparations and could be further reduced by advanced effluent treatments that mimic phosphate-fixing mechanisms of soil.

Though phosphate is the chief nutrient limiting productivity in the oceans and most estuaries and lakes, other nutrients—for example, nitrogen, iron, and molybdenum—can sometimes be identified as the primary limiter in certain isolated pristine lakes (such as Lake Tahoe in California and Nevada). Preserving the extraordinary clarity of such lakes calls for extraordinary prohibitions against soil disturbance, clearing, fertilizer use, and disposal of sewage effluent and other wastes within the watershed. There is even concern to limit escape of gaseous ammonia from sewage undergoing treatment.

Eutrophication may be a form of pollution of soil as well as water. Accidental nutrient inputs into valued natural landscapes may encourage unwanted growth or unwanted plants (i.e., weeds).

Haynes, R. J., and P. H. Williams. 1993. Nutrient cycling and soil fertility in the grazed pasture ecosystem. *Advances in Agronomy* 49:119–200.

Marschner, H. 1995. *Mineral nutrition of higher plants.* London: Academic Press.

Paul E. A., and F. E. Clark. 1996. *Soil microbiology and biochemistry.* San Diego: Academic Press.

Plant, R. E., A. Merner, G. S. Pettygrove, M. P. Vayssieres, J. A. Young, R. O. Miller, F. Jackson, R. F. Denison, and K. Phelps. 1999. Factors underlying grain yield spatial variability in three irrigated wheat fields. *Transactions American Society of Agricultural Engineers* 42(5):1187–1202.

Sharpley, A. N., and R. G. Menzel. 1987. The impact of soil and fertilizer phosphorus on the environment. *Advances in Agronomy* 41:297–324.

Stevenson, F. J. 1986. *Cycles of soil: C, N, S, P, micronutrients.* New York: Wiley.

Tisdale, S. L., W. L. Nelson, and J. D. Beaton. 1993. *Soil fertility and fertilizers.* New York: Macmillan.

Walsh, L. M., and J. D. Beaton, eds. 1973. *Soil testing and plant analysis.* Madison, Wis.: Soil Science Society of America.

## WEB RESOURCES

You can view nutrient deficiency and toxicity symptoms of plants at the following Web sites:
http://www1.iastate.edu/~rKillorn/nutrie~1.htm
http://phylogeny.arizona.edu/pubs/crops/az1007
   (citrus plants)

http://www.soil.ncsu.edu/lockers/Mikkelsen_R/
   Nutrientdeficiencies/plant_nutrient_deficiency_
   index.htm

http://www.agric.wa.gov.au/agency/Pubns/ff1/5c6
   4632.htm (canola)

A site without photos but with descriptions of deficiency symptoms in wheat, corn, soybean, and alfalfa can be found at http://www.agohio_ state.edu/~ohioline/b827/b827/index.html

## SUPPLEMENT 10–1   Fertilizers: Specifications and Calculations

### Physical Properties and Handling

Solid fertilizers are still the most widely used. They are easy to transport, store, and use with common equipment. Particle size and the method of mixing are important.

Particle size affects handling, spreading, and the rate of dissolution. Commonly, particle size ranges between 1 and 3 mm in diameter. Crystalline materials (e.g., rock, phosphate, KCl, ammonium sulfate)are ground and screened to the desired size. Fine powders (e.g., superphosphate, urea) are pelleted, or granulated, by moistening with steam or a water spray while the material is rolled in a rotating drum.

Solid fertilizers may be single compounds such as ammonium sulfate, S, calcium nitrate, and potassium chloride. Others, designed to provide two or more nutrients, are either homogeneous mixes or bulk blends. Homogeneous mixes, usually factory manufactured, have their component substances so closely compounded that every granule has the same composition (e.g., superphosphate is a homogeneous mix of calcium phosphate and calcium sulfate). Bulk blends, often prepared by a distributor to suit local soil and crop needs, result from the simple mixing of two or more granular materials, so that the blend contains granules of different composition. To prevent segregation during handling, the granules should be of similar size. Micronutrients are most conveniently applied along with other fertilizers; incorporation into a homogeneous mix is the most satisfactory method (e.g., Mo superphosphate, Zn superphosphate), but micronutrients can be added to a bulk blend as a small amount of solution sprayed on during mixing.

Liquid fertilizers, which are becoming more popular, include solutions and suspensions. Solutions should resist **salting out** (precipitating) caused by oversaturation when the temperature drops. Unanticipated salting out clogs equipment and causes uneven application. The fertilizer solutions most used are ammonia, ammonium phosphates, and ammonium polyphosphates. Polyphosphoric acid, made by condensing phosphoric acid molecules into chainlike molecules at high temperature, can be ammoniated to produce high N and P concentrations (e.g., 10 to 11 percent N, 15 percent P). Higher concentrations and the inclusion of Ca, Mg, and K are possible in suspensions of fine crystals. Agitation and the addition of clay keep the crystals suspended and reduce clogging of machinery. Liquid fertilizers can be easily metered into irrigation water.

The corrosiveness of liquid fertilizers limits the use of copper, brass, and iron equipment. Less corrodible are stainless steel, some aluminum alloys, glass, polyethylene, vinyl, rubber, and neoprene.

### Description: Analysis and Labeling

Most states and countries require that fertilizer labels clearly indicate the analytical values for all nutrients claimed to be present. In the United States the label usually at least indicates total N, soluble P, and soluble K. Additional information might include S, Mg, Ca, and micronutrients, if claimed; water-insoluble N, if slow N release is claimed; separate values for ammonium and nitrate; and the potential acidity or basicity, commonly expressed as the amount of calcium carbonate that would either neutralize the fertilizer's acidity or yield the same amount of alkalinity (e.g., "pounds $CaCO_3$ equivalent" per ton of fertilizer).

Nitrogen analyses are expressed as a percentage (by weight) of the element N, but the U.S. fertilizer trade expresses P, K, Ca, and Mg as if they were present as the oxides $P_2O_5$, $K_2O$, CaO, and MgO, respectively. The oxide basis is a relic from the nineteenth century. Conversions between the oxide and elemental basis are simply derived from the atomic weights (O = 16, P = 31, K = 39, Ca = 40, Mg = 24). To convert values of nutrient mass, weight, or percentage (by weight) from the element basis to the corresponding oxide basis, multiply by 2.29 for $P_2O_5$, 1.205 for $K_2O$, 1.40 for CaO, and 1.67 for MgO. To convert from oxide to element basis, divide by the same factors. Nutrient quantities and analytical values always look bigger on the oxide basis.

Listing N, $P_2O_5$, and $K_2O$ values in that order has led to useful jargon to identify fertilizers by the three numbers. Thus a blend with 10 percent each of N, $P_2O_5$, and $K_2O$ is called 10–10–10; an ammonium phosphate with 11 percent N and 48 percent $P_2O_5$ is 11–48–0; and superphosphate with 20 percent $P_2O_5$ is 0–20–0.

Fertilizer rate recommendations are often expressed as amounts of nutrient per unit area (or per plant in orchard and vineyard practice). The recommendation should also specify whether the elemental or oxide basis is intended, except in countries where the oxide basis has been dropped. Farmers calculate the required amount of the chosen fertilizer from the recommendation and the fertilizer's analysis. Other agriculturalists should also be able to do this arithmetic and convert between American and metric units (1 kg = 2.2 lb, 1 ha = 2.47 ac, 1 kg/ha = 0.90 lb/ac, 1 lb/ac = 1.12 kg/ha).

The following is a sample calculation:

The recommended phosphate addition is 30 lb P per acre. How much phosphate fertilizer with 20 percent $P_2O_5$ is needed?

*Procedure:* Convert the recommendation from element basis to oxide basis and then multiply by the amount of fertilizer per unit of $P_2O_5$:

$$fertilizer = \frac{30\ lb\ P}{1\ ac} \times \frac{2.29\ lb\ oxide}{1.0\ lb\ P} \times \frac{100\ lb\ fertilizer}{20\ lb\ ozide}$$

$$= 343.5\ lb/ac\ of\ fertilizer$$

## Other Examples to Check for Correctness

| Recommendation | Application[a] |
|---|---|
| 20 lb/ac $P_2O_5$ | 100 lb/ac 0–20–0 |
| | or 42 lb/ac 11–48–0 |
| 100 kg/ha N | 300 kg/ha 33–0–0 |
| | or 910 kg/ha 11–48–0 |
| 30 lb/ac $K_2O$ | 50 lb/ac 0–0–60 |
| 30 lb/ac K | 61 lb/ac 0–0–60 |

[a]Values are rounded up to the nearest whole number to ensure that adequate nutrients are added.

# 11

# ACIDITY AND SALINITY

## OVERVIEW

*Acidic soils are most common where high rainfall and free drainage favor leaching and the biological production of acids. Saline, sodic (high Na), and alkaline soils are most common where low rainfall, high evaporation, and limited drainage inhibit leaching and promote the accumulation of salt from incoming water. Both acid-related and salt-related constraints are more costly to remedy and control than are nutrient deficiencies.*

*Acid-related factors limiting plant growth include acidity itself, Al toxicity, Mn toxicity, and Ca deficiency. These tend to occur together, and they interact. They all are corrected by adding large quantities of liming materials (alkaline soil amendments containing Ca) or by flooding. Changes in fertility management and the use of tolerant crops sometimes help.*

*Salt-related factors limiting plant growth include salinity itself (i.e., a high concentration of salts), toxicities of individual components (e.g., Cl, Na, boron), poor soil permeability due to sodicity (high exchangeable Na), and micronutrient deficiencies associated with a high pH. These factors are managed by avoiding sensitive crops and improving water quality, irrigation management, and leaching. Effective leaching often calls for amendments to raise soluble Ca and for provision of artificial drainage and water disposal.*

**233**

**Acidification** and **salinization** are major natural soil-forming processes that can also become major land use problems. The two have similarities: Each is a complex of several related properties and processes. Each is characteristically expensive to remedy, compared with nutrient deficiencies, and in both cases the common remedies involve soil amendments containing Ca. Each involves soil cation exchange properties. Each is related to the soil-forming factors through effects on processes of solute leaching and accumulation. But acidity and salinity also differ in many respects. Most important, acidity arises from biological activity coupled with the vigorous leaching of salts, whereas salinity and alkalinity arise from the accumulation of salts and bases with inadequate leaching. Consequently, acidity is typically associated with high rainfall and good drainage, whereas salinity and alkalinity are associated with aridity and inadequate drainage.

More than most topics, these two have complexities important to some readers and not to others, and so this chapter has several supplements. Supplement 11–1 is a background on the chemistry of acidity and alkalinity, and the other supplements elaborate on parts of the main story.

## 11.1  ACIDITY AND RELATED PROBLEMS

### 11.1.1  Nature and Causes

Soils become acidic because of prolonged recurrent leaching, coupled with the input of acids (substances capable of releasing $H^+$). Which acids are involved? Water itself is an extremely weak acid (see Supplement 11–1), but soil water contains stronger acids, mostly of biological origin. Carbonic acid is the most abundant and is formed when $CO_2$ gas from respiration and fermentation dissolves in water.

$$CO_2 + H_2O \rightarrow H_2CO_3 \text{ (carbonic acid)} \qquad (1)$$

$$H_2CO_3 \rightarrow H^+ + HCO_3^- \text{ (bicarbonate)} \qquad (2)$$

Even as rainwater drops through unpolluted air, it absorbs enough $CO_2$ to lower its pH from 7.0 to about 5.5. In soil, microbes and roots greatly increase the $CO_2$ partial pressure and concentration of carbonic acid. Nitric and sulfuric acids, though less abundant than carbonic acid, are stronger (i.e., more fully dissociated). Both occur in air polluted by volcanoes and the burning of vegetation and fossil fuels, but even more of them are produced in soil by the bacterial oxidation of ammonium ions and sulfur (see Chapter 8). Finally, roots and soil microbes release organic acids or their anions. (Supplement 11–3 lists soil reactions and processes that produce $H^+$.)

But mere leaching and acid input do not explain cumulative acidification. Why do hydrogen ions ($H^+$) accumulate in the soil, lowering pH? Why do they not leach out? To explain this, two other things are important, the anions of the acids (their conjugate bases) and cation exchange (Figure 11–1). With every addition of acid to soil, some $H^+$ is adsorbed onto colloids, by its attachment to pH-dependent charge sites or by the exchange of $H^+$ with Ca, Mg, K, or Na. The retention of $H^+$ is aided when leaching removes the anions of the acids along with the displaced Ca, Mg, Na, and K that would otherwise compete with $H^+$ for attachment to the colloids. Progressively, over the years or centuries, the soil's **exchange complex** and the soil solution together become more acidic.

This mechanism explains why acidic soils develop mostly under conditions of vigorous leaching and biological activity. It also explains why acidic soils often contain little exchangeable and soluble Ca, Mg, K, and Na and why soil acidity develops most readily on parent materials that contain little of these elements. Further, it is the leached metal cations and base anions, moving in rivers and groundwater, that eventually accumulate in the alkaline salty soils and waters of poorly leached drainage basins.

As a natural process, acidification is favored by high rainfall, low evaporation (cold, humid climate), free drainage, ample oxidative biological activity that produces acids, and limited input of Ca, Mg, K, and Na from the soil parent material. The most extensive acidic soils are under upland humid forest or shrub land (tropical, temperate, or boreal) on high-silica rocks such as granite and quartz sandstones.

Acidification is accelerated by certain agricultural practices. Some are deliberate, such as the application of acidifying agents to neutral and alkaline soils (see Section 11.2). Other practices selectively promote $H^+$-producing processes in the cycling of N and other plant nutrients. The most important of these processes are nitrification,

$$NH_4^+ + 2O_2 \rightarrow NO_3^- + 2H^+ + H_2O \qquad (3)$$

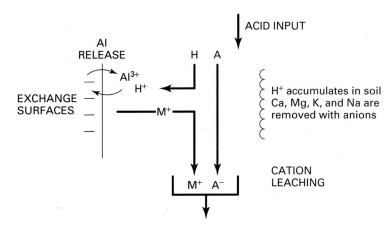

FIGURE 11–1 Roles of acid input, cation exchange, and leaching in soil acidification.

$HA$ = acids ($H_2CO_3$, $HNO_3$, $H_2SO_4$, organics)
$H^+$ = hydrogen ion
$A^-$ = anions ($HCO_3^-$, $NO_3^-$, $SO_4^{2-}$, organic)
$M^+$ = exchangeable metal cations ($Ca^{2+}$, $Mg^{2+}$, $K^+$, $Na^+$)

and the excessive removal of cationic plant nutrients—$NH_4^+$, $K^+$, $Ca^{2+}$, and $Mg^{2+}$—for which the plant exchanges $H^+$ into the soil. Practices that promote these acidifying processes include the production and removal of heavy harvests, the accumulation of soil organic matter, and the input of large amounts of N by means of ammonium fertilizers or biological nitrogen fixation (see Supplement 11–3).

Acidic soils normally develop high levels of dissolved and exchangeable Al. Aluminum is a major element in the aluminosilicates and aluminum oxides of the clay fraction. As the soil becomes more acidic, these minerals become more soluble, releasing Al into solution (Figure 11–2). As its concentration rises, $Al^{3+}$ displaces $H^+$ and other exchangeable cations, occupying much of the exchange capacity. More than 15 or 20 percent of the exchange capacity occupied by Al spells trouble for many crop species. Exchangeable $H^+$ itself remains minor, even in acidic soils.

Manganese toxicity is likewise connected with soil acidity, though less closely than is Al toxicity. Manganese toxicity occurs in acidic soils that are also wet and high in organic matter. These conditions favor the reduction of solid manganese oxide to produce $Mn^{2+}$ ions (Supplement 11–2). In acidic flooded soils, ferrous ($Fe^{2+}$) iron also reaches high concentrations in solution and may sometimes become toxic to rice crops.

## 11.1.2 Effects of Acidity, Al, and Mn

Acidity itself, at pH 5.5 or lower, can inhibit the growth of sensitive plant species, although it has little effect on insensitive species even at pH as low as 4. This pH effect is compounded and often overshadowed by Al toxicity, Mn toxicity, Ca deficiency, and Mo deficiency. These **soil acidity factors** tend to occur together, and they also interact, enhancing one another's effect. For instance, H, Al, or Mn will cause worse damage if calcium is at levels that, although low, would still be sufficient in a neutral soil. Figure 11–3 shows examples of such interactions.

Roots are commonly the first organs to show injury due to acid, Al, or lack of Ca. They become stunted, stubby, and sometimes brown (Figure 11–4). The stubbiness, most distinct in Al toxicity, is connected with the inhibition of cell division in the meristems of root tips. Stunted roots have difficulty getting immobile nutrients (e.g., P), which are frequently deficient in acidic soils. Because of poor root growth and reactions between Al and P, Al toxicity in plants resembles P deficiency, and the correction of Al toxicity can reduce the need for fertilizer P (Figure 11–5).

Molybdenum deficiency inhibits nitrate reduction in ordinary plants and nitrogen fixation in legumes, because Mo is an essential constituent of key enzymes in both processes. Leaf scorch caused by

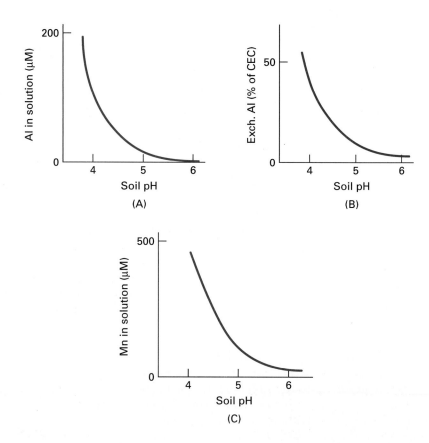

**FIGURE 11–2** The effect of soil pH on availability of aluminum and manganese. (A) Al concentration in soil solution. (B) Exchangeable Al. (C) Mn concentration in solution.

nitrate accumulation is one of the symptoms of Mo deficiency in nonlegumes, and chlorosis caused by N starvation is the common symptom in legumes.

Manganese toxicity produces varied patterns of leaf distortion, yellowing, and necrosis, depending on the plant species.

Calcium deficiency aggravates effects of acidity, Al, and Mn. Occasionally, plants also show symptoms of simple Ca deficiency itself. Leaf petioles collapse so that the leaf hangs wilted and dead. In certain fruit and vegetable crops, Ca deficiency causes localized death of tissues during storage after harvest.

Plant species and varieties have different sensitivity to soil acidity factors (Table 11–1 and Figure 11–6). For example, when given adequate fertilizer N, maize tolerates soil acidity well, but this tolerance is little help in crop rotations in which maize depends on nitrogen fixation by alfalfa and soybean, which are sensitive to soil acidity (Figure 11–7). Successful acid-tolerant cultivars have been developed in rice, wheat, and forage legumes.

### 11.1.3 Management of Soil Acidity

Many soils require no management of soil acidity. Their pH remains between 5.5 and 8.5, a level at which plants do well unless they are unusually sensitive. Soil pH remains steady if acid inputs are approximately balanced by alkalinity inputs. Alkalinity comes from several sources: weathering of carbonate and silicate minerals, microbial reduction reactions, and bicarbonate in irrigation water. Alkaline soils may need acidification (see Supplement 11–3). Even neutral or slightly acidic soils need acidification for crops especially sensitive to alkali-related conditions such as Fe deficiency. Usually, however, crop plants and many others grow better if the soil pH is raised or kept above pH 5 or 5.5.

One way to correct acidity is to flood the soil; microbial reduction processes usually raise the pH to 7 (see Section 8.3), where it remains as long as the soil stays flooded. This method suits only crops such as rice and taro that tolerate waterlogging. For most

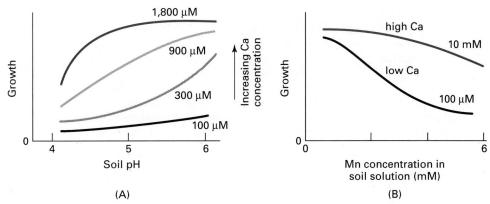

(A)                                           (B)

**FIGURE 11–3** Typical interactions between Ca and other acidic soil factors on plant growth. (A) Growth and nodulation of a legume dependent on N fixation becomes less sensitive to acidity (low pH) if Ca concentration is raised. (B) Growth of several species is less sensitive to high Mn if Ca is also high. These pH values refer to measurements in water-saturated soil pastes or suspensions in dilute salt solution (e.g., 0.01 M $CaCl_2$).

**FIGURE 11–4** Stunting of roots by Al toxicity. Healthy (left) and Al-affected (right) plants of *Trifolium subterraneum*. Calcium deficiency and extremely low pH can have similar effects. (Photo: Keith Helyar.)

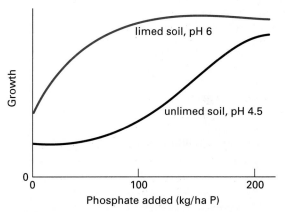

**FIGURE 11–5** Lime and phosphate interaction affecting growth of alfalfa on an acidic soil with severe Al toxicity. Liming corrects Al toxicity and reduces the apparent P requirement. Very high levels of P precipitate Al, producing the same benefit as the combined effect of lime with moderate P levels.

## TABLE 11–1
Relative Acid Sensitivity in Some Agricultural Plants

| | |
|---|---|
| Highly sensitive | Alfalfa, common bean, pea, red clover, crown vetch, *Leucaena,* spinach, cotton |
| Sensitive | Cabbage, wheat, soybean, white clover, sorghum |
| Moderately tolerant | Peanut, potato, oats, rice, rye, corn |
| Tolerant | *Stylosanthes,* kudzu, pineapple, tea, coffee, blueberry |

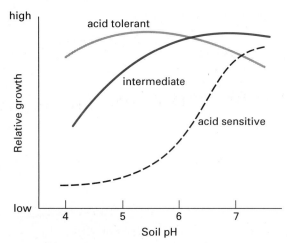

**FIGURE 11–6** Differences in acidic soil tolerance among species. Typical growth responses of sensitive and tolerant species on mineral acidic soils.

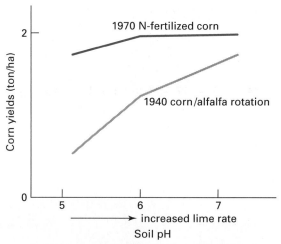

**FIGURE 11–7** Reduction of lime requirements following switch from N-fixing alfalfa to fertilizers as the source of N for maize.

crops, adding an **amendment** such as lime (liming) is the suitable correction.

**Liming**   Liming is the addition of alkaline material to soil. An alkaline substance provides a conjugate base, such as carbonate, hydroxide, or silicate, that reacts with $H^+$. Commercial liming materials must be cheap, available, and safe. In particular, the cation that accompanies the conjugate base should be beneficial or harmless in the large quantities required. Favored liming materials are crushed limestone, containing calcite ($CaCO_3$), and dolomitic limestone, containing calcium, magnesium, and carbonate. Plant ash and several kinds of furnace slag are also used. Calcium and magnesium hydroxide are usually too costly; potassium and sodium carbonates and hydroxides are too costly, caustic, and damaging to soil structure; and apatite and calcium silicate are too costly and slow to act.

Many things happen when lime is added to acidic soil (Supplements 11–3 and 11–5 describe some of the details). Primarily, the base—that is, hydroxide, carbonate, or silicate—reacts with $H^+$, thus:

$$OH^- + H^+ \rightarrow H_2O \qquad (4)$$

$$CO_3{}^{2-} + 2H^+ \rightarrow CO_2 + H_2O \qquad (5)$$

$$SiO_3{}^2 + 2H^+ + H_2O \rightarrow H_4SiO_4 \qquad (6)$$

As $H^+$ is reacted, the pH rises. One consequence is that the higher concentration of $OH^-$ immobilizes free Al, by precipitation of aluminum hydroxide:

$$Al^{3+} + 3OH^- \rightarrow Al(OH)_3 \text{ precipitate} \qquad (7)$$

This is not all. The reserve acidity must be reduced as the pH of the soil solution is raised, just as the reserves of a nutrient must change if its availability is changed.

**pH buffering**   Reserve acidity comes from colloid surface reactions that yield $H^+$ and $Al^{3+}$ to react with the added base ($OH^-$). These surface reactions are

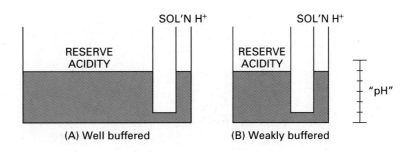

FIGURE 11–8 A model for reserve acidity of the soil's solid phase (tank) in equilibrium with the small amount of H in solution (sidearm gauge), in well-buffered soil (A) and weakly buffered soil (B).

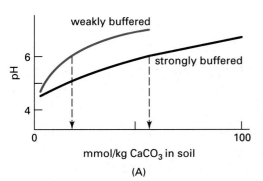

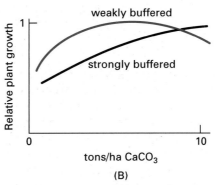

FIGURE 11–9 (A) Typical buffering curves. Vertical arrows indicate lime requirement to adjust the soils to pH 6. (B) Lime responses by plants on soils of different buffering capacity.

1. The release of protons from pH-dependent charge sites:

$$\text{surface} - \text{H} + \text{OH}^- = \text{surface negative sites} + \text{H}_2\text{O} \quad (8)$$

2. The displacement and precipitation of exchangeable Al:

$$\text{Al}^{3+} \text{ (exch)} + 3\text{OH}^- = \text{Al(OH)}_3 \text{ precipitate} \quad (9)$$

The $\text{H}^+$ and $\text{Al}^{3+}$ ions in the soil solution are important because they directly influence roots, microbes, and soil reactions, but their quantity is negligible relative to the amounts of acidity that the soil solid phase can release as the soil pH is raised. The solid-phase reserve acidity acts as a **pH buffer** system (Figures 11–8 and 11–9). It must partially react if the whole system is to come to equilibrium at a higher pH. The buffering also works in reverse: Solid-phase acidity must increase if the solution pH is to be dropped. The capacity to release and take up Al and H is the soil's *buffering* capacity, which determines the amount of lime (or acid) needed to bring about a given shift in pH.

In general, an acidic soil is better pH buffered if it has much clay and humus and if these colloids have high reactivity, cation exchange capacity, and pH-dependent charge.

Surface reactions on colloids provide fast buffering. The dissolution and precipitation of carbonate and silicate soil minerals provide slower buffering, so that in the long run a soil can resist acidification better if it is "young"—rich in weatherable silicate or carbonates. Unfortunately, most naturally acidic soils are not rich in these materials.

**Diagnosis of lime needs** As with the nutrient deficiencies, field trials with the relevant plants form the basis for diagnosing whether lime is needed and in what quantity. The field trials provide the basis for interpreting and calibrating other diagnostic tests.

The need for lime can be tested quickly and easily by measuring the pH of representative topsoil samples, with indicator dyes or by inserting the electrodes of a glass-electrode pH meter into a soil paste or suspension (see Supplement 11–5). Different procedures give different values, so the procedure must

## TABLE 11–2
Approximate Lime Requirements in Soils of Different Textures

| Soil texture | Amount of lime to raise pH from 4.5 to 5.5 |
| --- | --- |
| Sand | 1,000 kg/ha |
| Sandy loam | 2,000 kg/ha |
| Clay loam | 4,000 kg/ha |
| Organic soil (peat) | 8,000 kg/ha |

Note: 1 kg/ha equals approximately 1 lb/ac.
Source: Adapted from California Fertilizer Association, 1975.

be specified. Some testing services also use exchangeable Al or available Mn data. Plant analyses and plant symptoms help diagnose Mn toxicity and Mo deficiency but have little use in diagnosing other acidic soil problems.

To determine the amount of lime required, one can mix samples of soil with different measured amounts of lime, add water, let them react, measure the resulting pH values, and interpolate the lime quantity that gives the desired pH. The desirable pH is indicated by field trials with the crop and typically lies between 5 and 6 (Figure 11–9). There are less laborious and more arbitrary methods (see Supplement 11–4), and reasonable guesses can be made from soil texture and color, which indicate buffer capacity because they reflect clay and humus content (Table 11–2).

The lime requirement over the years depends little on the soil buffer capacity; rather, it depends on the annual rates of acid and alkali inputs from biological sources, water, weathering, and fertilizer. What does depend on the buffering capacity is the frequency of liming and the size of the dose. Thus well-buffered soils need larger but less frequent additions than weakly buffered soils, and a lime quantity suitable for a well-buffered soil would overlime a weakly buffered soil.

Subsoil acidity is difficult to manage. The clay-rich B horizons of Ultisols and Alfisols sometimes have a pH below 5, high exchangeable Al, and low exchangeable Ca. The combination of density and acidity prevents root access to subsoil water and causes severe drought stress between rains. Deep placement of lime is expensive, and the neutralizing effect of surface applications penetrates to subsoil exceedingly slowly. Gypsum (calcium sulfate) helps; although it has little effect on the topsoil pH, it increases free Ca

and sulfate so that together they leach down and ameliorate the subsoil's Ca deficiency.

**Management to reduce lime requirements**   In many agricultural systems, liming is clearly cost-effective, but this is not true where the returns are small per unit land area or where distance and terrain make it costly to transport lime, which is needed in large amounts. When lime is expensive and crop values limited, very acidic soils are left uncropped or are used only in ways that remove fewer cation nutrients and acidify soil more slowly (e.g., animal production versus crops). Nitrogen management can be modified to reduce acidification. For example, sparing use of $NH_4^+$-releasing fertilizers will reduce the rate of acidification, and, unlike biological $N_2$ fixation and the common N fertilizers, calcium nitrate does not lead to acidification or Ca depletion.

Certain species or varieties tolerate acidity better than others do (Table 11–1). Tolerance is more likely to be exploited in pasture or forestry with low inputs than in crop agriculture.

Alternative alkali sources might be exploited. Many irrigation waters contain enough bicarbonate to keep acidification in check. Ash from the burning of forest or savanna controls soil acidity in slash-and-burn or shifting cultivation, still an extensive kind of agriculture. Burning to dispose of crop residues also produces alkali.

Lime, like fertilizer, can be used more sparingly by several placement techniques. Drilling lime with seed, at only a few hundred kilograms per hectare, sometimes produces benefits similar to those of a ton of lime broadcast. Even cheaper, pelleting seed with a coat of lime helps pasture legumes through a critical stage of seedling establishment. Although soybean is acid-sensitive, it is grown successfully on soils with pH below 4.1, such as in Vietnam's Plain of Reeds, following the burning of trash from the previous rice crop. The people scrape the alkaline ash into the soybean planting holes to concentrate it for maximum effect.

**Special cases of extreme acidity** Mining and drainage of wetlands sometimes results in extreme acidity (pH < 4) in the resulting soils and drainage water. The extreme acidity is caused by oxidation of sulfide containing minerals in the mine wastes or waterlogged soils.

Sulfide-bearing minerals exposed to the atmosphere by mining include pyrite ($FeS_2$), pyrrhotite (FeS), chalcopyrite ($CuFeS_2$), arsenopyrite (FeAsS), and cobalite (CoAsS). Pyrite, the most common in drained soils, forms in strongly reduced (redox potential of $-400$ mV) waterlogged soils. Upon exposure to air, the iron and sulfur oxidize to form secondary minerals and sulfuric acid. Examples of the oxidation reactions are shown in Equations 10 and 11:

$$2FeS_2 + \frac{15}{2} O_2 + 7H_2O = 2Fe(OH)_3 + 4H_2SO_4 \quad (10)$$

$$FeS_2 + 14Fe^{3+} + 8H_2O = 15Fe^{2+} + \quad (11)$$
$$2SO_4^{2-} + 16H^+$$

The oxidation of pyrite by oxygen shown in Equation 10 is much slower than the oxidation by $Fe^{3+}$ shown in Equation 11. The reaction in Equation 11 is catalyzed at pH values of <4 by the bacteria *Thiobacillus ferrooxidans*, which rapidly oxidize $Fe^{2+}$ to $Fe^{3+}$, regenerating the supply of $Fe^{3+}$ for further oxidation of pyrite. The sulfuric acid produced is a strong acid—that is, it fully dissociates, contributing large amounts of $H^+$ to soils, surface runoff water, or groundwater through leaching. Under severely acidic, oxidizing conditions, pale yellow jarosite, $KFe_3(OH)_6(SO_4)_2$, is formed. This mineral is a diagnostic feature of acidic sulfate soils.

The toxicity, increased solubility of aluminum, unavailability of phosphate, low base status, and nutrient deficiencies common to acidic soils are also common to soils and mine tailings with extreme acidity. In addition, the environmental problems that accompany severe acidity, especially those associated with mining debris, include increased solubility of toxic metals, including Cu, Pb, Zn, Ni, and Mn. The combination of extreme acidity and toxicity of drainage waters from mine tailings is a serious environmental problem on millions of hectares around the world.

## 11.2 SALINITY AND RELATED PROBLEMS

### 11.2.1 Nature and Causes

**Saline** soils are soils that contain large amounts of soluble salts, appreciably more soluble than calcium sulfate. Most commonly, these are salts of Na, Ca, and Mg, with chloride, sulfate, and bicarbonate. An accepted measure and criterion of salinity is the electrical conductivity (EC) of the soil's saturation paste or saturation extract (see Supplement 11–5). The EC is easy to measure, and it relates closely to the concentration of salt (Figure 11–10) because electricity moves through solutions mostly by way of ion movement, water itself being a poor conductor. Conventionally, soils are considered saline if their EC exceeds 4 dS m$^{-1}$. Many plants suffer at this level.

**Sodic** soils contain Na as a significant proportion of their total exchangeable cations (more than 10 percent or so). Saline soils tend to be sodic as well, because soluble Na salts are more abundant than others. However, not all saline soils are sodic—they might have salts of Ca and Mg, for instance—nor are all sodic soils saline. Useful criteria of sodicity are the **exchangeable sodium percentage (ESP)** and the **sodium adsorption ratio (SAR)** (see Supplement 11–5). The ESP is the exchangeable Na expressed as a percentage of the total exchangeable cations. The SAR is a modified ratio of Na to other major cations (Ca and Mg) in the saturation extract (representing the soil solution). A soil's ESP and SAR are positively related (Figure 11–10C) because a soil's solution cations and exchange cations are nearly always in equilibrium with each other. Critical values for ESP vary from 15 percent down to 5 percent, depending on soil salinity.

Sodic and saline soils are usually **alkaline.** The pH is commonly above 8.3 to 8.5, at which $CaCO_3$ precipitates at atmospheric $CO_2$ pressures. It sometimes approaches 10 in sodic soils, especially if their salinity is low. (Salts keep the pH down somewhat by pushing $H^+$ off exchange sites into solution.) Acid exceptions include reclaimed salt marsh soils in which the oxidation of sulfides, in the absence of $CaCO_3$ from shells, produces sulfuric acid.

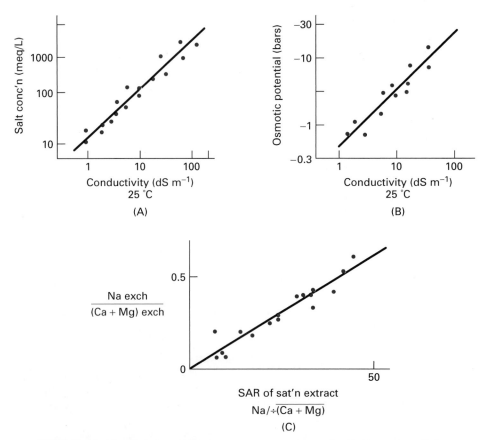

**FIGURE 11–10** Relationships between measures of salinity and sodicity. (A) Salt concentration in water (or soil extract) is directly related to the easily measured electrical conductivity. (B) The osmotic water potential is decreased as salt concentration increases. (C) The ratio of Na to other cations in the exchange phase relates directly to the ratio in the soil solution.

Alkalinity, sodium, and salts originate in mineral weathering. They reach high concentrations wherever water, carrying dissolved salts, collects, evaporates, and leaves the salts behind (Figures 11–11, 11–12). Water evaporates, but salts do not.

This idea, so obvious and easy to overlook, explains almost everything important about the development and control of soil salinity. It is the reason that saline and sodic soils develop in semiarid basins of limited drainage; the reason that irrigation often adds more salt than it removes, even when the irrigation water seems harmlessly low in salt; and much of the reason that the sea is saline.

## 11.2.2 Effects of Salinity, Sodicity, and Alkalinity

**Salinity** Damaging levels of salts in solution inhibit the growth and development of plants. The effects differ depending on climate, soil water, salt composition, kind of plant (Figure 11–13 and Table 11–3), and the plant's stage of development.

Salts and other solutes lower the osmotic component of the water potential (see Chapter 5). That is, water tends to move into solutions of high solute concentration and becomes harder for plants to remove from the soil as the concentration of solutes

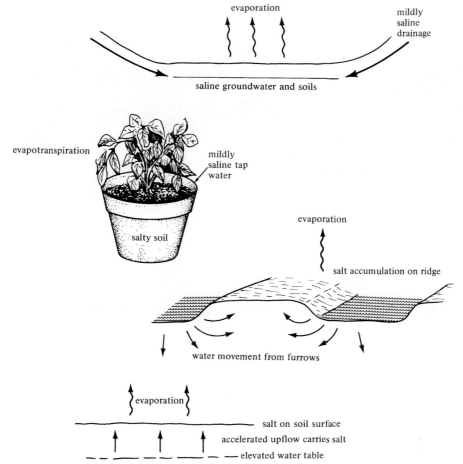

**FIGURE 11–11**   Salt accumulation: four examples of the movement and evaporation of water. Salts move in water. Water evaporates; salts do not.

rises. Water enters plants only if the water potential is lower inside the roots than outside them. To absorb water from a salty soil solution, the plant must increase the concentration of solutes inside its cells. The solutes used for this osmotic regulation are either organic compounds made by the plant or salt ions taken from the soil. Salt accumulation is a satisfactory means of osmotic regulation if the plant's cells can tolerate internal salt.

Plants that readily accumulate the common salt ions are called **halophytes,** which are common among the salt-tolerant grasses, herbs, and shrubs of deserts, shorelines, and salt marshes. They protect saline lands from erosion and they feed wildlife and livestock, but few are crop species. Crops with a degree of halophytic behavior include beet, date palm, jojoba, and barley. However, most crop plants are not halophytic; instead, they lower their internal water potential by accumulating organic solutes, and they attempt, with varying success, to exclude salt ions.

Evidently salt does not stop plants from acquiring water; they all can regulate their osmotic potential. Why, then, does salt inhibit growth? One reason is that osmotic regulation costs energy to make organic solutes or to take up ions and accumulate them in safe compartments. Another reason is the toxicity of one or more of the ions of the salt.

Plants suffering from salt damage may show leaf scorching, indicating sodium or chloride toxicity. Or leaves may prematurely yellow, wither, and drop from

**FIGURE 11–12** Salt (light-colored areas) frequently accumulates in patches where water accumulates and evaporates, often at the ends of an irrigated field. Note also the accumulation of salt on the ridges between irrigation furrows and the failure of plants to establish, survive, and grow. (Photo: D. M. Westphal, U.S. Bureau of Reclamation.)

**TABLE 11–3**

Relative Salt Sensitivity of Some Agricultural Plants

| | |
|---|---|
| Highly sensitive | Common bean, chickpea, white clover, red clover, celery |
| Sensitive | Alfalfa, corn, oats, rice, Sudan grass, ryegrass, carrot, lettuce, cabbage |
| Moderately tolerant | Rye, wheat, melilotus (sweet clover), tomato, cotton |
| Tolerant | Barley, sugarbeet, beet, spinach, Bermuda grass, date |

the plant. More often, the plants look dull and dark or show no sign except stunting. Frequently, salt damage is more severe in patches or at the low end of a field (Figure 11–12).

**Sodicity** Sodium is toxic to certain plants, especially if Ca concentrations are low. But the overriding effects of high Na are on the soil's physical properties. Sodic soils readily lose their structure, becoming impermeable, unless the salt concentration in the soil solution is kept high (Figure 11–14). Maintaining high salt concentrations is neither conducive to crop productivity nor possible at critical times when leaching just the topmost centimeter of soil disrupts that layer's structure and makes it an almost complete barrier to the further entry of water (see Chapter 15). Uncorrected sodicity makes most soils inhospitable to vegetation and unmanageable for agriculture.

Why does high Na combined with lowered salt concentration disrupt soil structure? It allows soil colloids—clay and humus—to disperse into individual hydrated particles instead of remaining **flocculated**—that is, stuck together into the clusters or packages from which stable structural units are built. (Supplement 11–6 outlines the physical chemistry of flocculation and dispersion.) Briefly, the hydrated exchangeable cations surrounding each colloid particle electrically repel their counterparts around the neighboring particles. If the repulsion is strong enough, it will overcome the short-range cohesive forces that hold particles together. High salt concentration helps the particles cohere by pushing the exchangeable cations against the clay surfaces, so that they interfere little with the cohesion between the particles. Exchangeable Ca, Mg, and Al ions stay close to the colloid surfaces, even at the low levels of salinity present in leached soils. By contrast, Na is so loosely held and so ready to hydrate that sodic colloids will disperse unless the total salt concentration is very high, comparable with that of seawater (about 500 mM).

Dispersion also explains the old terms *black alkali,* referring to nonsaline sodic soils, and *white alkali,* referring to saline soils. White alkali soils are so called because of the white deposit of salt on their surface. Their high salinity keeps humus flocculated and immobile. By contrast, in nonsaline, sodic black alkali soils, humus is dispersed, moves with water, and forms a black coating on the ground surface when the water evaporates.

**Alkalinity**   Extremely high pH (above 9) may itself directly injure some plants, but more important are nutrient deficiencies or toxicities induced by high pH. Calcium is immobilized because high pH promotes the formation of carbonate from $CO_2$, and carbonate precipitates with Ca, as $CaCO_3$. Micronutrient disorders related to alkalinity are Zn and Fe deficiencies and, less commonly, Mn deficiency, Cu deficiency, and boron toxicity. A high pH adversely affects the sorption behavior of these elements in the soil. These problems may be compounded by excess boron or bicarbonate in irrigation water and by exposure of alkaline calcareous subsoils when land is leveled for irrigation.

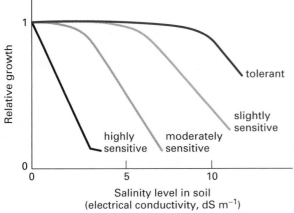

**FIGURE 11–13**   Plant growth responses to salinity.

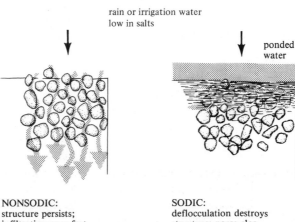

NONSODIC:
structure persists;
infiltration goes fast

SODIC:
deflocculation destroys
structure; pores clog
at surface

**FIGURE 11–14**   Effect of sodium on surface soil permeability.

## 11.2.3   Management of Saline and Sodic Soils

**Diagnosis of problems**   How much of a yield problem is due to salinity, sodium, alkalinity, or something completely unrelated? How bad is it? These questions are largely answered by the tests described in Supplement 11–5 and outlined here.

Salt and sodium often produce no clear symptoms in plants. The main symptom, poor growth, might not be evident unless healthy plants are nearby for comparison. Furthermore, plant-response trials, which are the best way of determining needs for nutrients and lime, are less useful in diagnosing salt and sodium problems. Plant analyses for Na and Cl and soil analyses for total salts (electrical conductivity) are useful indicators of salt problems, and high analytical values for exchangeable Na coupled with evidence of slow water infiltration and pH > 8.3 are good evidence that sodicity is a problem.

Water tests are helpful because irrigation water is a major source of salt, sodium, and alkalinity. The criteria for water appear to be more stringent than those for soil; for example, water is rated unacceptable with a salinity level (EC or total salts) that would be quite acceptable in soil. The reason is that water, applied again and again, mostly evaporates, leaving the salts to accumulate at increasing concentrations. Water quality measurements predict the likelihood of trouble developing from long-term use of the water. "Trouble" need not mean disaster; more often it means additional operating cost or less yield. If the water is poor, more must be done to prevent the accumulation of salts, Na, or alkalinity. Thus, poor water and soils make irrigation more expensive.

**Reclamation and control**   **Reclamation** of soil means treatment to correct a severe excess of salinity, sodicity, or alkalinity. It is the drastic fix needed if things are bad. *Control* is the preventive practices and management needed to keep problems from developing, recurring, or worsening. Reclamation and control depend on the same principles and use similar processes and practices. They both require

- *Amendments* that provide cations to displace exchangeable Na and flocculate soil colloids
- *Leaching* of salts and displaced Na
- *Drainage* for leaching
- An acceptable *sink* to receive drainage water

With luck, these are provided naturally. But more often, deliberate action is necessary to correct and control salt and Na in irrigated soils.

**Amendments for sodic soils**   Exchangeable sodium cannot be leached unless something is done to displace it into solution and to keep soil flocculated. Both objectives are achieved by maintaining a sufficient concentration of other cations in solution. In practice, the critical cation is calcium. It is abundant, and being divalent and little hydrated, it replaces Na readily and flocculates clay at quite low total salt concentrations.

**Gypsum**—$CaSO_4 \cdot 2H_2O$—is soluble enough to maintain Ca at a useful concentration. Saline soils whose surface horizons already contain gypsum need no amendments to leach readily. To other soils, gypsum is added, either by spreading it on the surface or dissolving it in the water applied for irrigation or leaching. Calcium chloride and calcium nitrate work at least as well as gypsum, but they cost more and their anions are harmful in large amounts. Calcium carbonate is too insoluble at high pH to be a worthwhile amendment for most sodic soils, but if the soil has $CaCO_3$ in the surface layer, acidifying agents (see Supplement 11–3) will dissolve the $CaCO_3$ and produce the same beneficial effect as gypsum does.

Reclamation of sodic soil requires large amounts of amendment, sometimes several tons per hectare. Even the periodic additions to control Na accumulation can be costly if the irrigation water has a high salinity and sodium adsorption ratio.

**Leaching methods**   The purpose of leaching is to lower the salt concentration and sodicity in the root zone and keep it low. Drip irrigation (see Chapter 6) offers a temporary, local control of salt; water trickling from properly placed emitters keeps salts moving away from the plant (Figure 11–15). Frequent conventional irrigation achieves the same thing, albeit less well. But the salts accumulated around the wet zone tend to move back into it, especially when the soil surface and root zone dry out. Accordingly, conventional leaching practice aims at deep leaching to move salts well below the root zone.

Deep leaching requires satisfactory natural or artificial drainage to remove salty water. It is hindered by impermeable horizons or **substrata** or by saline groundwater close to the surface. The leaching is commonly done by holding water **ponded** on the

(A) DURING IRRIGATION

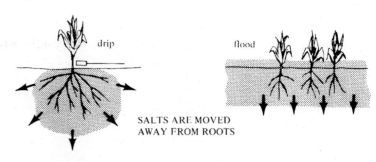

SALTS ARE MOVED
AWAY FROM ROOTS

(B) AFTER IRRIGATION

SALTS MOVE BACK

**FIGURE 11–15** Salt movement during and after irrigation. Shaded area shows wetted soil, and arrows indicate direction of salt movement.

land until enough has passed through (Figure 11–16). The time varies from days to months, depending on the soil's permeability. The depth of water needed ($Dw$) to remove salts by continuous ponding is a function of the depth of soil ($Ds$) to be reclaimed, the soil texture, and the fraction of initial salinity ($C/Co$) to be removed (Figure 11–17). To remove 90 percent of the salt from the top meter of a sandy loam–textured soil requires about 1 m of water. To remove the same fraction of salt from clay loam–textured soil requires more than 2.5 m of water (Figure 11–17).

Intermittent ponding, in which several smaller applications of ponded water are applied to the soil for shorter time periods, is another reclamation method. About one-third to two-thirds less water is needed for reclamation using intermittent ponding compared with continuous ponding, because the wetting and drying draw salts from the fine pores to the surface of aggregates, where the salts are removed by the leaching water. Alternatively, intermittent flooding or sprinkling can keep water moving down without ponding. Such "unsaturated" leaching consumes less water because it involves less salt retention in small pores bypassed by the large channels that carry most saturated flow (Section 3.4.3). Land

can be used during leaching for rice or salt-tolerant crops such as barley, but it often turns out to be both efficient and convenient to leach during a cool, wet, noncropping season. Floods or wet-season rains reduce the need for deliberate leaching and, if regular, improve the chance of supporting permanent irrigation agriculture without salinization.

**Leaching requirement (leaching fraction)**   The long-term or steady-state leaching requirement ($LR$) is important to planning irrigation and drainage systems, because it suggests how much water in addition to crop needs will be needed to control salinity. The leaching requirement ($LR$) is the amount of drainage water that must be produced ($Dw$), as a fraction of the amount of water applied to the land ($Iw$). Under the simplest conditions, $LR$ equals the ratio of salinity (salt concentration) in the irrigation water ($Si$) and the *acceptable* salinity in the soil solution as measured in the drainwater ($Sd$), so that

$$LR = Dw/Iw = Si/Sd \qquad (12)$$

The acceptable salinity level depends on the cost of control, the value of the crop product, and the crop's response curve (Figure 11–13). Tolerant crops permit the use of poorer and less water.

(A)

FIGURE 11–16  Leaching. (A) Leaching in progress. The plot on the right has ponded water standing on the surface. As this water infiltrates, it removes salt from the root zone. The plot on the left is not being leached. It is a control plot. (B) Results of a leaching treatment. The plots in the second row are continuously dark, indicating good, continuous plant cover, unlike the plots in the foreground, where plant growth is sparse due to inadequate salt removal. (Photos: Bert Krantz.)

(B)

The *LR* becomes an oversimplification if there are varying amounts of water of different quality. River water, for example, commonly becomes more saline as the stream level drops during a dry season or drought. Much land receives both salty irrigation water and salt-free rain. Furthermore, variability in land level and soil makes the average *LR* an underestimate of the amount of water needed to leach the whole field adequately.

Figure 11–18 (upper) shows chloride concentration (a measure of salinity) patterns for three leaching fractions applied using drip irrigation. Chloride is lowest at the emitter, directly beneath the plant row (R) and highest midway between emitters (M). The size of the chloride-free zone is largest with the highest leaching fraction (0.25) and smallest with the lowest leaching fraction (0.05). Throughout the profile, chloride (and salinity) decreases with depth. Observers have found that most of the plant roots are near the emitter. This is an example of salt control with proper irrigation.

After a rain (Figure 11–18, lower), the low-chloride zone is enlarged because the rainwater dissolves salt and carries it downward, farther from plant roots.

**Drainage and disposal**  Long-term salt control requires drainage into a suitable sink. Sometimes natural drainage and disposal are satisfactory. For example, the Nile Valley, cultivated since the beginning of agriculture, was leached annually by floods and drained through permeable sediments to deep groundwater systems extending out into hundreds of kilometers of desert east and west. Most of California's Sacramento Valley receives winter rains that leach salts into deep substrata. Areas less fortunate require artificial drainage (see Chapter 6), commonly with buried perforated pipe leading into a river, lake, bay, or ocean.

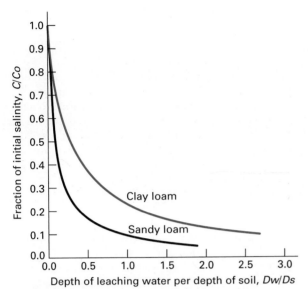

**FIGURE 11–17** Depth of water in excess of soil moisture depletion per depth of soil needed for reclamation by continuous ponding. (From Hanson, B., and S. Grattan. 1991. Reclaiming saline soils. *Soil and Water Newsletter,* Cooperative Extension, University of California, Davis, p. 5.)

Artificial drainage is expensive, and impermeable layers or excessive clay in any horizon (requiring extremely closely spaced drains) can raise the cost of drainage enough to make irrigation impractical.

Geographic and political realities of drainwater disposal provide a final constraint. Ancient Sumeria's agriculture declined, and modern Iraq's agriculture struggles in the same region, partly because of lack of a natural drainage outlet. Political, legal, and ecological constraints on drainwater disposal can also inhibit the use of saline land and water. Disposal into rivers adds salinity and sodium to afflict other users downstream. Nitrate, pesticides, or toxic substances such as selenium, leached with the salt, further limit the dumping of drainage into natural waters sensitive to such pollutants. The agricultural future of much of California's San Joaquin Valley is doubtful because of public hesitance to accept the costs and consequences of drainage into the San Joaquin River and San Francisco Bay.

## 11.3 SUMMARY

Acidification is a natural soil-forming process in humid climates, as is salinization in arid and semiarid climates. Acidity (excess hydrogen ions), salinity

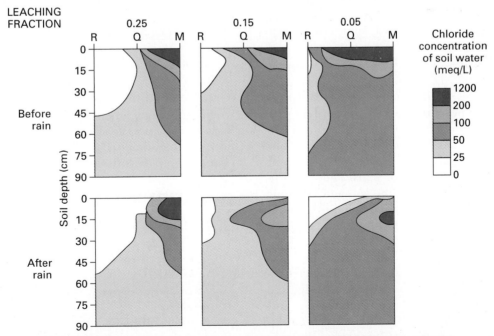

**FIGURE 11–18** Salt patterns under trickle irrigation. (From Hanson, B. 1991. Defining salt distribution patterns under trickle irrigation. *Soil and Water Newsletter,* Cooperative Extension, University of California, Davis, p. 4.)

# SOIL AND ENVIRONMENT
## *Acid Precipitation*

Normal rain and fog falling from a clean atmosphere is at pH 5.5 to 6.0 because of dissolved $CO_2$. Rain of lower pH, "acid rain," is increasingly prevalent due to atmospheric pollution with sulfuric and nitric acids from burning coal and petroleum. Acid rain and fog damage forests, mountain streams and lakes, buildings, artworks, and perhaps humans. It may measurably acidify some weakly buffered soils (low in clay, humus, and weatherable minerals), but in normally buffered soils any important effects are likely to be indirect consequences of damage to vegetation. Acid inputs from precipitation seldom approach the magnitude of biological production of nitric and sulfuric acid in fertile soils (see Supplement 11–7).

### Soil Acidification

Soil acidification by biological processes coupled with leaching is continuous in most well-drained soils with high rainfall. Since agriculture in these conditions typically accelerates acidification, it is not sustainable without continued monitoring and correction of acidity. Acidification might be slowed by managing fertility and vegetation so as to cut percolation losses of water and nitrate.

### Acidic Soils as Environmental Hazards

Mine spoils, open cuts, tailings from mineral processing works, road cuts, and the like often expose strongly acidified wastes or acidic subsoils of low fertility and poor physical structure. These visual scars on the landscape threaten neighboring land with their runoff and eroded sediments. Stabilizing them with vegetation requires fertilizer and liming treatments; it often calls for applied research to adapt established knowledge to unusual soils and plant species. How the substantial costs are distributed is important politics.

### Salinization and Salt Disposal

Salinization, sodification, and waterlogging have been common environmental outcomes throughout 4,000 years of irrigated agriculture in arid regions. Control of these problems requires disposal of drain water carrying salts and, frequently, nitrate, boron, selenium, or pesticides. Nearly always, disposal is into a river, lake, sea, or ocean. Oceans and open seas are generally acceptable sinks because they are big and salty and nobody owns them. Lakes and enclosed seas become less acceptable as salinization and eutrophication become obvious. Rivers are seldom acceptable sinks if vocal people downstream use them as water supplies. Toxic compounds aggravate disposal problems. Dumping salty water into a saline estuary might cease to be acceptable when the salt is laced with insecticides and nitrate; salt disposal into ponds becomes unacceptable when the selenium concentration becomes lethal to waterfowl.

(excess soluble salt), and sodicity (excess exchangeable sodium) are potential constraints to plant production. Acid-related limitations include the $H^+$ itself, Al and Mn toxicity, and Ca deficiency. Soils become acidic because of prolonged recurrent leaching, coupled with input of acids. They are usually corrected by adding alkaline soil amendments containing Ca. Salt-related factors limiting plant growth are high concentrations of soluble salts; toxicities of individual components such as $Cl^-$, $Na^+$, and boron; poor soil permeability due to high exchangeable Na; and micronutrient deficiencies associated with high pH. Salts and sodium originate from mineral weathering, rising groundwater, and irrigation water. Sodium and soluble salts accumulate where there is insufficient leaching. Salinity is managed by avoiding sensitive crops and improving irrigation management, drainage, and leaching. Sodicity requires the same management, plus addition of amendments to replace $Na^+$ on the exchange complex with $Ca^{2+}$.

## QUESTIONS

1. Explain the meaning of *acid, base, acidity, pH, buffering, lime requirement, salinity, sodicity, alkalinity, electrical conductivity, sodium adsorption ratio (SAR), exchangeable sodium percentage (ESP), flocculation, dispersion, halophyte,* and *leaching requirement.*

2. Neither soil acidity nor soil salinity results from one process alone. Construct a scheme to show

how the important processes work together in each case.

3. Suppose you had geographic data for a region you had never seen. Which combinations of climate, relief, soil parent materials, and vegetation would indicate the likelihood of (a) soil acidity problems and (b) soil salinity problems?

4. Suppose the only information you had for a region was a general map showing the soil orders (see Chapters 1 and 14). How helpful would that be in forecasting acid and salt problems? Which orders are likely to contain (a) acidic soils and (b) saline soils?

5. Why does Al become a dominant exchangeable cation in acidic soils? Why is it not a dominant cation in neutral soils?

6. Why does productive agriculture tend to acidify soils in humid areas?

7. Distinguish among (a) pH and acidity and (b) alkaline, saline, and sodic.

8. Explain what lime does to soil pH, cation exchange capacity, and exchangeable cations.

9. Why do some soils need more lime to raise their pH than others do? How can you tell how much lime will be needed?

10. Both SAR and EC are important attributes of irrigation water. What do these measures indicate? What is the danger of irrigating with water of (a) a high SAR and (b) a high EC?

11. How can you tell whether salt is affecting your plants?

12. Which conditions would you suspect if water infiltrated slowly into your alkaline soil? How would you confirm your suspicions? How could you correct the problem?

13. Why is drainage critical to irrigated agriculture?

14. Which of the following would be useful amendments for an acidic soil and which would be useful for a sodic soil? Explain why.

$CaCO_3$     $CaSO_4$     $Ca(NO_3)_2$     $CaSiO_3$

# FURTHER READING

## General Background

McBride, M. P. 1994. *Environmental chemistry of soils.* New York: Oxford University Press.

Rechcigl, J. C., ed. 1995. *Soil amendments and environmental quality.* Boca Raton, Fla.: CRC/Lewis.

Tan, K. H. 1993. *Principles of soil chemistry.* New York: Dekker.

## Soil Acidity

Adams, F. 1984. Soil acidity and liming. *Agronomy Monograph Series 12.* 2d ed. Madison, Wis.: American Society of Agronomy.

Dent, D. 1992. Reclamation of acid sulfate soils. *Advances in Soil Science* 17:79–122.

Hosner, L. R., and F. M. Hons. 1992. Reclamation of mine tailings. *Advances in Soil Science* 17:311–350.

Robarge, W. P., and W. D. Johnson. 1992. Effects of acidic deposition on forested soils. *Advances in Agronomy* 47:1–84.

Robson, A. D., and W. M. Porter. 1989. *Acid soils and plant growth.* San Diego: Academic Press.

Sutton, P., and W. A. Dick. 1987. Reclamation of acidic mined lands. *Advances in Agronomy* 41:377–437.

## Salinity

Ayers, R. S., and D. W. Westcot. 1976. Water quality for agriculture. *FAO Irrigation and Drainage Paper 29.* Rome: United Nations.

Bernstein, L. 1974. Crop growth and salinity. In *Drainage for agriculture, Agronomy Monograph Series 17,* (pp. 39–54), ed. J. van Schilfgaarde. Madison, Wis.: American Society of Agronomy.

California Fertilizer Association. 1975. *Western fertilizer handbook.* Danville, Ill.: Interstate Printers and Publishers.

Hagan, R. M., H. R. Haise, and T. W. Edminster, eds. 1967. *Irrigation of agricultural lands, Agronomy Monograph Series 11.* Madison, Wis.: American Society of Agronomy.

Rhoades, D. J. 1993. Electrical conductance methods for measuring and mapping soil salinity. *Advances in Agronomy* 49:201–253.

Richards, L. A., ed. 1964. Diagnosis and improvement of saline and alkali soils. *USDA Handbook 60.* Washington, D.C.: U.S. Government Printing Office. (A brief and comprehensive handbook. Out of date, but still valuable.)

van Schilfgaarde, J., ed. 1974. *Drainage for agriculture, Agronomy Monograph Series 17.* Madison, Wis.: American Society of Agronomy.

# SUPPLEMENT 11–1  Acids, Bases, and Exchangeable Bases

## Acids and Bases

An **acid** is a substance that releases $H^+$, and a **base** is a substance that combines with $H^+$. Thus an acid dissociates in solution, producing $H^+$ and a base (the conjugate base of the acid). These reversible reactions may be written in the following generalized form:

$$\text{acid} = \text{H ion} + \text{conjugate base}$$

For example,

$$HCl = H^+ + Cl^-$$

$$H_2SO_4 = 2H^+ + SO_4{}^{2-}$$

These examples, $HCl$ and $H_2SO_4$, are *strong acids*— that is, acids that dissociate almost completely in ordinary dilute solutions. Their bases, the chloride and sulfate anions, have little tendency to associate with $H^+$. *Weak acids,* by contrast, dissociate incompletely. In solution they form equilibrium mixtures of the undissociated acids, the anion base, and $H^+$. Examples are oxalic acid, orthophosphate, carbonic acid, and water:

$$HOOC(CH_2)_2COOH = H^+ + HOOC(CH_2)_2COO^-$$

$$H_2PO_4{}^- = H^+ + HPO_4{}^{2-}$$

$$H_2CO_3 = H^+ + HCO_3{}^-$$

$$H_2O = H^+ + OH^-$$

The anions in these cases have a strong tendency to react with $H^+$. They are *strong bases*.

## Exchangeable Bases

Soil scientists use the terms **exchangeable bases** or *basic cations* to mean those exchangeable cations that are not especially abundant in acidic soil (Na, K, Ca, Mg, Sr), as distinct from hydrogen and aluminum cations. This terminology is confusing. None of these cations is truly a base in the chemists' sense that we have been discussing. They do not combine with $H^+$, and their salts do not raise the pH of soils or solutions unless the anion involved is a true base, such as carbonate or bicarbonate. Nevertheless, the terminology persists in soil science.

Likewise, the term *base saturation*, or **exchangeable base saturation,** meaning the percentage of the cation exchange capacity (CEC) occupied by Na, K, Mg, Ca, Sr, and so forth, persists as an index of soil acidity. Base saturation is nearly 100 percent in neutral and alkaline soils. It becomes less than 100 percent in most acidic soils, for two reasons: (1) some of the exchangeable cations ($Al^{3+}$ and $H^+$) are now "nonbases," and (2) conventional methods of measuring CEC (at pH 7 or 8) greatly overestimate the actual exchange capacity of acidic soils that have pH-dependent charge.

Base saturation data are still used in soil classification but seldom in soil chemistry or fertility. Base saturation indicates plant response no better than do simpler, more direct measures such as pH, exchangeable Al, or total exchangeable acidity.

# SUPPLEMENT 11–2  Aluminum and Manganese Reactions

Why does the solubility of Al increase sharply as the pH drops? The dissolution of Al-containing soil minerals releases not only Al but also $OH^-$ (or consumes $H^+$), so that the dissolution equilibria depend on the solution pH. An example is the dissolution of gibbsite and its relatives (members of the hydroxyoxide group of clay minerals):

$$Al(OH)_3 = Al^{3+} + 3(OH)^-$$

This reaction proceeds to the right if its equilibrium is disturbed by a decrease in (OH). That is, lowering the pH allows more aluminum hydroxide to dissolve. The effect is large; in the ideal system represented by the equation, $Al^{3+}$ increases 1,000-fold for each 10-fold decrease in (OH) (i.e., for each one-unit drop in pH). The dissolution and formation of kaolinite and other silicate clays follow similar equilibria, except that $(Al^{3+})$ in solution is governed by levels of soluble silica as well as pH.

Trivalent $Al^{3+}$ is not the only form of Al in solution. Aluminum also forms soluble ion complexes with many anions, including hydroxyl. Its affinity for $OH^-$ causes $Al^{3+}$ to react with water, releasing $H^+$ and therefore behaving like an acid. Thus Al not only forms $Al(OH)_3$ solid but also partially hydroxylated cations, such as

$$Al^{3+} + H_2O \rightarrow Al(OH)^{2+} + H^+$$

There is less $Al(OH)^{2+}$ than $Al^{3+}$ at very low pH, but they are about equal in concentration at pH 5. Other Al complexes in soil solutions include complexes with $F^-$, $SO_4{}^{2-}$, and organic anions such as citrate and oxalate. Organic complexes (chelates) account for most of the dissolved Al in Histosols and many forest soils.

Manganese solubility in acidic soils is also influenced by chelate complex formation and by acid dissolution of oxides and hydroxyoxides. An example of the latter is the reduction and dissolution of the purplish black solid manganese dioxide, $MnO_2$, to form manganous ion:

$$MnO_2 + 4H^+ + 2e^- \rightarrow Mn^{2+} + 2H_2O$$

Because it is a reduction, this reaction in soil is favored by wetness, compaction, vigorous respiration, or any other condition that leads to oxygen deficiency. As you can deduce from the input of $H^+$, the dissolution is strongly favored by acidity.

## Note on Solubility Products: Gibbsite and Other Minerals

Like most minerals, gibbsite is only slightly soluble. It stops dissolving when the mathematical product of its molar ion activities, $(Al)(OH)^3$, approaches gibbsite's very small solubility product (Ksp). (The activity product relates to the probability of ions colliding in solution and combining.) In a solution saturated with respect to gibbsite,

$$(Al)(OH)^3 = Ksp_{gibbsite} = 10^{-34.1}$$

At pH 7, for instance, where $(H) = (OH) = 10^{-7}$ mol, we could estimate the $Al^{3+}$ concentration (ignoring activity coefficients) thus:

$$(Al) = Ksp/(OH)^3 = 10^{-34}/(10^{-7})^3 = 10^{-13}$$
$$mol/L$$

That concentration ($10^{-13}$ mol/L) is harmless. But any decrease in (OH) allows more gibbsite to dissolve. At pH 4, for instance, $(OH) = 10^{-10}$, giving $(Al) = 10^{-4}$ mol/L, toxic to most plants and fish.

Solubility products of important minerals are listed elsewhere (e.g., in W. L. Lindsay. 1979. *Chemical equilibria in soils*. New York: Wiley).

## SUPPLEMENT 11–3  Acidifying and Alkalinizing Reactions

Figure 11–19 indicates how soil pH can be affected by major steps in nutrient cycling. Table 11–4 lists soil reactions that can lower or raise the pH. Few readers will benefit from trying to memorize all these reactions; we list them only for reference.

Several of the soil's acidifying and alkalinizing processes oppose one another. For example, anion uptake and cation uptake by organisms cancel each other's effect on soil pH; only the imbalance between the two has a net effect on pH, and this depends mainly on the relative assimilation of $N_2$, $NO_3^-$, and $NH_4^+$.

Likewise, individual steps in nutrient cycling alter the pH, but complete, closed nutrient cycles have little net effect on it. Marked pH shifts result when nutrient cycling is incomplete; that is, nutrients are being added in one form and removed in others (Figure 11–19). The agricultural acidification of soils is an example.

A shift from oxidative to anoxic conditions usually neutralizes acidic soils, and the reverse shift drops the pH.

## SUPPLEMENT 11–4  pH Buffering and Lime Requirements

Adding an acid or a base changes the pH. It also influences reactions that involve $H^+$ or $OH^-$, generally in such a way that the reaction partly counteracts the imposed pH change. If an acid is added, the reaction will consume $H^+$; if a base is added, the reaction will release $H^+$. Such reactions are said to provide pH buffering. Most buffering reactions work best in a particular pH range. Carbonate and bicarbonate, for instance, buffer pH most strongly at pH above 7, whereas Al reactions with OH buffer most effectively at pH below 5.

Some buffering reactions operate slowly. The dissolution of oxide, silicate, and carbonate minerals in rock and coarse soil fractions (weathering) might be barely fast enough to keep up with natural acidification over a time scale of years or decades. Grinding minerals to a powder speeds their reaction enormously, and ground agricultural limestone or calcium silicate reacts within a few days or weeks.

Rapid buffering over most of the soil pH range is provided by colloid surface reactions that involve $H^+$ or $OH^-$ either directly or through effects on Al. These reactions are (1) cation exchange on humus and aluminosilicate clays and (2) proton addition and proton removal on humus and oxides. These reactions determine the magnitude of the lime requirement, as it is usually measured in the laboratory.

Buffering or lime requirement curves relate the soil pH to the amount of base added (Figure 11–9). The buffering capacity is the amount of base needed to produce a certain pH increase or the amount of

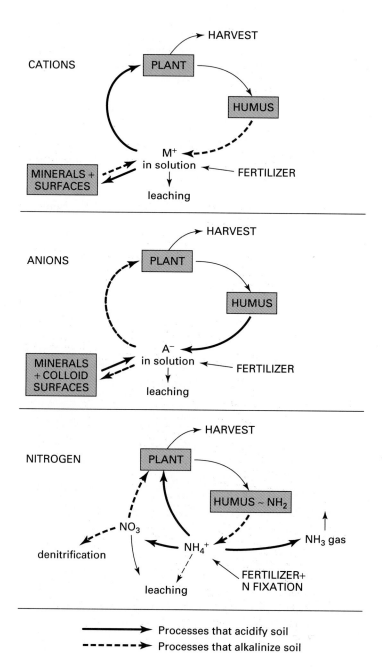

FIGURE 11–19 Soil acidification in relation to nutrient cycling. Complete nutrient cycles have no long-term effect on soil pH, but individual steps in each cycle release or consume acid. Net acidification or alkalinization can therefore result when nutrients are put into an ecosystem in one form and are lost or taken out in other forms.

acid needed to produce a certain pH decrease. Soils high in reactive colloids have large buffer capacities; sands have low buffer capacities. The buffering capacity often increases as the pH drops below 5 and rises above 7. Frequently the buffer curve changes with increasing reaction time over a period of months, because of slow reactions. Normal quick diagnostic tests of lime requirement essentially ignore slow reactions, but this can be considered when interpreting the test data.

**TABLE 11–4**
Acidifying and Alkalinizing Reactions

| PROCESSES THAT LOWER pH | | PROCESSES THAT RAISE pH |
|---|---|---|
| $\rightarrow$ | **GENERAL REACTION** | $\leftarrow$ |

$$\text{acid} \quad\quad \text{proton} + \text{anion}$$
$$\text{(base)}$$
$$HA \rightleftharpoons H^+ + A^-$$

**COLLOID SURFACE REACTIONS**

| | | |
|---|---|---|
| displacement of exchange Al | $Al^{3+} \text{ exch} + 3M^+ + 3OH^- \rightleftharpoons Al(OH)_3 \text{ solid} + 3M^+ \text{ exch}$ | exchange adsorption of Al |
| deprotonation of variable-charge surfaces | $\text{surface} - H \rightleftharpoons H^+ + \text{surface}^-$ | protonation of variable-charge surfaces |

**ION INTAKE BY PLANTS & MICROBES**

| | | |
|---|---|---|
| "excess cation intake" (e.g., $NH_4^+$ − supplied plant) | $\begin{array}{c} \text{IN} \end{array} \begin{array}{l} \leftarrow M \\ \leftarrow A \\ \rightarrow H^+ \quad \text{OUT} \end{array}$ | (M = cation, A = anion) |
| | $\begin{array}{c} \text{IN} \end{array} \begin{array}{l} \leftarrow M \\ \leftarrow A \\ \rightarrow OH^- \quad \text{OUT} \end{array}$ | "excess anion intake" (e.g., $NO_3^-$ − supplied plant) |

**ORGANIC ACIDS**

| | | |
|---|---|---|
| biological release of organic acids $\rightarrow$ | $\begin{array}{ccc} \text{COOH} & & \text{COO}^- + H^+ \\ \mid & \rightleftharpoons & \mid \\ -C & & -C \end{array}$ | biological release of $\leftarrow$ organic anion |

**MICROBIAL REDOX REACTIONS**

| | | |
|---|---|---|
| nitrification $\rightarrow$ | $NH_4 + 2O_2 \rightarrow 2H^+ + NO_3^- + H_2O$ | flooded-soil processes |
| | $(N_2O, N_2) + H_2O \leftarrow H^+ + NO_3^- + ne'$ | $\leftarrow$ denitrification |
| oxidation of S & sulfides $\rightarrow$ | $S + O_2 + 2H_2O \rightleftharpoons 4H^+ + SO_4^{2-} + 2e'$ | sulfate reduction |
| oxidation of ferrous compounds $\rightarrow$ | $Fe^{2+} + 3H_2O \rightleftharpoons 3H^+ + Fe(OH)_3 + e'$ | reduction of Fe oxides |

**PRECIPITATION/DISSOLUTION REACTIONS**

| | | |
|---|---|---|
| precipitation of Fe and Al added as soluble compounds (e.g., $Fe_2(SO_4)_3$) | $Fe^{3+} + 3H_2O \rightleftharpoons 3H^+ + Fe(OH)_3\downarrow$ | |
| [Commercial soil acidifying agents— | $Al^{3+} + 3H_2O \rightleftharpoons 3H^+ + Al(OH)_3\downarrow$ | weathering of silicate minerals |
| S, FeS, and ferrous, ferric, & Al | $Na^+ + Al(OH)_3 + 3SiO_2 \rightleftharpoons H^+ + NaAlSi_3O_8 + H_2O$ | Na − feldspar |
| sulfate and chlorides] | $Ca^{2+} + 2Al(OH)_3 + SiO_2 \rightleftharpoons 2H^+ + CaAl_2SiO_6 + 2H_2O$ | pyroxene |

**CO₂/CARBONATE SYSTEM**

| | | |
|---|---|---|
| Input of $CO_2$ or carbonic acid | $CO_2 + H_2O \rightleftharpoons H_2CO_3 \rightleftharpoons H^+ + HCO_3^-$ | input of bicarbonates |
| | $HCO_3^- \rightleftharpoons H^+ + CO_3^{2-}$ | $\leftarrow$ carbonates |
| | $CO_3^{2-} + Ca^{2+} \rightleftharpoons CaCO_3\downarrow$ | |

## SUPPLEMENT 11–5  Diagnostic Tests

### Acidic Soils

1. *Soil pH.* Soil pH is easily measured with a glass-electrode pH meter. The soil pH suggests the likelihood of several problems:

   pH < 5: Al and Mn toxicity, Ca and Mo deficiency

   pH < 5.5: Mo, Zn, K, and S deficiency

   pH > 7.5: salinity, Zn and Fe deficiency

   pH > 8.5: sodicity, salinity, Zn and Fe deficiency

   The most common deficiencies—N and P—are likely to exist regardless of the pH.

   Salts in solution lower the measured pH, because of the displacement of $H^+$ into solution from the colloid surfaces. This effect causes the field pH to fluctuate with fertilization and seasonal changes of ion concentrations in the soil solution. Laboratory procedures must take into account this salt effect; the water content or salt concentration must be controlled or specified during soil pH measurement.

   Common procedures measure pH on samples prepared in the following ways:

   **a.** Saturated paste: soil mixed with just enough distilled water to saturate it and left for about an hour to equilibrate.

   **b.** Dilute aqueous suspensions (1:2, 1:5, or 1:10 soil-to-water ratio) shaken for an hour or so. The soil-to-water ratio must be specified because additional water dilutes the soil's salts and raises the pH reading.

   **c.** Dilute suspensions in 0.01 mol $CaCl_2$. The 0.01 mol $CaCl_2$ is supposed to resemble an average soil solution. The pH measured this way is often close to the saturated-paste pH and varies little with the ratio of water to soil. Most of the soil pH values in this book are based on methods a and c.

2. *Lime requirement or buffering capacity.* Titration methods: Samples with different quantities of added base are wetted and allowed to react before the pH is measured. The lime requirement is read from the curve relating the soil's pH to the amount of added base (Figure 11–9). Buffer methods: The soil is mixed with a measured volume of buffer solution, and the lime require-ment is taken from a graph relating the final pH to the amount of acidity released by the soil (equivalent to the amount of lime required).

3. *Exchangeable Al.* Extracted with a large volume of concentrated (1 mol) KCl or $NH_4Cl$ (or other neutral salts), exchangeable Al can be determined by direct colorimetric or flame methods or by being titrated with standard alkali.

### Saline Soils and Irrigation Waters

1. *Salinity.* Electrical conductivity (EC) in water and saturated soil extracts is measured following the insertion of a standardized pair of metal electrodes. The salinity relates to EC, as shown in Figure 11–10. The total soluble salts in filtered water samples can also be weighed after evaporating the water. EC is reported in decisiemens per meter or millimhos per centimeter. The two units are numerically equivalent; for example, $10\ dS\ m^{-1} = 10\ mmho\ cm^{-1}$.

2. *Sodicity.* Exchangeable sodium percentage (ESP) is conventionally calculated from exchangeable Na, Ca, and Mg measured in molar ammonium acetate extracts from soil samples. The ratio is

$$ESP = 100\,[\text{exch Na}]/[\text{exch Na} + \text{exch Ca} + \text{exch Mg}]$$

The cation amounts are expressed in moles of charge (gram equivalents).

The sodium adsorption ratio (SAR) of water or soil solution extracts is calculated from direct analyses of Na, Ca, and Mg as

$$SAR = [\text{Na}]/\sqrt{[\text{Ca} + \text{Mg}]}$$

The cation concentrations are expressed in molarity.

## SUPPLEMENT 11–6  Salt, Cations, and Colloid Dispersion

Colloid particles may be flocculated—that is, stuck together into tight clusters or flocs—or they may be **deflocculated,** able to disperse readily into a suspension of separate particles. Flocculation of clay and humus is a prerequisite for aggregate structure. Dispersion, or deflocculation, destroys aggregation and therefore eliminates large pores and slows infiltration.

Dispersion is likely when the total salt concentration is low, especially when the cations are monovalent and highly hydrated ($Na^+$, $K^+$, $NH_4^+$). These conditions lead to dispersion because they allow the exchangeable cations to spread out loosely instead of being held tightly in a compact layer near the colloid surface (Figure 11–20).

## Effects of Cations and Salinity on the Exchangeable Cation Layer

Exchangeable cations are attracted to a negative colloid particle, but they also tend to diffuse randomly through the liquid. The exchange layer becomes more diffuse (less compact) if there are few cations in the solution and if the predominant cations are of kinds that are held weakly (Figure 11–20A). Increasing the salt concentration in solution (salinity) decreases the net likelihood of cations diffusing outward; it compresses the exchange layer. Compression of the diffuse layer will require less salinity if the predominant cations are tightly held ($Mg^{2+}$, $Ca^{2+}$, $Al^{3+}$) but high salinity if the predominant cations are loosely held ($Na^+$, $K^+$, $NH_4^+$).

## Effects of the Cation Layer on Flocculation

The diffuseness of the cation layer affects flocculation because it interferes with the weak cohesion between the colloid particles themselves (Figure 11–20B). Particles of clay or humus cohere because of van der Waals interatomic forces. These short-range forces become strong *only* when two particles are close enough for individual atoms to interact; just a little separation weakens the cohesion. Repulsion and separation arise from (1) the tendency of the cations to hydrate, drawing water in between the colloid particles, and (2) the tendency of the positive layer of ions around one particle to repel the positive diffuse layer around neighboring particles.

If the diffuse layers remain compressed and dehydrated, short-range forces can keep colloid particles together, and stable soil structure remains possible. But if the salt concentration in a sodic soil is lowered, the diffuse layer hydration and repulsion allow the colloid particles to disperse.

## Dispersion for Particle Size Analysis

Dispersion, undesirable in field soils, is required for particle size analysis in the laboratory (see Chapter 2). The same principles hold. Dispersing agents add $Na^+$ and eliminate $Ca^{2+}$, $Mg^{2+}$, and $Al^{3+}$, while adding little to the total electrolyte concentration. One such agent is sodium metaphosphate (its alkalinity precipitates Al, and the metaphosphate forms complexes with Ca and Mg, so they cannot compete with Na for exchange sites).

# SUPPLEMENT 11–7   Acid Precipitation in Cool, Humid Forests

Acid rains and fog can damage plants and affect soils indirectly as a result. Direct effects on soil are generally discounted because rains add negligible amounts of acid compared with most soils' biological acid production and buffering capacity. Some Northern Hemisphere forests are exceptions because of very high acid rain inputs, low soil fertility, and disuniformity of water flow through the forest system.

Typical humid forest soils are extremely acidic, and $H^+$ and $Al^{3+}$ are major exchangeable cations (Figure 11–21). The prominent O horizon's humus accounts for most of the cation retention and much of the large pH-buffering capacity. Strong buffering and soil variation together make it almost impossible to measure acid budgets, and shifts of pH, if they have happened, are probably small. But there are other effects, apparent in stream chemistry and tree nutrition.

The acid rain that reaches most of the forest floor's area becomes changed as it falls through the forest canopy (Figure 11–22). The leaves absorb some $H^+$ from the water and add some leached nutrient cations. This does not improve tree health but partly shields most of the forest floor from acid input.

However, in addition to crown drip (or throughfall), much of the rainfall is channeled down the stem system, with minimal leaf contact, so as to concentrate inputs of water, acid, and anions at the bases of tree trunks. Here some $H^+$ is swallowed in buffering reactions—protonation of humus and dissolution of minerals—which release Al, Mn, and exchange cations to join with the other solutes in the drainage flowing to streams. Thus the forest loses more nutrients than it normally would, and the

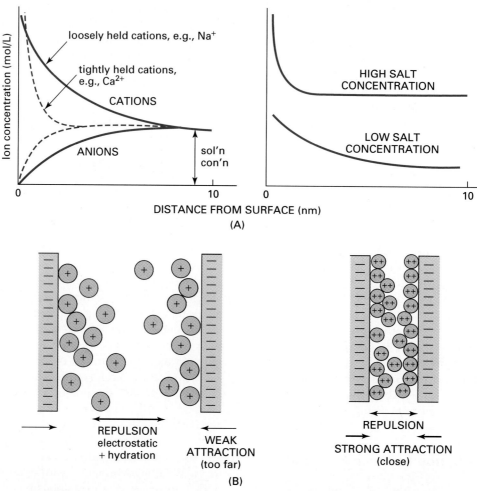

**FIGURE 11–20** Dispersing and flocculating effects of salt concentration and lyotropic properties of cations. (A) The cation concentration in the soil solution is highest near a negatively charged surface (clay) because cations are attracted to the surface, and the anion concentration is lowest because the anions are repelled from the surface. The small di- or trivalent cations are held more tightly than the large, monovalent cations. High salt concentration enhances the effect. (B) Clay dispersion is increased by repulsion of monovalent cations such as Na$^+$ in the solution around the clay particles. Divalent cations such as Ca$^{2+}$ help to reduce dispersion and increase flocculation.

streams gain them, along with acidity, Al, and Mn, which are potentially toxic to aquatic life.

The acidic forest soils of warm climates (Ultisols and their relatives) are less prone to lose nitrate, sul-fate, and accompanying cations because the hydrous oxides that abound in their lower horizons provide anion exchange capacity, which increases with acidi-fication (see Robarge and Johnson, 1992).

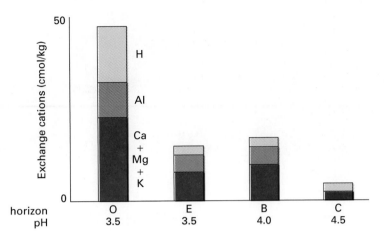

**FIGURE 11–21** Composition and concentration of exchangeable cations in horizons with different pH.

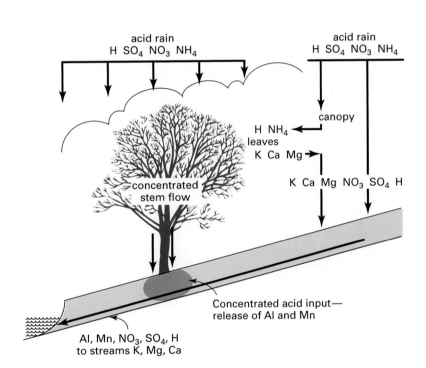

**FIGURE 11–22** Interactions of acid rain with the tree canopy and soil. Note the importance of stem flow to acid input to the soil.

# 12

# SOIL GENESIS

## OVERVIEW

*All kinds of geological materials such as hard-rock, lake, river, and glacial deposits, windblown silt, and sand are the starting point (the parent material) from which soil formation begins. The five soil-forming factors—parent material, climate, topography, biota, and time—interact by controlling the physical and chemical processes that change the parent materials into soils—soil genesis.*

    *Soils are closely related to the landscape on which they are found: The landscape influences the amount of water that enters the parent material. The length and degree of slope affect the balance between soil formation and erosion that destroys the soil. On steep slopes, the balance between*

*creation and destruction may be only slightly in favor of creation, and thin soils are formed. The processes that form soils from parent material act slowly and continuously.*

    *Water is necessary for all soil-forming processes, as it supports life, dissolves rock, and removes weathering products. The chemical process of hydrolysis, hydration, acidification, oxidation-reduction, chelation, solution, and the physical processes of transport and deposition—which are discussed in this chapter—cannot happen without water. All operate together within the framework of the five soil-forming factors to form a multitude of soils arrayed on landscapes over the earth's surface. This array is not a random assortment of pedons; each forms in response to the factors and reflects their interaction.*

## 12.1 FACTORS OF SOIL FORMATION

Soils are not rocks. This was a controversial statement 100 years ago when V.V. Dokuchaev, a Russian soil scientist, was probably the first to recognize soils as natural bodies formed at the earth's surface by processes different from those that form rocks. Soils, unlike rocks, form under atmospheric pressure and at normal earth surface temperatures. Our ideas about soils have changed further, especially with the recognition of soil-forming factors and processes. The purpose of this chapter is to describe the factors of soil formation, the processes that produce soils, and the modern concepts of soil genesis.

Soils are a key part of the geological cycle, which includes rock formation, landscape evolution, and weathering to produce soil, followed by erosion, deposition, and, depending on the depositional en-

vironment, either new rock or additional soil formation (Figure 12–1). Rocks form when hot **magma** cools or when eroded materials are compressed by the pressure of overlying sediments into new sedimentary rocks. Both sedimentary and igneous rocks may be changed by additional heat or pressure to form metamorphic rocks. Rocks are one kind of soil parent material.

**Soil genesis** is the process of creating soil from parent material. Genesis includes reducing the size of the parent material particles, rearranging the mineral particles, adding organic matter, changing the kinds of minerals, creating horizons, and producing clays. It is a continuous, slow process. We use the term *weathering* to include all the physical, biological, and chemical processes that change the size and chemistry of soil minerals.

These factors combine in infinite combinations to produce an infinite array of soils (Figure 12–2).

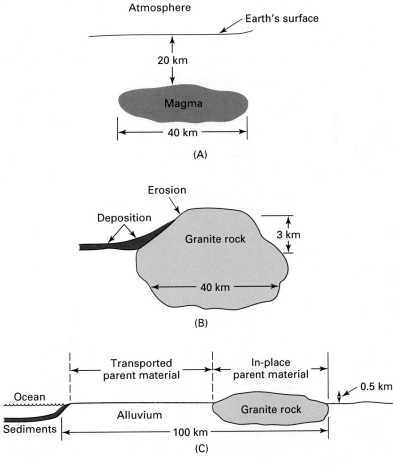

**FIGURE 12–1** The geological cycle. (A) A mass of molten rock rises slowly toward the surface and cools and solidifies. (B) Over millions of years the hard rock reaches the surface as erosion removes the overlying materials. At this scale, the soil formed on the rock is too thin to be shown. (C) Over additional millions of years the mountains are worn away, forming alluvial parent material for alluvial soils. Alluvium in the ocean becomes parent material for new sedimentary rocks, which someday will be uplifted to renew the process.

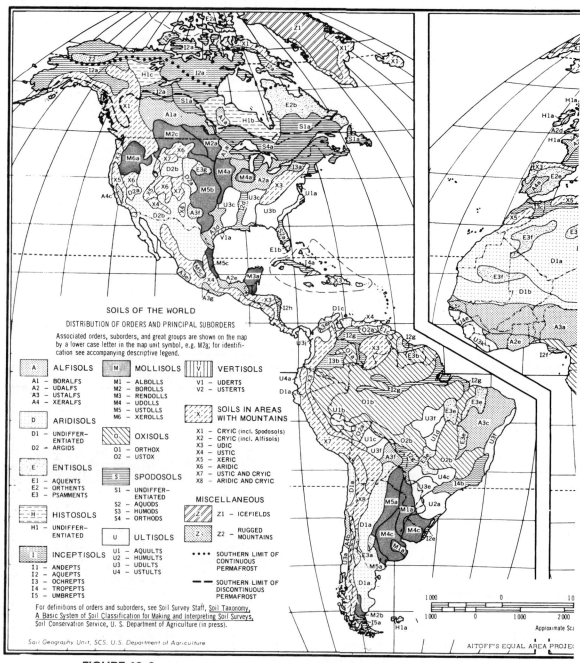

**FIGURE 12–2** Map of world soil distribution. (Natural Resources Conservation Service, U.S. Department of Agriculture.)

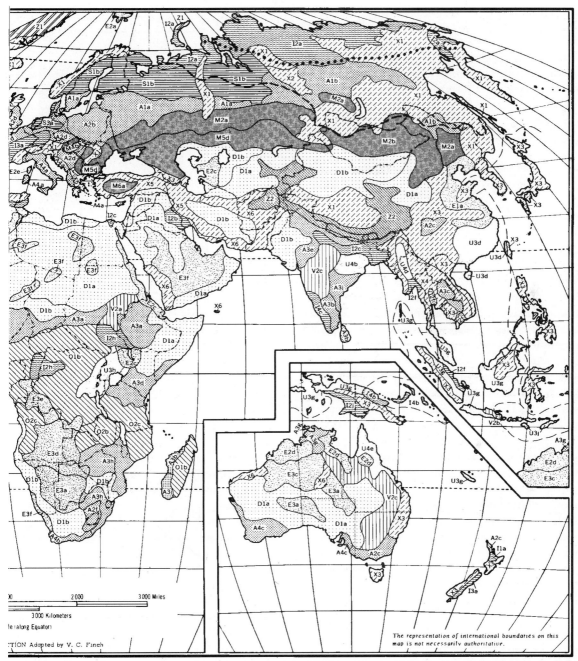

**FIGURE 12–2** Continued.

Physical, chemical, and biological processes transform parent materials into soils, reduce the parent material size, and convert the individual minerals into new minerals or soluble products that are carried out of the soil. These processes all act within the framework of the five soil-forming factors.

## 12.1.1  Parent Material

*Parent material* is the starting stuff of soils. Parent material mineralogy, hardness, and porosity determine weathering rates within climate and biotic regimes. Minerals weather at different rates; the weathering results in different products and hence different soils. The size distribution of minerals in rocks, or the sizes of the rocks themselves if the parent material is unconsolidated, determines the surface area per unit weight of material. The higher the surface area, the faster weathering will be. Finer-grained materials (compared to coarser ones) wet to shallower depths, retain water longer, and weather more rapidly. Parent material mineralogy determines the kinds and amounts of fertility elements (e.g., Ca, Mg, Fe, Na, K) released to the soil solution and secondary (pedogenic) minerals that will form over time under each climatic regime.

The parent material of Figure 12–3 is transported material from which the soil formed. Transport may be by water, wind, or gravity. In this case, transport and deposition by water of eroded material produced parent material for the soil. In Figure 12–3, the C2 horizon is unweathered (unchanged) soft transported geological parent material, and the C1 horizon has undergone some changes but continues to contain properties more like the parent material than like the solum. The solum is the portion of the profile that is distinctly soil, the A and B horizons.

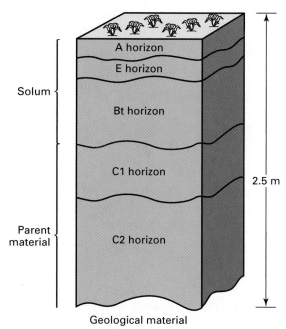

**FIGURE 12–3**  This pedon illustrates the concept of parent material. The A, E, and Bt horizons are formed from material with characteristics similar to those of the C2 horizon.

The factors controlling the rate of soil formation and the final product are
- Parent material
- Climate
- Topography
- Biota
- Time

Oxygen, silicon, aluminum, and iron contribute 96 percent (by volume) of the elemental composition of the earth's crust (Table 12–1); all the other elements contribute the remaining 4 percent. The elemental and mineralogical composition of the parent material in turn determines the soil's elemental and mineralogical composition. For example, a soil formed solely from quartz sand ($SiO_2$) does not contain clay minerals because the Al and associated cations necessary to form clays are not present. Furthermore, the soil's fertility is poor because nutrient cations are missing from the quartz. A soil formed from a basic igneous rock, such as basalt, can have abundant clay minerals and good fertility because the building blocks and cations are present in the rock.

Water is the main weathering agent that dissolves original minerals. Water must penetrate the parent material to weather it. Hard, consolidated igneous rocks such as basalt and granite are slowly permeable, whereas sedimentary rocks such as sandstone and shale are more permeable than are most igneous rocks and so generally weather more quickly. Transported parent materials such as river alluvium and glacial till are unconsolidated when they are deposited and weather rapidly compared with consolidated rocks.

## TABLE 12–1
Elemental Composition of the Earth's Crust

| Element | Mass (%) | Volume (%) |
|---------|----------|------------|
| O | 47 | 94 |
| Si | 28 | 1 |
| Al + Fe | 13 | 1 |
| All others | 11 | 4 |

**Examples: in-place materials** We separate parent materials into in-place (residual) and transported materials. Rocks are **residual** or **in-place parent material** because they weather to soils without first being moved by wind or water. Rocks are divided into igneous, sedimentary, and metamorphic groups, which are further subdivided by mineralogy, particle size, crystallinity, and specific mode of formation (Table 12–2).

Igneous rocks form when a hot mixture of elements (magma) cools. The earth itself was formed by this process, which is continuing today. Magma that cools at the earth's surface forms **extrusive** igneous rocks, and magma that cools below the surface forms **intrusive** igneous rocks. Rapid cooling produces small crystals in extrusive rocks, and slow cooling produces large crystals in intrusive rocks. Crystal size and composition determine the igneous rock type (Table 12–3), and crystal size is one factor that determines rock porosity and rock weathering rate. Rock composition also determines weathering rate and largely determines the soil's chemical composition. Sialic igneous rocks, such as granite, contain mostly quartz and K feldspars. Those containing less quartz and more Na or Ca feldspars are *mafic* igneous rocks, regardless of where they cool.

Sedimentary rocks consolidate from material deposited in lakes and oceans. Over time and under the pressure of the overlying material, sediments become sedimentary rocks. They vary in hardness, depending on formation pressure, and in composition, depending on composition of the original deposit. Sandstone originates from mostly sand-sized sediments, and shale is from mostly clay-sized sediments. Limestone is rich in carbonates from the shells of lake or ocean organisms. Most sedimentary rocks are softer and more porous than are igneous rocks.

## TABLE 12–2
Major Rock Types: Their Origins and Properties

| Rock type | Origin | Examples | Properties |
|-----------|--------|----------|------------|
| Igneous | Cooling of magma | Granite | Light colored, coarse grained |
| | | Basalt | Dark colored, fine grained |
| Sedimentary | Deposition and compaction | Shale | Any color, fine grained |
| | | Sandstone | Any color, coarse grained |
| | | Limestone | Light colored, shells or $CaCO_3$ present |
| Metamorphic | Change in igneous or sedimentary | Slate | Any color, hardened shale |
| | | Marble | Any color, changed limestone |

## TABLE 12–3
Simple Classification of Some Igneous Rocks by Crystal Size, Color, and Dominant Mineralogy

| Crystal size | High in quartz, high in K feldspar (light colored) | Low in quartz, high in Ca/Na feldspar (dark colored) | No quartz (dark colored) |
|--------------|---------|---------|---------|
| Large (intrusive) | Granite | Gabbro | Peridotite |
| Small (extrusive) | Rhyolite | Basalt | Limburgite |

## SOIL AND ENVIRONMENT
### Soil Stability, Nuclear Plants, and Sustainable Agriculture

One important use of soil genesis studies is determining the age and stability of landscapes for the siting of power plants, dams, and nuclear waste disposal sites. Our laws and good sense require that such facilities be sited where earthquake faults have no record of recent movement. *Recent* is a relative term, but in the law it is interpreted as 150,000 years. There are no good radiometric dating techniques for quantifying the age of rocks or soils this old. Relative ages based on landscape position and soil morphology are calculated to determine whether a landscape that contains faults has been stable for at least 150,000 years. If a fault is found in a landscape, and there is no displacement of the soil on either side of the fault, and if the soil can be shown to be more than 150,000 years old based on its morphology, a location is deemed stable for purposes of site selection. Soils are like living museums in that they reflect the sum of an area's climate and vegetation history. Fossil soils and buried soils are used to infer past climates and vegetation.

Sustainable land use, particularly in agriculture, requires the ability to predict and control long-term changes in soil, a critical new area for application of the principles and research methods of soil genesis.

Metamorphic rocks form when either igneous or sedimentary rocks are heated or put under additional pressure or both. If there is sufficient heat and pressure, the original minerals will melt, and as the melt cools, new minerals form. Examples of metamorphic rocks are marble formed from limestone, slate from shale, and gneiss from granite.

**Examples: transported materials**  Many of the most productive soils in the world have weathered from **transported parent materials.** The transporting agents include wind, liquid water, ice, and gravity. Transported materials are generally more porous than are in-place materials.

Wind lifts and carries clay, silt, and fine sand particles for varying distances, depending on the wind speed and the particle size. Deposits of windblown sand are **aeolian** deposits, and deposits of windblown silts and clay are **loess.** Sand dunes are the best-known form of aeolian deposit. Loess covers much of the midsection of the United States, and thick deposits are found in many agricultural regions of the world, including the Ukraine, Argentina, and the vast loess plateau of the People's Republic of China (Figure 12–4). Some windblown deposits are **stratified,** and they all decrease in thickness away from their source. Wind sorts the material by size, dropping the largest particles nearest the source and carrying the smallest particles great distances. Satellite photos show great clouds of dust blowing west across the Atlantic Ocean from West Africa, and dust from China has been identified in Hawaii. Continuous dust addition to soils is a major contributor to soil genesis in some regions. The particle size distribution of a loess deposit is often vertically uniform for meters, and the high porosity of windblown deposits causes them to weather rapidly to thick soils.

Liquid water and ice are important material transporters. All the great and small rivers of the world carry eroded soil particles, which when deposited and drained become parent material for soils. The Nile, Euphrates, and Mississippi deltas are familiar deposits of water-transported materials, or **alluvium.** In addition to deltas, rivers deposit materials along their courses as floodplains and terraces. In arid and semiarid environments, rivers build alluvial fans and basins with transported materials (Figure 12–5).

During the past 2 million years, glacial ice 3 or more kilometers thick has advanced across and retreated from Canada, the northern United States, and Europe many times. These **glaciations** created a large assortment of parent materials. **Glacial till** is an unsorted, unstratified mixture of particles from clay to boulder size, either carried on the surface or within the ice or bulldozed ahead of or under it (Figure 12–6). Water melting from the glacier deposits materials that are indistinguishable from other water-deposited material. These deposits are sorted, stratified **glacial outwash** (Figure 12–7). In addition, glaciers create their own landscapes with unique names (Figure 12–8).

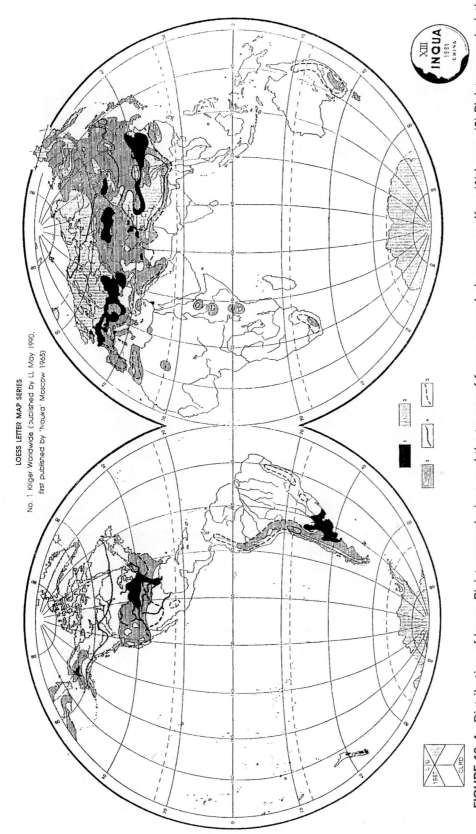

**FIGURE 12–4** Distribution of loess, Pleistocene glaciers, and the traces of frozen ground phenomena. Key: (1) Loess, (2) Pleistocene glaciation, (3) extent of Pleistocene frozen ground, (4) boundary of contemporary perennial frozen ground, and (5) seasonal frozen ground.

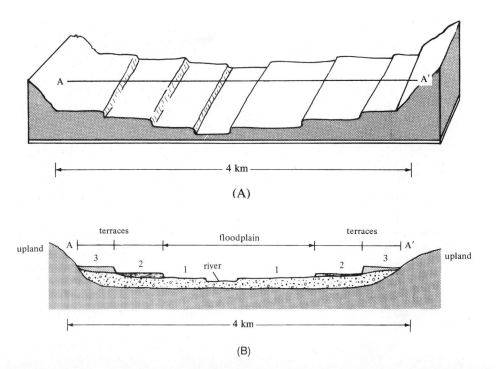

(A)

(B)

(C)

**FIGURE 12–5** (A) An aerial view of a river floodplain and terraces. (B) Cross section along the line A to A', showing the vertical arrangement of river floodplain and terraces. The floodplain may flood frequently, and the soils are young. On terraces, land surfaces and soils are older, and they do not flood often. The oldest parent material and soil are on terrace 3. (C) In arid environments such as Death Valley, California, streams create alluvial fans such as the one in this photo. The source of the alluvium is the mountain range in the background. (Photo © Mike Andrews.)

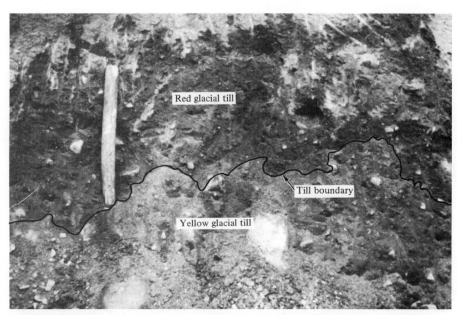

**FIGURE 12–6** Two unweathered glacial tills from Minnesota. The upper till is red, and the lower till is yellow or tan. The glaciers that deposited these tills came from different directions over different kinds of rocks. Note the coarse fragments in the till and the lack of stratification. The stick is about 15 cm long.

**FIGURE 12–7** Glacial outwash. Note the sorting of material and depositional features (layering) common to water-deposited material. Compare this with Figure 12–6.

Gravity carries material short distances down slopes. Such deposits are **colluvium,** which is found as erosional deposits on hillsides. It is often difficult to distinguish between colluvium and rock weathered in place when the colluvium sits on fresh rock.

**Organic soils** form from plants in various stages of decay. The types of plants partly determine the degree of decomposition of the organic material and the characteristics of the organic soil, but little attempt has been made to classify the different organic parent materials.

### 12.1.2 Climate

Precipitation and temperature are the two principal climatic factors influencing soil formation. They vary with elevation and latitude. Water is the main agent in weathering. The amount of water entering the soil depends on how much falls and how much runs off the soil surface. The amount of precipitation, its form (rain, snow, or both), and its distribution during the year are major factors influencing the kind of soil formed and the rate of soil formation. Increasing precipitation generally results in increasing rates of soil

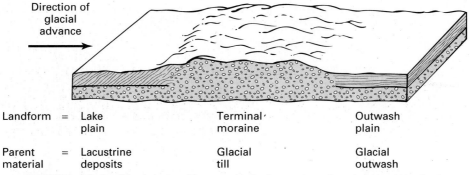

Direction of glacial advance →

| Landform | = | Lake plain | Terminal moraine | Outwash plain |
| Parent material | = | Lacustrine deposits | Glacial till | Glacial outwash |

**FIGURE 12–8** Glacial landforms include outwash plains, terminal moraines, and till plains. Each feature has a unique deposit that is a parent material for soil.

formation and thick soils. Rainfall amount and intensity affect the amount of runoff, erosion, and soil destruction. Seasonal rainfall distribution is important because the rain that falls when the soil is cold is less effective in weathering than is rain on warm soil.

Climate determines which soil-forming processes dominate in a profile. In cool, humid climates, organic matter accumulation, removal of weathering products, and acidification dominate. In arid and semiarid climates where evapotranspiration exceeds precipitation, organic matter accumulates more slowly; weathering products, especially soluble salts and carbonate, accumulate rapidly; there is slow removal of weathering products; and acidification takes much longer than in humid climates.

Another "climate" to consider is the local moisture regime of soils that have poor drainage or a high water table. Free water in a profile so greatly influences soil formation that the regional climate, characterized by mean annual temperature and precipitation, is much less important than it is in well-drained soils. In poorly drained soils, reduction and oxidation—but primarily reduction (See Supplement 8–2)—are the dominant processes that result in **gleying** and mottles.

Gley colors are the dark gray or unusual bright blue and green colors that are produced by prolonged reducing conditions. These colors are often described as having neutral hues. A designation such as N 2/0 is a common indication of gleying (see Section 12.2.2). Mottles are bright yellow and red or dull gray spots of color that indicate a fluctuating water table.

Precipitation and temperature determine the kind and amount of vegetation, another factor of soil formation. Temperature also determines how much precipitation falls as rain or snow. Generally, snowmelt is less effective than rain is in weathering, probably because the soil is cold when snow melts, and chemical weathering proceeds slowly at low temperatures. Low temperature decreases evaporation and increases leaching efficiency. Clay formation increases with a rising mean annual temperature. Clay type also is influenced by climate. Smectite is usually found in greater quantities in drier climates than is kaolinite. (Kandites and hydrous oxides accumulate as residual products of intense weathering in wet, warm climates [Figure 12–21].) Figure 12–9 illustrates some other effects of precipitation and temperature on soil formation.

### 12.1.3 Topography

Topography is a landscape characteristic. Slope angle and slope length are features of topography that govern the amount of water that runs off or enters a soil (the **effective precipitation**). Three levels of effective precipitation are shown in Figure 12–10. All the water that falls at point A enters the soil, and there is neither runoff nor run-on. At point B, only 50 percent of the precipitation enters the soil, and at point C, all the precipitation plus the 50 percent that runs off B enters the soil. As the slope becomes steeper, erosion increases, and soil development (pedon thickness and horizon differentiation) decreases. Water flowing through the soil adds to the wetness at the foot slope and toe slope positions. Figure 12–10 oversimplifies what happens on a slope because it does not show water that moves through the profile from the position of pedon A to pedon C. This interflow

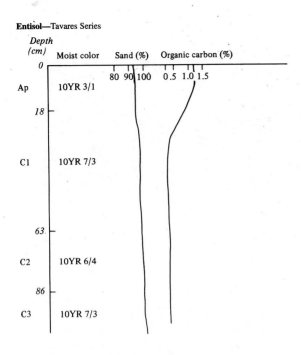

Entisol—Tavares Series

**PLATE 1.** Tavares Sand, a moderately well-drained Entisol from Pasco County, Florida, is formed in sandy marine (ocean) sediments. The simple profile consists of an accumulation of organic matter (1.2 percent organic carbon) in the A horizon over C horizons that extend to more than 2 m. Soil pH in water is 3.1 in the A and around 5 for the remainder of the pedon.

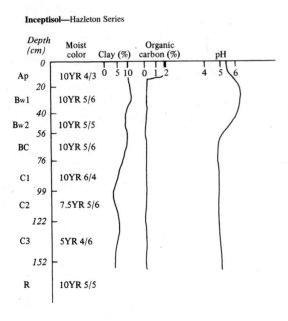

Inceptisol—Hazleton Series

**PLATE 2.** Hazleton channery loam, a well-drained to somewhat excessively drained Inceptisol from Butler County, Pennsylvania, is formed from micaceous sandstone. The channers (large flat stones) are inherited from the parent material. Few patchy clay films in the Bw horizons indicate soil development. Accumulation of bases, indicated by increasing pH, is another indication of development.

**Vertisol**—Houston Black Series

PLATE 3. Houston Black clay, a moderately well-drained Vertisol from Tarrant County, Texas, is formed from marine clay and shale. It has inherited the high (more than 55 percent) clay content from the parent material. The uniform carbon content is produced by natural mixing of the soil as the clay shrinks and swells. The C horizon is at 1.6 m.

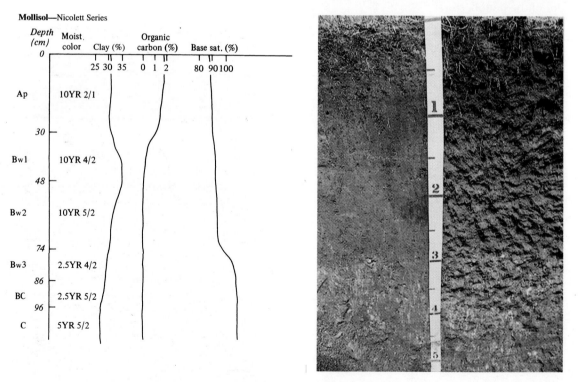

**Mollisol**—Nicolett Series

PLATE 4. Nicolett clay loam, a moderately well-drained to somewhat poorly drained Mollisol from Waseca County, Minnesota, is formed in calcareous glacial till. It has insufficient clay accumulation to have an argillic horizon. Calcium carbonate has been leached from the solum but is found in the BC and C horizons (white streaks).

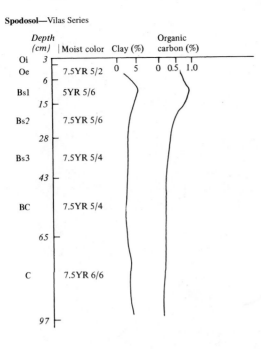

**Spodosol**—Vilas Series

**PLATE 5.** Vilas loamy sand, a somewhat excessively drained Spodosol from Oneida County, Wisconsin, is formed from glacial outwash. Organic matter has accumulated at the surface and has been leached with iron into the Bs horizon. The bleached white Oe horizon is obvious but is not needed to have a Spodosol. Some depositional layering can be seen in the C horizon.

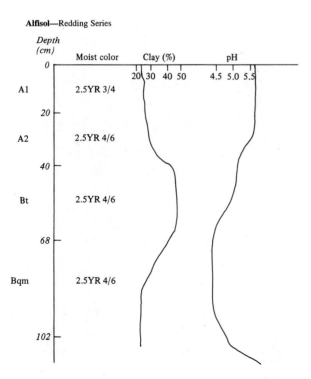

**Alfisol**—Redding Series

**PLATE 6.** Redding gravely loam, a well-drained Alfisol from Merced County, California, is formed in granitic alluvium. Clay has accumulated in the Bt horizon, and silica has accumulated and cemented the Bqm horizon. The Bqm is indurated; it does not soften when wetted.

**Ultisol**—Exum Series

Depth (cm) / Moist color / Clay (%)

| Horizon | Depth (cm) | Moist color |
|---|---|---|
| Ap | 0–25 | 10YR 5/2 |
| E | 25–41 | 2.5Y 6/4 |
| BE | 41–56 | 10YR 6/4 |
| Bt1 | 56–91 | 10YR 6/3 |
| Bt2 | 91–142 | 10YR 6/3 |
| BC | 142–183 | 7.5YR 4/6 |
| C | 183– | |

Clay (%): 0 10 20 30

**PLATE 7.** Exum very fine sandy loam, a well-drained Ultisol from Edgecombe County, North Carolina, is formed in siliceous and kaolinitic Coastal Plain sediments. It is deeply weathered. The thin E horizon overlies a transitional BE and thick BT horizon. Note the distinct mottles in the lower portion of the pedon.

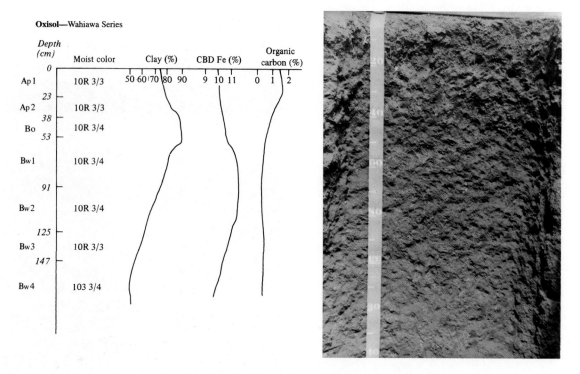

**Oxisol**—Wahiawa Series

Depth (cm) / Moist color / Clay (%) / CBD Fe (%) / Organic carbon (%)

| Horizon | Depth (cm) | Moist color |
|---|---|---|
| Ap1 | 0–23 | 10R 3/3 |
| Ap2 | 23–38 | 10R 3/3 |
| Bo | 38–53 | 10R 3/4 |
| Bw1 | 53–91 | 10R 3/4 |
| Bw2 | 91–125 | 10R 3/4 |
| Bw3 | 125–147 | 10R 3/3 |
| Bw4 | 147– | 10 3/4 |

Clay (%): 50 60 70 80 90
CBD Fe (%): 9 10 11
Organic carbon (%): 0 1 2

**PLATE 8.** Wahiawa silty clay, a well-drained Oxisol from Hawaii, is formed from siltstone. The Bw horizon extends below 1.8 m. Note the dark colors and the iron and clay contents throughout the pedon. CBD iron is a measure of the amount of iron that has weathered from the parent material.

**Aridisol**—Dona Ana Series

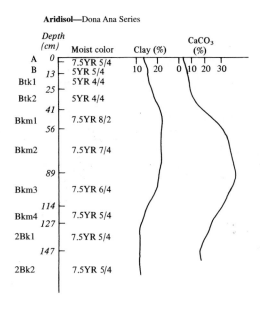

| Depth (cm) | | Moist color | Clay (%) | CaCO₃ (%) |
|---|---|---|---|---|
| A | 0 | 7.5YR 5/4 | | |
| B | 13 | 5YR 5/4 | 10  20 | 0  10  20  30 |
| Btk1 | | 5YR 4/4 | | |
| Btk2 | 25 | 5YR 4/4 | | |
| | 41 | | | |
| Bkm1 | 56 | 7.5YR 8/2 | | |
| Bkm2 | | 7.5YR 7/4 | | |
| | 89 | | | |
| Bkm3 | 114 | 7.5YR 6/4 | | |
| Bkm4 | 127 | 7.5YR 5/4 | | |
| 2Bk1 | 147 | 7.5YR 5/4 | | |
| 2Bk2 | | 7.5YR 5/4 | | |

**PLATE 9.** Dona Ana fine sandy loam, a well-drained Aridisol from Dona Ana County, New Mexico, is formed from sedimentary rock alluvium. It has an accumulation of clay and calcium carbonate. Like the Alfisol in Plate 6, the Bkm horizon is indurated, but the cementing material is calcium carbonate, not silica.

**Histosol**—Kina Series

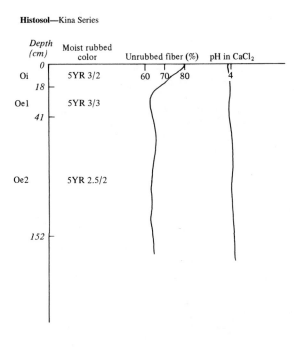

| Depth (cm) | Moist rubbed color | Unrubbed fiber (%) | pH in CaCl₂ |
|---|---|---|---|
| Oi | 0 | 5YR 3/2 | 60  70  80 | 4 |
| | 18 | | | |
| Oe1 | 41 | 5YR 3/3 | | |
| Oe2 | | 5YR 2.5/2 | | |
| | 152 | | | |

**PLATE 10.** Kina peat, a poorly drained Histosol from Ketchikan, Alaska, is formed from sedges. Note the high organic matter and low pH of this organic soil.

**PLATE 11.** Color aerial photograph of Modesto, California. Note that living vegetation is green, soils are brown, and the urban areas are light gray.

**PLATE 12.** False-color infrared aerial photograph of the same area as Plate 11. Living vegetation is red or pink, soils are gray or green, and urban areas are blue-gray.

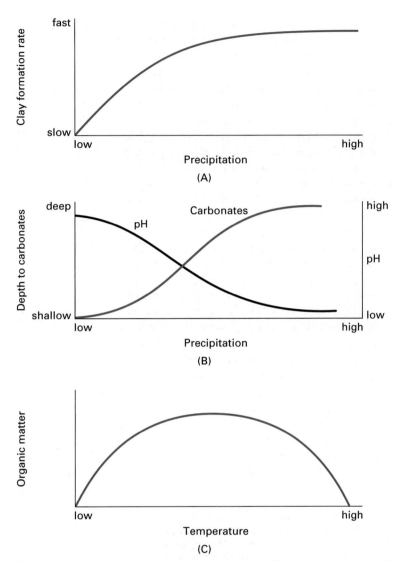

**FIGURE 12–9**   Precipitation and temperature effects on soil formation. (A) The rate of clay formation increases as water becomes more abundant. The curve levels off as other factors become limiting. (B) Depth to which inherited carbonates ($CaCO_3$, for example) have been leached increases and pH decreases as precipitation increases. (C) Organic matter accumulation is a balance between biological production and decomposition. The maximum accumulation is at moderate temperatures.

does not produce erosion and is far slower than overland flow, but it contributes to soil formation. Soil formation is considered further in Section 12.3.1 later in the chapter.

The direction in which a slope faces is the slope **aspect.** In the Northern Hemisphere, north-facing slopes are cooler than south-facing slopes are be-

cause they receive less direct solar radiation. Because they are cooler and temperature limits development, soils are shallower on north-facing slopes. In arid climates where moisture is limiting, cooler soil temperatures and less evaporation are conducive to deeper soils on north slopes. The opposite is true in the Southern Hemisphere.

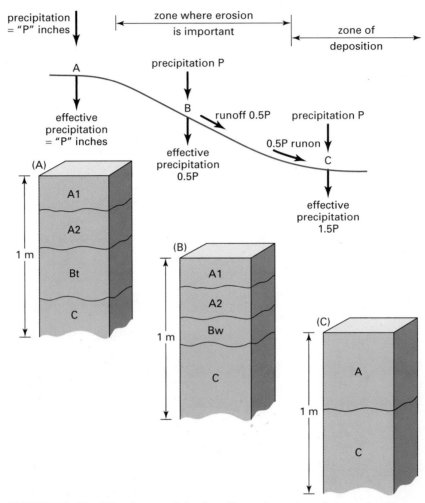

**FIGURE 12–10** Effective precipitation (from the point of view of soil genesis) is water that enters the pedon. These three pedons illustrate that the effective precipitation is greatest where runoff water accumulates. Position A is the summit, the convex portion between A and B is the shoulder, the long straight portion of the hill is the backslope, and the concave portion that connects to the toe slope at the base of the hill is the foot slope.

### 12.1.4   Biota

The **biotic factor** includes both plants and animals. Plants fix atmospheric carbon and add it to the soil, which is often the first step in soil genesis and certainly one of the most noticeable. Organic matter is added to the soil surface by plant tops and to the subsurface by roots. Plants do not have to die to add organic matter to the soil, because their roots exude material during plant growth. Root exudates and the sloughing of cells as roots push through soil are sources of nutrients and energy for soil microorganisms, which are also part of the biotic component.

Organic matter is important to weathering. As it is decomposed by soil microbes, compounds are produced that can chelate cations, and the soil's acidity is increased by humus production and destruction. All these influence the rate of soil formation and the kinds of products formed.

Both aerial (aboveground) and belowground plant parts affect the composition of the soil solution. Cations and anions are added to water when it is intercepted by aboveground vegetation. This **crown drip** reacts with the solid phase of the soil differently than pure rainwater does because it contains cations, anions, and dissolved organic compounds. Plant roots both remove ions, preventing them from being lost by leaching, and add ions during growth. Root and microorganism respiration releases $CO_2$ that contributes to lowering the pH of the soil solution. Leaves also drop to the ground and decay. This uptake and return of ions is **nutrient cycling.**

Roots and soil animals create pores and structure as they work their way through the soil. Soil animals come in all sizes, from microscopic single cells to gophers and wombats. Each plays a role in organic matter decomposition, the creation of pores, and bioturbation. **Bioturbation** is the biological mixing of surface organic matter into the subsurface and the mixing of subsoil horizons with the surface. In some ancient soils, bioturbation is the only agent of soil renewal. Some people consider humans part of soil biota (see Section 12.1.6).

## 12.1.5 Time

Soil scientists agree that soils are not all the same age and that the factors of soil formation act continuously over time, but not all soil scientists agree on what the starting point of soil genesis is or how genesis progresses from the starting point. There may be more than one expected "end point" for a soil in a region or under a climate. This makes generalizations about time as a factor of soil formation difficult. If clay accumulation is used as a measure of soil formation, it will take longer to "form" a soil than if salt accumulation is used as a measure of soil formation. Salt accumulation in arid climates is more rapid than clay accumulation in humid climates.

The eruption of Mount Saint Helens in Washington State in May 1980 provided a unique opportunity for soil scientists to study the initial stages of soil weathering and to determine rates of weathering with absolute certainty of when weathering began. Observations of the volcanic ash parent material began as soon after the eruption as safety permitted, and they continue to this day. Results of research published in 1991 (Ugolini et al., 1991) showed that horizons had not yet formed in the 1980 ash flow but that mineral weathering had begun. Smectite and vermiculite deposited with the ash were being de-

graded. Sufficient poorly crystallized minerals and kaolinite had formed to be detected by standard methods. Soil solution chemistry, on samples collected with lysimeters (see Chapter 3), also indicated that weathering of the inherited minerals was occurring. As time passes and the soils continue to form in the volcanic ash, the rates of formation of horizons and clay will be determined.

Complicating the study of time as a soil-forming factor is that the other factors change as the time period of soil formation increases. Thus, many soils are polygenetic; the factors of soil formation change over time. Another complication is that soil development can be progressive or regressive with time. Soils become more differentiated in progressive development; there are more horizons, and contrasts among the horizons are greater. The opposite occurs with regressive development; if soils are eroded or material is constantly added to the surface, progressive development is slowed or reversed.

There have been attempts to determine soil ages. In some locations, soils can be found vertically stacked like a layer cake, with the oldest soil buried by younger ones. A maximum age thus can be estimated by carbon 14 dating of the organic matter in the buried soils. In other locations, soils may be found that contain datable artifacts or volcanic ash.

Soil formation rates are too slow to measure directly, and so ages must be determined (or estimated) indirectly. When maximum soil ages are known, the pedon thickness is divided by the age to give an estimate of the soil depth formed per year. This estimate assumes that soil formation occurs at a constant rate, that soil thickness is the major indication of age, and that the major processes of soil formation act uniformly over time. Not all soil scientists agree with these ideas, though they do agree that the rates depend on the five soil-forming factors and are slow.

## 12.1.6 People as Soil Formers

The five soil-forming factors interact to produce all natural soils. As human populations have grown, they have indirectly or directly created or altered soils. For example, in Europe, where farms have occupied the land for thousands of years, regular additions of organic matter have produced soils with very thick A horizons and a high phosphorus content. Archaeologists rely on high soil organic matter and phosphorus to identify human habitation sites.

With urban and suburban expansion, landscapes are changed by cutting away steep slopes and filling over slopes to create level areas on which to build houses. The fill becomes new parent material for soil, as does the surface of the cut area. In many urban areas, fill material includes soil material mixed with human rubbish, old brick and concrete, coal ash from power plants, asphalt from roads, and myriad other materials. Are these areas soil? They have properties very different from soils formed without the human influence, but they have horizons and many properties similar to those of natural soils.

Within agricultural areas, land leveling for irrigation, deep tillage or ripping, and deep cultivation disturb soil horizons and dramatically change landscapes. Frequently the natural sequence of soil horizons is completely destroyed. In this case, the human influence is to reset the soil formation clock to zero when soil formation begins anew.

When natural soils are not suitable, people create or extensively modify soils to suit their needs. The processes and products are very different from natural soils. For example, landscape and golf course designers create soils and landscapes to suit their needs, and, on a smaller scale, horticulturists create potting soils for indoor and outdoor plants. Environmental laws require that strip-mined areas be left in a condition that will allow the regrowth of vegetation, which often requires reconstructing the soils that have been removed to gain access to the minerals. In low-lying coastal countries short of agricultural land, dikes have been built to hold back the ocean. The seawater is drained, and "new" soils are created. (These **polders** and other new soils are discussed in Chapter 16). We can also create soils on a small scale, although one purpose of this book is to show that it is not easy to create soils and that protecting the existing ones is necessary.

## 12.1.7   A Soil Formation Equation

Hans Jenny (1941, 1980) was the first American soil scientist to combine the five soil-forming factors into an equation, which states that the dependent variable, soil, depends on the five independent soil-forming factors:

$$S = f(P_M, C, R, B, T) \qquad (1)$$

In Equation 1, $S$ is a soil property that is a function of the parent material, $P_M$, climate, $C$, relief (topography), $R$, biota, $B$, and time, $T$. Jenny and others demonstrated that these factors are not truly independent, because climate influences biota, and relief affects climate and biota, but it is useful to isolate individual factors to study how each influences soil formation. It is sometimes possible to separate the factors so that one varies while the others are nearly constant. Groups of soils that reflect one soil-forming factor more than the other four are studied to determine the effects of the dominant factor. These groups are lithosequences, climosequences, biosequences, toposequences, and chronosequences.

A *lithosequence* is a group of soils formed on different parent materials under the same (or very similar) climate, relief, biota, and time conditions. By carefully selecting the sampling sites, the effect of the parent material on soil properties can be isolated from the other four factors. A **climosequence** is a group of soils of the same age, formed on the same parent material and under similar biotic and topographic conditions, that differ because of the climate under which they formed. An example of a climosequence is a group of soils formed on the same age loess, under grass vegetation, on the same slopes, in Iowa and Minnesota. Traveling from south to north, the mean annual temperature decreases and more of the precipitation is snow. The effect on soils is less development of soil characteristics such as pedon and Bt thickness. Similar sequences can be found for *toposequences, biosequences,* and *chronosequences.* In each the factors are not identical, but one dominates over the other four.

## 12.1.8   Soil-Forming Factors: Some Interactions

Climate and time: age and maturity    Soils have transient fluctuating properties such as water content and temperature profiles. But most properties take years, centuries, or millennia to develop, become evident, and perhaps approach a steady state with negligible further change. Different characteristics of a pedon develop at different rates. For instance, a pedon developing in a calcareous rock or sediment exposed to moderate rainfall might show an early increase in organic matter, approaching a steady level within a few decades following the establishment of vegetation.

Calcite leaching and clay formation normally would take longer to produce steady levels (Figure 12–11A).

At the same time, materials will be moving within the developing pedon. Some of these movements are suggested in Figure 12–11B, to be read from left to right as a sort of slow-motion view of the changing profile. The simple increase in organic matter shown in Figure 12–11A can now be seen as a growing thickness of the A horizon, followed by increasing concentration of organic matter into a shallower A (and O) as prairie is replaced by forest. Calcite is not just leached out; some of it precipitates in a calcareous B horizon, which gradually shifts down, and out of, the pedon. Clay not only forms but also moves from the A and developing E horizons, depositing or reforming lower down, so that most of the increase in total clay shows up as developing and thickening of the Bt horizon.

The slowly developed features are indicators of advanced profile development or maturity. They can also be interpreted as indicators of age, but their development depends on other soil-forming factors as well as time. In fact, Figure 12–11A and 12–11B could represent a climosequence if we were to substitute increasing rainfall for increasing time on the horizontal axis. However, on sites of similar topography, within a limited regional climate zone, mature profile features can be reasonable indicators of relative age.

These indicators of maturity, ranked approximately in order (faster to slower), include the following:

1. Subsoil deposits of slightly soluble $CaSO_4$ and $CaCO_3$
2. Thick, dark A horizons
3. Movement of clay or clay-forming constituents, producing Bt horizons
4. Severe weathering to produce accumulations of kandites and hydrous oxides (oxic horizons) in warm, wet conditions
5. Migration of weathering products to produce bleached E horizons and Bs horizons in cool, wet conditions
6. Subsurface hardened layers cemented by Fe oxides or $SiO_2$

Most of these will never develop if climate and topography are unfavorable.

**Time and topography: active and stable land surfaces**   Mature soil profiles normally develop only on a stable land surface, relatively undisturbed by erosion or deposition. Figure 12–11C depicts a

# SOIL AND ENVIRONMENT
## *Wetlands*

People act as a soil-forming factor when they modify environments. From the early days of European settlement until the 1980s in the United States, areas of poorly drained soils were considered undesirable. They were too wet to be farmed productively, and they harbored mosquitoes and other insects. A solution was to drain the swamps and wetlands by installing drain tile. Draining swamps and wetlands was not only considered ethical and smart, but the federal government had programs to help finance it. This was a form of creating "new" soils. Soil itself was not created, but by removing impediments to drainage, soils that were often anaerobic were made aerobic. Processes and soil characteristics were changed.

People change their minds about the value of soils. Recognition of wetlands as areas of high biological diversity and special ecological value has caused a reversal of the policy that encouraged wetland drainage. The same government that paid farmers to drain wetlands now forbids their draining. Through environmental laws, every hectare of wetland now being drained must be replaced by the creation of additional new hectares. Once again, soils are being created by human intervention.

What is a wetland? The official definition provides for three characteristics that describe a wetland: hydric soils, hydrophytic (water-loving) vegetation, and wetland hydrology. An important characteristic of wetlands is that they must be saturated with water at or near the soil surface for prolonged periods when the soil temperature is sufficiently high to result in anaerobic conditions. The temperature must be high enough to maintain an active microbiological community that helps to deplete the soil of oxygen, thus producing the anaerobic conditions that are the key requirement for a wetland. More details on hydric soil definitions and characteristics are found in the text.

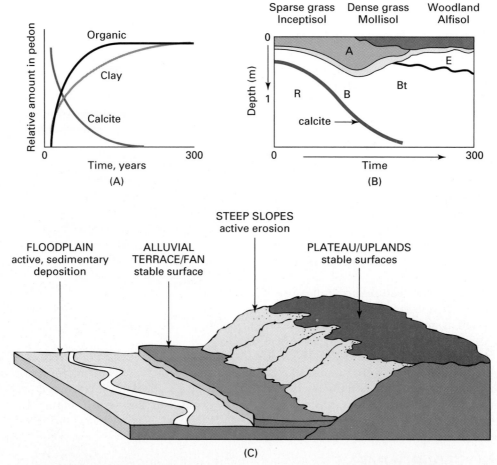

FIGURE 12–11  (A) Differing rates of development processes—organic matter production, calcite leaching, and clay formation—in a hypothetical pedon developing in a calcareous parent material under moderate rainfall on a stable land surface. (B) Distribution of materials, horizon formation, and vegetation change during development of the pedon represented in part A. (C) Examples of stable land surfaces (shaded), which favor mature profile development, and unstable surfaces subject to active erosion or deposition (unshaded), which favor development of undifferentiated profiles.

set of stable (shaded) and active (unshaded) surfaces. Soil development on active surfaces is frequently interrupted; floods deposit new parent material on top of the developing A horizon, or episodes of erosion strip topsoil away. In both cases, the soils develop only weak A horizons and little other profile differentiation. Strata in the parent material often remains as a clue to its origin. Steep hill slope soils may be shallow, infertile, and gravelly and floodplain soils may be deep, productive, and texture-sorted (sand near the river, clay farther

away), but all have immature profiles with weakly differentiated horizons and mostly fall in the same orders of soil taxonomy (Entisols and Inceptisols—see Chapter 14) regardless of climate, parent material, and the passage of time.

Stable land surfaces, neither steep nor flood prone, undergo little erosion or deposition (unless exposed to severe wind action). On such surfaces we expect mature soils fully expressing the influence of the other soil-forming factors. Uplands with low local relief are the principal areas where Oxisols and Ultisols occur

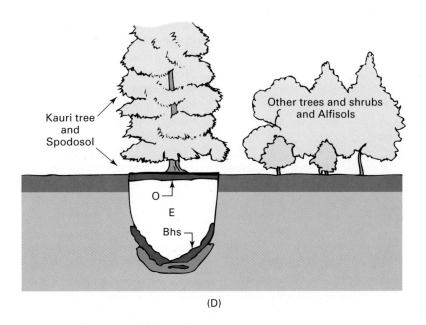

(D)

Vegetation and soil chronosequence on marine terraces, Mendocino Coast, California

Age:       current                     200,000 y                    400,000 y
Vegetation:   grass                        forest                       pygmy forest
Soil feature:  deep dark A horizon     low base sat'n & argillic B     Fe & humus transport
                                                                                   &oxide hardpan

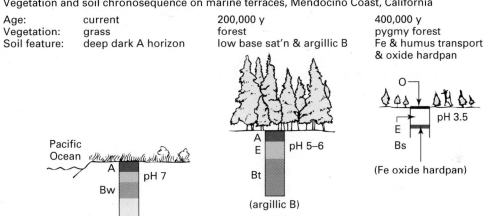

900 mm rain, dry summer with fog; sandstone underlying beach and windblown sands; 16 percent weatherable minerals (feldspar, etc.)

(E)

FIGURE 12–11   Continued. (D) A biotic effect. Heavy input of organic matter, acid, and chelating agents from the giant kauri (*Agathis australis*) produces a spectacular O horizon, a bleached (albic) E horizon, and illuviation of Fe and humic matter into a spodic Bhs horizon. Surrounding forest and shrub vegetation produces more mundane Alfisols with A horizon over textural B horizon. New Zealand, North Island: soft sedimentary rocks, moderate-low relief, high rainfall. (E) A chronosequence of vegetation and soils on terraces cut by wave action in Franciscan graywacke (feldspar-rich sandstone) during successive coastal uplifts, Mendocino County, California. The most mature soil on the highest terrace has 10 to 15 cm of O and E horizon overlying an almost impenetrable layer cemented by iron oxide (Bs). Acute acidity, nutrient deficiencies, winter waterlogging, and summer desiccation support dwarfed (<2 m) forest.

277

in warm, wet climates, Spodosols in cool, wet climates, and Alfisols in all except the most arid or frigid climates. Also, very mature soils often occur on the slightly elevated fringes of floodplains (Figure 12–11C), where old surfaces of alluvial terraces or fans have been stranded above current flood levels due to geological uplift or down cutting by the river. An example is the chronosequence we describe in Figure 12–19.

### Vegetation and soil as codependent variables

In discussing the independent factors of soil formation, we used the terms *biotic factor* and *biota* rather than *vegetation,* for two reasons. First, animals and microbes, as well as plants, affect soils. Second, the biota—the list of species with access to the site—is an independent variable, whereas the vegetation—the community of plants we see growing in the soil—depends on parent material, relief, climate, biota, and time, just as much as most soil properties do.

In Figure 12–18, we discuss the transition from forest to grassland in Minnesota, which we call a biosequence because relief, climate, parent material, and time are similar across the transect and because the soil changes can be explained as effects of the vegetation change and not vice versa. Figure 12–11D shows another example where such an argument could hold. Such judgments are seldom possible when soils and vegetation develop and vary in tandem, interacting on each other but starting with the same parent material and biota and directed by the same climate, relief, and chronology. A celebrated example is the sequence of soils and vegetations on a series of six firmly dated marine terraces developed on a feldspar-poor sandstone in Mendocino County, California. Figure 12–11E gives a simplified description.

## 12.2 PROCESSES OF SOIL FORMATION

### 12.2.1 Physical Weathering

The first process in the formation of soils from rocks is the reduction in particle size by physical weathering. Freezing and thawing, uneven heating, abrasion, and shrinking and swelling break large particles into small ones. Liquid water is more dense than ice, which is why ice floats and water exerts large outward forces when it freezes—a full, tightly closed bottle of water in the freezer will explode because of the ice

pressure. Likewise, the formation of ice within cracks in rocks helps shatter the rocks.

Uneven heating causes expansion and eventual breakage of the rock at weak points, and wetting and drying cause shrinking and swelling of rocks and mineral grains. These changes in volume shatter rocks and minerals along planes of weakness. Even roots growing into thin cracks can force cracks open until the rocks break, and growing salt crystals can do the same.

### 12.2.2 Chemical Weathering

Chemical weathering or, more precisely, biogeochemical weathering—because the biota interacts chemically with the geological material to weather the minerals—is the main process converting parent material into soil. Rates of soil formation are largely determined by the rates at which silicate minerals weather. Weathering is not mysterious (Figure 12–12). It is a process that changes minerals from their original composition to a new composition, through the addition of hydrogen to the structure (**hydrolysis**), addition of water (**hydration**), gain or loss of electrons (**oxidation** or **reduction**), dissolution, and carbonation. Each of these fundamental processes occur wherever water is in contact with mineral particles.

Water is the key ingredient, dissolving minerals and transporting reactants and products. Water enters the parent material (in the earliest stages of soil formation) or the soil and is altered by the biota and by interaction with the minerals. The water may continue farther into the developing profile, carrying soluble weathering products or solid particles. It may then stop within the profile or proceed upward or downward out of the soil (Figure 12–13). But water does not act alone; it contains organic and inorganic acids, especially carbonic acid, which speed mineral weathering. Each of these processes is influenced by the soil-forming factors.

The parent material determines the starting mineral composition. Climate and topography determine the amount of water entering the profile, and climate determines the temperature at which the reactions take place. Biota, particularly plants, determines the types and amounts of humus in the profile, and, finally, time determines the period over which the sum of all the processes acts.

**Hydrolysis**  Polar water molecules (see Chapter 3) are attracted to the surfaces of minerals. The $H^+$ in

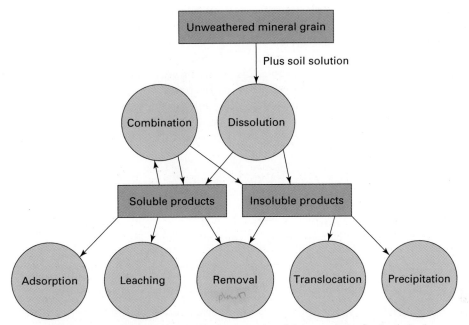

**FIGURE 12–12** Weathering includes a cycle of destruction of minerals, formation of new minerals, and destruction again. The cycle is shown here with processes in rectangles and products in circles. Biochemical weathering produces soluble and insoluble products. The insoluble ones may precipitate and remain where they precipitate, or, if they are small enough, they may be translocated (carried) by the soil solution to another part of the profile, where they accumulate. Soluble and insoluble products may be removed completely from the profile by soil solution that leaches (moves) beyond the solum or by plants that use the elements for nutrition or by overland flow if they are at the soil surface. They may be adsorbed to inorganic or organic surfaces, translocated to lower parts of the profile, or combine with other soluble weathering products to form new minerals. The new minerals may join unweathered minerals in this weathering cycle.

$H_2O$ can replace $K^+$, $Ca^{2+}$, and $Na^+$ in the mineral structure (Figure 12–14). In the two examples in Figure 12–14, the clay mineral and the feldspar release a $K^+$ to the soil solution and add an $H^+$ to their structure. The $H^+$ causes a physical strain in the structure because it is smaller than the cations it replaces, and this accelerates weathering. The $K^+$ may (1) recombine with other ions in solution to form new minerals, (2) be adsorbed by a clay mineral and participate in cation exchange, (3) be used by plants, or (4) be leached from the soil (Figure 12–12). The mineral continues to weather, and its component parts react to form new minerals, are used by plants, or are leached.

Lowering the pH increases hydrolysis because the concentration of $H^+$ ions in solution increases. Organic matter decomposition adds $H^+$ and speeds hydrolysis, as do other biological processes such as nutrient uptake, nitrification, and S oxidation. Warm temperatures increase the **dissociation** of water molecules, and the additional $H^+$ may participate in hydrolysis. Carbonation, or *acidification*, is similar to hydrolysis as a $H^+$ is added to a mineral's structure. (For more about acidification and solution, see Supplements 12–1 and 12–2.)

**Hydration**  Hydration adds water molecules to a mineral's structure. It differs from hydrolysis because the water molecule does not dissociate. Many minerals expand on hydration; for instance, smectite hydrates and dehydrates readily when water enters and leaves its interlayer. Mica also hydrates and expands. In the expanded condition, most minerals are more

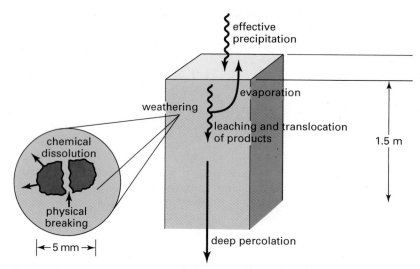

**FIGURE 12–13** The flow of water in a pedon not only weathers minerals but also carries products up and down the pedon, as illustrated here.

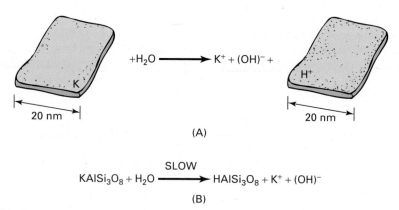

**FIGURE 12–14** (A) Hydrolysis of a clay mineral releases $K^+$ from the clay structure and adds an $H^+$ from solution. (B) Hydrolysis reactions written as a chemical equation look like this. Microcline (a feldspar) reacts with water and loses a $K^+$ as explained in part A.

porous and more susceptible to additional weathering by other processes.

Other examples of hydration are the change of boehmite to gibbsite or of red hematite to yellow or brown hydrated hematite:

$$\overset{150°C}{AlOOH + H_2O \rightleftharpoons AlOH_3}$$
$$\text{boehmite} \qquad \text{gibbsite}$$

*or*

$$Fe_2O_3 + 3H_2O \rightarrow 2Fe(OH)_3$$

These minerals contribute to the red, brown, and yellow color of soils.

**Oxidation reduction**    Oxidation reduction was described in Chapter 8, and it is important in weathering of minerals such as biotite mica, that have $Fe^{2+}$ in the mineral structure. When the oxidation of $Fe^{2+}$ to $Fe^{3+}$ occurs while the Fe is part of the mineral, the smaller $Fe^{3+}$ will strain the crystal, which accelerates the weathering rate. The red color of many soils is due to the "free" iron oxide including hematite ($Fe_2O_3$) that coat other mineral particles. (They are free because they are not part of a silicate mineral.) These oxides also are important to soil structure (see Chapter 2) and soil fertility (see Chapter 9).

Reduction is important to weathering because some cations are more mobile when reduced than when oxidized. For example, iron and manganese are more easily leached when in reduced forms. The formation of colored splotches called **mottles** is a result of Fe and Mn reduction, mobilization, concentration, and reoxidation.

### 12.2.3 Translocations

Water carries particles and soluble constituents through the soil. The downward translocation of ions, clay, and mineral fragments such as silica tetrahedra that recombine to form clay helps form subsoil horizons. Leaching or deep leaching occurs when translocation carries soluble constituents and clay out of the pedon.

The upward translocation of salts helps form the saline and sodic soils of arid and semiarid regions. Upward translocation stops at the soil surface when water evaporates, and downward translocation stops where the wetting front stops. In arid and semiarid regimes, this may be at shallow depths, and in humid areas, translocation may be mostly out of the profile. Climate is the main determinant. Water continues to move into and below the C horizons, but clays and oxides remain because of precipitation and other reactions.

Chelation also helps in translocation. Metal ions such as Fe and Al combine with organic molecules released during humus formation. The organic metal chelate is more soluble than the Fe or Al minerals are and is readily transported through the pedon.

### 12.2.4 Clay Formation

Clay formation and translocation are processes that differentiate soils from rocks, and they have attracted considerable attention from soil scientists. Examples of clay formation are offered in Supplement 12–3.

Smectite (2:1) and kandite (1:1) minerals are found in soils throughout the world. Smectite dominates the mineralogy of the high shrink-swell soils (Vertisols). Smectite clays form in environments that have high Si and sufficient $Ca^{2+}$, $Mg^{2+}$, and $Na^+$. Thus,

smectites are often found in basins where leaching is minimal, on parent materials rich in these cations.

Smectite can be inherited directly from some parent materials. Under well-drained conditions where leaching dominates soil formation, smectites are not expected to be a permanent part of the clay mineralogy because Si necessary for smectite formation and preservation is leached away, as are the accessory cations. Smectite weathering is enhanced in soils low in organic matter and with moderate acidity (pH 5).

Kaolin formation is favored by hot, humid, well-drained conditions and is commonly found in humid tropics and humid temperate zones. Kaolinite, the most common kaolin, is a weathering product of feldspars, mica, and other sources of Si and Al. Kaolinite is often found in small amounts in most soil-forming environments. It is a minor constituent in young soils but a major constituent of highly weathered soils throughout humid climates.

> Clays form (1) in place (without translocation) from the alteration of minerals or (2) either in place or after translocation, by the decomposition of minerals and recombination (recrystallization) of the weathering products.

## 12.3 PRODUCTS

This section discusses examples of the different kinds of horizons, first mentioned in Chapter 1. Combinations of these horizons may be found in mineral and organic soils.

### 12.3.1 Mineral Soils

Horizons    The **master horizons** are the A, O, E, B, C, and R horizons. W is used to designate a layer of water within the soil profile. The subscript f is added if the layer is permanently frozen (see Table 12–4). In addition to the master horizons, **transitional horizons** are found between master horizons (Table 12–4). One capital letter is used to designate master horizons and two are used for transitional horizons. Lowercase subscripts are used to subdivide master and transitional horizons (creating subordinate horizons) and to designate important horizon properties. Two or more horizons with identical letter identification are numbered

## TABLE 12–4

Master and Subordinate Horizon Symbols

| Symbol | Horizon property or characteristic |
|---|---|
| **Master horizons** | |
| O | Surface layer dominated by organic material |
| A | Mineral horizon formed at the soil surface or below an O horizon; it has accumulated humus |
| E | Mineral horizon in which the main feature is loss of silicate clay, Fe, or Al, leaving a concentration of resistant sand and silt particles |
| B | Horizon formed below an A, E, or O and dominated by the obliteration of original rock structure and by the accumulation of silicate clay, Fe, Al, humus, carbonate, gypsum, or Si |
| C | Horizon, excluding hard rock, little affected by soil genesis |
| R | Hard bedrock such as basalt, granite, or sandstone |
| **Subordinate horizons** | |
| a | Highly decomposed organic material |
| b | Buried genetic horizon |
| c | Concretions or nodules |
| d | Physical root restriction |
| e | Intermediately decomposed organic material |
| f | Frozen soil (permafrost) or water |
| ff | Dry permafrost |
| g | Strong gleying indicative of high water table |
| h | Accumulation of illuvial organic material |
| i | Slightly decomposed organic material |
| j | Accumulation of Jarosite |
| jj | Evidence of cryoturbation |
| k | Accumulation of carbonates |
| m | Cementation |
| n | Accumulation of sodium |
| o | Residual accumulation of sesquioxides |
| p | Tillage or other disturbance |
| q | Accumulation of Si |
| r | Soft or weathered bedrock |
| s | Illuvial accumulation of sesquioxides and organic matter |
| ss | Presence of slickensides |
| t | Accumulation of silicate clay |
| v | Plinthite |
| w | Development of structure or color |
| x | Fragipan character |
| y | Accumulation of gypsum |
| z | Accumulation of salts more soluble than gypsum |

Note: Transitional horizons have designations such as AB, BA, EB, BE, BC, and CB. The first letter represents the horizon to which the transitional horizon is most similar.

Source: *USDA Soil Survey Manual,* Chapter 3. Soil Survey Division Staff. 1993. Soil Survey Manual. Superintendent of Documents US Govt. Printing Office, P.O. Box 371954, Pittsburg, PA. http://www.statlab.iastate.edu/soils/ssm/gen_cont.html

consecutively—for example, A1, A2 and Bt1, Bt2. Examine the color plates for additional examples.

The A horizon is a mineral soil layer found at the earth's surface, unless it is buried by younger material. The high biological activity within the A horizon results in humus accumulation, the rapid weathering of minerals, leaching of soluble products, and eluviation. **Eluviation** is the translocation in suspension (movement by water) of weathering products out of one horizon to a lower horizon. Humus accumulation results in dark soil colors and a strong soil structure. Examine the A horizons of the profiles shown in the color plates to see the range of colors and thicknesses of A horizons. If there is more than one A horizon in a soil, each is numbered consecutively as A1, A2, A3, and so forth. Ap horizons are surface horizons that have been cultivated.

The O horizon is a predominantly organic rather than mineral horizon. Note the organic matter content of the O horizons in Plate 10 compared with the A horizons in the other plates. Mineral soils may have O horizons, as shown in Plate 5. A soil composed entirely of O horizons is an organic soil; O horizons are numbered consecutively from the surface. For example, the Kina series (Plate 10) has three consecutive O horizons.

The E horizon is typically found below an A or O horizon, but it may be found at the soil surface. It is a mineral horizon of strong eluviation, formed when organic acids from the A or O horizon combine with leaching waters to weather and translocate silicate clay, iron, and aluminum, leaving a residue of resistant minerals such as quartz. E horizons are usually lighter colored than are either the overlying A or underlying B horizons. The Vilas profile (Plate 5) has an E horizon. Some E horizons are characterized by platy structure.

Below an A, E, or O horizon are the B horizons, which are zones of illuviation. **Illuviation** is the accumulation of eluviated materials. Eluviated clays, iron and aluminum, $CaCO_3$, humus, silicates, and salts, alone or in various combinations, accumulate in B horizons. The kind of B horizon formed depends on the soil-forming factors. As water percolates through a soil, it may become saturated, at which time some of what it carries will precipitate. Changing chemical conditions (pH, oxidation reduction) in the soil also cause materials in solution or suspension to precipitate.

In humid environments, such as the northeastern and north-central United States and tropical and subtropical areas, leaching is a common process.

Cations, anions, and humus are eluviated from O and A horizons and accumulate in the subsoil to form new horizons. If Fe and Al are the dominant cations eluviated, Bs horizons form; Bh horizons form when eluviated humus accumulates in the subsoil. Large pores and much rain facilitate eluviation and subsoil horizon formation. If clay is the main material accumulated, Bt horizons form. Horizons of carbonate (Bk), gypsum (By), and salt (Bz) accumulation are common in arid and semiarid climates where there is insufficient salt leaching. Bt horizons are found in most climates.

Some soils do not have zones of accumulation but have zones in which a color or structure differentiates the horizon from the overlying A and underlying parent material. This is a Bw horizon.

Parent material is recognized as the C, R, and Cr horizons. The C horizon is little affected by soil formation and lacks the properties of the other horizons. It is not hard bedrock, for which the symbol R is reserved. Cr is used to designate soft or weathered bedrock.

Transitional horizons exist when there is a gradual change from one horizon to the next. The AB, EB, BE, and BC horizons are examples of transitional horizons A to B, E to B (which looks more like E than B), B to E (which looks more like B than E), and B to C.

**Example pedons** The example pedon we created in Figure 1–5 and have discussed throughout the book can now exist on a landscape. The pedon has five horizons: an A, E, Bt1, Bt2, and C (Figure 12–15). It has a well-developed profile because it has a thick, dark A horizon, a thin, eluviated E horizon, and a thick Bt horizon that has been subdivided into two horizons, over the C. Such pedons are common in the American Midwest.

*A lithosequence* Figure 12–16 illustrates a three-pedon **lithosequence.** Each soil has formed on level ground and under grass vegetation, 100 cm precipitation, and a mean annual temperature of 15°C. All three are the same age; only the parent material differences have caused the differences in the soils. The thin soil (Figure 12–16A) formed on basalt (a hard, extrusive, fine-grained, mafic igneous rock). Because it is thin, it does not support as much grass as does the limestone-derived soil.

The soils derived from limestone and granite (Figure 12–16B, and 12–16C) are considerably thicker than the basalt-derived soil. Limestone is a soft sedimentary rock into which water penetrates

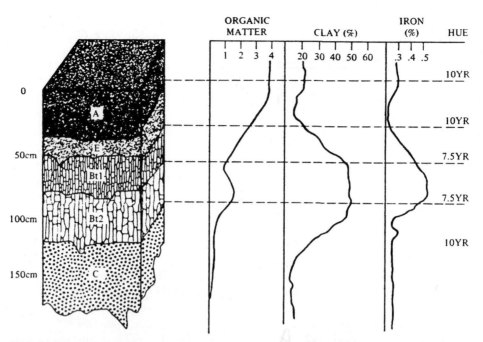

**FIGURE 12–15** This is the pedon we have considered in previous chapters. Starting in Chapter 1 with an empty three-dimensional rectangle, we have added properties and complexity to produce a complex pedon. What processes formed the horizons?

The A horizon has accumulated the most organic matter from the decay of grass roots. The E horizon has been leached of organic matter, clay, and iron oxides, which have accumulated in the thick B horizon. There is sufficient clay accumulation that the B has the subordinate horizon designation *t*. In addition, the upper part of the Bt, the Bt1, has accumulated some humus and iron. These all may have moved together. The C horizon is presumed to be the material from which the solum developed. We say presumed, because considerable data are needed to prove that the C is the parent material. It may represent a different geological deposit. In some complex soils, it is impossible to be sure the C is the parent material.

easily, and granite is a hard, coarse-grained, intrusive igneous rock. Granite's coarse grains make it more permeable than the fine-grained basalt. Granite is mostly quartz, muscovite mica, and potassium feldspar, and so it has low fertility. It holds more water than does the soil formed from basalt, but more than water is needed to grow the grass.

*A toposequence* The soils in a **toposequence** are formed from the same parent material, in the same general climate, with the same vegetation, and over the same length of time. They differ only in the position on the landscape where they are found (Figure 12–17). The position influences the microclimate and amount of water entering the soil. The toposequence illustrated in Figure 12–17 shows three soils formed on Pleistocene glacial till in the northeastern United States.

The soil at position 1 is moderately deep and developed. Because the soil at position 2 is on a steep hillside, erosion removes the A horizon at about the same rate as the parent material weathers to form new soil. In addition, although the same amount of rain falls on the surface, 50 percent of the water runs off, and only half as much is available for soil formation. At position 3, a deep, undeveloped profile shows the effects of prolonged wetness. This soil has a constant input of water and sediment eroded from position 2. The water-loving grasses, reeds, and sedges growing here added more humus to the soil than did the trees at positions 1 and 2. Vegetation has affected the morphology of the soils in the toposequence (Figure 12–17), but topography has been the dominant soil-forming factor.

The catena concept, first explained by G. Milne, who worked in East Africa, embodies the idea of how landscape hydrology can produce soils of different

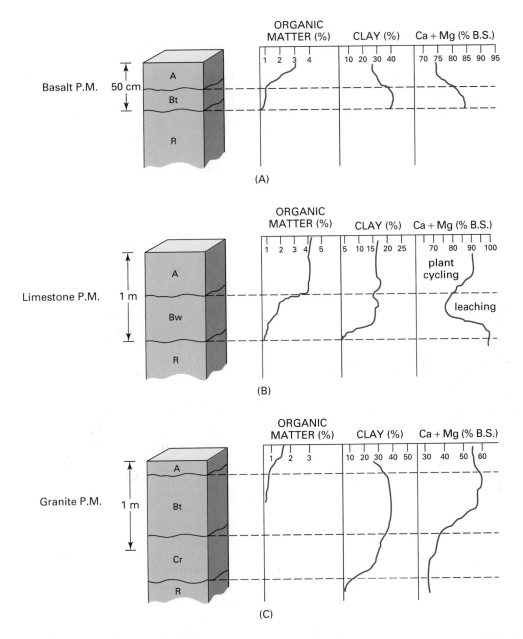

**FIGURE 12–16** This is a three-pedon lithosequence. The soil formed on basalt (A) has accumulated enough organic matter to have an A horizon, and sufficient clay has formed from rock weathering to have a Bt horizon. Soils A and B have much exchangeable Ca and Mg (high base saturation) because basalt and limestone contain more of these elements than does granite (soil C).

The limestone-derived soil (B) is rich in nutrients and supports luxuriant grass growth that cycles the nutrients as it did on the basalt. There is no Bt because insufficient clay has moved in the presence of $CaCO_3$. There is sufficient development of structure and a small clay increase for the soil to have a Bw horizon.

The soil formed from granite (C) has the thinnest A horizon because it has the poorest fertility and supports the poorest grass growth. There has been some recycling of nutrients by the grass, and much of the mica and some of the feldspar have weathered to clay. Deep penetration of water has helped develop a thicker Bt than in the basalt-derived soil. Some of the rock (R) has weathered to a paralithic horizon, an intermediate stage between hard rock and B horizon. This soft, partly weathered rock is a Cr horizon because the parent material has weathered in place.

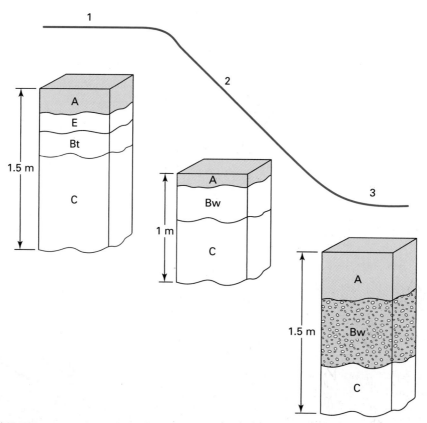

**FIGURE 12–17** A toposequence from a humid climate. The soil at position 1 has a thin A horizon, reflecting the mixed deciduous hardwood vegetation under which it formed. The E and Bt horizons also are thin, reflecting the cool temperatures and short weathering season. The soil at position 2 has a thin A over a thin Bw, which reflects the steep slope, increased erosion, and decrease in effective precipitation. At position 3, prolonged wetness has slowed the rate of clay formation, and so there is no Bt. The Bw is mottled with red and yellow splotches of color that indicate recurring cycles of wetting and drying during the year. The soils of the toposequence in this figure show the effect of changes in vegetation, but topography and its influence on profile hydrology are the major factors of soil formation influencing the pedon characteristics at each slope position.

drainage classes and different properties. The soils shown diagrammatically in Figure 12–17 form a drainage catena. Water moving over the surface contributes eroded soil, and water moving through the soil horizons contributes soluble weathering products as well as additional water to the soil lower on the slope. The soil forming at the top of the slope remains well oxidized. The soil at the bottom of the slope is somewhat poorly or poorly drained. It undergoes many periods of reduction, resulting in a different microbial population and different weathering processes and products than the well-drained soil up slope. The lower end of a drainage catena or toposequence is where hydric soils are found (see Section 12.3.3).

*A biosequence* In a **biosequence** (Figure 12–18), the soil-forming factors other than vegetation are "constant." In north-central Minnesota, soils formed under grass, mixed forest–grassland, and forest can be found on glacial parent material and topography of a similar age. The climate in Minnesota, particularly the precipitation, decreases from east to west, but the predominant difference in the soil-forming factors is the vegetation, which changes from forest to grassland and has left a distinct imprint on each of these pedons.

*A chronosequence* Time is isolated as a soil-forming factor in a **chronosequence.** The four pedons in

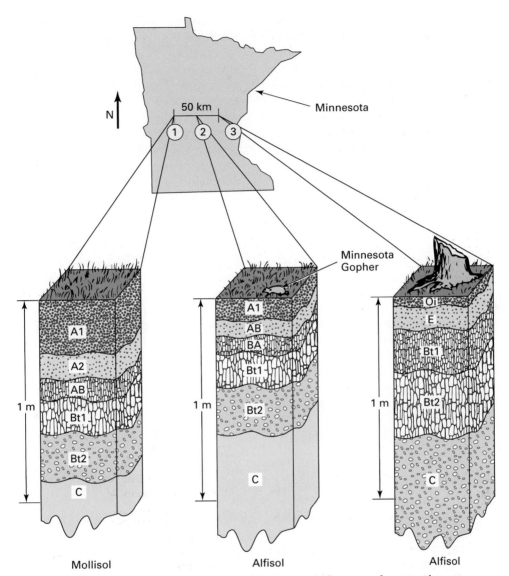

**FIGURE 12–18** A biosequence illustrates the influence of vegetation on soil morphology. The soil formed under grass (site 1) has a thick, dark A horizon. It formed under sufficient rainfall to form a prominent Bt horizon in the 10,000 years since the glaciers receded from the area. The soil formed under forest (site 3) has an Oi horizon of leaves and twigs. Beneath the Oi is an E horizon formed by the translocation of chelated compounds. The chelating agents formed from the decomposition of the forest litter. Below the E is a strongly developed Bt horizon.

Between the prairie and forest soils is a pedon with properties intermediate between the two (site 2). The A horizon is thinner, and the Bt is separated from the A by several transition horizons with characteristics intermediate between A, E, and B horizons.

Figure 12–19 are a chronosequence formed from river alluvium from the Sierra Nevada mountains in the central valley of California. The stair-step fan terraces were formed by a cycle of landscape construction and erosion. The process of deposition and erosion occurred repeatedly, but the river height progressively declined, so the younger surfaces were inset below the older surfaces. During this time, the older parent materials and surfaces were stable, and soil formation could proceed. The result is a series of soils of different ages that formed on the same alluvium and under the same semiarid climate, vegetation, and topography.

## 12.3.2   Organic Soils

Mineral soils are more common than organic soils, but organic soils still are important agriculturally. An organic soil consists predominantly of decomposed plant material. Soil taxonomy, which is discussed in Chapter 14, defines an organic soil as containing a minimum of 20 percent organic carbon. If the clay content is greater than 60 percent, an organic soil must have at least 18 percent organic carbon. If the clay content is zero, the soil must have at least 12 percent organic carbon. Organic soils usually contain at least 80 percent organic matter.

**Horizons**   Organic soil horizons are differentiated by three degrees of organic matter decomposition: slightly (Oi), intermediately (Oe), and highly (Oa) decomposed. The degree of decomposition is determined by the kind of vegetation and climate and is identified by sight and feel. Slightly decomposed organic horizons have plant remains that can be easily distinguished, whereas highly decomposed horizons do not. Intermediate between the two extremes are the Oe horizons. Rubbing highly decomposed material results in a slippery black product. The same treatment of slightly decomposed organic matter results in little black material and many observable plant parts.

**Example pedon**   Figure 12–20 and Plate 10 illustrate a pedon with a sequence of O horizons. The horizons may be in any order and are numbered consecutively; for example, the third Oa horizon is the Oa3, even though it is the sixth horizon. The organic material may rest on different kinds of mineral material, but the mineral material is not the parent material for the soil, even though it is a C horizon.

Organic soils are found in those areas on the landscape that are permanently wet, and they are most common in cold and wet locations where the combination of wetness and cold slows the decomposition of plant materials. In the United States, large areas are found in Alaska, glacial areas of Michigan and Minnesota (Figure 12–21A), and low-lying coastal areas in Florida and California (Figure 12–21B). The presence of water year-round permits the growth of reeds, sedges, and other water-loving plants. When the plants die, they decompose slowly under water, and, over time, the decomposed organic material accumulates. When an organic soil is drained for agriculture, the rate of decomposition greatly increases (see Chapter 15).

The physical and chemical properties of organic soils are different from those of their inorganic counterparts. They have low bulk densities that may range from less than 0.1 to 0.3 $Mg/m^3$, and their color is usually dark brown to black throughout the pedon. The low density means that organic soils have a high porosity and a water-holding capacity many times their dry weight. This high water-holding capacity is one reason that organic soils are valued for nursery and container growth of plants.

Decayed organic material is rich in colloids that have a high surface area and charge. The cation exchange capacity (CEC) of organic soils far exceeds the CEC of mineral soils. Depending on the conditions under which the organic soil formed, the base saturation may be high—or low if the organic soil is acidic. In general, the nitrogen-, phosphorus-, and sulfur-supplying capacity of organic soils is high.

## 12.3.3   Hydric Soils

One component of the definition of a wetland is the presence of hydric soils. Hydric soils form when the hydrology of the profile dominates the soil-forming factors. Thus, hydric soils can be found in any general climate zone. Conditions required to form a hydric soil are those that promote reduction and mobilization of Fe and Mn with or without complete removal of these elements. Iron and manganese play important roles in the definition of hydric soils because they contribute greatly to soil color and both elements are readily reduced and reoxidized in soil. Conditions include sufficiently warm temperatures during periods of wetness to have biological activity, abundant water, and Fe and Mn in soil.

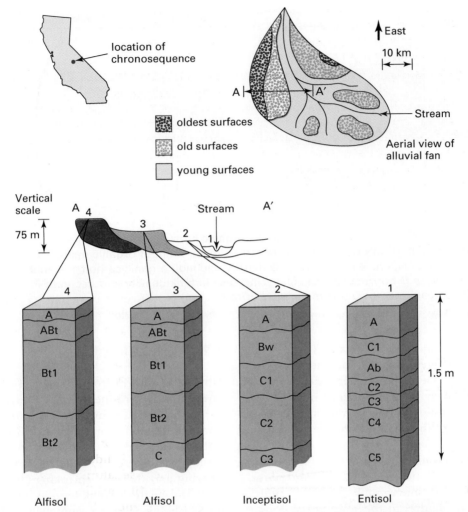

**FIGURE 12–19** This chronosequence from California illustrates the effect of 250,000 years on soil properties. Pedon 1, found in the river floodplain, is flooded in winter and spring but is otherwise well drained. The A is the only genetic horizon. Stratification is evident, and in the top meter is a buried A horizon, indicating that the soil surface was once at 60 cm for long enough to accumulate organic matter. The A was then buried by subsequent deposition. Continuing deposition may bury the present A horizon, making it another Ab.

Pedon 2 is above the stream and is not flooded frequently. It has been in this position for enough time that a Bw horizon has formed, and much of the stratification has been lost through biological mixing. The B has insufficient clay to be a Bt, but it does have some structure and red colors that differentiate it from the A and C horizons.

Pedon 3 is found on much older surfaces. It has never flooded, all stratification is lost, and the parent material has been deeply weathered. The A horizon is thin because less vegetation grows here than on site 1. The Bt1 and Bt2 are distinguished by structure, clay film thickness, and color.

The oldest pedon is found on the highest landscape position. The Bt horizons extend beyond 1.5 m. Iron and clay were probably translocated together into the Bt horizons, making them red and rich in clay, with thick clay films.

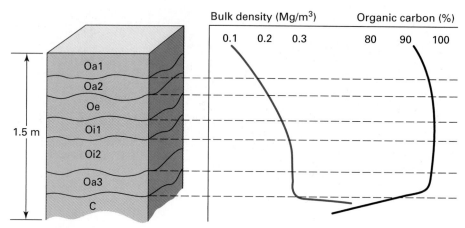

**FIGURE 12–20** This pedon has six organic horizons over a mineral layer. Each horizon's designation is based on the amount of organic matter decomposition. Note the low bulk density and high organic carbon contents of the O horizons and the sharp increase in bulk density and decrease in organic carbon in the mineral C horizon.

Hydric soils are mineral soils that have features produced by prolonged wetness including redox concentrations and depletions and organic soils. Organic soils or overly thick A horizons in mineral soils form when wetness inhibits organic matter decomposition. Both of these conditions are also indicators of hydric soils.

Redox concentrations in mineral soils are Fe and Mn nodules and concretions, masses, and pore linings. Nodules and concretions are small (millimeter size), hard bodies in the soil matrix. Nodules have no regular concentric pattern when cut in half; concretions do. Masses are soft (usually iron) bodies with reddish or brown colors. These are often called mottles because the reddish brown or yellow brown colors are spots or splotches of color on the soil matrix. Pore linings are zones of iron or Mn accumulation along pore surfaces. These concentrations are interpreted as indicating cyclic wetting (reduction) and drying (oxidation) that result in redistribution of Fe and Mn but not complete removal from the profile.

Redox depletions are zones of low chroma and neutral hue (gray colors) where iron and manganese have become mobilized and removed. Under prolonged saturation, Fe and Mn are completely removed from the soil, resulting in gray, low-chroma matrix colors throughout the reduced zone. These are referred to as gley colors.

## 12.4  SUMMARY

Mineral soils form from geological parent material. Five factors determine the characteristics of a pedon: parent material, climate, topography, biota, and time. These soil-forming factors influence the chemical and physical processes that convert the parent geological material to soil. Physical weathering by freezing and thawing, shrinking and swelling, abrasion, and heating break large particles into smaller ones. Chemical weathering such as hydrolysis, hydration, acidification, oxidation reduction, chelation, and solution change individual minerals into new minerals or into molecules that can recombine to form new minerals or that are removed from the pedon by leaching.

There is an almost infinite number of combinations of the factors and processes, resulting in a wide array of kinds of soils.

## QUESTIONS

1. What are the five soil-forming factors?
2. What is the difference between a soil-forming factor and a soil-forming process?
3. How do rocks and soils differ?
4. How do in-place parent materials differ from transported parent materials?

(A)

(B)

**FIGURE 12–21**  (A) Black spruce forest on raised bog from northern Minnesota. (Photo: D. Grigal.) (B) In the confluence of the Sacramento and San Joaquin Rivers is an area of organic soils that is valued for agriculture (top) and recreation (center). Undrained land is wildlife habitat.

5. Describe what soils you would find if you walked completely around a hillside in the Northern Hemisphere.
6. What are the differences between physical and chemical weathering products?
7. How do the horizon designations differ for mineral and organic soils?
8. What are the soil-forming factors expressed in the area where you live? Give the details of each. For example, how much rain falls in your area each year?

## FURTHER READING

Amundson, R., J. Harden, and M. J. Singer. 1994. *Factors of soil formation: A 50th anniversary retrospective.* Soil Science Society of America, Special Publication 33.

Birkeland, P. W. 1984. *Soils and geomorphology.* New York: Oxford University Press.

Dixon, J. B., and S. B. Weed, eds. 1989. *Minerals in soil environments.* 2d ed. Madison, Wis.: Soil Science Society of America.

Fanning, D. S., and M. C. B. Fanning. 1989. *Soil: Morphology, genesis, and classification.* New York: Wiley.

Jenny, H. 1941. *Factors of soil formation.* New York: McGraw-Hill.

Jenny, H. 1980. *The soil resource—origin and behavior.* New York: Springer-Verlag.

Smalley, I. J., and C. Vita-Finzi. 1968. The formation of fine particles in sandy deserts and the nature of "desert" loess. *Journal of Sedimentary Petrology* 38:766–774.

Ugolini, F. C., R. Dahlgren, J. LaManna, W. Nuhn, and J. Zachara. 1991. Mineralogy and weathering processes in recent and holocene tephra deposits of the Pacific Northwest, USA. *Geoderma* 51:277–299.

## SUPPLEMENT 12–1   Acidification

Carbonic acid is common in soils because of the abundance of $CO_2$ dissolved in water:

$$H_2O + CO_2 \rightarrow H_2CO_3 \qquad (3)$$

The weak carbonic acid formed helps disintegrate minerals. A familiar example is the dissolution of limestone by an acidic soil solution. The major mineral in limestone is calcite, which reacts with carbonic acid to form soluble calcium bicarbonate, which in turn is leached into groundwater:

$$CaCO_3 + H_2CO_3 \rightarrow Ca(HCO_3)_2 \qquad (4)$$

Other sources of $H^+$ ions are organic acids, microbiologically produced $HNO_3$ and $H_2SO_4$, and exchangeable $H^+$ from clay (see also Figure 8–8B).

## SUPPLEMENT 12–2   Solution

The different weathering mechanisms contribute to the dissolution of minerals in the soil. Most common minerals have limited solubility in pure water, but the acidification of the soil solution by $CO_2$, the dissociation of $H^+$, and the release of organic acids by means of organic-matter decomposition increase the solubility of most minerals. Over the tens or hundreds of thousands of years of soil formation, the repeated removal of the dissolution products permits further dissolution.

## SUPPLEMENT 12–3   Clay Formation

Feldspar, mica, amphibole, and pyroxene minerals change into clays through the processes of hydrolysis, hydration, and oxidation. For example, in biotite mica, $Fe^{2+}$ can oxidize, $K^+$ leaves the structure to maintain electrical neutrality, and the structure is weakened. Next, soluble $Ca^{2+}$, $Mg^{2+}$, and $Na^+$ in the soil solution replace the remaining biotite $K^+$ to form vermiculite or montmorillonite. All this may take place without any movement of the mineral.

Mica and other aluminosilicates can also slowly dissolve into individual silica molecules and Al, Mg, K, and Fe ions, which can recombine to form clay in the same location, or the various ions can be translocated to another location, where they recombine to form clay. A generalized scheme of clay formation is shown in Figure 12–22.

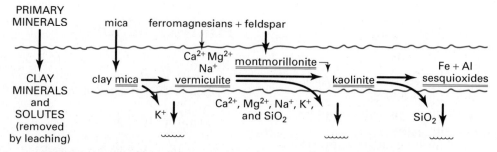

MINERAL WEATHERING: A GENERALIZED SEQUENCE

**FIGURE 12–22** Generalized scheme of mineral weathering to clay and clay transformations. Primary minerals weather to form clays that weather to solutes and to other clays.

# 13

# SOIL INFORMATION

## OVERVIEW

*Soil information includes data on horizons, a pedon, groups of pedons in a field, or all pedons in the world. Maps display geographic soils information, and reports offer information about single or multiple properties of pedons. Geographic information includes the distribution of soils on the landscape, current land use, and potential uses or suitabilities. Soil map units describe soil pedons as they exist on the landscape. Map scale determines what can be displayed on a map. Detailed maps on large scales are useful for detailed planning, and generalized maps on small scales are useful for broad overviews of an area.*

*Soil survey reports are accessible sources of point data and geographic data in the United States. The reports consist of tabular data and interpretations and generalized and detailed soil maps. Soil maps are made using satellite imagery, aerial photographs, and traditional techniques such as walking across the land digging holes and observing pedon characteristics. The map's level of detail is determined*

*by the number of observations made by the soil surveyor, by the proposed use of the information, and by the money available for a survey.*

*Remote sensing is gathering information about an object from a distance. Aerial photography with a camera–film–filter combination is routinely used to obtain soil survey information. Stereo photography is particularly useful, because it allows the observer to "see" a three-dimensional view of the earth's surface on a flat photo. Color and color-infrared film extend a surveyor's ability to interpret imagery. These and other forms of remote sensing allow the soil surveyor to preview the area to be mapped before going to the field. On-site observations—ground truth—must always be obtained for evaluating remotely sensed information.*

*Maps are a form of geographic data. Computerized storage, retrieval, and interpretation of geographic data eliminate the laborious task of drawing and redrawing maps when they are needed. Information may be entered into a computer file just once, and new maps may be displayed on the computer monitor and printed as needed.*

## 13.1  KINDS OF SOIL INFORMATION

Soil information would be useless if there were no way to transmit it to interested individuals. Farmers, ranchers, foresters, gardeners, home owners, engineers, soil scientists, tax assessors, and others want to know what kind of soil is on their land or on land for which they are planning some use. Often people ask questions such as, "What soils do I have?" or "What are the properties of the soil or soils on my land?"

Maps are the best way to answer the first question, because soil maps show the distribution of soils on the land. This information is **geographic data** because it describes a place's physical features and is easily displayed on maps. The properties of a soil or soils are identified in reports with data tables, or point data. **Point data** refer to the morphological, chemical, physical, mineralogical, or biological characteristics of the pedon. These characteristics can be displayed on maps, but it is difficult to display all the characteristics that describe and define a pedon.

Geographic and point data may be either observed or inferred. Soils are mapped as they are observed in the field, or their inferred properties are

mapped. Inferred properties are those that are not observed but are interpreted from observed properties.

### 13.1.1  Geographic Data

Soil maps are not too different from other kinds of maps, though several concepts are important to knowing how to use a soil map, including the kind of soil map unit and scale.

**Soil map units**  Pedons can be mapped, but they are too small to be useful units for general planning purposes. The largest pedons are less than 8 m on a side, and the smallest are only 1 m². Polypedons, however, are mapped. A **polypedon** is a contiguous group of similar pedons encircled by a line on a map, forming a polygon. Polygons are also called map unit delineations or simply delineations. A **map unit** defines the contents of the area inside a polygon on a soil map. The dark lines surrounding the map unit symbols in Figure 13–1 define a polygon. The degree of similarity of the pedons inside the map unit is a matter of choice and scale and leads to five commonly used kinds of soil map units.

**FIGURE 13–1**  Wabasha County, Minnesota. The original map scale is 1:20,000, or 3.2 in/mile, or 50 mm/km. The 1:20,000 scale is a *relative fraction,* which allows us to calculate the distance in any units. One inch on the map is 20,000 inches on the earth's surface, or 1 cm on the map is 20,000 cm on the earth's surface. This is an enlarged reproduction of part of a map sheet from the USDA Natural Resources Conservation Service Soil Survey of Wabasha County. A soil map unit is identified by a letter code within each polygon (e.g., BtA, MdA).

The **consociation** is the most detailed soil map unit. About 75 percent of the pedons in a consociation are a named soil series. The remaining 25 percent of the polypedon may include similar or dissimilar soils, but no single dissimilar soil may occupy more than 10 percent of the map unit. The 25 percent are **inclusions** within the map unit. Soil series and categories of soil taxonomy higher than the series may be consociations. An example of a consociation name is "Miami silt loam, 0 to 3 percent slopes." *Miami* is the named soil series, which ideally occupies 75 percent of the polypedon in the map unit. The consociation name includes a soil series and one or more phases such as surface texture, slope, or drainage.

A **soil complex** is a map unit that consists of two or more kinds of soils that occur in a regular repeating pattern, but the pattern is so intricate that the two components cannot be separated at the selected map scale. The major soils of the map unit provide the name for the unit. For example, "A-B complex, 5 to 15 percent slopes" identifies a complex in which series A and B are the major components.

**Soil associations** are map units that contain two or more kinds of soils, generally associated in a regularly repeating pattern and individually large enough to be separated on soil survey maps. Soil associations are used to reduce the cost of making a survey, without reducing its usefulness. Soils within an association are geographically associated but may have no similarities.

The **undifferentiated group** is used as a map unit when the management of two or more soils is very similar for common uses. For example, if two or more kinds of soil, each on very steep slopes, have the same probable uses and management, they may be combined into an undifferentiated group. Other common features used to determine an undifferentiated group are stoniness, shallowness, or flooding. An example of an undifferentiated group name is "A and B clays, shallow." The word *and* distinguishes an undifferentiated group from either an association or a complex.

Areas with little natural soil are mapped as **miscellaneous areas.** Examples of miscellaneous areas are rough, rocky land, gullied land, and dune land. Generally, the variability in these areas is so great that no "typical" pedon can be found.

**Soil maps**   Soil maps may be of any scale or level of generalization. The map in Figure 13–1 is a portion of a detailed soil map from the Wabasha County, Minnesota, soil survey report published in 1965. The map scale is 1 to 20,000. One unit of length on the map is 20,000 units of length on the ground. For example, 1 mm on the map is 20,000 mm, or 20 m, on the ground. There are three kinds of map units on this portion of the map: consociations such as CaB–Chaseburg silt loam, undifferentiated groups such as DhC2–Downs and Mt. Carroll silt loams, and miscellaneous areas such as Sr–steep stony and rocky land. In addition to the soil boundaries that separate the polypedons, cultural features such as roads and towns, bodies of water such as Gorman Creek, and geographical markers such as section, township, and range lines help precisely locate a place on the map. The section–range system is explained in Supplement 13–1.

Figure 13–2 is a generalized soil map from Lauderdale County, Alabama. The scale is 1 to 190,080, or about 2 km/cm (3 miles/in), much less detailed than the Minnesota map. It has only four map units, all of which are associations of two or more soils. A few major geographic features are shown on this map to help orient the user.

The general soil map presents the broad pattern of soils on the landscape, which is useful in general planning, whereas the detailed map provides information on the intricacies, which are valuable for farm and city planning. Both are important to understanding soil distribution, but neither gives sufficient detail for site-specific planning, such as where to plant a tree or build a house. Scale determines, to a large extent, what can be seen on a map. Table 13–1 illustrates the effect of scale on the minimum-size map delineation. A minimum-size delineation is a square 6 mm on a side because it is about the smallest area in which an identifying symbol can be printed and read; 6 mm is about 1/4 in.

Both Figures 13–1 and 13–2 are *basic maps* because they describe the kinds and distribution of soils without interpreting their use. *Interpretive maps,* on the other hand, group areas into similar use categories or limitations. For example, Figure 13–3 shows slope groups, rather than soil map units, for the area in Figure 13–1. Another example of an interpretive map (Figure 13–4) shows severe, moderate, or slight limitations for septic tank filter fields. In this map, the soils' properties, rather than polypedons, have been interpreted, and the interpretive classes are displayed.

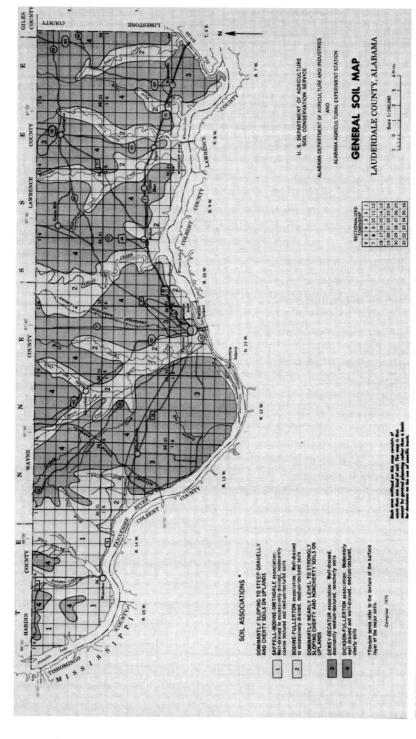

FIGURE 13–2 Lauderdale County, Alabama. This is a general soil map. The scale is 1:190,080, and only soil associations are shown. This is a reproduction of the general soil map from the USDA Natural Resources Conservation Service Soil Survey of Lauderdale County.

**TABLE 13–1**

Examples of Map Scales and the Minimum Delineation Normally Found

| Map scale | Minimum delineation | |
|---|---|---|
| | Acres | Hectares |
| 1:500 | 0.0025 | 0.001 |
| 1:5,000 | 0.25 | 0.10 |
| 1:10,000 | 1.00 | 0.41 |
| 1:15,840 | 2.50 | 1.00 |
| 1:24,000 | 5.7 | 2.3 |
| 1:62,500 | 39 | 15.8 |
| 1:100,000 | 100 | 40.5 |
| 1:250,000 | 623 | 252 |
| 1:500,000 | 2,500 | 1,000 |
| 1:1,000,000 | 10,000 | 4,000 |

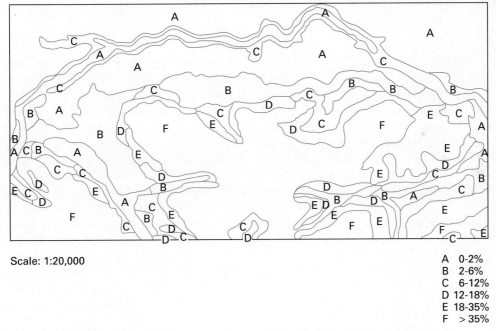

Scale: 1:20,000

A  0-2%
B  2-6%
C  6-12%
D  12-18%
E  18-35%
F  > 35%

**FIGURE 13–3**  Slope groups are an interpretive grouping of soils. The soil map units from Figure 13–1 have been grouped into slope groups for this figure. Note that the number of lines, the soil groups, and the complexity of the map have decreased compared with those of Figure 13–1. North is at the top.

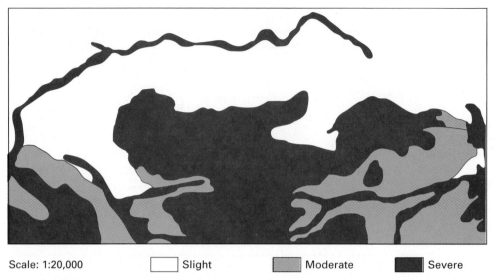

Scale: 1:20,000    □ Slight    ▨ Moderate    ■ Severe

**FIGURE 13–4**  Septic leach field suitability is another interpretive grouping of soils. The soil map units of Figure 13–1 have been grouped into
slight, moderate, and severe limitations for septic leach fields.

## 13.1.2  Point Data

Soil maps show how soils are arrayed on the land surface but give little information about pedon properties or detailed information about the polypedon
represented on the map. Rather, the point data, ex-
trapolated from several soil pedons, provide that information. In the United States, the single most accessible source of point data is the soil survey report.

**Soil survey reports**   Soil survey reports, such as
those illustrated in Figure 13–5, are the cooperative
product of the U.S. Department of Agriculture
(USDA) Natural Resources Conservation Service, or
NRCS (formerly the Soil Conservation Service, or
SCS), other federal agencies such as the U.S. Forest
Service, and cooperating agencies such as the states'
agricultural experiment stations. These reports consist of a map section with generalized and detailed
soil maps and a text section with point information
and interpretations.

The text section of a soil survey report is filled
with detailed pedon and map unit descriptions, laboratory data, and interpretations. The typical soil survey report text begins with a discussion of the general
nature of the area that was mapped, including history
and development and an overview of the agriculture,

physiography, relief and drainage, and climate. Reports also contain a brief description of how the maps
and report were prepared.

Next comes a description of the soils in the soil
map units on the general soil map. The general soil
map units are associations of major soils or major
soils and miscellaneous areas. The general soils map
is typically at too small a scale for planning or management purposes but gives an overview of the distribution of soils in the survey area.

Following the general soil map unit descriptions are map unit descriptions for the detailed soil
maps. These are typically listed alphabetically. Each
map unit description contains information on the
landform where the map unit is commonly found, including slopes, elevation, and mean annual precipitation. The composition of the map unit is also described. As noted earlier, a consociation ideally
contains 75 percent of the named unit and 25 percent inclusions. In this section of the soil survey report, the actual composition of each map unit is described, including details about the named soils and
inclusions.

The discussion is supported with data tables on
climate, acreage of each soil map unit, and soil map
units that meet definitions of prime agricultural
land. Permeability, plant-available water, effective

**FIGURE 13–5**  Soil survey reports from the United States.

rooting depth, runoff rate, erosion hazard, suitability for typical uses, and USDA capability units are part of each map unit description. Tables of yields of commonly grown crops, USDA Land Capability, Storie Index (in California), rangeland productivity, and characteristic plant communities are part of the soil survey.

The morphology of a typical profile of each soil series and the allowable range of the profile's characteristics are part of every soil survey report (Figure 13–6). The series descriptions include the soil classification (see Chapter 14) and the exact location where the profile was described. Some soil surveys also include a description of soil formation and how it was influenced by the climate, organisms, and geomorphology.

In addition to the agricultural data, soil survey reports provide nonagricultural interpretations. Chapters 15 and 16 discuss some of the interpretations in more detail. Interpretations include suitability or limitations of each map unit for recreational development, including camp and picnic areas, playgrounds, paths and trails, and golf fairways; wildlife habitat; building-site development, including shallow excavations, dwellings with and without basements, small commercial buildings, local roads and streets, and lawns and landscaping; and construction materials, including road fill, sand, gravel, and topsoil. These interpretations are supported by engineering test data, physical and chemical properties of the soils, and estimated values of engineering and soil properties based on the test data.

Interpretations are also provided for sanitary facilities, including septic tank absorption fields, sewage lagoon areas, sanitary landfills, and daily cover for landfills. Altogether, the soil survey report is the single best available source of information on distribution, morphology, and interpretation of soils.

**Other sources**  Soil survey maps and reports are made for privately owned agricultural land by the NRCS. Other federal agencies and some state agencies map soils in their jurisdictions and generally work closely with the NRCS. The U.S. Forest Service makes soil surveys of the national forests that are less detailed than the NRCS surveys are. The U.S. Bureau of Land Management (BLM) makes soil surveys of nonforested public lands to help in their planning and management, but neither the BLM nor the U.S. Forest Service maps are prepared for distribution to the general public. Perhaps the most ambitious project is the soil map of the world made by the Food and Agricultural Organization of the United Nations.

**Yolo Series**

The Yolo series consists of nearly level, well-drained soils on alluvial fans. These soils formed in mixed alluvium derived from sedimentary rocks. Where these soils are not cultivated, the vegetation is annual grasses and forbs. The average annual temperature is 60° to 62° F., the average annual rainfall is 18 to 25 inches, and the frost-free season is 240 to 260 days. Elevation ranges from 25 to 150 feet.

In a representative profile, the surface layer is dark grayish-brown silty clay loam about 28 inches thick. The next layer is brown clay loam about 8 inches thick. The substratum is brown loam that extends to a depth of more than 60 inches.

The effective rooting depth is more than 60 inches.

Yolo soils are used for orchards, irrigated row crops, forage crops, truck crops, dryfarmed small grain, wildlife habitat, and recreation.

Following is a representative profile of Yolo silty clay loam:

Ap—0 to 9 inches, dark grayish-brown (10YR 4/2) silty clay loam, very dark grayish brown (10YR 3/2) when moist; massive; hard, friable, sticky, plastic; many very fine roots, few fine roots; common very fine pores; mildly alkaline; abrupt, wavy boundary.

A11—9 to 18 inches, dark grayish-brown (10YR 4/2) silty clay loam, very dark grayish brown (10YR 3/2) when moist; weak, medium, subangular blocky structure; hard, friable, sticky, plastic; few fine and very fine roots; few fine pores and common very fine pores; mildly alkaline; thin films on ped faces and in pores; gradual, wavy boundary.

A12—18 to 28 inches, dark grayish-brown (10YR 4/2) silty clay loam, very dark grayish brown (10YR 3/2) when moist; weak, medium, angular blocky structure; hard, friable, sticky, plastic; few fine and coarse roots; common fine pores; neutral; increase in films (may be organic staining) on ped faces; clear, wavy boundary.

AC—28 to 36 inches, brown (10YR 4/3) light clay loam, dark brown (10YR 3/3) when moist; weak, medium, subangular blocky structure; slightly hard, very friable, slightly sticky, slightly plastic; few very fine and medium roots; many very fine and fine pores; mildly alkaline; thin very dark grayish-brown (10YR 3/2) films on ped faces; gradual, wavy boundary.

C1—36 to 44 inches, brown (10YR 5/3) loam, brown (10YR 4/3) when moist; weak, medium, subangular blocky structure; slightly hard, very friable, slightly sticky, slightly plastic; few fine and coarse roots; common fine and very fine pores; mildly alkaline; thin films on ped faces; gradual, wavy boundary.

C2—44 to 60 inches, brown (10YR 5/3) loam, brown (10YR 4/3) when moist; weak, medium, subangular blocky stucture; slightly hard, very friable, slightly sticky, slightly plastic; few fine and coarse roots; common fine and very fine pores; mildly alkaline; thin films (may be organic staining) on ped faces.

The A horizon ranges from dark grayish brown to grayish brown in color, from silty clay loam to loam in texture, from slightly acid to moderately alkaline in reaction, and from 18 to 36 inches in thickness. The C horizon ranges from brown to yellowish brown in color. It is loam or silt loam in texture and neutral to moderately alkaline in reaction.

**Yolo loam** (Yo).—This soil has a profile similar to the profile described as representative for the series, except that it has a loam texture throughout. Included with this soil in mapping are small areas of Reiff fine sandy loam, Brentwood clay loam, Yolo silty clay loam, and Sycamore silty clay loam.

Permeability is moderate. Runoff is slow, and erosion is a slight hazard. The available water capacity is 9 to 11 inches.

The soil is used mostly for almonds, peaches, apricots, walnuts, sugar beets, corn, tomatoes, and alfalfa (fig.15). It is also used for dryfarmed barley, urban development, wildlife habitat, and recreation. Capability unit I–1 (17); not placed in a range site.

**Yolo loam, clay substratum** (Yr).—This soil has a profile similar to the one described as representative for the series, except that the surface layer is loam and a buried clay substratum is at a depth of 40 to 60 inches. Included with this soil in mapping are small areas of Reiff fine sandy loam, Yolo loam, Sycamore silty clay loam, Brentwood clay loam, and a soil that has a clay substratum at a depth of 20 to 40 inches.

Permeability is slow. Runoff is slow, and erosion is a slight hazard. The available water capacity is 9 to 11 inches.

This soil is used for irrigated sugar beets, tomatoes, grain sorghum, and alfalfa. It is also used for dryfarmed barley, wildlife habitat, and recreation. Capability unit IIs–3 (17); not placed in a range site.

**Yolo silty clay loam** (Ys).—This soil has the profile described as representative for the series. Included with it in mapping are small areas of Reiff fine sandy loam, Brentwood clay loam, and Sycamore silty clay loam.

Permeability is moderately slow. Runoff is slow, and erosion is a slight hazard. The available water capacity is 10 to 12 inches.

This soil is used mostly for almonds, peaches, sugar beets, tomatoes, alfalfa, walnuts, and dryfarmed barley. It is also used for urban development, wildlife habitat, and recreation. Capability unit I–1 (17); not placed in a range site.

**FIGURE 13–6** This is a typical pedon description and map unit descriptions for the Yolo series as found in the USDA Natural Resources Conservation Service Soil Survey Report for Solano County, California. Three map units are used in the survey, and each is described.

## 13.1.3  Soil Variability

Soil map units contain more than one soil series because soil distribution is too complex to separate into completely "pure" units. Variability is found in both geographic and point data and is a natural function of the soil-forming processes. Variability occurs at large and small spatial scales. This variability is accepted, but methods are needed to describe it quantitatively. Variation in point data is described using standard statistical techniques such as averages, deviations around the average, and range of values. However, standard statistical procedures assume that variations are random. Soil variations are usually not random; there is a high degree of spatial dependency, so that samples taken close together vary less than samples taken far apart (a glance at Figure 13–4 or Figure 13–13 makes this apparent). **Geostatistics** are used to describe this spatial variability.

Two frequently asked questions are, "How many soil samples are needed to know the value of a property?" and "Where is the best place in a field to collect samples?" Geostatistics help answer those questions. In geostatistics, many samples are taken, and each is compared with every other sample in relation to the sampling distance that separates them. The relationship between the variations of the sample values and the distance between them provides useful information on where to sample, how many samples are needed, and how they should be spaced to describe soil variability accurately.

## 13.2  COLLECTING AND DISPLAYING INFORMATION

### 13.2.1  Making a Soil Map

The USDA has a standard procedure for making a soil survey. Making a soil survey takes many years of hard work. The soil surveyor's tools are shown in Figure 13–7. Before walking through the countryside examining the soils, the soil surveyor collects whatever information is available on the five soil-forming factors, which may include old soil and geology maps of the area, vegetation information, and climate data. Similar information about the surrounding areas is also assembled and studied.

Modern soil maps are made on aerial photo bases. Old and new photos of the survey area are studied for clues to the kinds of soils that may exist on the landscape. Once the information has been assembled, the surveyor is ready to begin looking at the soils, by driving over and around the area. Once the general "lay of the land" is understood, the soil surveyor will (with permission of the landowners) walk over much of the survey area, digging holes, examining pedons, and constructing a legend. The legend contains names of the soil map units that the soil surveyor wants to use in the survey.

As the survey progresses, other soil surveyors from the state and region review the soil surveyor's work. These reviews improve the quality of the survey and ensure the correlation of the soil series in the mapping area with the same soil series in other areas. **Correlation** is the step that ensures that soil series names are applied consistently, so that all the pedons being called by a soil series name fit the series description. Any pedon, anywhere, called Yolo must fit within the defined limits of Yolo. A major responsibility of the soil surveyor is to match what is found in the field with the known concepts of soil series.

What happens if pedons are found that fit no existing soil series? New soil series names are created. Thousands of new series have been created since soil taxonomy was developed, because soil taxonomy de-

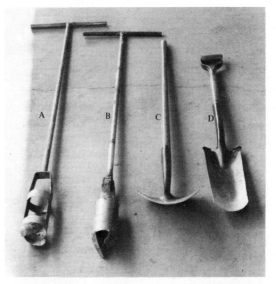

**FIGURE 13–7**  Tools of the soil scientist include (A) a mud auger, (B) a bucket auger, (C) a shovel, and (D) a tile spade. Also needed is a good pair of boots (not shown).

mands precise limits on any particular series. Over 23,000 soil series existed in the United States in 2000.

By walking, digging holes, drawing lines on aerial photos, reviewing, describing pedons, reviewing again, and walking more, the soil surveyor completes the soil survey. While the mapping is under way, information is gathered for the soil survey report. The soil surveyor in charge of the survey writes the survey report, which when completed is published and distributed through local NRCS offices.

There are five general levels or orders of soil surveys. The most detailed surveys are the first-order surveys. Most soil survey reports are second-order surveys, which are less detailed. Surveys become more general up to the fifth order, which consists of broad reconnaissance surveys that contain few observations. (More details are given in Supplement 13–2.)

Every map needs a suitable base onto which the information is drawn. Before the advent of aerial photography—until about 1926—soil maps were made by **plane table** mapping. The mapper would begin with a small, flat, level table on a stand, a compass, and a blank paper. The soil surveyor marked the location of landmarks on the paper, using the length of a step to measure the distance to each landmark and a compass to note its direction from the location of the table and from other landmarks. Trees, houses, roads, cemeteries, and other more or less permanent features were used to identify the location on the base map. The soil surveyor then walked over the area, digging holes and marking the location of the observations on the base map.

Since 1935 virtually all soil maps have been made on aerial photo bases, which offer the advantage of detailed information on cultural and natural features in the survey area. The soil surveyor can always know positions with certainty, which improves speed and accuracy. Today, a wide assortment of aerial imagery is available, including manned and unmanned satellite imagery and photography from various kinds of aircraft using color, color-infrared, and black-and-white film. Aerial photography is a form of remote sensing.

Global positioning systems (GPSs) are a tool for locating landmarks and locations accurately. The U.S. Department of Defense has made signals from its 24 global positioning satellites available for civilian use, and new applications of this technology are developed constantly. These earth-orbiting satellites continuously transmit signals. A ground-based receiver, which may be small enough to put in a pocket, locates itself (and the user) by receiving transmissions from three or more satellites. The positions of the GPS satellites relative to a location on the ground are changing constantly. The receiver measures the time it takes for the signal to travel, at the speed of light, from each satellite to the receiver. The distance from the receiver to each satellite is calculated by the receiver based on the fact that distance = time × velocity. The receiver's location is determined by triangulation of three or more signals. Triangulation works because the signals from three or four satellites intersect at only one unique point on the earth's surface. Advanced systems can locate a position within 1 m. This is valuable to soil surveyors, particularly when the terrain is featureless or covered by thick vegetation.

## 13.2.2   Remote Sensing

**Remote sensing** is the acquisition of information from a distance. More precisely, it is the detection, by means of a sensor in the atmosphere or space, of energy transmitted or reflected from natural sources on the earth. The term is more general than aerial photography and is used because some "pictures" cannot be obtained by traditional methods of photography.

**Basics**   With all methods of remote sensing, radiation from a part of the **electromagnetic spectrum** (Figure 13–8) is being sensed. When we see an object, our eyes are sensing reflected radiation in the

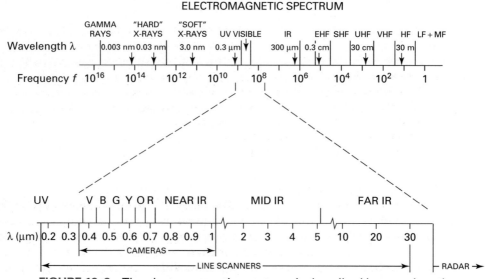

**FIGURE 13–8**   The electromagnetic spectrum is described by wavelength, frequency, common name, and appropriate remote sensing device.

visible portion of the electromagnetic spectrum, and our brains are converting the spectral information into recognition of the object. Radiation travels at the speed of light and is described by its energy and wavelength, which are inversely related. Short-wavelength radiation has more energy than does long-wavelength radiation. The source of energy may be natural (the sun), in which case the remote sensing is **passive,** or the source may be artificial, in which case the remote sensing is **active.** X-rays, radar, and microwave radiation are parts of active remote reconnaissance systems.

Energy (the sun's radiation) entering the atmosphere interacts in several ways with the atmosphere and objects (Figure 13–9). It may be **reflected,** returned unchanged to the atmosphere; **absorbed,** giving up energy by heating the object; **scattered,** deflected from the object; **transmitted,** passed through an object; or **reemitted** by the object or matter that absorbed the energy.

Radiation traveling from an object to our sensors, our eyes, or to mechanical or electrical sensors is partially absorbed by the atmosphere. Water, $CO_2$,

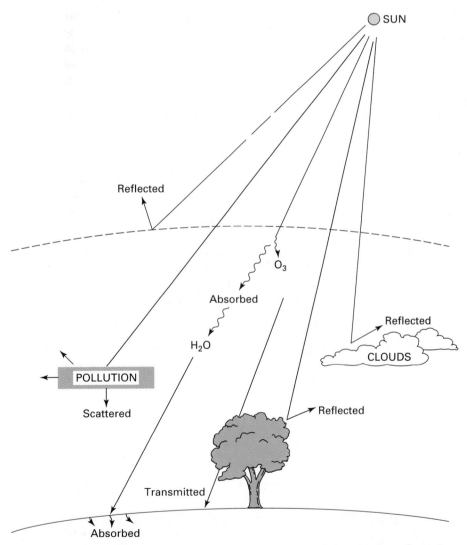

**FIGURE 13–9**  Before energy can be usefully sensed, it must be reflected from an object. Absorbed, scattered, and transmitted energy and energy reflected into space from the upper parts of the atmosphere cannot be photographically sensed.

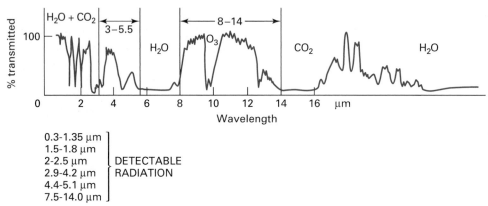

**FIGURE 13–10** Only energy transmitted through the atmosphere can be usefully sensed. The transmission characteristics of the atmosphere for a range of wavelengths is shown here. Note that $CO_2$ and $H_2O$ are responsible for much energy adsorption. The 0.3- to 1.35-μm wavelength range is the visible and near-infrared.

and ozone ($O_3$) do most of the absorbing. Because of absorption, there are only small "windows" in the atmosphere through which we can detect radiation. Figure 13–10 shows that out to 0.3 μm, there is almost total absorption of radiation; the visible region from 0.4 to 0.7μm is relatively free from absorption. There are several windows in the infrared to 25 μm, but beyond 25 μm the atmosphere absorbs most of the radiation. **Scattering** by atmospheric particles causes radiation to move in unpredictable directions and so makes it useless for remote sensing.

Reflected and reemitted energy are the most useful because they are "sensed" to obtain an image with information about the reflecting or emitting bodies. The direction of reflected radiation is predictable and therefore useful for remote sensing. Reemitted radiation has too long a wavelength to be detected optically, but it can be detected by instruments.

Each kind of energy interaction, whether reflected, transmitted, absorbed, or emitted, is specific for each object and energy wavelength. This is important because it enables us to remotely distinguish among objects on the ground. Objects look different when we recognize a distinctive shape, size, texture, or color. Sometimes we cannot distinguish among objects because the scale does not allow it or because the objects "look" too much alike. Even when our eyes cannot distinguish among objects, some detectors can do so because the objects' spectral signatures may be very different. A **spectral signature** is the reflectance or emittance characteristics of an object at

each wavelength in the electromagnetic spectrum (Figure 13–11). Thus, a sick plant may look different from a healthy plant to our eyes and also to a sensor (film or electromechanical) because of the plants' different spectral signatures.

Regardless of the energy source, we must have a detector that can sense the differences in the reflected or emitted energy. Most frequently, camera–film–filter combinations are used, but satellites and some high-flying aircraft use other instruments to detect radiation. The instrument readings are later converted into images that look like photographs.

**Verification: ground truth**    Although the topography, vegetation, geology, and surface soil properties depicted on an image can be useful to the soil mapper, one major disadvantage of remote sensing for soil science is that none of the methods provides a clear picture of the soil's properties with depth. We are limited to looking at the earth's surface.

Any form of remote sensing requires checking our interpretations with what is actually on the ground, **ground truth.** Ground truth is collected by visiting the area to be photographed and comparing what is actually on the ground with what the photograph indicates. Another method for obtaining ground truth is to determine in the laboratory the spectral signatures of materials such as soils and plants. The spectral signatures can then be compared with remotely sensed spectral signatures. Collection of ground truth is a means of quality control for remote sensing.

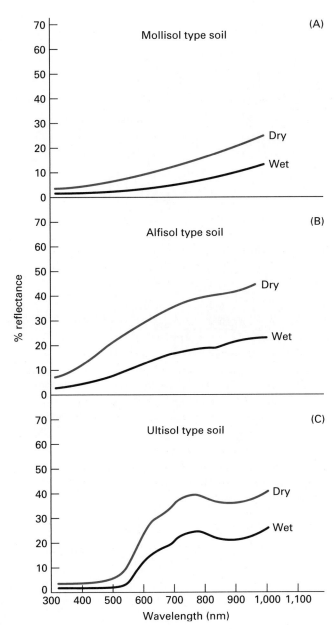

**FIGURE 13–11** These soils look different to our eyes, and the sensitive instrument that recorded these spectral signatures illustrates why. At each wavelength the percentage of energy reflected is different.

**Platforms and detectors** Several combinations of platforms and detectors are used to obtain imagery for soil surveys. The platforms are most often aircraft, but manned and unmanned satellites are also used. Satellites such as Landsat (Figure 13–12)

and Global Operational Environmental Satellite (GOES) weather satellites are well-known platforms for gathering remotely sensed data. Satellite imagery offers repeated observations of large areas. The imagery provides an overview of areas before the detailed soil surveying is begun.

Cameras are the most frequently used detectors for soil survey. When combined with different films and filters, they can be highly specific in the radiation detected. Radiation may be sensed as (1) broad wave band sensing, (2) narrow wave band sensing, (3) bispectral sensing, and (4) multispectral sensing. Filters are used to reduce the range of different radiation wavelengths that pass through the camera lens to the film. Cutoff and cuton filters transmit only shortwave and longwave radiation, respectively, and bandpass filters transmit only a limited band of radiation.

Nonspecific, broadband sensors such as black-and-white, or monochromatic, film may be used to sample a mixture of radiation across the visible portion of the electromagnetic spectrum, within the film's limits of sensitivity. Narrow wave band sensing records only a single selected wavelength, determined by the film or film–filter combination sensitive to the wavelength. When two nonadjacent wavelengths are sensed and compared, it is called *bispectral sensing.*

*Multispectral sensing* is a powerful remote sensing tool that arranges a series of optical or mechanical sensors to permit the simultaneous sensing of three or more wavelengths of radiation. When more than one camera is used simultaneously to take photos of the same object, different films and filters are employed to detect the object's spectral responses at the same instant. By comparing two or more images in which different wavelengths are detected, objects can be identified that might not have been detected with only one photo, using either a single narrow wavelength or a broad range of wavelengths.

Lens and film combinations may be chosen to control further what is being detected. Lenses vary in focal length and angular field. *Focal length* is essentially the distance from the lens to the film. Short-focal-length lenses are 15 cm (6 in), normal lenses are 15 to 30 cm (6 to 12 in), and long-focal-length lenses are > 30 cm (12 in). The focal length is one factor determining photographic scale. Short-focal-length, wide-angle lenses permit wide coverage from a low altitude. Wide-angle-lens photography is useful for topographic mapping that requires measurements of

**FIGURE 13–12**   Drawing of a Land-sat satellite. (NASA.)

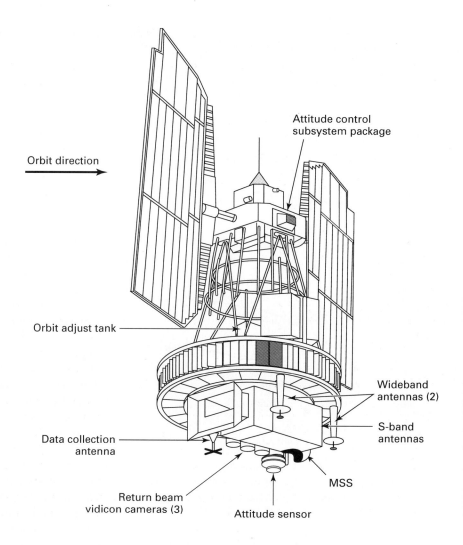

Orbit direction

Attitude control
subsystem package

Orbit adjust tank

Wideband
antennas (2)

S-band
antennas

Data collection
antenna

MSS

Return beam
vidicon cameras (3)

Attitude sensor

height and elevation, because three-dimensional stereoscopic effects are exaggerated. Long-focal-length lenses are preferred for large-scale photography, and medium-focal-length lenses are often the best on the average. The angular field determines the detection width. A normal lens has a field width of up to 75 degrees, a wide-angle lens has a field of 75 to 100 degrees, and a superwide-angle lens has a field of view of over 100 degrees.

The scale of the photograph is determined by the camera's focal length and height above the ground. By varying the lens focal length or the height of the platform, various scales of imagery can be obtained.

In addition, light-sensitive instruments are used to detect energy in the 0.35-to-0.5- and 0.8-to-1.4-μm wavelength range, which is not visible to an optical

detector such as a camera. The radiation detected is converted to a numerical value or to an image on a television-like device called a *cathode ray tube (CRT)*. The picture on the CRT is then recorded on film.

The advent of high-speed computers, advanced optics, and miniature data storage and manipulation devices has greatly advanced collection of digital imagery. Many digital cameras and recorders record the spectral response of an object pixel by pixel. Each pixel records the average radiance (brightness) of the ground area that it represents. Each image consists of a two-dimensional array of these discrete picture elements. The image can be viewed immediately, stored and enhanced electronically, and transferred electronically or printed.

Thermal infrared (IR) (0.8 to 1.4 μm) is detected in this way. Thermal IR is useful for detecting

**FIGURE 13–13**   Nighttime thermal image near Middleton, Wisconsin. Hot objects appear white and cool objects appear black in this photo. The white dots in the upper end of the field are cows. (Photo: Lillesand, T., and R. W. Kiefer. 1987. *Remote sensing and image interpretation.* 2d ed. New York: Wiley. Reprinted with the permission of the authors and John Wiley and Sons, Inc.)

fires, hot spots on the earth such as geothermal areas and volcanoes, and animals (Figure 13–13). It is also useful for detecting moisture differences in soils. Dry soils heat up more than wet soils do and therefore emit energy at a different intensity. Vegetation differences can be detected because different plants emit energy at different intensities, depending on their health.

**Film**   Optically detected radiation is recorded on film as an image. Choices of film include black-and-white, color, and **false-color infrared.** (See Plates 11 and 12.) Film differs in sensitivity, resolution, and speed depending on the kind of **emulsion.** Sensitivity is the range of wavelengths to which the film will respond; resolution is determined by the film's grain size; and grain size also determines the sharpness of an image and the **film speed,** or amount of light required to produce an image.

**Stereo imagery**   Photographs may be taken directly above an object, which produces a **vertical photo,** or they may be taken at an angle, which results in an **oblique photo** (Figure 13–14). Vertical photos have the advantage that the imagery is much like a map. Vertical photography is used when stereo imagery is desired.

Stereo photography is important to soil science because it provides a three-dimensional view of the earth's surface, from which heights and slopes can be measured. A stereo image is obtained by taking two photos of the same object from different angles. The apparent change in the object's position that results from a change in the viewer's position is **parallax.** Most frequently, an airplane flies over the area to be photographed, and pictures are taken at an interval that allows for overlap in the photos (Figure 13–15). When the photos are viewed at the same time through a stereoscope, a three-dimensional image appears. The two photos are called a **stereo pair** (Figure 13–16). Sophisticated mechanical and electro-mechanical instruments have been developed for making measurements on stereo images, which are used to make contour maps and accurately measure the height of objects.

The NRCS uses vertical black-and-white stereo pairs of aerial photographs to make soil maps. Pairs of photos are examined in the office, where the soil scientist makes preliminary soil and landscape delineations based on slope groups and tones on the photo. Tones represent differences in soil moisture status or vegetation. The photos are then taken into the field, where observations are used to confirm or refute the preliminary delineations. Good-quality aerial photos are an important tool in making soil maps.

### 13.2.3   Geographic Data Handling

As the amount of soils information increases and the cost of computing decreases, it is natural that soils and landscape information be stored, sorted, retrieved, and interpreted with the help of computers.

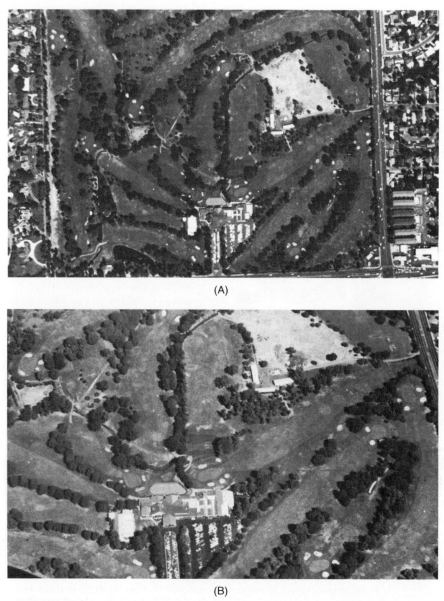

(A)

(B)

**FIGURE 13–14**   This pair of photos illustrates the differences between vertical (A) and oblique (B) views of the same scene. What is the land use? (Photos: W. E. Wildman.)

Any information that can be displayed on a map is geographic data, and numerous computer systems have been devised to store and recall this information.

Environmental problem solving and planning often require that many sources of information be merged and compared. Before computers, information was merged by a map overlay method. Individual maps were traced onto transparent material, and the maps were registered to a common geographic datum. By examining the overlays, a composite analysis can be made and new interpretive maps derived (Figure 13–17). Today, each map layer is entered into a computer database. The maps are georeferenced so that the overlays are geographically accurate. Unlike the former transparency method, the number of layers is limited only by the information available. The computer can store, sort, and display large amounts of information.

**FIGURE 13–15** Overlapping photographs fool our eyes by providing two views of the same object from slightly different positions. We see a three-dimensional view on a pair of two-dimensional photographs.

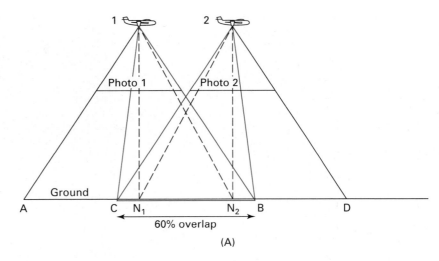

(A)

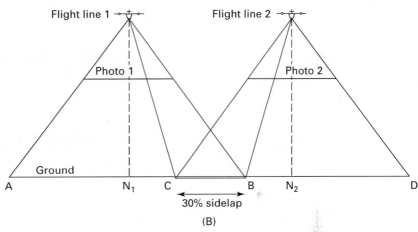

(B)

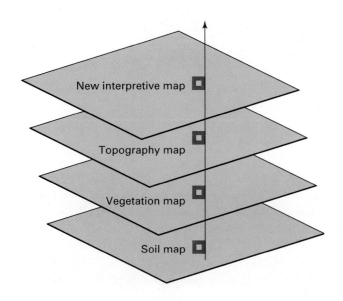

**FIGURE 13–16** Geographic information systems provide a technique for combining geographic information in new ways. Illustrated here is how a map showing the distribution of soils on a landscape can be combined with a vegetation distribution map and a topographic map to create an all new interpretive map. The arrow illustrates how information in each cell of each map can be combined to create new information in the final product.

**FIGURE 13–17** If you look at it through a stereoscope, this stereo pair of photos will give you a three-dimensional view of the mountains and valleys.

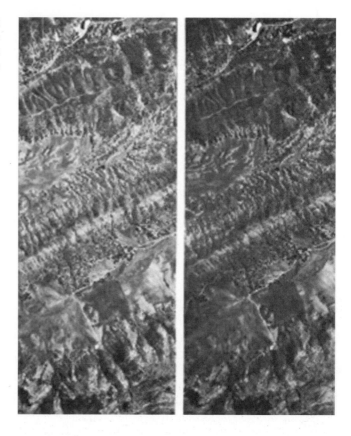

Geographic data are entered into computers by means of digitizing (Figure 13–18) or by computer scanning. Each point on the map must have a unique $(x, y)$ coordinate. Hand digitizing or manual encoding requires that each line on a map be traced with an electronic cursor that identifies the location of the lines. Scanners measure the intensity of light passing through an image and convert this information into a picture of the image in a computer. Scanners have increased the speed of image entry and reduced greatly the drudgery of hand digitizing. Once the lines have been digitized, errors are removed by editing programs.

Geographic data–handling systems are useful because very large amounts of information may be stored, manipulated, and retrieved in a variety of ways. Information can also be easily corrected as conditions on the ground change. The plastic transparency technique's limitations are that only a few combinations are possible and the drawing of such overlays is tedious. A computer can be programmed to "overlay" many combinations of properties to produce a new map. In addition, the various properties

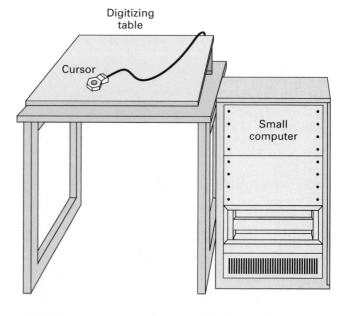

**FIGURE 13–18** The electronic digitizing table and cursor are used to enter geographic or point data into the computer.

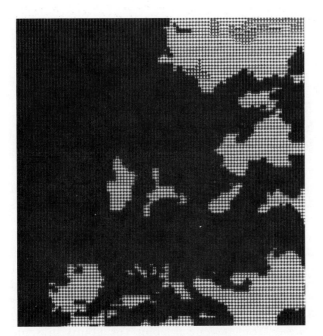

**FIGURE 13–19** This is a simple computer-drawn interpretive map based on a soil map. A soil survey report map was digitized, and the morphological characteristics of each soil series were entered into the computer. The computer was then instructed to classify the soils and print an interpretive map. (Used with permission of the Soil Conservation Society of America; reproduced from the *Journal of Soil and Water Conservation.*)

being combined can be weighted or valued to give a weighted combination map. The output from a geographic data–handling system may be tables of information or, more often, maps (Figure 13–19).

## 13.3 SUMMARY

There are two kinds of soil information: geographic data and point data. Both types are needed if soils are to be used properly. Geographic data are available as soil maps that are made on aerial photo bases. The maps portray the arrangement of soils on the landscape as seen by the soil surveyor. Point data are detailed information about individual pedons or parts of pedons and polypedons. Physical, chemical, and morphological properties of soil series are common point data.

Soil survey reports contain both geographic and point data and many interpretations of soil information. General and detailed soil maps show the distribution of soils on the landscape, and different soil map units are used to identify the soils on the maps. Data tables and soil pedon descriptions indicate the nature of soil individuals, map units, and potential uses of the soils.

Remote sensing is a technique for obtaining information at a distance. Radiation from the electromagnetic spectrum is sensed by a camera–film–filter

---

## SOIL AND ENVIRONMENT
### *Soil Survey as Resource Base*

Environmental or resource management planning begins with assessment of the resource base. Soil maps locate land resources and define areas with similar soils, vegetation, local climate, and topography. In the United States, the Natural Resources Conservation Service produces the most comprehensive and detailed soil maps and soil survey reports, showing the location of soil map units in most agricultural and many nonagricultural counties. In addition, the reports provide soil series descriptions and guidelines for different land uses, including waste disposal, building of small structures, growing plants, and locating and managing recreation and wildlife areas.

Soil surveys assist the extension and adaptation of research knowledge from one area to others. Problems

and management worked out for one site and soil are likely to relate to similar sites and soils elsewhere.

A soil survey contains much information, but it seldom indicates recent changes (due to clearing, deep cultivation, leveling, ripping, draining, catastrophic erosion, and so on). Nor does it specify conditions likely to be transient (e.g., fertilizer and lime requirements, salinity levels). Soil surveys emphasize the less changeable properties of pedon and site—namely, the physics and mineralogy of the pedon and its climate, topography, and landform. These characteristics are impractical or at best expensive to alter. Together they set constraints within which management systems have to operate. That is why they are essential data for deciding appropriate land uses.

combination or by a mechanical or electromechanical device. Either a photograph or an image constructed from the data collected by the device is interpreted to obtain information about an area.

## QUESTIONS

1. Give several examples of observed and inferred data that might be shown on a soil map.
2. Examine a soil survey report from your area and find the number of soil series and soil map units.
3. Name three different kinds of map units and explain how they are different or the same.
4. How does scale influence what is seen on a map?
5. What is soil correlation? Why is it important?
6. What is remote sensing? Why is it important to a soil survey?
7. Why are color, color-infrared, and stereo imagery useful?
8. What is stereo imagery?
9. What is the electromagnetic spectrum?
10. What is computer geographic data handling?

## FURTHER READING

Barrett, E. C., and L. F. Curtis. 1982. *Introduction to environmental remote sensing.* 2d ed. New York: Chapman & Hall.

Lillesand, T. M., and R. W. Kiefer. 1987. *Remote sensing and image interpretation.* 2d ed. New York: Wiley.

Petersen, G. W., J. C. Bell, K. McSweeney, G. A. Nielsen, and P. C. Robert. 1995. Geographic information systems. *Advances in Agronomy* 55:68–112.

Soil Survey Staff. 1951. *Soil survey manual.* Washington, D.C.: U.S. Department of Agriculture.

Soil Survey Staff. 1983. *Soil survey manual.* Washington, D.C.: U.S. Department of Agriculture.

## WEB RESOURCES

Following are three sources of remote sensing imagery on the World Wide Web:
NASA has a Web site at
http://www.nasa.gov/gallery/index.html
The Jet Propulsion Lab has information on remote sensing and many images of the earth taken from space at http://www.jpl.nasa.gov/

The U.S. Geological Survey has aerial photo coverage of the United States. See if you can find a photo of where you live at http://www.usgs.gov/

## SUPPLEMENT 13–1   Section–Range System

The **section–range system** was devised as a legal survey system to accurately locate a parcel of land. A section is a square, 1 mile on a side, with a total area of 640 ac. (We will discuss this system in American units because that is how it was devised and exists.) Thirty-six sections arranged in six rows of six each is a township (Figure 13–20). Each section is identified by a number in the center of the 640 ac. The number (1 to 36) located in the center of each small square in Figure 13–20 is the *section number.* The United States is divided into larger blocks called *townships* and *ranges.* From a central point (or points in some states), north–south and east–west lines are drawn. Range 1 east is the first 6 miles to the east of this central point, and Range 1 west is the first 6 miles to the west. Township 1 north is the first 6 miles to the north of the central point (Figure 13–20). The official township–range description of the square land parcel in Figure 13–20 is SW 1/4, SW 1/4, SW 1/4 Section 29, Township 1N, R1E. The SW 1/4 is the southwest one-fourth of the section. By quartering the section three times, we can locate an object or a soil to within 10 acres and, by describing the location in more detail, can find out exactly where something is.

## SUPPLEMENT 13–2   Orders of the Soil Survey

*First-order* surveys are made for intensive land uses in small areas, such as agricultural experiment station fields. Soil boundaries are identified by making transects of the area, and the map units are mostly consociations with some complexes. The delineations have a minimum size of about 1 ha or less. These are the most detailed soil surveys.

*Second-order* surveys are less detailed than the first-order surveys, and fewer observations are made. Although observations are made at closely spaced intervals, strict transects are not made. Map units are mostly consociations and complexes, and the delineations are variable in size, with a minimum of 0.6 to 4 ha.

**FIGURE 13–20** The township–range system illustrated here is used to locate land parcels in the United States. Each numbered square is 1 square mile, or 640 acres. Each may be subdivided into half, quarter, and sixteenth sections to precisely describe a location.

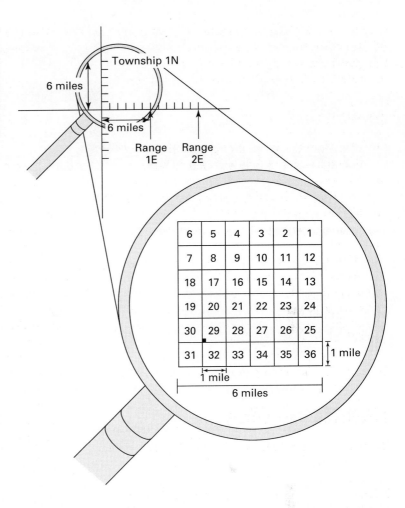

*Third-order* surveys are made for extensive land uses that do not require detailed soil information. Soil boundaries may be interpreted by remote sensing and verified by observation in the field, though fewer observations are made. The map units are mostly associations, but complexes, consociations, and undifferentiated groups can also be used. The delineations range in size from 1.6 to 256 ha.

*Fourth-order* surveys are very general. Most of the map units are associations, and most of the boundaries are determined by interpretation of remotely sensed data. The minimum size of the delineations is 40 to 4,000 ha, depending on the map size.

*Fifth-order* surveys are made in very large areas at a low level of detail. The minimum size of the delineations is 1,000 to 4,000 ha.

# 14

# ORGANIZING SOILS INFORMATION

## OVERVIEW

*Soils are classified to make information about them easier to understand, remember, and use. The natural systems classify what is seen, and the technical systems classify what can be inferred about the potential uses and limitations to uses of soil.*

*Soil taxonomy is a natural soil classification developed by the U.S. Department of Agriculture (USDA). It organizes soils into 12 broad groups called soil orders, which are further subdivided into suborders, great groups, subgroups, families, and soil series. Other countries, such as Canada, and international organizations, such as the Food and Agriculture Organization (FAO) of the United Nations, have soil classification systems.*

*The USDA Land Capability Classification is an interpretive, or technical, classification. Soils are grouped into eight classes (I to VIII) according to their usefulness for cultivated agriculture, range, and forestry. These classes are*

*then subdivided into subclasses and units that specify the reasons that a soil is limited for agriculture.*

*The Storie Index is another system for rating land for cultivated agriculture. It is a numerical index from 0 for nonagricultural soils to 100 for the best agricultural soils. There are also other systems for rating land for range, forestry, and nonagricultural uses.*

*Timber and range sites are classified according to their potential for growing timber or range species. These systems classify landscapes according to their soil properties, slopes, climate, and other factors that influence plant growth.*

*Nonagricultural classifications include drainage class, limitations for numerous special uses, and soil potential. Soil potential is a new system designed to rate soils for different uses based on the cost of improving poor soils for the desired use. Because this rating is related to the cost of correcting limitations, it is a more useful classification system than is the simple limitation rating.*

**314**

# 14.1 NATURAL SYSTEMS

## 14.1.1 Why Organize Information?

Information is organized to make it understandable and useful. Many thousands of kinds of soils exist in the United States and throughout the world. Soils are organized to help us remember their names and important properties and because it is inefficient to have to consider every soil to find information about one particular kind. Organizing soils into groups with similar properties minimizes the problem of locating information about any one soil. For classification systems to be useful, they must be designed for a specific purpose.

## 14.1.2 Kinds of Organization

Soils are organized according to observed or inferred properties. **Observed properties** are those that can be seen in the field or measured in the laboratory. Classification systems based on observed properties are therefore **natural classifications.** For example, in a natural classification, soils may be organized into those groups that have black (10YR2/1) surface horizons and those that do not. **Inferred properties** are not observed but are determined to exist based on observable or measured properties. We can observe that a soil has dark colors and loam textures, but we infer that it has moderate or high potential for plant growth. Dark colors and loam textures are observed or natural properties, whereas growth or yield is an inferred property (until it is measured). A classification based on inferred properties is therefore a **technical classification.** A potential yield classification is a technical system, as is soil drainage class, in which soils are organized into groups with similar wetness characteristics.

## 14.1.3 Soil Taxonomy

Classification systems for soils began in the nineteenth century with the Russian soil scientist V. V. Dokuchaev, who organized soils into groups with similar observable properties based on broad climatic zones. In the late 1920s and early 1930s, C. F. Marbut of the U.S. Department of Agriculture developed the first American system of soil classification based on the soil profile and soil as a natural body. This system, first published in 1938 in the *Yearbook of Agriculture,* classified soils according to their properties and the perceived processes of soil formation. This system had some limitations, however, because the processes were not well understood. The 1938 American system, with some revisions, was used until 1960, when the first version of the present system was introduced. The system now used, **soil taxonomy,** was first known as the **seventh approximation,** because it was the seventh version of the system. In 1975 soil taxonomy was published, and the current classification system is known by this name. The second hard-cover edition was published in 1998. It has undergone continuous revision since its introduction because it is not a static system. Every other year, a new paperback edition of *Keys to Soil Taxonomy* is published. Each edition includes additions, deletions, and corrections that have been suggested by soil scientists from around the world. The newest soil order, the Gelisols, was introduced to soil taxonomy in the 1998 version of the keys. This new order includes the soils that have permanently frozen layers. The purpose of soil taxonomy is to organize soils into groups with similar natural properties.

The **soil individual** is the object being classified in soil taxonomy. A soil is represented by a three-dimensional natural body, the pedon, the smallest volume that can be called a soil. You are already familiar with one pedon, the one that we started to construct in Chapter 1. The soil individual is defined by its morphological properties, including the number, kind, and arrangement of horizons, color, texture, structure, accumulation of clay, iron oxides, humus, silica, and carbonates. A *polypedon* is a group of similar pedons next to one another that can be identified on a map and used as a field classification or soil-mapping unit (see Chapter 13).

The **soil series** is the most detailed category in soil taxonomy. A soil series consists of a typical or modal soil and a defined range of properties. For example, the Yolo series is defined as having an A and C horizon and no B. The range of horizon thickness, color, pH, and structure is also stated. A soil with a single property outside the defined range for the Yolo belongs to a different series. As soil scientists have studied soils, the number of soil series has increased.

Soil taxonomy does not grant all soil characteristics equal importance when placing a soil series into taxonomic categories. Three special zones or layers have been defined in soil taxonomy for purposes of classification: epipedons, subsurface diagnostic horizons, and the control section. These layers may be either a single morphologic horizon or more than one.

**Epipedons**   An **epipedon** is a surface horizon in which most of the rock structure has been destroyed and that is either appreciably darkened by organic matter or is eluviated. Rock structure is either fine stratification (< 5 mm) in unconsolidated sediments or weathered rock (saprolite) in which the weathered minerals retain their positions relative to each other. In other words, the material looks like rock but is strongly weathered. An epipedon may include both A and B illuvial horizons, provided that darkening extends from the surface into or through the B horizon. The eight epipedons recognized in soil taxonomy are the mollic, umbric, anthropic, plaggen, folistic, melanic, ochric, and histic epipedons (Table 14–1). Figure 14–1 is a key to the epipedons.

**Diagnostic horizons**   All soils have an epipedon, but not all soils have subsurface diagnostic horizons. The **subsurface diagnostic horizons** may be part of the A or B horizon, although they are most commonly part of the B. Soil taxonomy lists 20 subsurface diagnostic horizons, 9 of which are listed in Table 14–2. The argillic, spodic, cambic, and albic horizons are most common in North America. The oxic horizon is

> **The advantages of soil taxonomy over earlier classifications are**
>
> 1. It is based on observed soil properties rather than inferred processes of soil formation.
> 2. It focuses on soils as natural bodies.
> 3. It permits the classification of soils of unknown origin.
> 4. It is flexible and open to modification as knowledge about soils increases.

found in Hawaii and is diagnostic (required) for the Oxisol order.

Much attention is given to the argillic horizon in soil taxonomy because clay is important to soil behavior. An argillic horizon is a subsurface zone (one or more horizons) of clay accumulation. Thus, for a zone to be an argillic horizon, it must be higher in clay than are those horizons above or below, and there must be evidence for clay illuviation. The evidence for clay illuviation is the presence of **clay films** on ped faces and the partial or complete filling of pores with translocated clay (Figure 14–2). Argillic horizons are found in all climates and under most vegetation and often control the soil's moisture relationships, rooting depth, and nutrient status.

## TABLE 14–1
Epipedon Names and Important Characteristics

| Name | Derivation | Important characteristics |
|---|---|---|
| Mollic | *Mollis,* soft (Latin) | Thick, well-structured, base saturation >50%, dark-colored mineral soil horizon |
| Umbric | *Umbra,* shade (Latin) | Like mollic but with base saturation <50% |
| Anthropic | *Anthropikos,* human (Greek) | Like mollic but with very high phosphorus content due to long period of cultivation and fertilization |
| Plaggen | *Plaggen,* sod (German) | Overly thick mollic (>50 cm) due to long, continued manure application |
| Folistic | *Folia,* leaf (Latin) | Organic soil material saturated less than 30 cumulative days during the year |
| Melanic | *Melas, melan,* black (Greek) | Thick black horizon containing high concentrations of organic carbon associated with poorly crystalline (amorphous) minerals |
| Ochric | *Ochros,* pale (Greek) | Surface mineral horizon that does not meet criteria for other epipedons |
| Histic | *Histos,* tissue (Greek) | Thin organic horizon saturated 30 consecutive days or more, unless drained; if mixed with mineral material, remains very high in organic matter |

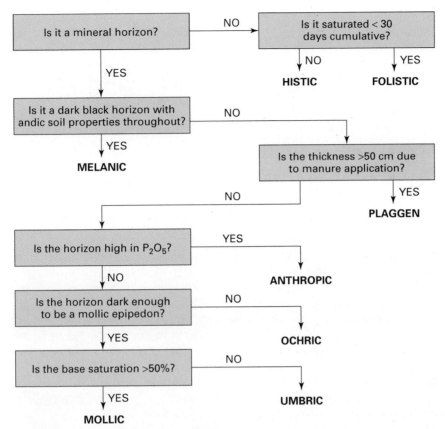

**FIGURE 14–1** Key to the epipedons in soil taxonomy. The words in bold-face are the names of the epipedons.

## TABLE 14–2

Some Subsurface Diagnostic Horizons

| Horizon name | Derivation | Major features |
|---|---|---|
| Argillic | *Argilla,* clay (Latin) | Accumulation of silicate clay, exemplified by clay films on ped surfaces |
| Natric | *Natrium,* sodium (Latin) | Special kind of argillic that includes accumulation of Na and clay |
| Spodic | *Spodos,* wood ash (Greek) | Accumulation of amorphous oxides of Al, Fe, and humus |
| Cambic | *Cambiare,* to change (Latin) | Altered horizon without illuvial or eluvial horizons |
| Albic | *Albus,* white (Latin) | Eluvial horizon from which Fe oxides and clay have been removed |
| Oxic | Modified from *oxide* | Accumulation of hydrated oxides of Fe, Al, and 1:1 clays |
| Placic | *Plax,* flat stone (Greek) | Thin black to dark reddish Fe, Mn, or Fe humus-cemented pan |
| Sombric | *Sombra,* shade, hence dark (Spanish) | Accumulation of illuvial humus with low base saturation and without Al or Na |
| Agric | *Ager,* field (Latin) | Accumulation of illuvial silt, clay, and humus due to longtime cultivation |

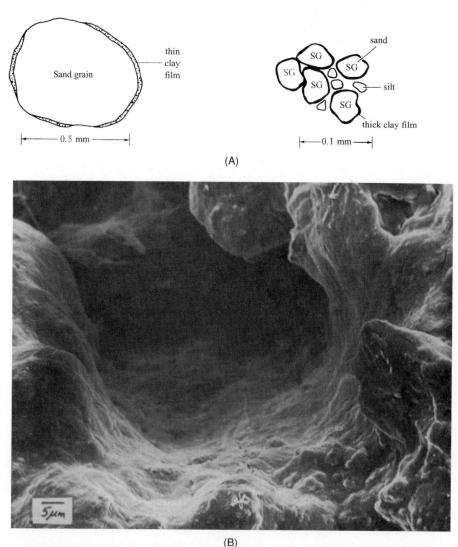

(A)

(B)

**FIGURE 14–2** (A) Clay films are layers of clay that coat sand and silt particles and pores. Thick clay films may completely fill fine pores. (B) Scanning electron micrograph of clay films coating pore walls. Note that the surfaces are smooth. Clay covers sand and silt grains. (Photo: R. J. Southard, reproduced from *Soil Science Society of American Journal* 49(1985): 167–171, by permission of the Soil Science Society of America.)

The spodic horizon is a zone of illuviation of Al and Fe combined with humus. Spodic horizons are sometimes hard and impermeable. The albic horizon is defined by its color, because it represents extreme eluviation and thus has the color of the uncoated sand and silt grains. An oxic horizon represents extreme weathering, but rather than being a horizon of leaching, it is a horizon of alteration. It consists of a mixture of hydrated oxides of iron or aluminum or both, 1:1 clay, and quartz sand.

Two subsurface diagnostic horizons that are always cemented are the duripan and petrocalcic horizons. The main cementing agent is $SiO_2$ in duripans and $CaCO_3$ in petrocalcic horizons. The fragipan is a

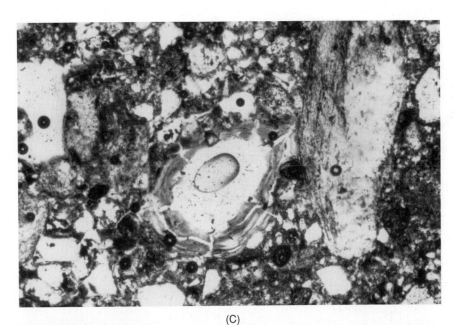

(C)

**FIGURE 14–2** *Continued.* (C) A thin section showing a pore with clay accumulation on the walls. The clay appears as light-colored, wavy, discontinuous bands around the pore walls. Thin sections are made by embedding soil samples with a liquid resin that hardens, followed by cutting the sample into thin slices that can be viewed with transmitted light. (Photo: Unpublished photo by R. J. Southard.)

subsurface diagnostic horizon of slow permeability that is not cemented. Calcic, gypsic, and salic horizons are diagnostic subsurface horizons usually found in arid and semiarid climates and are distinguished by their accumulations of $CaCO_3$ (calcic), $CaSO_4$ (gypsic), or Na, Ca, or Mg salts (salic).

Many soils have horizons between the epipedon and parent material that have been altered by soil formation but that do not qualify as illuvial horizons. Such horizons often meet the requirements for a cambic horizon, a horizon of "change." Examples are (1) a horizon of insufficient clay accumulation for the horizon to be an argillic, (2) sufficient alteration of the parent material so that the color of the cambic is redder than the parent material, and (3) rock weathering advanced to the point that the rock structure is no longer apparent.

Epipedons and subsurface diagnostic horizons are the building blocks for soil taxonomy and are defined so that they can be recognized by carefully examining soil horizons. Some also require laboratory data to quantify the important characteristics.

**Control section** The **control section** is the third part of the soil profile important to soil taxonomy. It is a defined thickness, not a genetic soil horizon, within the pedon used for defining the soil moisture regime, particle size classes, mineralogy, and cation exchange activity classes and for differentiating the series. The location of the particle size, mineralogy, and cation exchange activity control section for mineral soils without permafrost and with more than 36 cm to a root-limiting layer depends on its intended use and the presence or absence of diagnostic subsurface horizons. If there are no subsurface diagnostic horizons, the control section starts at 25 cm below the surface and extends to 100 cm. If there is a diagnostic subsurface horizon such as an argillic horizon, then the control section starts at the top of the diagnostic horizon and extends to a defined depth (Figure 14–3).

**Taxa** Soil taxonomy is a nested, or hierarchical, system containing six categories (taxa): order, suborder, great group, subgroup, family, and series. The order is the highest and most general level in the system, with all the soils in the world fitting into 12 orders (Table 14–3). The soil series is the lowest and most specific level; there are over 23,000 named and defined

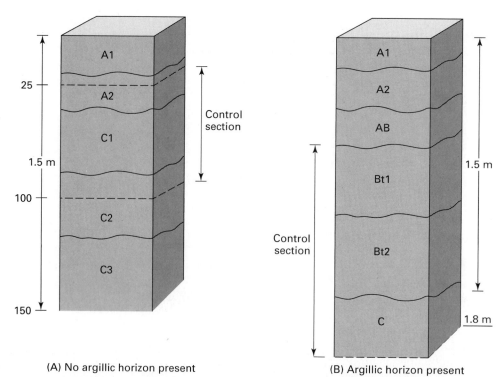

(A) No argillic horizon present                (B) Argillic horizon present

**FIGURE 14–3**    The control section is not a natural (pedogenic) horizon but is a designated part of a pedon used for placing soils into proper categories in soil taxonomy. (A) The control section in a pedon without an argillic horizon. (B) The control section in a pedon with an argillic horizon.

soil series in the United States alone. Each level above the series uses Greek, Latin, or nonderivative roots as formative elements in the classification terminology.

The soils in each order share one or more diagnostic horizons, or features formed by similar soil-forming processes. We do not need to know the specific processes to place a soil into an order, but we must be able to recognize the results of the processes. The results are mainly different kinds of horizons.

The degree of profile differentiation is one characteristic that separates the orders. Undifferentiated profiles that do not require subsurface diagnostic horizons are the Vertisols, Entisols, Mollisols, Aridisols, Inceptisols, Andisols, and Gelisols.

*Vertisols* form in clay-textured parent materials and have clay textures that swell when wet and shrink when dry so that large cracks, diagnostic of the order, are obvious. Vertisols are found in sufficiently dry climates that the soils dry regularly and form the wide cracks.

*Entisols* do not have the high shrink-swell of Vertisols. Entisols are found in all climates and under almost any vegetation. They are generally thought of as young soils.

*Mollisols* are soils with a large surface accumulation of organic matter. The mollic epipedon is diagnostic for the Mollisol order. Remember that a mollic epipedon must be dark, high in bases, and well structured. Mollisols may or may not have argillic horizons.

*Aridisols* are soils from arid regions and are the only order that places climate at the highest level of abstraction in soil taxonomy. Aridisols are separated into seven suborders: those with argillic horizons (argids), calcic horizons (calcids), gypsic horizons (gypsids), duripans (durids), salic horizons (salids), those without these diagnostic horizons (cambids), and cold aridisols (cryids). Other differences in aridisols are used in the system's lower categories.

**TABLE 14–3**

Order Names and Some Important Order Properties

| Order name | Formative element[a] | Diagnostic horizon | Other diagnostic properties |
|---|---|---|---|
| Entisol | Ent | None required | Ochric epipedon usually |
| Inceptisol | Ept | Cambic | No mollic epipedon or argillic horizon allowed |
| Vertisol | Ert | None required | Ochric epipedon usually, argillic not allowed |
| Aridisol | Id | Subsoil diagnostic required | Ochric or anthropic but not oxic or spodic |
| Alfisol | Alf | Argillic | May not have mollic epipedon, oxic, or spodic. |
| Mollisol | Oll | Mollic epipedon | May have argillic but not oxic or spodic |
| Ultisol | Ult | Argillic | May not have oxic or spodic |
| Oxisol | Ox | Oxic | May not have spodic or argillic |
| Spodosol | Od | Spodic | May not have oxic or argillic |
| Histosol | Ist | None | Organic soil has none of the mineral diagnostic horizons |
| Andisol | And | None required[b] | Colloidal fraction dominated by noncrystalline or poorly crystalline minerals |
| Gelisol | El | None required | Presence of permafrost |

[a]The formative element is used to form the taxonomic names of the soil. It is always the last two or three letters in the taxonomic name, and it uniquely identifies the order.
[b]The soil must exhibit "andic" properties, including high phosphate retention, low bulk density, and much noncrystalline or weakly crystallized mineral material.

*Inceptisols* have horizons of alteration, including some of the subsurface diagnostic horizons of the other orders. For example, they may have calcic, petrocalcic, gypsic, or petrogypsic horizons but may not be in an arid region. They have insufficient translocated clay to have an argillic horizon. The cambic horizon is diagnostic for Inceptisols.

*Andisols* are soils formed on volcanic parent materials such as volcanic ash, pumice, cinders, or lava. The clay fraction is dominated by poorly crystallized or noncrystalline (amorphous) minerals. Laboratory analyses are needed to measure the kinds and amounts of minerals and amount of volcanic glass in these soils. Three unusual properties of Andisols are their low bulk density (typical $<1.0$ g/cm$^3$), high water-holding capacity, and high phosphate retention. All three properties are due to the abundance of poorly crystalline minerals. Andisols may have any diagnostic epipedon, provided that the minimum requirements for the order are met in and/or below the epipedon (Figure 14–4).

*Gelisols* are soils that have permafrost within 100 cm of the soil surface or have "gelic" materials within 100 cm of the soil surface and permafrost within 200 cm of the soil surface. Permafrost indicates that the soil is frozen ($<0°C$) for 2 or more years in a row. Gelic materials are mineral or organic soil materials that show evidence of frost churning called cryoturbation. These soils occur in the Arctic, Antarctic, subarctic, boreal, and some alpine regions. A representative Gelisol from Alaska is shown in Figure 14–5. A second Gelisol from the Canadian Northwest Territories and an example of the vegetation found at this site are shown in Figures 14–6 and 14–7.

Examples of pedons from the Entisol, Vertisol, Aridisol, and Inceptisol orders are diagrammed in Figure 14–8. The Entisol and Vertisol have uniform clay distributions with depth, and both have some accumulation of organic matter in the A horizon. The organic matter in the Vertisol has been mixed to 60 cm by shrinking and swelling. The Aridisol has an accumulation of subsoil clay and a small accumulation of organic matter, and the Inceptisol has a subsoil

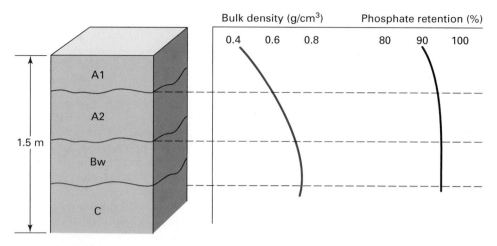

**FIGURE 14–4** Example Andisol pedon. Major properties of the Andisols are low bulk density and high phosphate retention, which are characteristic of the poorly crystallized clay minerals and volcanic glass that dominate the Andisols.

**FIGURE 14–5** Gelisol profile from Toolik Lake (north slope), Alaska. Permafrost begins at 30 cm and extends to the bottom of the photo. (Photo courtesy of USDA–Natural Resources Conservation Service [NRCS] soil scientist Dave Swanson.)

Horizons

— Oᵢ
— Oₐ
— $B_{w1}$
$B_{w2}$
$B_c$
— $A_f$
— $C_f$
—

**FIGURE 14–6** Gelisol formed on patterned ground (area of cryoturbation) in the Canadian Northwest Territory. The measuring tape is in the center of a hummock that is approximately 45 cm above the low point in the center of the photo. The parent material is fine-silty glacial till. (Photo courtesy of USDA-NRCS soil scientist Dave Swanson.)

**FIGURE 14–7** Mature black spruce–lichen forest growing on hummocks landscape in the Canadian subarctic. The trees are leaning in many directions, which is indicative of ground movement. Hummocks and swales also affect vegetation distribution. Earth hummocks are formed in areas of periodic cryoturbation. The Gelisol shown in Figure 14–6 was described in an area such as this. (Photo courtesy of USDA-NRCS soil scientist Dave Swanson.)

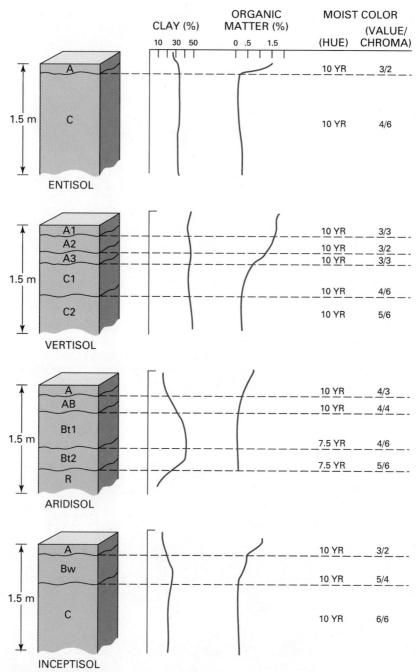

**FIGURE 14–8** Example Entisol, Vertisol, Aridisol, and Inceptisol pedons. Note the different kinds of horizons, the differences in clay and organic matter content, and different colors. (See Color Plates 1, 2, 3, and 9 for additional examples.)

<table>
<tr><td colspan="2">Soil Taxonomy has six categories:</td></tr>
<tr><td>Order</td><td>Subgroup</td></tr>
<tr><td>Suborder</td><td>Family</td></tr>
<tr><td>Great group</td><td>Series</td></tr>
</table>

clay accumulation that is too small to be an argillic horizon.

Alfisols, Ultisols, Oxisols, and Spodosols have accumulations of translocated weathering products and subsurface diagnostic horizons.

*Alfisols* have accumulations of translocated (illuviated) clay in the subsoil. The argillic horizon is diagnostic (required) for an Alfisol (Figure 14–9). In addition, Alfisols have at least 35 percent of

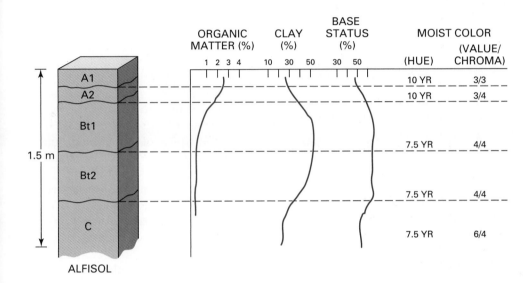

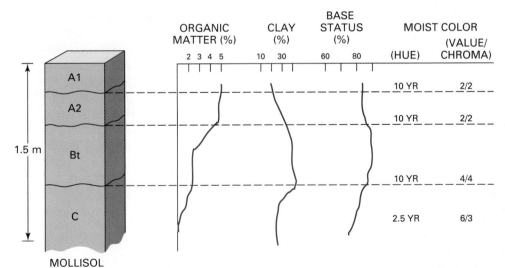

**FIGURE 14–9** Example Alfisol and Mollisol pedons. Note the accumulation of clay in the argillic horizon of the Alfisol and the organic matter in the mollic epipedon of the Mollisol. The Bt is not diagnostic (necessary) for a soil to be a Mollisol; the mollic epipedon is required. (See Color Plates 4 and 6 for other examples.)

their cation exchange capacity in the control section occupied by cations such as $Ca^{2+}$, $Mg^{2+}$, $Na^+$, and $K^+$ and little accumulation of organic matter in the surface.

*Ultisols* are similar to Alfisols except that the base saturation at a predetermined depth is less than 35 percent. Ultisols result from more intense weathering and leaching or a longer time of weathering than Alfisols. Often in addition to the low base saturation, the Ultisols have redder colors than the Alfisols, indicating the release of Fe from silicate minerals and subsequent oxidation.

*Oxisols* are most common in hot, humid climates where weathering and leaching are intense. Minerals in the profile are dominantly quartz, iron and aluminum oxides, and kaolinite. The diagnostic subsurface horizon is the oxic horizon. The combination of extreme weathering and residual mineralogy often makes these soils acidic and infertile, unless heavily fertilized.

*Spodosols* have a diagnostic spodic horizon, a subsoil accumulation of translocated humus combined with aluminum, or humus combined with aluminum and iron as amorphous materials. These soils are usually found in cool, humid regions and on coarse-textured parent material where leaching occurs rapidly.

Example pedons of the Ultisol, Oxisol, and Spodosol orders are illustrated in Figure 14–10. Note the different patterns of clay, humus, and base saturation in the three pedons.

*Histosols* include all organic soils. They must have at least 20 percent organic carbon if never saturated for more than a few days or at least 20 percent organic carbon, depending on the clay content of the mineral fraction, in more than half their thickness. The organic carbon is the remains of plant material in various states of decompostion.

Suborders are the next category in soil taxonomy. Each of the orders is divided into categories that emphasize climatic or morphological features of importance at the suborder level. There is no standard rule for how many suborders may be in each order or what differentiates the suborders. In general, climate, wetness, vegetation, and the presence of diagnostic horizons, when they are not diagnostic for the order, are the criteria that differentiate the suborders (see Supplement 14–1). The approximately 60 suborders are divided into over 300 *great groups*. Soils in a suborder that are similar in kind, arrangement, and degree of expression of horizons or that have similarities in soil moisture and temperature regimes or base status are placed into the same great group. Table 14–4 gives examples of the formative elements used in naming the suborders and great groups. For example, there are many wet Entisols. The Aquents suborder is separated into seven great groups based on properties such as temperature regime (Cryaquents, cold) or sandy wet Entisols (Psammaquents).

Great groups are further divided into subgroups, which indicate (1) the typical member of the great group, (2) members of a great group that are not typical but have properties of another great group, and (3) members of a great group that have properties that are not typical of that great group or any other great group. Thus we have a mechanism for including soils that are most typical of a great group and that are called *typic* and soils that are either transitional to another great group or do not seem to fit perfectly into any great group. Approximately 1,000 subgroups have been identified in the United States.

The subgroup, great group, suborder, and order are the higher categories in soil taxonomy. The two lower categories, the family and the soil series, are the most specific classifications of a soil. The family criteria group soils according to similar chemical and physical properties that influence a soil's use. Particle size distribution, mineralogy, cation exchange activity, temperature regime, and thickness of the soil penetrable by roots are five of the nine criteria used to place soils into families.

The series is the lowest (most detailed) category in the system. It specifies the nature of the soil individual and the allowable range of characteristics of an individual. As we already pointed out, if soil series A is defined as having a Bt horizon that ranges in thickness from 40 to 70 cm, and we are classifying a soil that has a Bt horizon 75 cm in thickness, it will not be soil series A.

This discussion has focused on mineral soils and mineral soil classification. Organic soils—the Histosols—because they all tend to be wet and have none of the mineral diagnostic horizons, are classified differently from the other 11 orders. Three

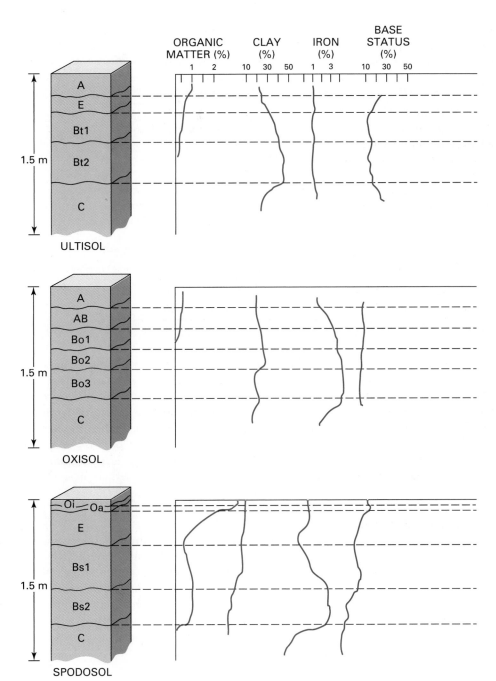

**FIGURE 14–10** Example Ultisol, Oxisol, and Spodosol pedons. Each of these orders requires a subsoil diagnostic horizon. The Ultisol has an argillic horizon (Bt1 and Bt2), the Oxisol an oxic horizon (Bo1, Bo2, and Bo3), and the Spodosol a spodic horizon (Bs1 and Bs2).

**TABLE 14–4**
Some Formative Elements for Suborder and Great Group Names

| Formative element | Derivation | Connotation |
|---|---|---|
| Alb | *Albus,* white (Latin) | Presence of albic horizon |
| Aqu | *Aqua,* water (Latin) | Wetness, aquic conditions |
| Cry | *Kryos,* icy cold (Greek) | Cold |
| Dur | *Durus,* hard (Latin) | Durable, a duripan |
| Fluv | *Fluvius,* river (Latin) | Formation in floodplain |
| Hapl | *Haplos,* simple (Greek) | Minimum horizon development |
| Natr | *Natrium,* sodium (Latin) | Presence of natric horizon |
| Ochr | *Ochros,* pale (Greek) | Presence of ochric epipedon |
| Orth | *Orthos,* true (Greek) | Common |
| Pale | *Paleos,* old (Greek) | Excessive profile development |
| Psamm | *Psammos,* sand (Greek) | Sand texture |
| Torr | *Torridus,* hot and dry (Latin) | Torric moisture regime |
| Ust | *Ustus,* burned (Latin) | Ustic moisture regime |
| Xer | *Xeros,* dry (Greek) | Xeric moisture regime |

suborders—the Fibrists, Hemists, and Saprists—designate the three levels of the organic matter's decomposition. The Fibrists are the least, the Hemists are intermediate, and the Saprists are the most decomposed. A fourth suborder, the Folists, designates organic soils that are never saturated with water for more than a few days following a heavy rain. The great groups divide the Histosols into climatic groups based on temperature and recognize special properties, such as the presence of sphagnum species as the main organic material. The subgroups are determined by the nature of the mineral material underlying the organic soil. Family criteria include the particle size and mineralogy of the underlying mineral material if the Histosol is thin, reaction, soil temperature, and soil depth.

Color Plates 1 through 10 show examples of soil series from each of the original 10 orders, and Table 14–5 classifies the pedons that the plates illustrate.

*Classification of our pedon*    Consider the classification of the pedon we have been examining throughout this book (Figure 14–11). It has three distinct horizons, two of which are diagnostic horizons: a mollic epipedon and an argillic horizon. The E horizon does not meet the color requirements for an albic horizon, and the mollic epipedon is diagnostic for the Mollisol order. This soil is found in humid regions where water is normally not limited and where

mean annual temperatures are not below 8°C, making it a Udoll. A Hapludoll is a simple Mollisol, but the pedon in Figure 14–11 has an argillic horizon. Udolls with argillic horizons are separated from the other Udolls at the great group level in the system. This pedon is an Argiudoll, and because it has no other special characteristics, it is a Typic Argiudoll.

### 14.1.4    Other Systems

Soil taxonomy has had a profound effect on how soil scientists look at and classify soils, though it is only one of many classification systems in the world. We offer in Supplement 14–2, for comparison, brief comments about the Canadian system and the United Nations FAO system. The purpose of both systems, like that of the American one, is to organize soils information to increase its usefulness.

## 14.2    INTERPRETIVE SYSTEMS

The natural systems described in the first part of this chapter classify observed soil properties. Technical, or interpretive, systems group soils according to their inferred properties, for particular interpretations or uses. For example, soils are classified according to their usefulness for growing crops, building houses,

**TABLE 14–5**

Classification of Pedons Illustrated in the Color Plates: Each Pedon Is Classified According to Soil Taxonomy, the Canadian System (Order), and the FAO System

| Plate | System | Classification |
|---|---|---|
| 1 | USDA | Hyperthermic, uncoated, Typic Quartzipsamments |
|   | Canada | Regosolic |
|   | FAO | Arenosol |
| 2 | USDA | Loamy-skeletal siliceous, subactive, mesic Typic Dystrochrepts |
|   | Canada | Brunisolic |
|   | FAO | Cambisol |
| 3 | USDA | Very fine, smectitic, thermic Oxyaquic Hapluderts |
|   | Canada | No equivalent |
|   | FAO | Vertisol |
| 4 | USDA | Fine-loamy, mixed superactive, mesic Aquic Hapludolls |
|   | Canada | Chernozemic |
|   | FAO | Phaeozem |
| 5 | USDA | Sandy, mixed, frigid Entic Haplorthods |
|   | Canada | Humic-ferric Podzolic |
|   | FAO | Podzol |
| 6 | USDA | Fine, mixed, active, thermic Abruptic Durixeralfs |
|   | Canada | Luvisolic |
|   | FAO | Luvisol |
| 7 | USDA | Fine-silty, siliceous, thermic Aquic Paleudults |
|   | Canada | No equivalent |
|   | FAO | Acrisol |
| 8 | USDA | Very fine, kaolinitic, isohyperthermic Rhodic Eutrustox |
|   | Canada | No equivalent |
|   | FAO | Ferralsol |
| 9 | USDA | Fine-loamy, mixed, superactive, thermic Typic Calciargids |
|   | Canada | No equivalent |
|   | FAO | Xerosol |
| 10 | USDA | Dysic Typic Cryohemists |
|   | Canada | Mesisolic |
|   | FAO | Histosol |

Source: USDA Web site: http://www.statlab.iastate.edu:80/soils/osd/

disposing of wastes, or providing gravel, topsoil, or sand. In each case, we deduce from the observed properties that a soil has qualities good or poor for the particular use.

## 14.2.1   USDA Land Capability Classification

The USDA Natural Resources Conservation Service developed the USDA Land Capability Classification as a means of grouping arable soils according to poten-

tials and limitations for the sustained production of common cultivated crops. (*Sustained production* means that the soil can be used for agriculture without excessive soil erosion.) Nonarable soils are grouped according to potentials and limitations and risks of soil damage. This system was devised to give conservationists and farm planners a tool for evaluating soils' agricultural potentials and limitations and risks of soil damage. It has 8 classes, 4 subclasses, and 10 units.

**Classes**   Each class contains those soils with similar broad limitations. Class I soils have few permanent

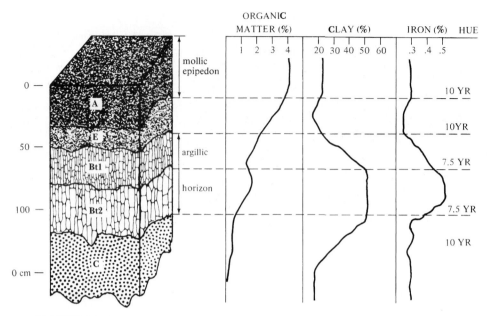

**FIGURE 14–11**   This is the pedon that began as an empty rectangle in Chapter 1. Here it is repeated with its mollic epipedon and argillic subsurface diagnostic horizon illustrated. Note the thick accumulation of organic matter in the A and the large increase in clay content between the E and Bt1 horizons.

limitations for cultivated agriculture. They are greater than 100 cm deep and well drained and have medium surface textures (loams and sandy loams), good water-holding capacity, no alkali or toxicity limitations, and no moderate or severe salinity. To be in Class I, soils must be on level parts of the landscape and in a climate where the rain is sufficient and the frost-free season is long enough for common cultivated crops to be grown. Class I soils require ordinary good management such as fertilizer, lime to reduce acidity, animal manures, or rotations to maintain productivity (Figure 14–12).

Class II, III, and IV soils have progressively more limitations for sustained crop growth. Class II soils have some limitations and require careful management and some special management to maintain productivity. For example, Class II soils may require erosion control practices such as contour cultivation or nontillage (see Chapter 15).

Class III soils have severe limitations that restrict crops and require special practices to maintain productivity. Soils in this classification may require erosion control practices such as contour cultivation, nontillage, strip cropping, or terrace construction. Water management is also important to soils in Class III, which may require artificial drainage.

Class IV soils have very severe limitations that require very careful management and restrict the choice of crops. Limitations such as shallowness to bedrock, steep slopes that increase the risk of soil erosion, low water-holding capacity, and wetness during the growing season increase from Class II to Class IV. Class IV soils are the lowest capability arable soils (Figure 14–13).

Class V soils are the best nonarable soils. They have fewer slope and soil thickness limitations than Class IV soils do but are excluded from cultivation because of limitations such as climate or wetness. Soils in Classes VI and VII are range soils with increasing limitations. Management practices may be successfully applied to Class VI soils, but Class VII soils are so severely restricted that no management is possible. Class VIII soils are unsuitable for cultivated agriculture and severely limited for range and pasture but are often important scenic, recreational, and watershed resources that require careful management.

**FIGURE 14–12**   The level, well-drained, medium-textured, fertile soil growing these vegetables is an example of Class I land.

**FIGURE 14–13**   This rolling landscape from Minnesota contains Classes II, III, and IV soils. Poor drainage (w) and potential or actual erosion (e) due to slope are two factors that reduce the capability class of these soils.

**Subclasses**   Within Classes II through VIII are the subclasses $e$ (erosion), $s$ (soil morphology), $w$ (wetness), and $c$ (climate), which specify a kind of limitation or hazard. The subclass is not usually used for Class I soils, but sometimes the $e$ subclass is included to indicate that there may be a potential erosion problem. An example of a IV$e$ classification is a deep (over 1 m), well-drained soil with slow permeability on slopes of over 15 percent. Because of slow permeability and slope, erosion is the dominant management problem.

## SOIL AND ENVIRONMENT
### *Appropriate Land Use*

Choosing appropriate uses for different types of land is a prerequisite if environmental damage and other costs are to be kept acceptably low. An appropriate use is one that is economically sound and suited to the land's characteristics, as judged from soil survey information and the natural and interpretive soil groupings derived from it.

### Agriculture, Urban Development, and Other Land Uses

As populations increase, so do demands for food, forest products, and water, while towns and cities extend mostly onto agricultural land. Deciding which land should become urban, which should remain agricultural, and which should be preserved for forest and watershed become critical environmental questions.

Prime agricultural land, that best suited for agriculture, has high priority to be kept under agriculture. One reason is that sustained crop production with minimum negative environmental impact requires special properties. These properties—long growing season,

level topography, deep pedons of uniform medium texture, lack of horizons that impede root growth or water movement, availability of water, adequate drainage, and freedom from flooding or severe salinity—are exemplified in agricultural land capability classifications. Capability classifications provide rational bases for definitions of "prime agricultural land." Prime agricultural land, a small fraction (5 to 10 percent) of the total land, is now almost wholly used for intensive agricultural systems whose increasing productivity has permitted this century's increases in world food production.

Another reason to preserve prime agricultural land is to avoid extending agriculture onto nonprime lands, which are more easily damaged by agriculture (erosion, waterlogging, salinization, and so on). Most nonprime lands should be valued for their own special qualities and preserved for watershed, range, forest, recreation, and wildland. For these areas, too, soils information underlies sound decisions about particular choices of land use and management.

The *s* subclass indicates that there is a morphological problem such as shallow rooting depth, very sandy or clayey textures, undesirable pH, or salinity or alkali problems in the root zone. Examples of III*s* classifications are (1) a soil with sandy textures throughout the pedon with low water-holding capacity and (2) a soil on a 0 to 2 percent slope that has a horizon or layer (such as bedrock) that limits rooting depth.

Wetness is designated with a *w*, which usually means that soil cultivation is limited by a high water table during the growing season. These may be somewhat poorly to poorly drained soils that are unlikely to erode and that have no root-limiting layers within the pedon.

When climatic limitations reduce the number of crops that can be grown or the success of growing crops, the *c* subclass is used. Soils in northern latitudes or in mountainous areas may have desirable physical and chemical properties, but the growing season may be too short for most crops. Such soils are placed in the *c* subclass.

**Units**   Units group soils with similar management responses and specify the most limiting characteristic

or predominant problem within the subclass. There is no standard for the numbering system, but 0 to 9 appears to be the common range of unit designations. The unit classifications used in California are shown in Table 14–6. Only one unit is used when classifying a soil. If there is more than one problem, the classifier must choose the most important one and use the appropriate unit.

The complete classification of a soil series includes the class, subclass, and unit, such as II*e*-1 or IV*s*-5. Our pedon in Figure 14–11 would be Class I if it was found on a slope of less than 2 percent and the Bt horizon were not limiting to roots. If there was sufficient clay to limit root penetration or to cause restricted drainage, the pedon would be Class II or III, depending on the degree of the limitation.

### 14.2.2   Timber and Range Site Indices

Less intensive and more extensive uses of soils include growing trees for wood products and growing animals for products such as meat, hides, and wool. These are less intensive uses because the land gen-

**TABLE 14–6**

Capability Units Used in California

| Capability unit | Principal limitation |
|---|---|
| 0 | Coarse sandy or very gravelly substrata limiting root penetration and moisture |
| 1 | Potential or actual wind or water erosion hazard |
| 2 | Drainage or overflow hazard |
| 3 | Slowly or very slowly permeable subsoils or substrata |
| 4 | Coarse or gravelly textures |
| 5 | Fine or very fine textures |
| 6 | Salinity or alkali sufficient to constitute a continuing limitation or hazard |
| 7 | Stones, cobbles, or rocks sufficient to interfere with tillage |
| 8 | Hardpan or hard, unweathered bedrock within the root zone |
| 9 | Low inherent fertility due to strong acidity or toxicity |

erally receives fewer inputs, such as fertilizer and cultivation, and produces less product value per unit area. Soils used for these purposes are often on steep slopes or are shallow and must be managed carefully to maintain their productivity. It is also important to evaluate the sites' potential productivity, most commonly by measuring the **site index** for timber and range.

The timber site index is a numerical measure of a site's quality for potential tree production. It is calculated according to the heights reached by selected tree species at an arbitrary age. In general, 100 years is used as a base for trees growing west of the Great Plains and 50 years for trees growing east of the Great Plains. For example, a site index of 150 indicates that the dominant and codominant trees growing in a normally stocked stand can be expected to reach a height of 45 m (150 ft) at age 100 years.

The **dominant** and **codominant** trees in a stand are the most abundant and the second most abundant trees, respectively. The number of trees per area is the stocking rate. The higher the stocking rate is, the greater will be the competition for nutrients and water. The site index is usually specific for tree species, climate, and stocking rate because these factors, in addition to soil qualities, influence productivity.

The site index is related to several soil properties, including the soil's depth to rock or an impermeable layer, gravel content, fertility, and permeability. The soil's position on the landscape also affects the site index, the two principal position characteristics being aspect and elevation.

Woodland suitability groups are used by the Natural Resources Conservation Service for woodland interpretations. Groups of soils with similar productivities of similar tree species that require the same general types of management are placed into a woodland suitability group. Each group is rated on the basis of productivity, plant competition, potential seedling mortality, and management factors such as equipment limitation, erosion, and wind-throw hazard.

Potential productivity is based on soil features such as drainage, texture, depth, slope, and erosion. Site index is often used to estimate potential productivity when making a woodland suitability classification. Plant competition is based on the estimated rate at which noncommercial species invade areas opened by cutting and thinning. Seedling mortality and site regeneration are affected by soil properties such as texture and moisture content.

The equipment for harvesting is limited by soil factors such as wetness, stoniness, clay and sand content, and slope steepness. For example, some kinds of equipment cannot be used on very steep land, and loose, coarse sands or excessively clayey textures limit the traction of wheeled equipment. Wind-throw hazard is the danger of trees being blown over by the wind.

A woodland suitability group is identified by arabic numbers and a letter symbol. The number represents the productivity level (1 = best), and the letter indicates the major limit or limits to its use. A third symbol, an arabic numeral, is often used to subdivide the rating further. Examples of symbols are $x$ for the presence of stones or rocks, $w$ for wetness, $t$ for toxic materials, and $d$ for rooting depth restriction. A 1$d$ rating indicates that the land has a high production potential but the rooting depth is restricted. A 4*r*3 rating indicates a low level of productivity (4) due to severe (3) steepness of relief (*r*).

   **Rangeland** is usually evaluated separately from cultivated land and forestland. Although it is considered in the USDA Land Capability Classification discussed earlier, there is at least one other method for evaluating rangeland. The USDA uses range site as an index of potential sustained vegetation production. Consequently, the potential for soil erosion damage and vegetation deterioration is considered. Range site is determined for land parcels rather than a single soil, as in the Land Capability Classification. A parcel may include several soil series that can be managed as a single unit. It is based on broad groups of soil properties, such as texture, water-holding capacity, and solum thickness, that influence total forage production. Although the composition of range vegetation is included, vegetation that is poisonous or not eaten by animals naturally detracts from the range site. Range site quality and woodland suitability are part of the interpretations made in soil survey reports.

## 14.2.3  Special-Purpose Groupings

Included in this section are (1) a frequently used interpretive grouping, soil drainage class; (2) an example of how nonagricultural soil evaluations are made; and (3) soil potential, which is a new kind of technical soil classification.

**Soil drainage class**   The climate and position of a pedon in the landscape and the properties that influence water flow through the pedon determine the **soil drainage class.** Soils that have slow internal drainage because of clayey textures or dense impermeable horizons have yellow and red mottles in the wet zone. *Mottles* are spots where Fe has been reduced and reoxidized, and they indicate a fluctuating water table, or temporary wetness. They are described by abundance, size, and contrast (Table 14–7). As mottles become more abundant, larger, and more contrasting, the interpretation is that the wetness has become more of a factor in the soil's genesis and management. Gray colors reflect permanent wetness.

   The USDA recognizes seven soil drainage classes:

*Very poorly drained soils* have a water table at or near the surface most of the year. They often have accumulated organic matter in the surface and may have histic epipedons. These soils are wet enough to prevent the growth of most important crops (except rice) without artificial drainage.

*Poorly drained soils* are usually wet, because the water table is commonly at or near the surface during a considerable part of the year. Both very poorly and poorly drained soils have gray colors, and they may or may not have mottles, depending on the length of time during the year when part of the pedon is free of water.

*Somewhat poorly or imperfectly drained soils* are also wet for significant periods. They tend to have mottles in the lower A, B, and C horizons. The A horizons are frequently overly thick, compared with those of well-drained soils in similar climates. Growing crops is possible but is often restricted by wetness unless there is artificial drainage.

*Moderately well-drained soils* are wet for a small but significant part of the year. This wetness influences the choice of crops and the timing of planting and tillage operations. Mottles are usu-

**TABLE 14–7**
Descriptive Terms for Mottles

| Abundance (% of pedsurface) | | Size (diameter) | | Contrast (color)[a] |
|---|---|---|---|---|
| Few | < 2 | Fine | < 5mm | Faint |
| Common | 2–20 | Medium | 5–15 mm | Distinct |
| Many | > 20 | Coarse | > 15 mm | Prominent |

[a]Contrast is determined by the difference in color between the mottle and the matrix colors. The greater the difference is, the more easily the mottles can be seen.

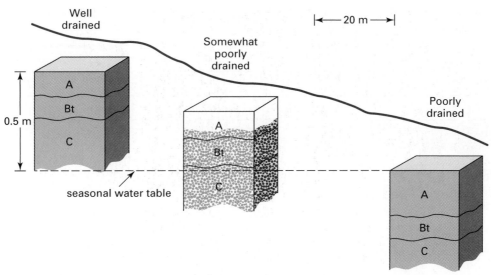

**FIGURE 14–14** The three pedons on this hillside illustrate morphological properties indicative of their drainage classes. The well-drained profile has no mottles in the solum. The somewhat poorly drained pedon is mottled throughout the Bt and C. The poorly drained pedon has a thick A horizon and is gray throughout because it remains wet for most of the year. It has few mottles. These drainage classes are most appropriate in humid climates.

ally not found above the lower B horizon in soils with B horizons, and overly thickened A horizons are not common.

*Well-drained soils* have no water table in the profile, are not adversely affected by seepage or overland flow, and have no mottles within the solum. Well-drained is the optimal condition for plant growth. Well-drained soils tend to be intermediate-textured soils without impermeable horizons and with good water-holding capacities (Figure 14–14).

The *somewhat excessively* and *excessively drained* classes are sandy and gravelly soils whose waterholding capacity is low and in which the water flow through the pedon is rapid or very rapid. Mottles are not found in either of these classes, but crop production is limited by the lack of water rather than its excess.

## Limitations for nonagricultural uses
Soils are often classified as having slight, moderate, or severe limitations for many nonagricultural uses. *Slight* is the rating given to soils that have properties favorable for use. *Moderate* is used when the limitation can

be overcome by special planning, design, or maintenance, but soils with one or more properties unfavorable for a use are rated *severe* for that use. This is sometimes called a "stop light" system because the colors used to represent the three classes on interpretive maps are green (slight), yellow (moderate), and red (severe).

The criteria used to place soils into these three groups depend on the particular interpretation. In general, they include soil texture, slope, chemical, and special soil characteristics that influence the use. Bedrock near the surface, flooding, high shrink-swell potential, and a seasonal high water table are additional properties used in the ratings. The ratings were designed to indicate the kind and degree of soil limitations, but they do not mean that the soil cannot or should not be used for a particular use.

Table 14–8 provides an example of the criteria used to rate soils for local roads and streets. A soil being evaluated for this use is judged by the 11 criteria in the left column. All soils are assumed to have slight limitations before the rating is made. If the soil does not meet one or more of the criteria for slight limitation, it is rated moderate, and if it does not meet one or more of the criteria for moderate, it is rated severe

**TABLE 14–8**

Limitations for Construction of Local Roads and Streets

| | Limits | | | Restrictive |
|---|---|---|---|---|
| Criterion | Slight | Moderate | Severe | feature |
| 1. USDA texture | — | — | Ice | Permafrost |
| 2 Depth to bedrock (cm) | | | | |
| Hard | >100 | 50–100 | <50 | Depth to rock |
| Soft | >50 | <50 | — | |
| 3. Depth to cemented horizon (cm) | | | | |
| Thick | >100 | 50–100 | <50 | Cemented pan |
| Thin | >50 | <50 | — | |
| 4. AASHTO Group Index | 0–4 | 5–8 | >8 | Low strength[a] |
| 5. AASHTO rating | — | A4, A5 | A6, A7, A8 | Low strength |
| 6. Depth to high water table (cm) | >75 | 30–75 | 0–30 | Wetness[b] |
| 7. Slope (%) | 0–8 | 8–15 | >15 | Slope |
| 8. Flooding | None | Rare | Common | Floods |
| 9. Potential frost action | Low | Moderate | High | Frost action |
| 10. Shrink-swell | Low | Moderate | High | Shrink-swell[c] |
| 11. Fraction >8 cm (wt. %) | >25 | 25–50 | >50 | Large stones[d] |

Source: *USDA National Soils Handbook,* Soil Survey staff. 1999. National Soil Survey Handbook Title 430-VI. USDA-NRCS US Government Printing Office. Washington, DC.
[a]American Association of State Highway and Transportation Officials rating of soil strength. If criterion 4 is not available, use criterion 5.
[b]If the soil has standing water on it (is ponded), then it is rated severe.
[c]Shrink-swell of the thickest horizon between 25 and 100 cm is rated.
[d]The stone content of the top 100 cm is rated.

for the particular use. The most restrictive feature is listed in the rightmost column.

Criterion 1 rates soils on the presence or absence of permanently frozen ground within the pedon. If it exists, the soil is rated severe. If the soil has hard bedrock or a cemented horizon within 1 m of the surface, it is given a moderate rating, and if the rock is at less than 50 cm, it is rated severe. Soft rock or a thin cemented zone is rated slight at 50 cm because both are more easily removed than is hard rock or a thick pan.

Criteria 4 and 5 rate the soil's strength. The other criteria are the presence of a high water table, which influences the soil's strength, and slope, which affects the cost of road building. It is more difficult to build on slopes greater than 15 percent than on level ground. The location of the road is rated for flooding potential, and as the potential for flooding increases, the limitations increase.

Frost can cause a street to buckle and crack, and so the next criterion rates the likelihood of frost action on the soil. Clay content and type of clay determine

the amount of a soil's shrinking and swelling. Low shrink-swell soils require less modification during construction, and roads built on low shrink-swell soils require less maintenance than do those built on high shrink-swell soils. The final criterion rates the weight percentage of stones in the pedon, because the presence of large stones increases road construction costs.

**Soil potential**  The "stop light" approach to interpretations has the advantage of ease of use: It implies that some uses may be difficult but provides no economic justification for the interpretation; that is, it does not tell the potential user what the cost of modifying a site may be for an intended use. Soil potential is intended to be an interpretive soil classification that indicates the cost of using a soil.

**Soil potential** is defined as the usefulness of a site for a specific purpose using available technology at a cost expressed in economic, social, or environmental units of value. Soil potential includes the concept that the use of a soil must be both cost-effective and environmentally sound. Both soil limitations and measures needed to modify the limitations are considered. The units of value may be dollars or potential environmental hazards, if a use is permitted. Potential is determined locally based on a relative ranking (best = 100) and is not expected to be a national rating such as Land Capability Classification or the "stop light" system.

Four steps are used to determine soil potential. First, the properties that influence each soil use are identified. Second, the practices used to overcome or minimize the unfavorable effects of the limiting soil properties are specified. Next, the relative cost or degree of difficulty of available practices or technologies is calculated. Finally, the continuing limitations for a particular use for each soil are determined. The soils in an area are then ranked from best to worst for each use. The best soil (or soils if two or more have an equal potential) is rated 100, and all others are rated lower, depending on their potential problems.

The soil potential index (SPI) is calculated as follows:

$$SPI = P - (CM + CL)$$

where $P$ is a locally derived index of performance or yield, $CM$ is an index of *costs of measures* to overcome or minimize the effects of soil limitations, and $CL$ is an index of *continuing costs* caused by limitations. For soils with $SPI = 100$, $CM$ and $CL$ are 0.

The best soils have (1) the lowest initial cost for the use of interest and (2) the lowest operating or maintenance cost while maintaining the environment. The worst soil naturally has the highest costs and lowest SPI. The index values $P$, $CM$, and $CL$ vary proportionately with the relative costs. The advantage of soil potential is clear: It gives a planner important information on *why* a soil may be limiting for a use and *how* to overcome any problems. The major disadvantage is that there are no data available for a multitude of soils and uses.

## 14.3  SUMMARY

There are tens of thousands of soils in the world; to remember their properties, it is necessary to organize them. Natural soil classification systems such as the USDA's soil taxonomy, the Canadian soil classification system, and the United Nations FAO's soil classification all group together soils with similar morphological properties. Pedologic processes leading to

observable or measurable characteristics are also used to group similar soils. In the USDA's soil taxonomy, soils are grouped at the order level according to the presence or absence of surface and subsurface diagnostic horizons that indicate the kinds and intensities of soil-forming processes. At the soil series level, specific properties—the number, order, thickness, and kind of horizons; color; structure; and texture—are needed to classify the soil properly.

Technical soil classification systems group together soils with similar potentials for specific uses. Examples are the USDA Land Capability Classification, the Storie Index (see Supplement 14–3), the timber site index, and ratings for nonagricultural uses of soils. Interpreted or inferred properties, rather than observed properties, are classified.

In the Land Capability Classification, Class I soils are deep, well drained, on level ground, and have few limitations for cultivated agriculture. Class II, III, and IV soils have increasing degrees of limitations for cultivated agriculture. Classes V through VII have increasing limitations for range and forestry and an increasing potential for degradation through soil erosion. Class VIII soils are reserved for wildlife and watershed.

Soil potential is a system that rates soils for specific local uses against a local standard. It not only rates a soil but also describes the general nature of any limitations.

## QUESTIONS

1. Why is soil information organized?
2. List some observed and inferred soil properties.
3. What features make soil taxonomy a hierarchical classification system? What are the advantages of such a system?
4. Name and describe the properties of several epipedons and subsurface diagnostic horizons. Can any epipedon occur with any subsurface horizon? Why?
5. How would you explain a subsurface horizon that had the properties of an epipedon? Or a surface horizon with the properties of a subsurface horizon?
6. What is a control section? Why does its position vary?
7. How do USDA Land Capability Class II soils differ from Class III, IV, and VII soils?
8. What is a Storie profile group?
9. How does soil potential differ from soil limitations?

10. For what purposes is a natural classification system not as good as a technical classification?

## FURTHER READING

Ahrens, R. J., and R. W. Arnold. 2000. Soil taxonomy. In *Handbook of soil science* (pp. E-117–E-135), ed. M. E. Sumner. Boca Raton, Fl.: CRC Press.

Bartelli, L. J. 1979. Interpreting soil data. In *Planning the uses and management of land, Agronomy Monograph 21* (pp. 91–116), ed. M. T. Beatty, G. W. Petersen, and L. D. Swindale. Madison, Wis.: American Society of Agronomy.

Fanning, D. S., and M. C. B. Fanning. 1989. *Soil: Morphology, genesis, and classification.* New York: Wiley.

Food and Agriculture Organization. 1988. FAO-UNESCO *Soil map of the world, revised legend. World Soil Resources Report 60.* Rome: Food and Agriculture Organization.

Highsmith, R. M., Jr., J. G. Jensen, and R. D. Rudd. 1971. *Conservation in the United States.* 2d ed. Chicago: Rand McNally.

Lloyd, W. J., and W. M. Clark. 1979. Timber production on intensive holdings. In *Planning the uses and management of land, Agronomy Monograph 21* (pp. 387–428), ed. M. T. Beatty, G. W. Petersen, and L. D. Swindale. Madison, Wis.: American Society of Agronomy.

Soil Survey Staff. 1983. *Soil survey manual.* Washington, D.C.: U.S. Department of Agriculture.

Soil Survey Staff. 1993. *Soil survey manual. USDA Agriculture Handbook 18.* Washington, D.C.: U.S. Government Printing Office.

Soil Survey Staff. 1998. *Keys to soil taxonomy.* 8th ed. Blacksburg, Va.: Pocahontas Press.

Soil Survey Staff. 1998. *Soil taxonomy: A basic system of soil classification for making and interpreting soil surveys. USDA Agriculture Handbook 436.* Washington, D.C.: U.S. Government Printing Office.

Storie, R. E. 1976. *Storie Index rating.* University of California Division of Agricultural Sciences Special Publication 3203.

## WEB RESOURCES

The U.S. Department of Agriculture, Natural Resources Conservation Service, has many Web resources for further exploration of soil taxonomy and soil survey. Following are some useful sites:

USDA NRCS Soil Survey Division home page:
http://www.statlab.iastate.edu/soils/index.html/

*National Soil Survey Handbook:*
http://www.statlab.iastate.edu/nssh/index.html/

National Soil Survey Center for access to soil data, soil taxonomy, photographs of soils and the *Soil Survey Handbook:*
http://www.statlab.iastate.edu/soils/nssc

World soil resources including access to the FAO soils map legend and classification systems of other nations:
http://www.nhq.nrcs.usda.gov/wsr/

## SUPPLEMENT 14–1    Suborders: An Example

Suborders are used to separate the orders into groups of broadly similar soils. For example, the suborders within Mollisols are shown in Table 14–9. Six of the suborders separate Mollisols of different climates or moisture regimes. Rendolls—Mollisols formed on limestone—may be found in many climatic regions.

*Aquolls* are Mollisols that have morphological features such as mottles or blue-gray reduced zones within the pedon that are produced by prolonged wetness. Aquolls may have a histic epipedon overlying the mollic epipedon.

*Cryolls* are Mollisols of cold climates where the mean annual soil temperature is equal to or below 8°C (47°F).

*Ustolls* are Mollisols found in seasonally dry climates such as North and South Dakota, where some moisture is present when conditions are suitable for plant growth but the total moisture is limited.

*Udolls* are found in regions where moisture is generally not limiting and the pedon is not dry in all parts for as long as 45 consecutive days between June and October. Southern Minnesota, Iowa, and parts of Wisconsin and Illinois have udic moisture regimes where Udolls, Udalfs, and Udults are found. These are the moist Mollisols, Alfisols, and Ultisols.

*Xerolls* occur in areas that have cool, moist winters and hot, dry summers. California's central valley and the area that borders the Mediterranean Sea are places with xeric moisture regimes where Xerolls, Xeralfs, and Xerults are found.

*Albolls* have an albic horizon that lies immediately under the mollic epipedon or that separates

**TABLE 14–9**

Example of Suborder Names of the Mollisol Order

| Suborder name | Major identifying characteristic |
|---|---|
| Albolls | Leached, presence of albic horizon below the mollic and argillic below the albic |
| Aquolls | Wet, presence of mottles, gray colors, and other signs of wetness |
| Rendolls | Mollisols of high carbonate content, often found on limestone parent material |
| Cryolls | Mollisols of cold climates; mean annual soil temperature colder than 8°C |
| Udolls | The moist Mollisols; udic soil moisture regime |
| Ustolls | The sometimes-dry Mollisols that do get summer rain; ustic soil moisture regime |
| Xerolls | The Mollisols of Mediterranean climates, with dry summers and cool, rainy winters |

horizons that together meet all the requirements of a mollic epipedon. Albolls are required to have an argillic horizon and indications of wetness in the albic or argillic horizons.

*Rendolls* are Mollisols formed on limestone. They must have at least 40 percent $CaCO_3$ below the mollic epipedon and must not have subsurface diagnostic horizons such as argillic or calic horizons.

## SUPPLEMENT 14–2   The Canadian and FAO Soil Classification Systems

### The Canadian System

The Canadian system is a natural classification system based on soil properties. Like the U.S. system, it is a hierarchical system of several categories into which all the soils of Canada must fit. Genetic processes are not assumed; rather, the taxa are concepts based on generalizations of soil properties and are differentiated according to observable soil properties.

Six categories (taxa) are recognized in this system: order, suborder great group, subgroup, family, and series. The nine groups at the order level are based on pedon properties that reflect the nature of the soil environment and the effects of the dominant soil-forming processes (Table 14–10). Great groups are used to differentiate soils within orders based on (1) properties that reflect differences in the strengths of the dominant processes and (2) a major contribution of a process in addition to the dominant

one. For example, clay translocation is a major process recognized at the great group level.

Subgroups within the great groups are determined by the arrangement and kinds of horizons, which indicate either the central concept of the great group or an intergrade toward the soils of another order. Families are differentiated on the basis of parent material characteristics such as texture and mineralogy, soil climatic factors, and soil reaction.

The soil series is the most detailed level in the classification, as it is in the U.S. system, and is defined in much the same way. The series is a group of pedons with a similar number and arrangement of horizons and similar properties such as color, texture, structure, consistence, and thickness. Each of the pedons on the color plates has been placed into the Canadian classification (Table 14–5).

### The FAO System

The FAO of the United Nations prepared a soils map for the world at a scale of 1 to 5 million. A legend was required for the map, and because no suitable classification was available for soils at that scale, the FAO designed a special soil classification system that contains elements of many of the soil classification systems existing in the world today.

The original monocategorical system was transformed into a multicategorical system in a 1988 revision (FAO, 1988). The number of major soil groupings (MSGs) was increased to 28 (Table 14–11). These are

**TABLE 14–10**

Soil Orders in the Canadian Classification System

| | |
|---|---|
| Brunosolic | Weak to moderately developed soils |
| Chernozemic | Soils with mollic-like A horizons |
| Cryosolic | Soils with permafrost; similar to the Gelisols |
| Gleysolic | Wet soils with reducing conditions |
| Luvisolic | Soils with subsoil accumulation of clay |
| Organic | Similar to Histosols |
| Podzolic | Soils have podzolic B horizon, similar to the Spodosols |
| Regosolic | Soils with little development |
| Solonetzic | Soils with subsurface horizons with prismatic or columnar structures and Ca/Na > 10 |

**TABLE 14–11**

Major Soil Groupings in the FAO System

| | |
|---|---|
| Histosols | Soils with organic matter accumulations |
| Anthrosols | Soils significantly modified by human activity |
| Leptosols | Thin soils limited by bedrock or highly calcareous material |
| Vertisols | Soils with high shrink-swell properties |
| Fluvisols | Weakly developed soils showing fluvic (layered) properties |
| Solonchaks | Weakly developed soils with salt accumulation |
| Gleysols | Weakly developed soils showing signs of saturation |
| Andosols | Soils with properties derived from volcanic ash |
| Arenosols | Weakly developed soils coarser than sandy loam textures |
| Regosols | Weakly developed soils with only ochric or umbric A horizon |
| Podzols | Soils with a spodic B horizon |
| Plinthosols | Strongly developed soils containing plinthite |
| Ferralsols | Soils with B horizons that have low cation exchange capacity |
| Planosols | Strongly developed soils with an albic horizon showing signs of wetness and abruptly overlying a slowly permeable B horizon |
| Solonetz | Soils with a natric B horizon |
| Greyzems | Soils with a mollic A horizon overlying a strongly developed B horizon |
| Chernozems | Soils with a mollic A horizon overlying a carbonate-rich B horizon |
| Kastanozems | Soils with a lighter-colored mollic A horizon overlying a carbonate-rich B horizon |
| Phaeozems | Soils with a mollic A horizon with a high base saturation |
| Podzoluvisols | Soils with strongly developed B horizons with A horizon material tonguing into the subsoil |
| Gypisols | Soils with a gypsum accumulation |
| Calcisols | Soils with a carbonate accumulation |
| Nitisols | Soils with strongly developed, overly thick B horizons |
| Alisols | Soils with low base saturation, high cation exchange capacity, and strongly developed B horizons |
| Acrisols | Soils with low base saturation, low cation exchange capacity, and strongly developed B horizons |
| Luvisols | Soils with high base saturation, high cation exchange capacity, and strongly developed B horizons |
| Lixisols | Soils with high base saturation, low cation exchange capacity, and strongly developed B horizons |
| Cambisols | Soils with cambic (weakly developed) B horizons |

equivalent to orders and some suborders in soil taxonomy. Subunits of the MSGs are the 153 soil units. Soil units are subdivided into soil subunits. Many diagnostic horizons and properties are used as part of the system.

## SUPPLEMENT 14–3   The Storie Index System

The USDA Land Capability Classification is a qualitative system because numbers are not used to quantify the classes. The Storie Index is an example of a quantitative system developed for use in California. It is named for R. E. Storie, who developed the system when he observed that yields of commonly grown crops in California depended on soil properties. The numerical value of a soil—from 0 for no agricultural value to 100 for excellent agricultural soil—was to be used as a guide for farmers and others in agriculture. The system is used in California and in several countries of the world where Storie taught the system. Although it is not in wide use in the United States, it is a good example of a quantitative system.

The four factors in Storie's system are *A,* the soil profile factor, based on the degree of profile development and the kind of parent material; *B,* the surface texture factor; *C,* the slope factor; and *X,* the soil management factor. The soil management factor, *X,* has several parts: drainage, salinity, sodicity, nutrient level, acidity, erosion, and microrelief. Each part is evaluated for a soil, and then all are multiplied together to determine the *X* factor.

Within each factor is a range of values from 0 to 100. The rating is divided by 100, and the four values are multiplied to obtain a product that in turn is multiplied by 100 to yield the Storie Index rating for the soil:

$$(A/100 \times B/100 \times C/100 \times X/100) \times 100 =$$
$$\text{Storie Index rating}$$

Because the system is multiplicative, soils can easily receive low ratings if they have several management problems. Storie also grouped the soil indices into groups called soil grades (Table 14–12).

### Profile Groups

Profile group I soils are young alluvial soils without Bt horizons. Profile groups II, III, and IV soils have increasing Bt horizon development. Storie viewed these four profile groups as a sequence from young to old as in a chronosequence (see Chapter 12). Profile group IV soils have an abrupt change in texture from overlying horizons to the Bt that greatly reduces the soil permeability. Profile group V soils are those that have indurated hardpans within the profile (Figure 14–15). The soil rating increases as the depth to the root-limiting layer increases, but the maximum *A* factor value for a PG V soil is 80, compared with 100 for a PG I soil. All five profile groups are reserved for soils formed from transported parent materials. The usefulness of the soil for agriculture decreases from PG I to PG V.

The remaining profile groups are reserved for soils formed from igneous (VII), hard sedimentary (VIII), and soft sedimentary (IX) rocks (Figure 14–16). Within each of these, soils are rated for depth to the rock, the shallow soils receiving lower ratings than the deep soils. Profile group VI is no longer used in the Storie system.

Surface texture is rated in factor *B.* Loam, silt loam, and fine sandy loam textures score 100 percent, and textures such as clay (50 to 70) and sand (60) are graded lower. Gravel and stones in the soil also reduce the value of the *B* factor. The rating is based on the fact that ease of tillage, seed germination, infiltration rate, and water-holding capacity depend on the soil texture.

Slope is rated in the *C* factor. As the slope increases from level (0 to 2 percent) to steep (30 to 45 percent) or very steep (more than 45 percent), the value of the *C* factor decreases. Nearly level slopes of 0 to 2 percent are rated 100 percent, and steep and very steep slopes are rated from 30 to 50 percent and 5 to 30 percent, respectively. As the slope increases, the problems of soil erosion increase, cultivation and other machine operations become more difficult, and land productivity is reduced.

**TABLE 14–12**
Storie Index Grades

| Storie grade | Storie Index | Land quality |
| --- | --- | --- |
| 1 | 80–100% | Excellent |
| 2 | 60–80% | Good |
| 3 | 40–60% | Fair |
| 4 | 20–40% | Low |
| 5 | 10–20% | Very poor |
| 6 | <10% | Nonagricultural |

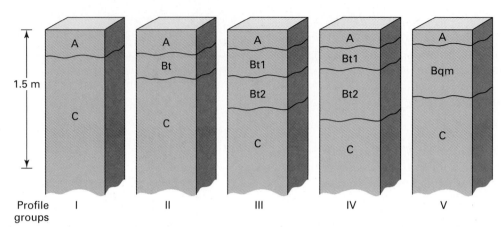

**FIGURE 14–15**  These pedons illustrate the morphological properties of Storie profile groups I through V. Note the increase in Bt thickness among profile groups II, III, and IV and the Bqm horizon in profile group V.

Well-drained (factor $X = 100$) soils have neither a high water table nor a fluctuating water table and are not subject to flooding. Those soils subject to frequent overflow are rated between 20 and 80 percent, depending on the flooding frequency and its duration. Soils without excess salts, sodium, or other toxicity problems are rated 100 percent, and those that suffer from such problems are given a lower rating according to the degree of the problem.

Nutrient level is inferred from soil characteristics such as organic matter level and clay content and type or it is measured by laboratory tests. Good, fair, poor, and very poor levels of fertility are recognized in the $X$ factor. Nutrient level is difficult to judge in the field and is easily improved, and so the lowest value of this part of the $X$ factor is 60 percent. Similarly, acidic soils are downgraded in the $X$ factor to 80 percent if they are extremely acidic.

The kind and severity of erosion are considered as part of the $X$ factor. For example, moderate sheet erosion is downgraded less than is severe gully erosion. The observer is left considerable freedom to evaluate the soils in this category. The microrelief portion of the $X$ factor is reserved for soils that have small relief features such as channels or mounds that can interfere with cultivation or that require earthmoving equipment to level them before cultivation can begin.

Our pedon in Figure 14–11 is moderately well developed on loess (profile group III), and so it receives 85 for factor $A$. The silt loam A horizon texture rates 100 percent for factor $B$, and a slope less than 2

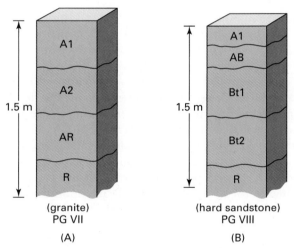

**FIGURE 14–16**  The parent material in profile groups (A) VII (hard igneous rock) and (B) VIII (hard sedimentary rocks) are used to separate soils in the two groups.

percent rates 100 percent for factor $C$. The soil is well drained (100 percent), has no alkali (100 percent) or pH problem (100 percent), and has high natural fertility (100 percent). There is no erosion (100 percent) or microrelief (100 percent). The total $X$ factor is the product of all the subparts and is thus 100 percent. The Storie Index rating for our pedon is 85 percent:

$$(85/100 \times 100/100 \times 100/100 \times 100/100) \times 100 = 85 \text{ percent}$$

# 15

# SOIL DEGRADATION

## OVERVIEW

Soils are neither permanent nor indestructible but may be altered or destroyed by mismanagement. The degree of alteration or mismanagement that "destroys" a soil must be decided within the context of our concept of soil and its intended uses.

Erosion by water, wind, and gravity occurs continuously, but soil formation rates exceed or equal the natural (or geological) rates of degradation. Therefore, when mismanagement increases erosion rates beyond the geological rates, a serious problem is created.

Water erosion occurs when raindrops strike bare soil, detaching particles, and overland flow carries the detached soil from the field. Water erosion takes several forms, depending on the concentration of water flow. Sheet or splash erosion occurs when there is little or no concentration of water flow over the soil surface. Rills form as water concentrates in small channels; when rills deepen, they become gullies.

Wind moves soil by rolling, bouncing, and lifting particles. Rolling and bouncing move large- and medium-sized particles short distances. Lifting suspends fine particles and may carry them long distances.

Mass wasting such as landslides and soil creep is caused by gravity, assisted by water that lubricates and adds weight to the soil being pulled downslope.

Soil crusting and sealing of the surface few millimeters reduce the rate of water penetration into soil. A seal or crust forms when water disperses clay that moves into the pores, plugging them. This may be a natural process or may be accelerated by management. Soils with a high exchangeable sodium percentage are most susceptible to this problem.

**343**

*Soil compaction also reduces water penetration. Compaction is often caused by excessive traffic or loading, especially when the soil is wet. While supporting the load, the soil particles can be rearranged to reduce the total pore space and pore size.*

*Mechanisms such as chelation, precipitation, and cation exchange, which help retain nutrients, are also important to retaining toxic compounds. Soil retention of toxic compounds reduces the amount of toxins reaching drinking water but increases soil pollution.*

*Protection and conservation of the soil are necessary because the rates of soil formation are slow. Erosion control includes keeping the soil surface covered with living or dead plant material, using engineering practices such as terraces and contour strips, and practicing conservation tillage. Wind erosion control includes cover, intercropping, and windbreaks to reduce wind speed over the ground surface.*

## 15.1   SOIL EROSION

Soil erosion by water is a major worldwide problem. By selectively removing organic matter and clay, **water erosion** not only removes nutrients but also may reduce the soil's chemical capacity to retain added nutrients (Table 15–1). Erosion reduces the pedon thickness and the volume of soil available for water storage and root exploration for nutrients. Under extreme gully erosion, enough soil is removed that farm machinery and animals cannot move from field to field across the gullies.

Soil "lost" from one field is often deposited on another, or it is carried into a stream, river, or lake, where the sediment becomes a pollutant. Many millions of dollars are spent annually cleaning rivers and harbors to maintain them for shipping. But sediment added to streams is not always pure; it may carry nutrients or pesticides to the receiving water. Thus, erosion is a two-part problem: a loss of fertility and soil depth in the eroding place (on-site problems) and the addition of unwanted sediment in the receiving place (off-site problems).

### 15.1.1   Water Erosion

Water flows over the ground in thin sheets or as concentrated flows in rills (Figure 15–1), gullies (Figure 15–2), and stream channels. As it flows over the ground, water dislodges and transports soil particles. Erosion and deposition are natural geological processes that wear down mountains and create spectacular vistas, such as the Grand Canyon of the Colorado River and (less noticeably perhaps) new parent material for soils.

**Geological erosion** at natural rates is not a management problem, and it would be foolish, or at least very expensive, to try to stop it entirely. But there are times and places where erosion rates are **accelerated** because of poor management. For example, on construction sites the soil may be left bare, or sloping farm fields may be plowed up and down the hill, caus-

**TABLE 15–1**

Selective Losses of Nutrients and Organic Matter Reflected in the Enrichment Ratio[a]

| Nutrient or property | Enrichment ratio |
|---|---|
| Organic matter | 1.1–4.7 |
| Nitrogen | 1–5 |
| Phosphorus | 1–4 |
| Potassium: total | 1–7.3 |
| available | 4.7–12.6 |
| Calcium | 2.1–2.4 |
| Magnesium | 1.4 |

[a]Calculated by dividing the amount of nutrient or soil property in runoff water by the amount found in the soil surface. A ratio greater than 1 illustrates that erosion rapidly removes nutrients, organic matter, and clay.
Source: Data from H. L. Barrows and V. J. Kilmer. (1963). Plant nutrient losses from soil by water erosion. *Advances in Agronomy* 15:303–316.

**FIGURE 15–1**   Rill erosion such as this near Pullman, Washington, removes many tons of soil from the field, lowers soil depth and fertility, and lowers water quality when the soil reaches a stream. (Photo: D. McCool.)

**FIGURE 15–2**   This erosion gully is about 10 m deep. It formed from uncontrolled runoff originating in a 51-ha watershed. Productivity is lost, and the eroded sediment becomes a pollutant when it enters a stream or lake. (Photo: U.S. Department of Agriculture [USDA] Natural Resources Conservation Service [NRCS].)

ing soil and water to run off the field rapidly. Such erosion problems need management.

**The erosion process**   Water erosion begins when a water drop strikes bare soil. The drop detaches a bit of soil from the pedon surface, and it may transport the soil downslope as it splashes or loosens the particle so that it may be transported by water flow over the soil surface as sediment in **overland flow** (runoff). Overland flow may detach particles and transport them downhill. Detachment and transport

are necessary for soil erosion to occur, and raindrop detachment and overland flow transport are the two main parts of the process.

Ponding must occur before water begins to flow over the soil surface. Water ponds when the rainfall rate exceeds the soil infiltration rate. Rainfall rates that are less than infiltration rates may cause ponding. If the soil is shallow to bedrock, the pedon may become saturated, or if the pedon has a very slowly permeable horizon, water flow into the pedon can be restricted, causing water to fill the soil pores to the

# SOIL AND ENVIRONMENT
## *Air and Water Quality*

Soil erosion is not just an agricultural problem. The on-site damage from erosion is indeed a problem to agriculture, but off-site problems are general to our society. Soil removed from an agricultural field by wind or water erosion may pollute the air and water. Dust in the air is unhealthy, reduces visibility, and, when it settles, makes things dirty. Sediment clogs rivers, lakes, and navigable waterways, reducing their potential for recreation and navigation. It reduces the water storage capacity of reservoirs and increases flooding. Agricultural fields are not the only locations where erosion exists. Soils along highway rights-of-way, pipeline and electrical line installations, new home developments, and nonagricultural watersheds erode at rates equal to or higher than that of agricultural erosion. Soil erosion in off-highway vehicle (OHV) recreation areas is a major management problem. Conservation practices in these locations may be very different from those in agricultural fields, but the methods for predicting agricultural wind and water erosion are useful in calculating urban and rural nonagricultural erosion.

Water erosion control practices along highways include rock mulches, artificial and natural netting, special plantings, and various chemicals to either hold mulch in place or seal the soil to protect it. All of these practices are too expensive for use in agriculture, but they work as agricultural conservation practices do by protecting the soil surface from raindrop impact energy, reducing the velocity of water running over the soil surface, and holding the soil in place.

soil surface. Surface conditions change during rainfall: raindrops beating on the soil surface can create a crust or seal that can greatly reduce the rate of water penetration into soil and cause ponding.

Once water has ponded in small depressions on the soil surface, they can overflow, and **runoff** will occur. If the runoff is in thin (a few millimeters) sheets, it is **sheet flow,** and the erosion it produces is **sheet erosion. Splash erosion** is produced when raindrops impact the soil surface, detaching particles and transporting them through the air. Most soil scientists agree that water seldom flows in thin sheets for more than a few meters on soil surfaces, and, at times, there may be no thin sheet flow at all. Instead, the flow is in many small channels a few millimeters wide and deep, called **rills.** A large rill may be too wide or deep (> 0.5 m) to fill by normal cultivation; then it is a **gully.** All forms of erosion require **particle detachment** and **particle transport.**

When most soil transport is by rills, the form of erosion is rill erosion. When most soil transport is by gullies, it is gully erosion. As water becomes concentrated in rills and gullies, it moves faster and detaches and transports soil particles more efficiently. Erosion rates increase as the process changes from sheet to rill and gully erosion.

Soils have different susceptibilities; that is, they respond differently to erosive forces. The relative susceptibility is the erodibility of a soil. For example, loess soils from the midwestern United States tend to be highly erodible. Surface texture and structure, infiltration rate, and permeability influence soil erodibility. Silt-sized particles (0.002 to 0.050 mm) tend to be the most easily detached and transported. Sands are too big to be transported easily, and clays cohere to one another and act like larger particles. Structure, particularly the size and water stability of aggregates, is important for the same reason. Particles that are bound together as stable aggregates resist detachment and transport. What binds these aggregates?

Organic matter, oxides of iron and aluminum, and clay are agents that bind particles together to form aggregates (see Chapter 2). Soils high in organic matter or oxides tend to be less erodible than those low in these aggregate-stabilizing materials. Good structure and medium texture provide for rapid infiltration rates, which reduce runoff and erosion. Soils with slow infiltration rates and permeabilities pond early in a storm and may have more runoff than do those with rapid infiltration and permeability.

The pedon we have been considering throughout this book has a strong surface structure because of its high organic matter content, but it also has a horizon of restricted permeability and silt loam textures, characteristics that make the soil erodible. It will become even more erodible if the surface is mismanaged and the organic matter content declines.

The water erosion process requires a slope. Level land erodes slowly because the runoff velocity is slow. But as the slope increases, runoff velocity and erosion increase. This fact and others have been combined into a relationship (the Universal Soil Loss Equation) useful for predicting the rates of soil erosion.

**Erosion prediction**   The **Universal Soil Loss Equation (USLE)** is widely used in the United States and other countries to predict the severity of erosion from farm fields. It is universal because the six factors are sufficient to describe the process:

$$A = RKLSCP \qquad (1)$$

In Equation 1

   $A$ is the long-term average annual soil loss for a field.

   $R$ is the long-term average rainfall–runoff erosivity factor.

   $K$ is a soil erodibility index.

   $L$ is a slope length factor.

   $S$ is a slope angle factor.

   $C$ is a soil cover factor.

   $P$ is an erosion control practice factor.

(For more detailed explanations, see Wischmeier and Smith, 1978, or Morgan, 1995.)

Rainfall impact and overland flow are responsible for soil erosion, but they are not easily measured. Rainfall amount (mm) and intensity (mm/hr) are easily measured, however, and these two factors are combined to estimate the long-term erosivity of rainfall and runoff for a location. The actual calculation of the $R$ factor and other factors in the USLE is in Wischmeier and Smith (1978).

The $K$ factor is a single number that combines the many soil properties that influence the soil's reaction to rainfall and runoff. Five properties have been combined to determine $K$: silt plus very fine sand, other sand, organic matter content, soil structure, and the soil permeability of the least permeable horizon. Silt and very fine sand are the most easily moved particles. Other sand (particles between 0.10 and 2.0 mm diameter) is less easily moved, and in-

creasing organic matter decreases erodibility because it helps strengthen soil aggregation. Soil structure and permeability relate to the rapidity of ponding and runoff initiation. The least permeable horizon controls the profile's hydrology. Soils with strong structure do not detach easily and so are less erodible than are soils with weak or no structure.

Water flowing downhill accelerates to a maximum velocity based on the hill's steepness and length. The general relationship between slope length and soil loss is shown in Figure 15–3. For short slopes the rate of increase in soil loss rises rapidly, but for long slopes the rate of increase is very small. This shows that there is a limit to the length of a slope that will influence soil loss.

Slope angle also influences soil loss (Figure 15–4). Unlike slope length, as slope angle increases, soil loss rate also rises at an increasing rate, demonstrating why slope is such an important variable in soil loss.

Cover is the principal erosion control practice. If a soil is completely covered with vegetation, raindrops and overland flow will not be able to detach particles. In the USLE, the cover factor ($C$) is the ratio (from 0 to 1) of soil loss from a plot with some level of cover to soil loss from a bare plot. The $C$ value is determined by experience rather than by an equation and includes the kind of cover and the amount of the soil surface covered. (See Section 15.4 for more on erosion control.)

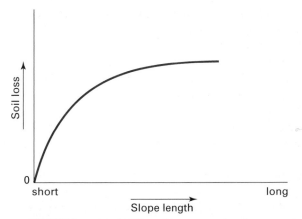

**FIGURE 15–3**  Soil loss increases as slope length increases, up to some maximum length depending on the slope angle. Most agricultural field lengths are in the steep portion of the curve.

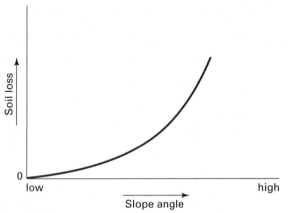

**FIGURE 15–4** Steep slopes are susceptible to severe soil erosion because soil loss
increases exponentially as the slope angle
increases.

The final factor in the USLE is the conservation
practices (*P*) factor. Like the *L, S,* and *C* factors, it com-
pares the soil loss from a soil with conservation practices with that of one plowed up and down the hill with
no erosion control practices. Conservation practices
include contour tillage, terraces, and grass waterways.
The no-practice condition gives a *P* factor of 1.

Once all of the factors have been determined,
they are multiplied together to yield the predicted average annual soil loss from the field or construction
site. This number can be used to determine whether
**conservation** practices are required and what level of
*C* or *P* factor is necessary to reduce the soil loss to the
desired level. The equation is not designed to predict
erosion from watersheds or individual storms.

The USLE has served the erosion community
well since its inception, but it had limitations, particularly in the western United States. The USDA has developed two new erosion models in recent years. Both
are products of the USDA Agricultural Research Service and the USDA National Soil Erosion Research
Laboratory. One, the revised USLE (RUSLE), modi-
fies each of the six factors in the USLE but does not
change the basic structure of the equation. Each of
the changes was made to improve the accuracy of pre-
dictions and to extend the usefulness of the equation
to the entire United States.

The RUSLE expands on the original USLE and
includes changes that make it more widely applicable in
the United States. It continues to rely on field plot and
rainfall simulation data and is thus an empirical equation much like the original USLE. The factors are defined in much the same way as in the USLE. The *A* factor continues to be defined as a long-term average
annual soil loss over a field slope, and losses at various
points on the slope may differ greatly from one another.

*R*   The *R* factor, with a few exceptions, is calculated for the RUSLE just as it is calculated for the
USLE. The changes that have been made are in how
rainfall energy is calculated. The formula that is now
recommended is as follows:

$$e = 1099[1 - 0.72 \, e^{(-1.27i)}] \qquad (2)$$

*i* has units of inches per hour, and *e* has units of foot-
tones (ft) per acre per inch.

The metric equivalent is

$$e_{\mathrm{m}} = 0.29[1 - 0.72 \, e^{(-0.05im)}] \qquad (3)$$

Units of $e_{\mathrm{m}}$ are megajoules per hectare per millimeter rain, and $i_{\mathrm{m}}$ has units of millimeters per hour.
One can calculate the *R* value for an area by taking
the average sum of *EI* values for all storms over all
years. The original *R* value was calculated for storms
that had more than 0.5 in. of rain. The *R* value in the
RUSLE includes all storm records for the western
states. This recognizes that in the western United
States, any storm, not just those producing over 1/2 in.
of rain, may have the potential to produce erosion.

Two other changes in the RUSLE are the inclusion of the effect of ponded water on the soil surface
and an *R* equivalent for cropland in the northwestern
wheat and range region. The *R* equivalent is used
where rain on frozen ground and snowmelt have been
shown to be very important in soil loss prediction.

*K*   The definition of *K* remains the same in
the RUSLE as in the USLE, but calculation methods
have changed. For example, surface rock fragments
are now incorporated into the RUSLE; surface rock
fragments protect the surface from direct raindrop
impact and influence interrill flow on the soil surface. Subsurface rocks influence infiltration by reducing the volume of soil available to transmit water.
In the RUSLE, surface rock fragments are considered within the *C* factor and subsurface rock fragments in sand- and loamy-sand–textured soils. The
effect is considered within the permeability portion
of the *K* factor calculation.

Seasonal variation in $K$ values is considered in the RUSLE. Variations in $K$ values appear to be due to soil freezing, soil texture, and soil water content. It is thought that freezing and thawing increase soil erodibility through the effect on soil structure, hydraulic conductivity, bulk density, aggregate stability, and soil strength. In areas that do not have many freeze-thaw cycles, it has been proposed that the $K$ value declines to a minimum over the growing season and then increases to its maximum some 6 months after the end of the growing season.

*L and S* The $LS$ factor represents the ratio of soil loss on a given slope length and steepness to soil loss from a slope that has a standard length and steepness. The equations were first developed for uniform slopes and then extended to include nonuniform slopes. The slope length factor, $L$, is calculated, as it was in the USLE, from the following equation:

$$L = \left(\frac{\lambda}{72.6}\right)^m \qquad (4)$$

where $\lambda$ is the actual slope length and $m$ is a variable slope length exponent based on slope steepness. All units are feet. A major change in the RUSLE is the way in which the slope steepness factor, $m$, is calculated. It is now related to rill and interrill erosion processes through a factor termed $\beta$, which is the ratio of rill to interrill erosion. If a soil is highly susceptible to rill erosion, the exponent $m$ should be increased. In some cases the value of $\beta$ needs to be doubled to adequately predict $m$. Similarly, if slopes are known to resist rill erosion, $\beta$ should be decreased. In the RUSLE manual, tables are given to assist in the selection of values for $\beta$ and $m$.

The slope steepness factor, $S$, is calculated using equations for different conditions as follows. Implicit in all of these equations is the concept that runoff is not a function of slope steepness.

$$S = 10.8 \sin \theta + 0.03 \text{ for } s < 9 \text{ percent} \qquad (5)$$

$$S = 16.8 \sin \theta - 0.50 \text{ for } s > 9 \text{ percent} \qquad (6)$$

$$S = 3.0 (\sin \theta)^{0.8} + 0.56 \text{ for} \atop \text{slopes shorter than 15 ft} \qquad (7)$$

Under the special conditions of the Pacific Northwest, where recently tilled soil is thawing, in a weakened state, and subjected primarily to surface flow, use

$$S = 10.8 \sin \theta + 0.03 \text{ for } s < 9 \text{ percent} \qquad (8)$$

$$S = (\sin \theta/0.0896)^{0.6} \text{ for } s > 9 \text{ percent} \qquad (9)$$

Both the USLE and RUSLE have a systematic method for calculating $LS$ for nonuniform slopes.

*C and P* The $C$ and $P$ factors are used in the RUSLE as they were used in the USLE, but some new twists have been added. For cultivated crops, the $C$ factor is calculated for time periods. In forestland and rangeland situations, the $C$ factor may not change appreciably over the season and a single $C$ factor may be used.

The terminology in the RUSLE changes a bit in that the soil loss ratio (SLR) for each crop stage is calculated and then multiplied by the fraction of rainfall and runoff erosivity ($EI$) for the period. These $EI$ weighted values are summed for the year to obtain an overall $C$ value. The time step for calculating the $C$ factor, like the $EI$ factor, was chosen as 15 days. The first half of the month consists of 15 days always, and the second half has a variable number of days depending on the length of the month. This provides 24 periods during the year for which SLRs are calculated. These periods may be subdivided if needed.

The calculation of the soil loss ratios (SLR) is based on five subfactors ($PLU*CC*SC*SR*SM$). Each is calculated separately, then multiplied together to equal the SLR for the given conditions. These subfactors are as follows:

$PLU$ is the prior land use subfactor

$CC$ is the canopy cover subfactor

$SC$ is the surface cover subfactor

$SR$ is the surface roughness subfactor

$SM$ is the soil moisture subfactor

The RUSLE uses a crop database to store values required to calculate the variables that go into the subfactors. These values include growth characteristics of vegetation, amount of residue produced, and residue characteristics. Another database within the RUSLE has temperature and rainfall data for locations so that residue decomposition rates can be calculated.

$PLU$ considers the influence on soil erosion of subsurface residual effects from previous crops and the effect of previous tillage practices on soil consolidation. $Bu$ accounts for the effect on erosion rates of live and dead roots and incorporated residue. Subfactor $CC$ accounts for interception of raindrops and reduction in the energy of raindrops that reach the soil surface. The *SC* subfactor accounts for the reduction in area of the soil surface that can be directly impacted by raindrops, thus reducing detachment. It also accounts for reduced capacity for transport by overland flow due to slowing of the overland flow rate and increased tortuosity of the flow path produced by the cover. Finally, cover produces ponding behind the cover elements, which results in deposition, and, if deep enough, it also reduces the effect of raindrop impact. This is perhaps the most important of the subfactors.

The surface roughness subfactor, *SR*, accounts for the effects of the random depressions and barriers that trap water and sediment on a rough surface. Surface roughness also reduces the velocity of overland flow, thus reducing its transport capacity and detachment. The soil moisture subfactor (*SM*) accounts for antecedent soil moisture and its effect on infiltration and runoff and their influence on soil erosion.

The support practice factor, *P*, has been defined for common practices including contouring, cross-slope strip-cropping, buffer strips, filter strips, terracing subsurface drains, diversions, and windrows. As with the other factors, *P* is the ratio of soil loss with a specific support practice to the corresponding loss with upslope and downslope tillage. An overall *P* factor is calculated as a product of *P* subfactors for individual practices when practices are used in combination. These practices are considered in Section 15.4.1.

A second new USDA erosion model is the Water Erosion Prediction Project (WEPP). The WEPP model is a process-based computer model that uses fundamental physical principles and basic understanding of water erosion processes to predict soil loss. Erosion is a complex process, and the WEPP models the complex interactions among raindrops, overland flow, topography, soil properties, and surface cover to predict amounts of soil loss. Because of the approach taken by the designers of the WEPP, it includes deposition, ephemeral gully erosion, sediment yield, and spatial and temporal variability. The hillslope profile version of the model estimates when and where on the hillslope erosion is occurring. The goal is to use the model to better design soil erosion control measures and reduce sediment yield from watersheds. To be used, the WEPP will require much more data and more computer resources than the RUSLE.

## 15.1.2  Wind Erosion

Wind erosion rarely produces the catastrophic dust storms that were well known in the dust bowl days of the 1930s, but wind erosion, like water erosion, damages the soil by removing organic matter and plant nutrients, and by decreasing soil thickness. Silt- and clay-sized particles are selectively removed from the surface soil by wind, resulting in reduced nutrient and water-holding capacity. Kimberlin, Hidlebaugh, and Grunewald (1977) reported that from 405,000 to 6,000,000 ha of land were damaged by wind erosion between 1935 and 1976. The greatest damage was in Texas. About 75 percent of all land damaged was in the southern Plains states. Damage is defined as lands that are losing soil at rates of approximately 34 metric tons per hectare per year. Many more hectares of land are losing soil at lower rates, but the cumulative effect is the same: a poorer medium for plant growth.

In addition, blowing soil particles can damage or kill plants, particularly young seedlings, by "sand blasting." Seedlings may be cut off at ground level, requiring reseeding. Vegetables are particularly susceptible to damage due to abrasion. Studies have shown that even at relatively low rates of soil loss, damage can be severe. Due to surface abrasion, plants may be damaged so that they do not grow as quickly and yield is reduced, or the quality of products to be marketed may be reduced. Wind-induced damage can also make plants more susceptible to damage by insects and disease.

In addition to these on-site problems, the off-site problems are air pollution and detrimental deposition. Soil particles in the atmosphere contribute to poor visibility and the suffering of people with allergies. Suspended particles may create visibility hazards for automobiles and aircraft. Newspaper accounts of traffic accidents caused by windblown particles are testimony to the danger of windblown soil. Windblown particles also pile up around plants, fence posts, and houses (Figure 15–5), and dunes may damage or destroy whatever they bury. Costs associated with the off-site problems include cleanup. Examples of cleanup costs may be removal of sediment from roadside ditches, car cleaning, and outdoor-table cleaning. Many costs are unaccounted for because it is difficult to determine costs for cleaning clothes or houses due to dust from wind erosion.

**Processes**  Wind, like water, is a fluid, and the process of wind erosion is similar to the process of wa-

**FIGURE 15–5** Wind erosion has moved the sand and deposited it on the fence. The fence has acted as a windbreak and has protected the road from burial. (Photo: USDA.)

ter erosion. When the wind force (a function of velocity) exceeds the forces holding a particle to the ground, it moves the particle. There are three forms of movement: saltation, creep, and suspension (Figure 15–6). **Saltation,** a lifting and bouncing of particles, is the most important of the three. It moves the largest number of particles and initiates creep and suspension. A wide range of particle sizes, from silts to sands, is moved by saltation, but particles of 0.1 to 0.5 mm diameter are most commonly transported by this process. Wind flowing over the ground surface exerts a force on particles like the lifting force exerted on an airplane wing as it flies through the air. This lifting force causes particles to jump from the ground, rise into the wind, and fall back to the ground where they may jump again or dislodge other particles. Saltating particles usually are not lifted more than 15 cm into the air, but particles do not have to be lifted high to be moved.

Coarse and very coarse sand grains move by **creep** because they are too large to be lifted by all but hurricane-force winds. Rather, they are rolled or pushed short distances across the ground surface.

**Suspension** is the lifting of clay- and silt-sized particles < 0.1 mm diameter high into the air, where they may be carried long distances. Although this

process does not carry as much soil as saltation does, it is the most visible mechanism of wind erosion, and it has received much public attention. The great dust storms of the 1930s, when suspended soil particles from the southwestern United States reached St. Louis and Chicago, drew public attention to the problem of soil erosion.

**Prediction** A wind erosion equation similar to the USLE summarizes the major factors important to wind erosion:

$$E = f(I, C, K, L, V) \qquad (10)$$

Equation 10 is solved both numerically and graphically, though its complete solution is not considered here. Briefly, the equation states that the potential average annual soil erosion, $E$, is a function of $I$, soil erodibility; $C$, local wind erosion climatic factor; $K$, soil surface roughness; $L$, unprotected width of field; and $V$, equivalent quantity of vegetative cover.

Soil erodibility ($I$) depends on the percentage of soil greater than 0.84 mm diameter. The local climatic factor ($C$) is the product of the average wind velocity and the average moisture of the soil surface. Moist soil moves less easily because it is heavier than dry soil, and soil particles cohere when moist. Soil

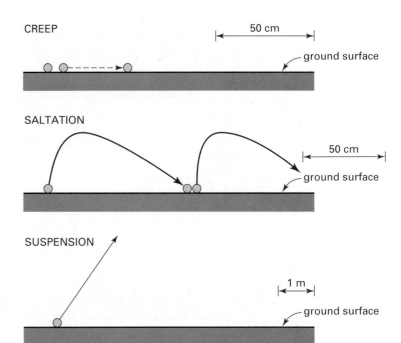

**FIGURE 15–6** Wind moves soil particles by creep, saltation, and suspension. Suspension attracts attention because it lifts small particles high into the air and carries them long distances. However, most particles are moved by saltation.

surface roughness ($K$) is expressed in height of ridges. More roughness (either more or higher ridges) increases the soil's resistance to wind erosion. The unprotected width of the field ($L$) is the mathematical difference between the total distance along the prevailing wind erosion direction across a field ($D_T$) and the protected distance along the prevailing wind erosion direction ($D_P$):

$$L = D_T - D_P \qquad (11)$$

Protection is by a wind barrier such as a windbreak. The quantity, kind, and orientation of the vegetative cover are combined into a single vegetative cover factor ($V$).

Note that Equation 10 states that $E = f$, a function of these parameters. It is not strictly equal to the mathematical product of the factors. The solution of the equation requires graphical methods.

### 15.1.3 Gravity Erosion (Mass Wasting)

**Gravity erosion** is generally termed **mass wasting** because large masses of soil move. Water usually assists, although flowing water is not the main moving force. The movement may be instantaneous, a **landslide,** or slow and persistent over many decades, **soil creep.**

Mass wasting is often spectacular (Figure 15–7), but it moves much less soil than the amount moved annually by wind and water.

**Types of mass wasting**   Various attempts have been made to describe or classify kinds of mass wasting by means of the direction or speed of soil movement. **Falls** occur when gravity pulls a part of a cliff or steep hillside vertically downward. **Rotational slides** or **slumps** actually rotate around a point in the center of the slump (Figure 15–8). A **compound slide** is a combination of a fall and a rotational slide, and a **transitional slide** occurs when a break causes a mass to slide along the face of the supporting mass. **Debris flows** and **avalanches** are instantaneous, whereas creep and **earth flows** are slow. *Slow* is a relative term, because earth flows may move rapidly once they get started. But usually they move a short distance and stop before moving again.

**Processes**   Mass wasting is a balance between the gravity pulling a mass downhill and the friction holding the mass in place. Anything that decreases the friction or increases the mass (weight) helps move the mass downhill. Water can do both. Dry soil masses may fall, but without water, they infrequently move as rotational slides or landslides. Water lubricates the sliding plane and adds mass to the block to be moved.

**FIGURE 15–7** The block of soil on the right broke off from the block on the left and instantaneously slid several meters downslope. Note the adult standing in the center of the photo and the gentle slope.

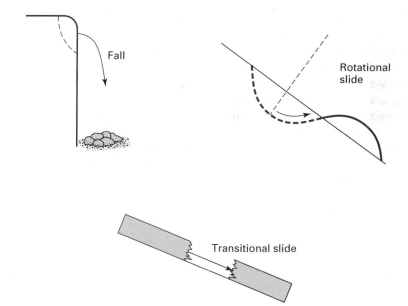

**FIGURE 15–8** Falls and rotational and transitional slides are forms of mass wasting. Saturated soil increases the danger of slides on steep slopes.

Saturated soils have a positive pore water pressure that enhances the possibility of slides. Water cannot be compressed (see Chapter 3). When a pore is full of water, the water exerts a pressure against the surrounding particles. If this is a positive pore water pressure (outward), the soil grains will not be in contact in many places (or, in extreme cases, not touching at all), and the soil will lose strength and move.

In addition to the amount of water, soil properties and vegetation influence mass wasting. Soils such as clays, which drain slowly and have high water-holding capacities, slide more easily than do sands, which drain quickly. Soils formed over weak or shattered bedrock that can absorb water also tend to fail easily. As slopes become steeper, mass wasting becomes more likely.

Vegetation, particularly shrubs and trees with deep roots, helps reduce landslides. Plants use water, thereby decreasing the pore water pressure and mass of the soil and increasing the friction among soil particles. Roots help bind the soil together. Thick taproots help anchor slopes in place. In addition, plant roots stretch and help slow movement.

**Prediction**     Although landslides and the conditions that favor them have been studied, there is no precise analytical or mathematical method for predicting where, when, or how they will occur. Some natural conditions are known to predispose a slope to mass wasting, including clay layers in the subsoil that act as sliding planes, shallow bedrock that also acts as a plane of weakness, and steep slopes (over 80 percent). Periods of intense rainfall often cause slopes to fail. Removal of the toe (downhill portion) of a slope in building and construction and removal of trees are management factors that predispose a slope to mass failure.

## 15.2  PHYSICAL DEGRADATION

### 15.2.1  Surface Crusts and Seals

The soil surface is permeable to water and air because of the pores that connect the surface and the subsurface. If surface pores clog or their average size is reduced, the rate water enters the soil will be reduced. This process is called surface sealing or crusting. Seals and crusts are similar, except that **seals** occur when the soil is wet and **crusts** develop when the soil dries (Figure 15–9).

A surface crust may be as thin as 1 mm or as thick as 1 cm. Particles in the top few tenths of a millimeter are sorted and may be layered. Clay separated from the top few tenths of a millimeter is washed into the next few tenths to clog the pores. This **washed-in zone** is the primary reason that water flow through crusts is very slow. McIntyre (1958) measured the flow of water through some crusts and found that it was 1,000 times slower than through the same soil uncrusted.

In irrigated agriculture, slow water penetration is a problem because it requires additional labor to irrigate more frequently at a slower application rate. In cases of severe crusting, insufficient water may reach the root zone, and plants can be water stressed. In addition, water may be ponded around sensitive trees or vines, resulting in their injury or death. On sloping ground, slow water penetration results in more runoff, more soil erosion, and less efficient water use.

In addition to slowing water penetration, crusts block seedling emergence, and those seedlings unable to emerge from the soil die, thereby reducing the number of plants per acre. This, of course, can ultimately lower the yield.

**Processes**     Crust formation is both a physical and a chemical process. Raindrops or sprinkler drops hitting the soil surface break aggregates, compact the surface, and sort the particles. As water enters aggregates, the air in the aggregates pushes outward, causing the aggregate to fall apart—to **slake.** Infiltrating water carries some surface particles downward, where they plug the small pores. The process is assisted when sodium is on the exchange complex, because sodium reduces aggregate stability and causes the clays to disperse (see Chapter 11). Dispersed clays are more mobile than flocculated clays are and therefore block pores more readily.

### 15.2.2  Compaction

Bulk density is a measure of the mass of soil per unit volume (see Chapter 3). When soil particles are pushed close together, increasing the mass per unit volume, the soil is *compacted*. There is no single bulk density above which all plant growth is reduced, but 50 percent porosity is usually satisfactory. A soil with minerals that average $2.65$ Mg m$^{-3}$ and 50 percent pore space has a bulk density of $1.32$ Mg m$^{-3}$. Compacted subsoils often have bulk densities in excess of $1.6$ Mg m$^{-3}$. The major problem for plants at high soil densities is difficulty in root extension, because plants do not readily push their roots through dense soil. Water flow through compacted soils is also restricted because the number of large pores is reduced.

**Processes**     Soil compaction occurs when a weight on the soil surface rearranges the soil particles. The weight is transmitted through the soil to a depth at which the particles support the load. The total load and the load per unit area are the main factors influencing the soil's behavior. Figure 15–10 illustrates

(A)                                                        (B)

**FIGURE 15–9**   (A) Uncrusted soil before raindrop impact. (B) The same soil after 30 minutes of simulated rainfall at 40 mm/hr. The surface structure is destroyed, and a crust has formed.

**FIGURE 15–10**   Noncohesive sands and loamy sands are compacted when a force rearranges particles into a denser packing. Note that both the size of the pores and the total volume of pores are less in the compacted sand.

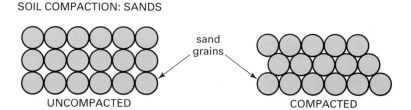

SOIL COMPACTION: SANDS

sand grains

UNCOMPACTED                    COMPACTED

SOIL COMPACTION: <2 mm

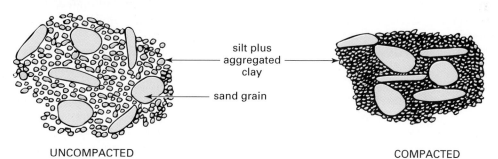

silt plus aggregated clay

sand grain

UNCOMPACTED                                    COMPACTED

**FIGURE 15–11**   Cohesive soils have sufficient clay to form some aggregates. When compacted, the fine particles are rearranged, and the larger particles become oriented.

what happens when a loose, sandy soil is compacted. Particles arranged in an open porous network are rearranged, primarily through sliding and rolling into less open, less porous, denser organizations. In fine-textured soils, the same process occurs, and in addition, the coarser particles in the fine matrix become oriented (Figure 15–11).

The dominant soil factor influencing compaction is the soil water content at the time the soil is loaded. The relationship between density and water

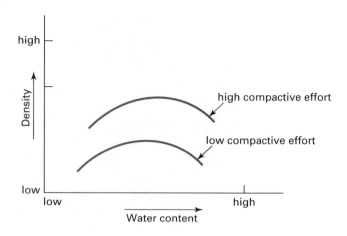

FIGURE 15–12 These are Proctor curves for a soil subjected to two compactive efforts at different water contents. As the water content increases, density (and compaction) increases to a maximum and then decreases.

content for two compactive efforts is shown in Figure 15–12. The upper curve is for the higher compactive effort (e.g., a heavier tractor or a smaller wheel tread). As the soil water content increases, the density of the compacted soil increases up to a maximum and then decreases. Dry soil particles do not readily slide or roll over one another. Water reduces friction and enhances movement. This does not continue beyond a maximum because water cannot be compressed; it takes up space. At a water content near saturation, the particles are pushed completely apart by the positive pore water pressure, causing the soil to lose strength completely.

## 15.3 CHEMICAL DEGRADATION

### 15.3.1 Nature of the Problem

Chemical soil degradation is most frequently a problem of soil contamination, which occurs when sufficient quantities or concentrations of harmful substances accumulate beyond their "natural" or background level in soils. Sources of soil contamination include industrial waste, dredge spoils, ore smelter wastes, sewage sludge, petroleum spills, fly ash disposal, landfills, and use and misuse of agricultural chemicals including fertilizers. Such contamination interferes with normal (routine) soil chemical and biological processes.

A constant danger is that contaminated soils may also contaminate drinking water supplies and food chains. Soil contamination occurs when useful chemicals are added to the soil in excess or through the accumulation of chemicals added in small amounts over time. We often use the soil, rather than water or air, for waste disposal. Such direct contamination may be tolerated as less dangerous than waste disposal in water or air (through burning). Although waste may not be dangerous when first buried, it may produce dangerous chemicals when it decomposes. Improper disposal and accidental spills and leaks may also contaminate soil.

Contamination may present a risk of explosion, fire, corrosion, or chemical toxicity. The toxicity reduces crop yields and the usefulness of soils for agriculture and is dangerous to humans and animals through direct contact, through drinking contaminated water, or through eating contaminated products grown on contaminated soils. Such contaminants include the following:

1. Elements such as lead, mercury, cadmium, copper, chromium, selenium, and arsenic, and their compounds
2. Organic chemicals such as pesticides, plastics, solvents, oils, and tars
3. Radioactive materials
4. Asbestos and other hazardous materials
5. Sodium, salts, and acids (see Chapter 11)

### 15.3.2 Contaminant Interactions with Soil

Few contaminants are so inert that they remain in soils unchanged. Elements and their inorganic compounds may be adsorbed on clay and organic matter surfaces as exchangeable ions, may be fixed in nonexchangeable forms, or may be precipitated in the soil (Figure 15–13). These processes may prevent

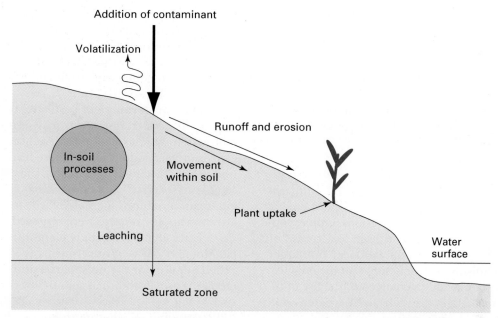

**FIGURE 15–13**  Possible fates of soil contaminants. Soil processes that retard movement or degrade contaminants include biodegradation, sorption, and chemical decomposition. Nutrients such as nitrogen and phosphorus are not usually thought of as soil contaminants, but when they leach to groundwater or erode or run off into surface water, they become water pollutants.

the movement and plant uptake of the contaminant but may not reduce its toxicity. Organic chemicals, including oils and tars, are frequently susceptible to microbiological decomposition in soils.

In addition to solid-phase interactions, contaminants in soils may be dissolved in the soil solution or simply occupy space in soil pores either as gas or as nonaqueous-phase liquids (liquids that do not dissolve in water), along with the soil solution and soil gases. In general, the solubility of hydrocarbons in water decreases with the increasing number of carbon atoms in the compounds. The mobility of contaminants that do not interact with the solid phase is a major concern because of the possibility that they will move from the soil into the groundwater.

*Pesticide* is the general term applied to chemicals discovered or developed to control a wide variety of pests that attack crops and humans. Their development and use have contributed to improvement in agricultural productivity and human health throughout the world. They include insecticides, used to control insects; herbicides, used to control weeds; fungi-

cides, used to control fungi; and rodenticides, used to control rats and other rodents. Because they are designed to kill organisms, it is of concern that they kill only the target organisms and do not spread beyond the point of application.

Pesticides reach the soil directly, through surface or subsurface application, or indirectly, as drip from foliage when sprayed directly on plants. They also reach soil through spills and improper disposal. Some portion of the chemicals remaining on plant parts or assimilated by plants may become available to soil organisms when the plant parts fall to the soil surface or are incorporated into the soil during tillage. The length of time a pesticide remains in the soil is the **persistence.** Pesticide persistence is important in determining the length of time the pesticide is effective and the potential for pollution. Chemical structure and environmental conditions such as soil temperature, soil moisture, and amount of organic matter and clay are factors that determine persistence.

Persistence may be either beneficial or problematic. Persistent pesticides are unlikely to be leached into groundwater because they are tightly held by the soil, but persistent pesticides are more likely to be transported with eroded soil particles than are pesticides with short residence times. The longer a pesticide remains in the soil, the more opportunity exists for plants and other soil organisms to assimilate it. This improves control of target organisms but is an environmental problem if nontarget organisms assimilate the pesticide.

Many tests have been undertaken to determine danger to nontarget microorganisms by use of pesticides. Although exceptions can always be found, evidence suggests that most populations are not greatly affected by low pesticide concentrations. One exception is fungicides and fumigants used for control of soilborne pathogens. The population of nontarget soil microorganisms is frequently significantly reduced and the composition changed by the use of fumigants. Some changes are short-lived, whereas others continue for long periods.

Concentration of pesticides over time can also lessen the subsequent usefulness of soils for agriculture. For example, some herbicide used to control weeds in one crop can carry over, thereby lowering the yield of the next crop. Changing weather and soil moisture conditions can cause carryover to occur even if the pesticide is applied properly and in the appropriate amounts.

Organic pesticides undergo several kinds of transformations that reduce their amounts in soil. They may be volatilized and transferred to the atmosphere, leached to groundwater, or chemically altered to nontoxic products within the soil. Microbial decomposition is the major mechanism of pesticide transformation. Many observations have been made of a wide range of genera of soil fungi and bacteria that use pesticides as substrates. Microorganisms metabolize pesticides and other organic compounds because the compound supports their growth, or they may decompose a compound as part of *cometabolism.* Cometabolism is metabolism of a compound that the cell cannot use as a source of energy or as an essential nutrient. In general, changes in environmental factors that enhance the growth and proliferation of the populations responsible for pesticide transformations also enhance the breakdown of pesticides. The pesticides that are not decomposed by microorganisms are **recalcitrant** molecules.

Examples of persistent or recalcitrant pesticides are chlordane, DDT, dieldrin, and toxaphene. These pesticides are detectable in soils for many years after application. Recalcitrance is largely a function of the molecular structure of the compound. Simple changes such as removal of a chlorine atom or replacement of a chlorine atom by a hydroxyl group can change a compound from recalcitrant to easily assimilated (Figure 15–14).

The long-term use of pesticides that cannot be degraded by microorganisms may also lead to contamination. One example occurred in the apple orchards in Washington State, where lead arsenate (an insecticide) accumulated in the soils. Arsenate resembles phosphate in its immobility in soil. The accumulation reduced the yield of wheat crops planted after the apple orchards were removed.

Trace elements include plant nutrients (e.g., B, Cl, Mo, and Zn) and trace metals (e.g., Co, Ni, and Pb). They are, for the most part, naturally occurring elements usually found in soils in low concentrations ($<100$ mg kg$^{-1}$). Trace elements act differently in soil depending on whether they exist as metals or as oxyanions. For example, metals tend to be more available as pH goes down, and oxyanions tend to be less available. This difference results because the pH influences the dependent charge sites on clays and humus, converting negatively charged sites to positively charged sites through protonation (addition of $H^+$ to charged sites).

Management of trace elements includes addition if the element is an essential element in short supply and reducing concentrations or immobilization of the element if its concentration is sufficient to be toxic. Soil with excess mobile trace elements is remediated by removal of the soluble constituent. Immobilization is the preferred management technique for less mobile trace elements. For example, raising pH above neutral is often effective in reducing the mobility of many metals.

### 15.3.3  Treating Contaminated Soils

Returning a soil to its "healthy," uncontaminated condition means reducing or eliminating the impacts that interfere with the normal chemical and biological processes of uncontaminated soils. There are two stages in reclamation: removing the source of contamination and treating the problems caused by the contaminant.

Reclamation includes containment, various on-site or in situ cleanup processes, and removal of soil.

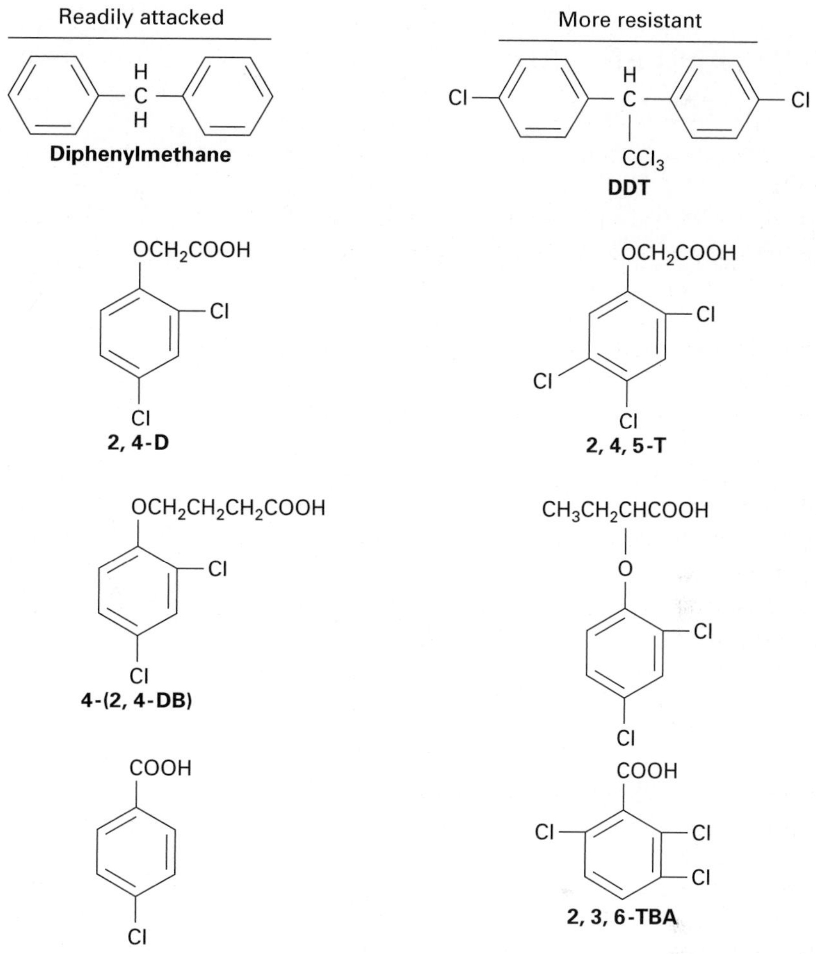

**FIGURE 15–14**   Compounds similar in structure but differing in biodegradability. (From Alexander, M. 1977. *Introduction to soil microbiology.* 2d ed. New York: John Wiley & Sons.)

Most are engineering solutions that require tens of millions of U.S. dollars per site to implement. The high cost of engineering solutions is often justified because they are frequently more rapid than biological solutions.

**Containment**   Contaminated soil may be "treated" by containing the waste so that it does not move from the contaminated site. Containment is attained by grading the site to reduce erosion, covering the waste with an impermeable cap of clay so that water cannot enter the waste and leach away the contaminants, and revegetating the site. Sometimes it is possible to pump cement or other materials into the soil under the contaminated areas to prevent the contaminants' downward movement. Such containment is very expensive.

The role of a cover cap is (1) to prevent human or animal exposure to contaminants through either blowing or erosion of contaminated soil, (2) to sustain vegetative growth on the cover, and (3) to fulfill any engineering function. A cover must control the movement of gas generated within the contaminated soil. This is particularly important in old sanitary landfills in which the decay of organic debris produces carbon dioxide and methane (which is explosive). The advantage of a soil cover is that it is relatively inexpensive, it may be installed rapidly, and contaminated material need not be excavated. Excavation leads to further contamination and greater costs.

The success of a cover material depends on its texture and mineralogy. Clay-rich material is preferred as a cap because it is very slowly permeable and can be compacted to lower even further any water and gas transmission. High shrink-swell materials are not useful because cracks enable gas to escape and water to enter. Well-constructed covers usually are

## SOIL AND ENVIRONMENT
### The MTBE Story

Methyl tertiary butyl ether (MTBE) is an additive to about 30 percent of the gasoline in the United States. It is an octane enhancer that helps to reduce carbon monoxide emissions from motor vehicles, thus reducing air pollution. MTBE smells bad and tastes bad, and it might cause cancer in humans. An unfortunate and unexpected side effect of MTBE addition to gasoline is widespread water pollution due to the high solubility of the chemical in water. MTBE enters groundwater when underground storage tanks leak and enters surface water from powered watercraft. Once MTBE gets into soil or groundwater, conventional remediation techniques have been shown to be only partly effective.

Biodegradation of MTBE in the laboratory has been clearly demonstrated, and both pure and mixed cultures of microorganisms that can degrade MTBE have been found. Soil microbiologists at the University of California–Davis (Hanson, Ackerman, and Scow, 1999) have demonstrated that a bacterial strain (strain *PM1*) originally isolated from compost uses MTBE as a sole carbon and energy source under aerobic conditions. The organism is a small gram-negative rod that requires oxygen for MTBE degradation. It is a strain of the β-1 subgroup of *Proteobacteria*. In field and laboratory tests, the organism consumes MTBE, converting it to harmless by-products. When the organism was added to MTBE in the laboratory and kept supplied with oxygen, it demonstrated its effectiveness by rapidly consuming the MTBE. Additions of cultures of the organism to groundwater at a polluted site in California have produced similar results. In this example of bioremediation, nature supplied the organism and soil scientists found it, cultured it, and are introducing it to new habitats to clean up our messes.

---

layers of fill materials. Alternating gravel layers with clay reduces the upward capillary rise of contaminated water as well as the entry of waste into contaminated ground. Soil may be mixed with up to 8 percent cement or lime to improve its usefulness as a cover material.

**In situ cleanup**   When the simple covering of a site is impractical or the contaminant must be removed from the soil, two choices are (1) in situ removal of the contaminant or (2) excavation of soil, off-site cleanup, and return of the soil to the site. Both of these are expensive, high-technology solutions.

Pumping and treating is one method of in situ cleanup of soil. Wells are drilled into soil and parent material around a contamination plume, and the contaminated groundwater is pumped from the ground, treated, and returned to the ground. If the contaminant is volatile, vacuum extraction is used. Among the other expensive alternatives are steam flooding and leaching with water, acid, organic solvents, surfactants, and chelating agents.

Another in situ cleanup method is microbiological. Microorganisms metabolize organic toxicants and mediate numerous kinds of reactions in soils. *Detoxication*, is the direct conversion of a toxic molecule to a nontoxic product without complete degradation of the original molecule. Complete conversion of the original toxic compound to simple products such as $CO_2$ and $H_2O$ is *degradation*. This statement greatly oversimplifies the process, which may include many steps and many enzymes. *Conjugation*, or *complex formation*, is a process that changes toxic compounds to nontoxic or less toxic compounds by the addition of an active group to the original compound. Active groups that are added include amino acids and methyl groups.

The advantages of microbial decomposition are as follows:

1. Low cost
2. Low energy consumption
3. Little hazard of air pollution from a "natural" process

Two major disadvantages are it is slow and favorable conditions for the microorganisms must be maintained. Twenty or more years may be necessary for microorganisms to finish decontaminating a site.

Favorable conditions for microorganisms include adequate nutrients, near-neutral pH, and sufficient water and oxygen. Nutrients can be added, the

pH can be adjusted, and the soil can be drained or irrigated as needed to provide sufficient water and oxygen. Organisms that can efficiently decompose and
decontaminate soils can be extracted and cultured to
enlarge their populations. Mixtures of these larger
populations can then be reinjected into the soil to
speed the decontamination.

Another in situ reclamation procedure is *phytoremediation.* Some plants live in soils that have concentrations of elements and compounds that are
toxic to most plants. They have evolved mechanisms
to exclude the toxic element or compound from
their cells or mechanisms to make the toxic material
ineffective if it enters the cells. Plants that accumulate
toxic materials without noticeable effect can be utilized for reclamation of contaminated soils. The
plants are grown on the contaminated site and harvested, thus removing the contaminant. This slow but
inexpensive reclamation technique is *phyto-* (plant)
remediation.

Contaminants may be immobilized on-site by
adding chemicals that enhance adsorption, precipitation, or complexation of the contaminants. Liming
of acidic soils increases adsorption and precipitation
of toxic metals and enhances complexation of metals
by soil organic matter. Adsorption of some oxyanions, such as arsenic ($AsO_4$), molybdenum ($MoO_4$),
chromium ($CrO_4$), and selenium ($SeO_4$), is reduced
with increasing pH, which complicates reclamation.

**Removal**    Contaminated soil can be excavated and
reburied, but this approach is expensive. And unless
the soil can be decontaminated off-site, it will simply
become a disposal problem at another site. Reburial
is typically in special landfills where leaching and
drainage are carefully controlled.

## 15.4    DEGRADATION CONTROL

### 15.4.1    Erosion Control

**Water**    Three kinds of erosion control for water
are cover, engineering or mechanical, and tillage
practices. Maintenance and enhancement of soil
cover is by far the best method of erosion control.
Cover includes a canopy effect, a mulch effect, and
an in-soil effect.

The canopy effect reduces rainfall energy
reaching the soil surface by interception of raindrops

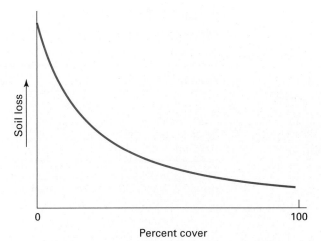

**FIGURE 15–15**  Small additions of cover
quickly reduce soil loss, as shown in the
curve. As the cover on the soil surface approaches 100 percent, the effectiveness of
each additional unit of cover decreases.
Note that even at 100 percent cover, some
soil erosion may occur.

(Figure 15–15). This effect was shown many years ago
by Norman Hudson, working in East Africa. Two sets
of field plots were constructed. In one, thin window
screen was suspended a centimeter above the soil,
and in the other there was no window screen. Both
sets of plots were bare, and the soil was exposed directly to natural rainfall for 10 years. The amount of
soil erosion from the window-screen-covered plot was
9.4 Mg/ha in 10 years; erosion from the uncovered
plot was 1265.7 Mg/ha in 10 years. The amount of
rainfall reaching the soil surface was the same on
both plots. The window screen effectively reduced
the rainfall energy reaching the plots, just as soil
cover does.

Canopy characteristics such as leaf size, orientation, inclination, and surface texture have been
considered important determining variables. The reduced volume and number of drops reaching the soil
surface is somewhat offset by the increased drop size.
This may be a serious problem in tall vegetation,
where the larger drops fall sufficient distance to regain their terminal velocity and high **kinetic energy.**
Trees control erosion rather poorly if they are widely
spaced and there is no other ground cover. Trees with
a ground cover of litter or grass provide effective erosion control (Table 15–2).

## TABLE 15–2

Magnitudes of Annual Soil Erosion Losses in Four Localities of West Africa (metric tons/ha)

| Location | Forestland | Cultivated land | Bare soil |
|----------|-----------|-----------------|-----------|
| Ouagadougou, Upper Volta | 0.1 | 0.6–8.0 | 10–20 |
| Sefa, Senegal | 0.2 | 7.3 | 21 |
| Bouake, Senegal | 0.1 | 0.1–26 | 18–30 |
| Abidjan, Ivory Coast | 0.3 | 0.1–90 | 108–170 |

These figures illustrate that forestland with an adequate ground cover reduces erosion compared with cultivated land, which, in turn, has less erosion than does bare soil. What are some reasons for the range in erosion amounts for cultivated land and bare soil?

Source: Data from P. A. Sanchez. (1976). *Properties and management of soil in the tropics.* New York: Wiley-Interscience.

**Mulch** is any material in contact with the soil surface. Conventionally, in farming systems, plant residue (corn stubble, wheat straw) left on the soil surface after harvest is mulch. However, a wide variety of materials are used as mulches, including gravel, wood chips, limestone, plastic, pine needles, and oak leaves. Preferred mulching materials are abundant, do not interfere with farming operations, and do not harbor disease organisms, weed seeds, or insects. In landscaping applications, preferred materials have all of these attributes and must look nice (Figure 15–16). Cover in contact with the soil surface is the most effective vegetative method of erosion control. The mulch intercepts raindrop energy, and in addition there is a filtering effect and reduction in overland flow velocity that reduces transport capacity. Velocity is reduced by increasing the tortuosity of the water flow and the roughness of the ground surface and by creating shallow pools into which flowing water deposits entrained sediment. To be effective, mulches must cover > 70 percent of the ground surface. Straw application rates of 5 Mg ha$^{-1}$ are sufficient to provide adequate soil protection.

In addition to reducing erosion by protecting the soil surface from raindrop impact, mulches reduce surface crusting, evaporative water losses, soil temperatures, and weed growth, and they enhance infiltration, soil water storage, and crop yield.

The in-soil effect is produced by living plants or incorporated biological material. Living plants, with their roots in the soil, bind soil particles together physically, and root exudates promote aggregation. After death, root decay adds humus to the soil, promoting aggregate stability directly, and stimulates production of microbial polysaccharides. Roots also create pores, which often are large and promote the rapid entry of water into soils, thus reducing overland flow. Living plants use soil water and help maintain infiltration rates.

Farmers and other conservationists have developed many strategies for reducing erosion by maintaining plant cover. Examples include rotations and **strip-cropping,** in which bare fallow or a row crop such as corn or soybeans is grown in alternating strips with grass or a legume hay crop (Figure 15–17). The grass or hay crop reduces soil loss by trapping eroded sediment as it moves downhill. Strip-cropping around the slope, on the contour, is particularly helpful because the furrows retain water and slow its flow down the slope. **Cover crops** are grown during the off-season or as a ground cover under trees in orchards or between vines in vineyards (Figure 15–18). If the cover crops include legumes, an added benefit is nitrogen fixed symbiotically by the plants and their bacterial symbionts. Cover crops grown during the normal growing season compete with the crops for water and nutrients and must be managed appropriately.

Farmers maintain soil cover by planting perennial crops such as grasses, by mulching between the crop rows, or by leaving crop residue on the field surface rather than plowing it into the ground. Erosion control along highways is often more difficult than control on farmland because road cuts are often deep and steep, and plant materials used for erosion control are difficult to establish in C horizons (Figure 15–19).

**FIGURE 15–16** Wood chips used as mulch in landscaping beautify, reduce evaporation, keep the soil cool, and protect the soil from wind and water erosion. (Photo: USDA.)

**FIGURE 15–17** Strip-cropping near Loganville, Wisconsin. (Photo: USDA/NRCS.)

Erosion control materials have been developed for special situations such as roadside stabilization and construction sites. These materials fulfill the basic need of soil cover, but they are often quite different from natural plants. Until plants are established, slopes can be stabilized with netting made of jute (a natural plant fiber) (Figure 15–20) or plastic. Asphalt, latex, and plastic sprays are applied to road cuts to stabilize or waterproof the soil. Often they are applied with fiber, seed, and fertilizer. The fiber (straw, bits of paper, wood chips, or shredded wood fiber) acts as a mulch to retain soil moisture until the

**FIGURE 15–18** Photograph taken in April of a newly established vineyard in the steep hills overlooking the Napa Valley in California. The clover and grass cover crop in the foreground protects the soil between the newly planted vines (inside the plastic tubes) from erosion. In the background, a grass cover crop is planted between older, established vines.

**FIGURE 15–19** Soil along this highway was left unprotected after the highway was constructed, and it is severely eroded. The productive capacity of the soil is being lost, and the eroded soil is expensive to remove from the highway. (Photo: USDA.)

seed germinates, and the chemical sticks the fiber to the side of the hill. The plants, usually grass, are the long-term soil conservation treatment. The technique of spraying this slurry of fiber, seed, fertilizer, and chemical on the roadside is called **hydromulching** (Figure 15–21).

Mechanical conservation practices such as conservation terraces and contour cropping are commonly used to reduce slope lengths and steepness. Conservation **terraces** (Figure 15–22) have many different names, depending on how and where they are built. They are sometimes called **graded terraces** because the ground surface is shaped or graded to control the direction and velocity of water flow. *Infiltration terraces,* or level terraces, are created by grading a slope into several level segments separated by short, steep ridges. Water is ponded on the level segments, where it infiltrates into the soil. These terraces have the combined advantage of reducing soil erosion and conserving water. *Back-sloped* or *reverse-sloped terraces* are graded such that the slope of the terrace is opposite the general slope of the ground. Thus, water is prevented from running off the outer edge of the terrace. Conservation terraces do not look like the terraces on steep mountain sides in Southeast Asia, built over many years with great expense of human labor. Those in Asia are built to conserve soil and to create new places to grow crops where land is in very short supply.

In all cases, terraces reduce the average land slope. This may be done by clever plowing across the slope for years, but more frequently, in the United States, conservation terraces are designed by engineers and built with earth-moving equipment. Figure 15–23 shows the profile of a farm field before and after terracing. The steep upper part of the field was cut, and the soil was moved to the lower part, resulting in a lower average slope.

Terraces must be designed and maintained carefully. Most accumulate water, and this accumulated water must be safely removed from the terraces either by drains and underground pipes or by directing the water along the terrace contour to a grass waterway. In either case, the water is discharged at low velocity into a stream, where its remaining energy cannot erode the soil. If the drains are poorly designed—for example, if they are not placed at the lowest point on the terrace, or the terrace backslopes are not well vegetated and carefully maintained—the accumulated

water will overflow the terrace and cause severe gully erosion. Terraces are often combined with grass waterways to remove excess water. A **grass waterway** is often a natural drainage **channel,** but it may be an engineered channel that is planted and left in grass permanently. The grass prevents erosion in the channel when water is flowing through it. Design and maintenance of a grass waterway require that it be sufficiently wide to avoid water overflow. If the water volume exceeds the grass width, water can seep into the soil at the sides and beneath the grass, causing severe erosion.

Tillage is the practice of mixing or inverting the surface soil to prepare a seedbed for planting or to mechanically remove weeds. Tillage includes plowing and various cultivation activities that break large clods and mix soil to uproot weeds. Conventional cultivation typically leaves the soil surface free of vegetation and exposed to the full energy of water drops or wind. **Conservation tillage** broadly describes farming systems that minimize **tillage** and increase residual soil cover, in contrast to *clean tillage,* the object of which is to incorporate all residues. **Minimum tillage** and **no-tillage** are examples of conservation tillage. Conservation tillage ranges from eliminating one or more tillage operations during the growing season to eliminating all plowing and most other tillage operations. No-till management leaves the soil undisturbed from harvest to planting except for nutrient injection. Seeds are planted or drilled in narrow seedbeds or slots created by tools that leave surface residue mostly undisturbed. Weed control is accomplished primarily with herbicides. Minimum tillage is like no-tillage except that some forms of tillage are practiced.

No-till farming restricts the number of equipment trips over the field to only those needed for planting, spraying, and harvesting. After harvest, the unharvested plant parts (crop residue) are left on the soil surface as a mulch, and the next crop is planted directly into the residue. The soil surface is seldom bare, and erosion is greatly reduced.

Crop residue management through conservation tillage is the preferred form of conservation practiced by the majority of farmers across the United States. In addition to conserving soil, benefits include reduced fuel use, prolonged life of equipment because of fewer trips across each field, increased capture of runoff due to maintenance of infiltration rates and an increase in macropores, and

**FIGURE 15–20**  Jute netting temporarily protects the soil on the slope above the channel until vegetation can be established for long-term erosion control. The rocks also protect the soil from scour by water flowing in the channel. (Photo: USDA.)

**FIGURE 15–21**  This hydromulcher is spraying a mixture of seed, fertilizer, and chemical mulch on the road cut. The chemical mulch holds the seed in place and protects the slope from erosion until the grass germinates and becomes established. (Photo: B. Kay.)

increased organic matter and nitrogen. Actual fuel use depends on size of equipment, number of operations, and soil type. Planting directly in untilled soil requires more energy than planting into tilled soil, but savings come from fewer total operations due mostly to fewer operations to control weeds. Infiltration rates in wheat fields that had been in no-till and conventional tillage for 4 years in California were five times as high in a no-till field compared with a conventionally tilled field during sprinkler application of water at 5 cm hr$^{-1}$ intensity. Even at 10 cm hr$^{-1}$, twice as much water infiltrated (Pettygrove et al., 1995).

Conservation tillage requires new thinking about old practices. For example, no-till requires that

**FIGURE 15–22**   This aerial view shows the extensive use of conservation terraces in the Midwest. Note the gully erosion in the center foreground. Improper disposal of water can cause erosion control practices to fail. (Photo: W. E. Wildman.)

**FIGURE 15–23**   Typical cross sections illustrating how the land surface slope is changed when a terrace is built. Slopes are exaggerated. According to Beasley (1963), maximum slopes of terraces range from 2 percent for short (less than 30 m) to 0.3 percent for long (greater than 366 m) terraces.

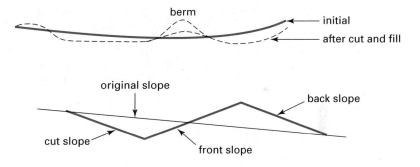

crop stubble (straw) not be baled or heavily grazed by animals. It also requires that residues be uniformly distributed across a field rather than left in rows. Lack of tillage may have unanticipated effects. In some studies, weed populations changed from annual species to perennials because buried annual weed seeds were no longer returned to the soil surface by tillage operations.

Conservation tillage often includes the use of cover crops that provide habitat and food for wildlife, especially birds. Use of legume cover crops is an additional nitrogen source for the crop. Thus there can be many potential benefits to conservation tillage, but there can be disadvantages, too. First, special machinery is required to plant into residue. Second, the

tillage control of weeds is replaced by additional chemical treatments. Third, additional nitrogen fertilizer is sometimes required, depending on the C/N ratio of the residue. In some areas of the United States, farmers have found that conservation tillage reduces yields. Disease and insect damage to crops is sometimes higher with conservation tillage. In many situations it is an appropriate conservation technique, but generalization is difficult; farmers need to weigh the benefits, risks, and disadvantages for their individual situations.

**Wind**   The control of wind erosion is primarily by means of windbreaks and maintenance of soil cover.

**FIGURE 15–24** This aerial photograph shows numerous windbreaks. Each reduces wind erosion and crop damage. (Photo: USDA.)

Other methods include the control of soil surface roughness and moisture. Windbreaks reduce the speed and turbulence of wind near the ground. Trees and shrubs are common windbreaks because, in addition to reducing soil erosion, they produce food, firewood, and wildlife habitats (Figure 15–24). They also retain snow during the winter and early spring, and the melting snow recharges the soil moisture storage. Sometimes wooden, plastic, or metal fencing is used as a windbreak.

Figure 15–25 illustrates some important windbreak concepts. *H* is the height of the windbreak. The sheltered distance on the downwind side of the windbreak is a multiple of *H,* often up to 20 *H. W* is the width of the windbreak. To a certain degree, the length of the sheltered distance is a function of *W,* but as the windbreak becomes wider, more land must be taken from production. Few windbreaks are solid walls, because a porous windbreak reduces wind turbulence on the downwind side more than a dense windbreak does.

Windbreaks have a few disadvantages. They occupy land and lower crop productivity nearby because they shade the crop and compete for water. However, these effects are usually more than compensated for by the higher yield in the rest of the field.

Increasing the field roughness by cultivation also decreases wind erosion, which is particularly useful if cultivation is done perpendicular to the prevailing wind direction. Mulching and minimum tillage are also useful techniques for reducing wind erosion. Mulching roughens the surface, minimizes the amount of bare soil available for transport, and helps retain soil moisture. The greater soil moisture increases particle cohesion and weight so that stronger winds are required to initiate movement.

Wind erosion can be diminished by not plowing or cultivating after harvest. Crop stubble left on the ground surface increases the surface roughness, and the plants' old root systems help hold the soil in place. An added benefit in some regions is the retention of snow and increased soil moisture in the spring.

**Mass wasting**   There is little one can do to stop a landslide once it has started, but careful planning can reduce the likelihood of mass wasting. Altering the natural slope by undercutting during road building and home construction leaves a steep, often unstable hillside that does not support the soil on the slope above the cut. Minimizing the size of cuts or eliminating them entirely helps prevent mass wasting.

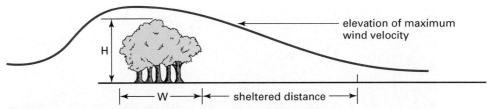

elevation of maximum wind velocity

H

W

sheltered distance

**FIGURE 15–25** Windbreaks reduce the wind speed near the ground. The height (H) and width (W) determine the sheltered distance. The wind is blowing from left to right in this figure.

**FIGURE 15–26** Soil compaction in this orchard was caused by tractor traffic when the soil was wet. Where on the curve in Figure 15–12 do you think the soil moisture content was when the tractor tires made these ruts? (Photo: W. E. Wildman.)

Most control efforts aim at removing water from potential slip areas by means of diversion ditches, plants, and perforated pipes. Diversion ditches placed at the top of the slope or along the slope lessen the amount of water flowing into the soil. In addition, perforated pipes placed horizontally into slopes can improve subsurface drainage.

## 15.4.2 Control of Physical Degradation

Soil compaction and crusting can be ameliorated through soil management. The most common soil management technique is cultivation to physically disrupt compacted soil or surface crusts. A better management technique is to reduce the load applied to the soil by using smaller equipment, spreading the load over larger tires or treads, and making fewer trips across each field. These techniques lower the total compactive effort that the soil must support. One management technique, not in widespread use, is the use of permanent traffic lanes. Equipment wheels are restricted to specific paths in the field. The remainder of the field remains untrafficked.

The timing of farming operations is often critical, and thus the soil may have to be plowed or driven on when it is too wet to support the load without compaction (Figure 15–26). If compaction has occurred, it can be reduced temporarily through ripping (Figure 15–27), deep plowing, and slip plowing (Figure 15–28). These mechanical treatments break up or mix the soil below the depth of normal tillage operations. They are necessary because there is little natural reduction in high soil densities from either freezing and thawing or shrinking and swelling.

**FIGURE 15–27**  These ripper shanks are pulled through the soil to break cemented pans. The person is about 2 m tall. (Photo: E. Ciolkosz.)

**FIGURE 15–28**  This giant slip plow mixes soil horizons as it is pulled through the soil. Soil is lifted up the inclined blade to the top, where it falls back into the silt made by the plow. This is an effective tool for destroying stratification. (Photo: W. E. Wildman.)

Mixing the soil surface by means of light cultivation helps break up surface crusting, and adding gypsum also helps improve infiltration by reducing the dispersion of clays. Mulches curtail crusting by weakening the effect of raindrops or sprinkler irrigation drops on the soil surface.

> Three methods to lessen compaction are (1) lowering the machinery's weight, (2) decreasing the number of trips the machinery makes, or (3) keeping off the soil when it is wet, by careful timing of field operations.

## 15.5 SPECIAL PROBLEMS

### 15.5.1 Desertification

Many African and Asian nations have experienced a measurable expansion of deserts. The causes of the change from arid but useful agricultural land to desert are not known with certainty, but the results are obvious. The United Nations reported in 1984 that nearly 3.5 billion ha of the world's range, rain-fed, and irrigated cropland is affected by desertification.

As the desert expands, it covers grazing land and farmland with sand. Elsewhere, the removal of vegetation greatly accelerates erosion, resulting in lowered crop productivity. The people, short of food, migrate to other parts of their country or to a neighboring country, putting additional pressures on medical, educational, and other services. The result has been catastrophic for many nations.

Some scientists suggest that desertification is a natural process brought about by climatic change. If this is the case, no human intervention will be useful. Other scientists argue that desertification is caused, at least in part, by poor management of the natural resources, including soils. Overgrazing, overcultivation, poor irrigation practices, and deforestation have been blamed for exacerbating the problem. In addition, both dry animal manure and almost all types of trees and shrubs are burned as fuel. This use of animal manure as fuel removes the farmer's one source of fertilizer, further diminishing plant productivity. Burning trees, litter, and shrubs directly removes plant cover that stabilizes the soil against erosion.

Attempts are under way in some African countries to restabilize sand dunes with vegetation, with some success, but the problem is large. Where poor management has accelerated desertification, the lesson to be learned is that resources have limits; once they are exploited beyond those limits, great problems will ensue.

### 15.5.2 Organic Soil Degradation

Organic soils (Histosols) have several management problems, at least two of which are peculiar to organic soils: wind erosion and microbial decomposition.

Because organic soils have low bulk densities—in the range of 0.1 to 0.3 $Mg\ m^{-3}$—once dry, they are highly susceptible to wind erosion.

The topography where organic soils are found is normally level, providing long open areas over which winds travel unimpeded. As already shown in Section 15.1.2, as the length of uninterrupted distance over which wind flows increases, wind erosion increases. In California, more than 150,000 ha of organic soils in the Sacramento–San Joaquin Delta region have been farmed for more than a century. Wind erosion was a serious on-site problem because it reduced the thickness of farmable soil. Off-site, the blowing peat dust reduced visibility in nearby cities and coated cars, laundry, and people downwind. One solution was to grow windbreaks of grain in between rows of asparagus. This technique reduced both the loss of Histosols and the poor visibility caused by the blowing peat dust. Careful water management, to maintain sufficient moisture in the peat, can also help make it less susceptible to wind erosion.

There are solutions for wind erosion, but there is no reasonable solution for microbial decomposition, a problem unique to organic soils when they are drained. Organic soils exist because of the slow accumulation of plant remains in anaerobic conditions. Lowering the water table and aerating organic soils cause soil microorganisms to convert the organic matter into energy, $CO_2$, and $H_2O$. The result is a rapid decrease in Histosol thickness (Figure 15–29). One partial solution is to flood the Histosols during periods when no crops are planted, slowing the rate of microbial decomposition by excluding $O_2$.

**FIGURE 15–29**  The floor of this building was at the ground surface seventy years ago. The organic soil has oxidized and been eroded by wind, leaving the floor more than three meters above the present ground surface. (Photo: E. L. Begg.)

## 15.6  SUMMARY

Erosion by water, wind, and gravity removes valuable topsoil, with its organic matter and nutrients, and deposits it as a pollutant. The processes of water and wind erosion are similar. Particles must be detached from the soil mass and transported from the site of detachment.

In water erosion, raindrops do most of the detachment, and water flowing over the soil surface does most of the transportation. Wind does both the detachment and transport in wind erosion. Gravity moves blocks of soil instantaneously as landslides or slowly as soil creep. Water assists gravity by lubricating the slip plane, adding mass to the soil, and weakening the strength of the soil.

Soil conservation interrupts the forces moving soil. Cover on the soil surface is the most effective method of erosion control. Vegetation or mulch on the soil surface intercepts raindrops, cuts the velocity of overland flow, and stops the wind from contacting the soil surface. Windbreaks are a major method of wind erosion control.

Soil compaction and surface crusting are physical soil management problems. Compaction occurs when the load on the soil exceeds the soil's strength and the particles rearrange themselves into a more tightly packed arrangement. The higher density and strength of the compacted soil reduce (1) water entry and flow through the soil, (2) the rate of water and air exchange, and (3) the rate and amount of root penetration.

Crusting is a surface phenomenon that includes compaction by raindrops or sprinkler drops, slaking, sorting of particles, and washing of dispersed clays into pore spaces. Crusting reduces seedling emergence and the penetration of water into soils. The most common methods of control are using mulch to alleviate the water drop impact and using gypsum to reduce clay dispersion.

Desertification is a special problem in arid and semiarid lands that lowers soil productivity. It is accelerated by mismanagement of soils and vegetation.

To use organic soils for agriculture, they must be drained, but when drained they decompose rapidly. Wind erosion can be reduced by proper management, but subsidence due to microbial decomposition can be diminished only by reflooding during fallow periods.

## QUESTIONS

1. How do sheet, rill, and gully erosion differ?
2. What are detachment and transport processes in soil erosion?
3. What are the three processes of wind erosion?
4. Explain each of the factors in the Universal Soil Loss Equation.
5. Compare the Universal Soil Loss Equation with the wind erosion equation. Why are the terms different?
6. What is mass wasting?
7. How does water influence mass wasting and water and wind erosion?
8. What are the differences between soil compaction and soil crusting?
9. What are the major factors affecting soil compaction and crusting?

10. Why is soil crusting both a physical and a chemical problem?
11. Why do organic soils "subside"?
12. Why is it undesirable to eliminate these degradation processes entirely?

## FURTHER READING

Beasley, R. P. 1963. A New Method of Terracing. *Missouri Agricultural Experiment Station Bulletin* 699. 22p.

Blevins, R. L., and W. W. Frye. 1993. Conservation tillage: An ecological approach to soil management. *Advances in Agronomy* 51:33–78.

Hanson, J. R., C. E. Ackerman, and K. M. Scow. 1999. Biodegradation of methyl tert-butyl ether by a bacterial pure culture. *Applied and Environmental Microbiology* 65:4788–4792.

Kimberlin, L. W., A. L. Hidlebaugh, and A. R. Grunewald. 1977. The potential wind erosion problem in the United States. *Transactions American Society of Agricultural Engineers* 20:873–879.

Laflen, J. M., L. J. Lane, and G. R. Foster. 1991. WEPP: A new generation of erosion prediction technology. *Journal of Soil and Water Conservation* 46(1):34–38.

Logan, T. J. 1992. Reclamation of chemically degraded soils. *Advances in Soil Science* 17:13–35.

McIntyre, D. S. 1958. Permeability measurements of soil crust formed by raindrop impact. *Soil Science* 85:185–189.

Morgan, R. P. C. 1995. *Soil erosion and conservation,* 2d ed. London: Longman Group.

Pettygrove, G. S., M. J. Smith, T. E. Kearney, and L. F. Jackson. 1995. *No-till wheat and barley production in California.* Sacramento, Calif.: California Energy Commission.

Renard, K. G., G. R. Foster, G. A. Weesies, D. K. McCool and D. C. Yoder, coordinators. 1997. *Predicting soil erosion by water: A guide to conservation planning with the Revised Universal Soil Loss Equation (RUSLE).* U.S. Department of Agriculture Handbook 703. Washington, D.C.: U.S. Department of Agriculture.

Renard, K. G., G. R. Foster, G. A. Weesies, and J. P. Porter. 1991. RUSLE: Revised Universal Soil Loss Equation. *Journal of Soil and Water Conservation* 46(1):30–33.

Renard, K. G., G. R. Foster, D. C. Yoder, and D. K. McCool. 1994. RUSLE revisited: Status, questions, answers, and the future. *Journal of Soil and Water Conservation* 49(3):213–220.

Schwab, G. O., R. K. Frevert, T. W. Edminster, and K. K. Barnes. 1981. *Soil and water conservation engineering.* New York: Wiley.

Stallings, J. H., 1957. *Soil conservation.* Englewood Cliffs, N.J.: Prentice Hall.

Wischmeier, W. H., C. B. Johnson, and B. V. Cross. 1971. A soil erodibility nomograph for farmland and construction sites. *Journal of Soil and Water Conservation* 26:189–193.

Wischmeier, W. H., and D. D. Smith. 1978. *Predicting rainfall erosion losses; A guide to conservation planning.* U.S. Department of Agriculture Handbook 537. Washington, D.C.: U.S. Department of Agriculture.

## WEB RESOURCES

The National Soil Erosion Research Laboratory has a Web site with information about the national soil conservation and erosion prediction efforts. Included are links to sites for downloading software. The site is at http://topsoil.nserl.purdue.edu/nserlweb

Another site specifically for more information on RUSLE is at http://www.sedlab.olemiss.edu/rusle

Wind erosion information can be found at http://topsoil.nserl.purdue.edu/nserlweb/weppmain/jhtml/weps.html

# 16

# NONAGRICULTURAL USES OF SOILS

## OVERVIEW

Soil disposal is the final step in the on-site purification of household sewage. The soil filters and treats the effluent as it passes through the soil. Soil microorganisms use the organic matter and nutrients in waste and deactivate disease organisms, purifying the effluent before it reaches the groundwater or surface water supplies. The success, capacity, and longevity of a septic system primarily depend on the soil in which it is placed. Soil properties also determine the success or failure of sludge and effluent disposal on land. The treatment is successful when the waste is safely disposed of or utilized without the contamination of surface water or groundwater.

Shear strength is the most important measure of soil strength. It determines, in part, whether a soil will erode by water, fail catastrophically under the pull of gravity, or compact under a load.

Sometimes there is not enough soil or it is the "wrong" soil for our needs. In some special cases, soils can be made. These "soils" are not what we have described throughout the book; they have no pedogenic horizons formed from the five soil-forming factors. Examples are polders and mine soils.

When the soil is not quite right for our intended use, such as for special lawn areas, golf courses, and horticultural areas, one is created by blending different materials. In greenhouse culture, for example, soil materials need to be well drained and sterilized so that seedlings can get a good start.

# 16.1 WASTE DISPOSAL ON SOILS

## 16.1.1 On-Site Disposal

Many rural and suburban homes are not served by central wastewater (sewage) treatment plants. Waste from sinks, showers, and toilets thus must be treated where it is produced. This is **on-site disposal.** The most common on-site disposal is through a septic tank and leach line into the soil (Figure 16–1).

**Septic system suitability** Principles that apply to household waste disposal also apply to disposal from feedlots and agricultural and urban industries. The purpose of waste treatment is to dispose of wastewater without polluting surface water or groundwater. Both must be protected from the nutrients, organic matter, toxins, and disease organisms present in waste. The soil acts as the final filter before wastewater enters surface water or groundwater. The soil's suitability for a septic system is determined by the soil's ability to treat waste, as well as the difficulty and expense of installing the system.

The septic tank accumulates and retains waste from the house, allowing most of the solids to settle to the bottom. Up to 60 percent of the biological oxygen demand and 70 percent of the suspended solids are removed in the septic tank. The microorganisms in the tank build a large anaerobic population that decomposes some of the organic matter, producing $CO_2$, $CH_4$, and $NH_4^+$. The remaining (suspended) solids and liquid, the **effluent,** enter the soil through leach lines. *Leach lines* are pipes, with holes along either side, that are buried in the soil. The effluent is pumped or flows by gravity into the leach lines. Effluent flows from the holes in the leach lines into the soil.

The **biological oxygen demand (BOD)** is a measure of the amount of microbially decomposable material in the waste, and the amount of oxygen the microorganisms require is one measure of waste quality. If waste with a high BOD is put into water, the microorganisms will use the dissolved oxygen to decompose it, leaving little or no oxygen in the water for desirable fish and plant species.

A septic system's volume capacity is the amount of effluent that can be treated by the soil each day. This capacity depends mainly on the soil's saturated hydraulic conductivity (see Chapter 5). The soil's texture and structure determine its saturated hydraulic conductivity and capacity. Figure 16–2 illustrates the

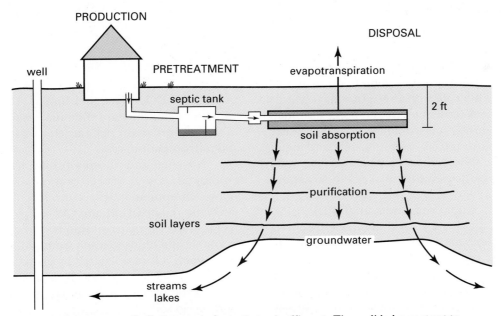

FIGURE 16–1 Soil disposal of septic tank effluent. The soil is important to purifying the effluent before it reaches the water table. (From J. Bouma, W. A. Ziebell, W. G. Walker, P. G. Olcott, E. McCoy, and F. D. Hole 1972. Soil absorption of septic tank effluent. *Information Circular 20.* Reproduced with permission of the University of Wisconsin–Extension, Geological and Natural History Survey.)

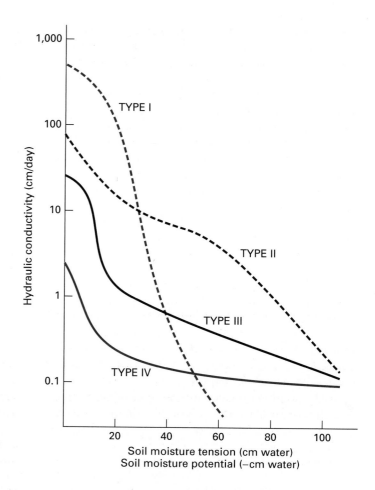

**FIGURE 16–2** Hydraulic conductivity curves for four major types of soil. (From J. Bouma. 1974. *Innovative on-site soil disposal and treatment systems for septic tank effluent.* National Home Sewage Symposium, American Society of Agricultural Engineers, Chicago.)

effect of soil texture and water content on hydraulic conductivity. Sandy soils with large pores have a large capacity to accept wastewater. Large pores also help maintain an aerobic environment good for waste treatment. The drawback to sandy soils and rapid flow rates is the short **contact time** between the waste and the solid particles, which makes the treatment less effective.

Medium-textured soils, types II and III in Figure 16–2, have moderate water flow rates and contact times. Type IV soils have slow hydraulic conductivities that limit their capacity, but they have long contact times.

For a system to be successful, the waste must be treated sufficiently to lower the BOD, reduce the suspended solids content, retain nutrients such as nitrogen and phosphorus, and deactivate disease organisms. This is done through the interaction of the waste with the soil surfaces and the microorganisms that live in the soil. Organic matter is decomposed by the native soil microorganisms, and under aerobic

conditions these microorganisms convert the organic matter into $CO_2$, $H_2O$, and humus. The pathogens (disease organisms) are then **inactivated** by the lack of $O_2$, adsorption on clay particles, or attack by native soil microorganisms.

Nitrogen and phosphorus are used by soil microorganisms and plants growing over the leach lines. Ammonium ($NH_4^+$) produced in the septic tank is abundant in effluent and is rapidly oxidized to $NO_3^-$ under aerobic conditions. Nitrate ion ($NO_3^-$) is soluble and tends to leach from soil. In aerobic environments it is rapidly used by plants and microbes. Phosphate ($PO_4$)$^{3-}$ is used by plants, but most of it forms insoluble or slowly soluble compounds that are retained in soil.

In addition to insoluble compounds precipitating in soils, cations in the waste are held on the soil's cation exchange complex. In general, the amount and kind of chemical transformation occurring in the soil depend on the amount and kind of clay. Soils with moderate amounts of smectite clay have high chemical retention capacities and moderate waste intake capacities. Soils too high in clay have high capacities to clean waste but a slow intake rate and low volume capacity.

Suspended solids and microorganisms are filtered from the waste. The wastewater flows along a tortuous path around the soil particles, which act like a sieve. The suspended particles and organisms are retained while the water passes through the soil.

Soil properties influence the cost of installing a septic system, because the soil determines the length of leach line necessary for adequate treatment of the wastewater.

> Soil properties determine
> - Capacity to accept waste
> - Success in treating waste
> - Cost of septic system

**The perc test**   The area required for a septic leach line is calculated according to the rate at which the water flows through the saturated soil. The rate is measured by a simple percolation test, or **perc test,** a standard test required by many states before a septic system may be installed. The perc test is run at the site where the leach field is to be installed and at the soil depth at which effluent will enter the soil (Figure 16–3).

**Septic system failure**   A septic system fails when its water intake capacity is too low for the needed volume or when it no longer cleans the waste. Clogging is the main cause of failure, and clay soils are particularly susceptible. In addition to reduced flow rates and therefore the system's decreased capacity, clogging can allow untreated effluent to reach the soil surface. This is both a health hazard and unsightly. **Clogging** is the biological, chemical, or physical plug-ging of soil pores so that water cannot move through them at the desired rate.

Biological clogging is caused by the production of microbial biomass and slime under prolonged anaerobic conditions.

Chemical clogging occurs in three ways:

1. Under anaerobic conditions, iron sulfide (FeS), a black precipitate, blocks the pores.
2. Dispersed clay moves into and blocks the pores.
3. Swelling reduces the pore size by pushing mineral particles into the pore spaces.

Dispersion and swelling are accelerated by the presence of sodium in the water. Sodium is added to water in soaps, detergents, and home water softeners that require large amounts of salt (NaCl) to recharge the exchange system.

Physical clogging occurs during the construction of septic leach lines. Compaction by construction equipment reduces the pore size, and the smearing of clays seals over the pores just as butter covers the holes in bread. Finally, vibration during construction causes particles to move and repack into a denser arrangement.

Clogging can be reduced, avoided, or eliminated by taking care during construction and by keeping the soil aerobic. Avoiding clogging is most easily accomplished by not overloading the system. **Dosing,** the intermittent addition of waste to the leach lines, is another method for maintaining aerobic conditions.

**Soil properties**   Published by the U.S. Department of Agriculture (USDA), the *USDA National Soils Handbook* lists eight different soil properties used to determine septic tank absorption field suitability (see Table 16–1). Most of the properties (1, 3, 4, 5, 6, and 7) relate directly or indirectly to the rate of water movement in

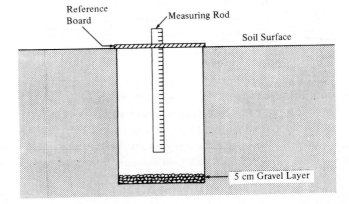

**FIGURE 16–3**   The percolation test hole is excavated to the depth of the proposed leach line, and the rate of water entry into the saturated soil is measured.

**TABLE 16–1**

Suitability for Septic Tank Absorption Fields

| Property | Limits | | | Restrictive feature |
|---|---|---|---|---|
| | Slight | Moderate | Severe | |
| 1. USDA texture | — | — | Ice | Permafrost |
| 2. Flooding | None, protected | Rare | Common | Floods |
| 3. Depth to bedrock (cm) | >183 | 100–183 | <100 | Shallow soil |
| 4. Depth to cemented pan (cm) | >183 | 100–183 | <100 | Cemented pan |
| 5. Depth to high water table (cm) | >183 | 120–183 | <120 | Wetness |
| 6. Permeability (mm/hr) (60–183cm) | 50–150 | 15–50 | <15 | Percs slowly |
| 7. Permeability (mm/hr) all layers > 60 cm | — | — | >150 | Poor filter |
| 8. Slope (%) | 0–8 | 8–15 | >15 | Steepness |
| 9. Fraction >75 mm (wt. %) | <25 | 25–50 | >50 | Large stones |

Source: Adapted from *USDA National Soils Handbook,* Part II, Table 403.1(a). Soil Survey Staff, Natural Resources Conservation Service. 1999. National Soil Survey Handbook. Title 430-VI Washington, D.C. United States Government Printing Office.

soil. A too-rapid movement of fluid results in poor treatment of waste, which is as serious as a low water intake capacity. Flooding (2) increases the chance of untreated or inadequately treated waste contacting water. Slope (8) relates to the difficulty of installation, the possibility of a too-rapid flow of effluent through the soil, and the possibility of untreated effluent reaching the soil surface. The coarse fraction (9) interferes with the system's installation.

## 16.1.2  Off-Site Disposal

**Off-site disposal** occurs in a central treatment plant away from the site where the waste is produced. Three kinds of household waste are treated off-site: effluent, sludge, and solid waste. In addition to household waste, various industrial, agricultural, energy-production, and mining wastes are generally treated off-site. These wastes will not be considered here.

The quality of the sewage determines how it is treated by soil and how it affects soil and water quality. Sewage quality depends on what goes into the treatment plant and how it is treated. Sewage may go through primary, secondary, or advanced wastewater treatment (Figure 16–4). Each type of treatment suc-

cessively improves the quality of the liquid discharge (the effluent) from the treatment plant. At the same time, the treatment reduces the effluent's usefulness as a fertilizer.

As the effluent becomes cleaner, the **sludge** (biosolids) becomes dirtier. The chemical content of the sewage entering the plant particularly affects the quality of the biosolids leaving the plant. Sewage from industrial areas may be heavily contaminated with metals such as zinc (Zn), cadmium (Cd), and lead (Pb), all of which are poisons concentrated in biosolids.

**Sewage effluent**  Sewage entering a treatment plant is filtered and allowed to settle. In a secondary-treatment plant, oxygen is added to the effluent from the primary treatment to reduce the BOD, and additional solids are removed by settling. The effluent from primary- and secondary-treatment plants can be disposed of on soils and can be either beneficial or harmful.

A major benefit of effluent is its water content. Although effluent is rarely used for irrigation of crops directly consumed by humans because of public resistance, golf courses, parks, and other public areas have been successfully irrigated with effluent. Effluent rarely contains heavy metals or

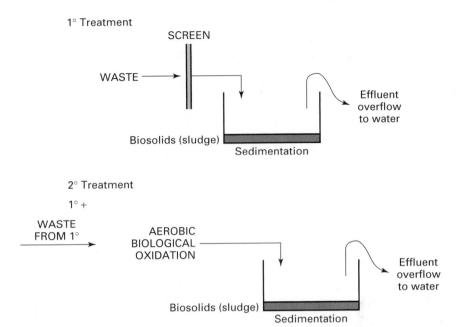

**FIGURE 16–4**   Primary (1°) waste treatment is a single sedimentation after coarse filtering. Secondary (2°) treatment includes a second sedimentation and an aerobic oxidation. Advanced wastewater treatment (not shown) comprises a large number of sophisticated (and expensive) treatments of 2° effluent, including aerobic and anaerobic biological decomposition and special filtration.

toxins. It usually has some nitrogen, potassium, and phosphorus but is limited as a fertilizer because it has a low nutrient content, and the nutrients are not in the ratio preferred by plants (Table 16–2). Concentrations of soluble salts such as NaCl are often high in effluent, limiting its usefulness in arid and semiarid areas, where salt accumulation is a problem. Special care needs to be taken when using effluent to irrigate crops to ensure that an adequate volume of water is added to leach salts below the plant root zone.

Effluent is added to soil as overland flow or through sprinklers. The overland flow method requires a slope. As waste flows across the surface, sunlight deactivates microorganisms, vegetation filters the liquid, and there is some evaporation. Thus when the effluent reaches the stream, it has been "treated." The sprinkler disposal of effluent does not require a slope, because the effluent percolates into the soil and is treated just as septic effluent is.

Soil properties influencing effluent disposal are the same as those for septic leach lines. Medium-textured soils with moderate infiltration rates are preferred for sprinkler disposal. Overland flow disposal requires soils with more clay so as to ensure that there is adequate runoff. In either case, it is desirable to retain aerobic conditions in the soil.

**Sewage sludge**   An alternative term to *sewage sludge* is *biosolids*. This term hides the source of this material but does not reduce the concern most people have with its use in agriculture. Sludge at a treatment plant is about 5 percent solid and 95 percent liquid (Figure 16–5). The solids generally contain low levels of plant nutrients such as nitrogen and phosphorus and may, depending on source and treatment, contain toxic elements. Typical sludge disposal is by drying and burning, burial in landfills, or dumping in water. Disposal or utilization in soil by spreading on the soil surface or by injection into the subsurface is becoming more common. Drying and burning are expensive in dollars and energy; landfills are filling very quickly, and dumping in water is neither environmentally sound nor legal. Adding sludge to soil involves some risk of surface water pollution resulting from erosion of surface-applied sludge. If sludge is applied to level ground, with proper erosion control practices, this problem is minimized.

Disposal is more common than sludge utilization on crops because the public is wary of the potential health hazards of sludge. In addition, sludge is expensive to transport and must be dewatered to be useful. Large amounts are needed for it to be an effective fertilizer, but the popularity of organic gardening has increased the market for dried, deodorized, and

**TABLE 16–2**

Composition of Sewage Effluent (from Several Sources)

| Source | Composition (mg/L) | | | | | | | EC[a] (dS/m) |
| | N | P | K | Na | Ca | Mg | pH | |
| --- | --- | --- | --- | --- | --- | --- | --- | --- |
| 1 | 17–27 | 4–12 | 9–13 | — | — | — | 8–8.2 | 1.7–2.1 |
| 2 | 5 | 8 | — | 45 | 40 | — | — | — |
| 3 | — | — | 20 | 236 | 87 | 53 | — | — |

[a]EC—electrical conductivity
1. C. E. Clapp, T. C. Newman, G. C. Marten, and W. E. Larson. Effects of Municipal Wastewater Effluent and Cutting Management on Root Growth of Perennial Forage Grasses. (1984). *Agronomy Journal* 76(4):642–647.
2. D. E. Hill. Waste Water Renovation in Connecticut Soils. (1972). *Journal of Environmental Quality* 1:167.
3. E. DeJong. The Movement of Sewage Effluent Through Soil Columns: The Major Ions (Na, Ca, Mg, Cl, and SO₄). (1978). *Journal of Environmental Quality* 7:133–136.

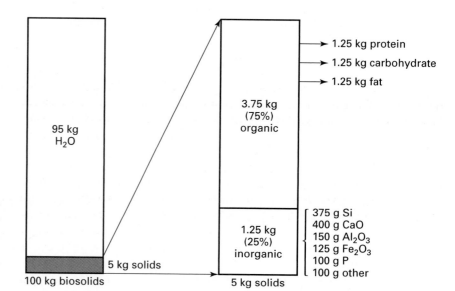

**FIGURE 16–5** Biosolids are mostly liquid. The solid portion is mostly organic. The small inorganic part is most limiting to sludge usefulness because it often contains many toxic elements.

composted sludge. You might find some in a bag at your local nursery supply store. Benefits of sludge additions to soil include increased organic matter and subsequent improvement in aggregate stability, infiltration rate, and water-holding capacity. The increase in aggregate stability is largely a function of the increased microbial activity produced by C and N addition from the biosolids. Both macro- and micronutrients are added in small but useful amounts. There is some evidence that biosolids addition, with their large microbial populations, may be helpful in suppressing plant disease organisms, and we should not discount

the positive value of the liquid portion in semiarid and arid climates.

Heavy-metal contaminants have caused the most concern among agriculturalists and health officials. Such elements as Pb and Cd are known toxicants of people. Small levels of these elements can be concentrated in soils and plants (such as leafy vegetables) and further concentrated in milk if cows are fed contaminated grain or forage. Biosolids addition to soil also has potential negative impacts on the soil, especially to soil organisms, from metals such as Zn, Cu, Ni, Cd, Pb, and Cr. There is conflicting evidence on the magnitude of the impact of biosolids on microorganisms. Because microorganisms have remarkable adaptability, essential soil processes appear to continue even in heavily contaminated sites. The nitrogen fixers and nitrifiers are most sensitive to the effects of these trace metals, which appear to reduce the effectiveness of infection by *Rhizobium* spp. and subsequent nodulation. Once infection has occurred, the bacteroids are protected by the root nodule. The impact on the large group of heterotrophic microorganisms responsible for N mineralization is highly soil dependent. Some mineralization is reduced by the presence of these metals in soils, but high cation exchange capacity (CEC) and pH reduce toxicity to the microorganisms.

Biosolids also contain soluble salts that may accumulate in soils, especially in locations where precipitation is too low to leach the salts out of the root zone. The salt content may have the effect of making soils saline and/or sodic.

Biosolids utilization has more potential in forests than in agronomic crops. Most forest soils are deficient in N, P, and S and, being acidic, benefit from the Ca and high pH of most biosolids. Any heavy metals are contained within the wood of growing trees or forest litter.

Another potential use of biosolids is for mine spoil and wasteland reclamation. Coal, sand, gravel, stone, clay, copper, iron ore, phosphate rock, and other minerals are mined in open-pit mines that require reclamation. In addition, wasteland is created in the United States where soils are used for deposition of dredge materials and fly ash from coal-fired power plants. Biosolids help in the establishment of vegetation on mine spoils and wasteland, especially when high-pH material is added to extremely acidic mine tailings. Waste that has been limed during the treatment process has neutral to alkaline pH and can be beneficial in raising the pH of mine spoils. Most important, mine spoils and wastelands frequently have very low organic matter content. Additions of biosolids increase soil humus. This has the important benefit of increasing the microbial population and reestablishing the spoil's microbiological community. Organic matter in biosolids improves the physical properties of mine spoils, including increased water-holding capacity, decreased bulk density, increased aggregate stability, increased water infiltration and hydraulic conductivity, and decreased surface temperature if left on the soil surface.

Soil properties important to biosolids disposal are those that affect the retention of nitrogen and heavy metals. Clay and humus content are the most important. Soils with low CECs are less useful for biosolids disposal than are those with medium and high CECs. Soils with a high pH are more suitable than are those with an acidic pH. At too high a pH, sludge makes nutrients unavailable to plants, and revegetation becomes difficult, but at pH <7 trace element availability increases.

The U.S. Environmental Protection Agency (EPA) has developed guidelines (not regulations) for the maximum amounts of Pb, Zn, Cu, Ni, and Cd allowed on agricultural land used for growing crops that enter the human diet with or without processing, based on soil CEC (Table 16–3). Not all federal agencies or

## TABLE 16–3

EPA Recommended Maximum Amounts of Trace-Metal Loadings for Agricultural Cropland

| Metal | Soil cation exchange capacity (cmol kg$^{-1}$) | | |
| | <5 | 5–15 | >15 |
| | Amount of metal (kg ha$^{-1}$) | | |
| --- | --- | --- | --- |
| Pb | 560 | 1,120 | 2,240 |
| Zn | 280 | 560 | 1,120 |
| Cu | 140 | 280 | 560 |
| Ni | 140 | 280 | 560 |
| Cd | 6 | 11 | 22 |

Source: U.S. EPA, 1983.

## SOIL AND ENVIRONMENT
### Waste Disposal

Garbage, sewage, toxic waste, manufacturing waste, and chemical waste are disposal problems in our society. Clean air and clean water laws have dictated that much waste disposal by burning and dilution in water have been replaced by land (soil) disposal. Some soils have more capacity than others to detoxify wastes and purify wastewater. Rainwater passing through buried waste can dissolve toxic materials and carry pollutants into groundwater. Soil physical and chemical properties are important when selecting waste disposal sites to prevent groundwater pollution, because waste disposal adds toxic chemicals and high concentrations of chemicals to soils, making them unsuitable for other uses.

Many homes are not connected to wastewater treatment plants and so use septic tanks and leach lines in soil to treat waste. The soil determines the installation and operation cost, the leach line longevity, and success or failure. The soil microbial community utilizes the partially treated waste that enters the soil from the septic tank through the leach line. This bioremediation occurs when the soil environment is hospitable to microbial activity. The soil atmosphere usually contains sufficient oxygen for efficient waste disposal.

---

state authorities agree with these guidelines, but all agree that the toxic-metal addition to soils by sludge must be limited to protect consumers.

**Solid waste**   Household and industrial waste that is not treated by a sewage treatment plant is **solid waste.** Solid waste is picked up by local sanitation workers and hauled away. Its final resting place is in the soil of a sanitary landfill. A **sanitary landfill** differs from a garbage dump because the waste is covered with soil every day, groundwater is protected from leachates, and gas is trapped before it is vented to the atmosphere.

Soils at the base of the landfill must prevent percolating water from entering the groundwater, and the soil used for the daily cover and ultimate final cover must be easy to move. Clay soils make the best lining for sanitary landfills, and medium-textured soils are best for the daily and final cover materials. Often the base of a sanitary landfill is below the usual depth investigated by a soil scientist. In such cases, knowledge of the parent material is helpful in selecting landfill sites.

Sites for landfills must be accessible to large trucks throughout the year, and during dry periods they should not be too dusty or susceptible to wind erosion. Soils information can be useful in making site assessments based on these criteria. Improper site selection can lead to soil, air, and water pollution.

Soils are useful for waste disposal because they retain nutrients, organic materials such as herbicides and insecticides, and metals. While protecting the water supply, the soils become contaminated with the waste materials. It is possible to contaminate a soil permanently so that it cannot be used for agriculture or anything else. There are a growing number of examples in the United States in which waste products from industries have been "permanently" disposed of in soils, only to appear many years later when the sites were to be used for other purposes. Today, the many industrial and agricultural chemicals that require safe disposal complicate the job of finding appropriate sites for disposal. (Chapter 15 has additional information on soil contamination.)

## 16.2   ENGINEERING PROPERTIES

### 16.2.1   Soil Strength

The soil's ability to support a load or respond to stress is a measure of its strength. It is possible to **stress** soils in three ways (Figure 16–6). **Compressive strength** is the soil's resistance to compressive (squeezing) stress, **tensile strength** is the soil's resistance to tensile (pulling) stress, and **shear strength** is the soil's resistance to shearing (sliding) stress.

Two factors determine soil strength. The first is **frictional resistance,** which is determined by the total surface area of the soil particles (primarily dependent on texture) and the particles' shape. Flat particles contact one another more than do round particles

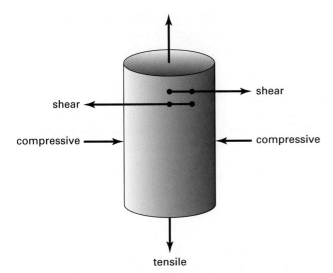

**FIGURE 16–6**   This soil core is being stressed by squeezing (compressive stress), pulling (tensile stress), and shearing (sliding stress).

(Figure 16–7A), and angular or rough particles interlock more than do perfectly smooth particles. Angular and flat particles therefore resist moving past one another more than smooth or round particles do (Figure 16–7B). In addition to these soil properties, the amount of frictional resistance increases as the force (weight) on the soil that pushes the particles together increases. Frictional resistance is reduced by water in the soil, which lubricates the particles and helps them slide or roll over one another. As the soil moisture content rises, the soil strength falls.

A second force holding soils together and contributing to their strength is cohesion. Cohesion between clay particles and between the water layers surrounding the soil particles adds to the soil's strength. Cohesion between layers of water is termed *apparent* cohesion because it is temporary, unlike true cohesion between clay particles and other charged surfaces. Apparent cohesion is well known to people who frequent beaches. At the beach, wet sand is used to build sand castles because dry, cohesionless sand grains fall apart but wet sand can be shaped and retains the shape until dry. Soils without clay are **cohesionless,** and those

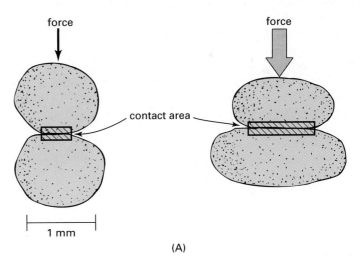

(A)

**FIGURE 16–7** Particles resist forces (stress) by interlocking and friction. (A) As the weight on a soil increases, the particles of sand change shape slightly, increasing the contact area and friction. (B) Particle shape influences interlocking.

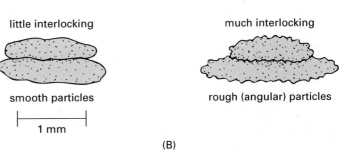

(B)

with clay are **cohesive.** In general, cohesive soils have greater strength than cohesionless soils do when dry. However, as the clay content of cohesive soils increases, the water-holding capacity increases and the strength decreases. The best materials to build on are well-drained sands.

## 16.2.2 Atterberg Limits

An engineer named Albert Atterberg developed a classification of soils based on their mechanical properties at different water contents. The classification is used as a guide to predict the engineering behavior of soil. Each **Atterberg limit** is the water content (mass %) at which the soil's mechanical behavior changes (Figure 16–8). The first Atterberg limit is the moisture content at which clays in the soil begin to swell. Below this shrinkage limit, a soil can dry without changing volume. Above the shrinkage limit, the soil swells. Soils act as hard, brittle solids when they are dry. A very hard, brittle soil clod will shatter if hit with a hammer, but the same soil, when moist can be deformed and will hold its shape. A soil is said to act as a plastic over the moisture content range in which the soil can be deformed and still holds its shape. The lowest moisture content at which the soil first begins to act in this plastic manner is the **plastic limit.** At the **liquid limit,** a soil behaves as a viscous liquid. It can be deformed, but it will not hold its shape. At the liquid limit a soil has no strength. A load applied to a soil at a water content at or above the liquid limit will not be supported. An easy test of a soil's inability to support a load at its liquid limit is to walk across a wet soil and feel yourself sink. There are special tools for determining the liquid limit in the laboratory (Figure 16–9).

Engineers use the Atterberg limits to determine whether soils are suitable for different uses. Nonplastic soils are preferred for building materials, compared with soils with large plastic indices. The **plastic index** is the difference between the liquid limit and the plastic limit (Figure 16–8). Soils with high liquid limits are clay soils, which are not as useful for building as are well-drained, coarse-textured soils.

## 16.2.3 Shrink-Swell

Most clay-textured soils change volume as their moisture content changes. The amount of volume change depends on the type and amount of clay in the soil. The 2:1 lattice clays such as smectite have the greatest shrink-swell because water molecules enter the interlayer between the clay unit cells, causing them to expand (Figure 16–10). This is repeated over thousands of unit cells. At the soil surface, shrinking and swelling are seen as cracks that develop when the soil dries (Figure 16–11). This is manifested in cracked streets and sidewalks, walls, and window frames; doors that do not close in winter; and, in severe cases, the destruction of walls and foundations (Figure 16–12).

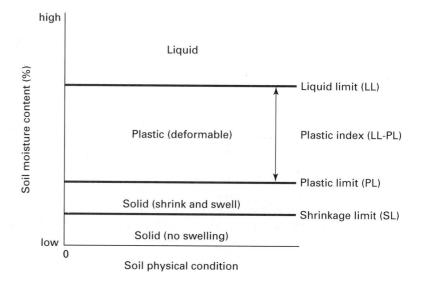

**FIGURE 16–8** The Atterberg limits: As the soil moisture increases from oven dry, the soil's consistency changes from a brittle solid, to a deformable plastic that holds its shape when molded, to a deformable liquid that does not hold its shape when molded.

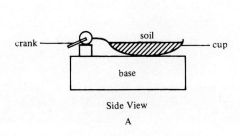

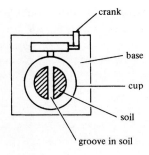

Side View

A

Top View Start of Test

B

Top View Completion of Test

C

**FIGURE 16–9** The liquid-limit device (A) is used to determine the moisture content of a soil at the liquid limit. The test begins with an open groove in the soil (B) and ends when the groove sides touch along the bottom of the cup (C). The soil "flows" because the cup is dropped repeatedly on the hard base by turning the crank.

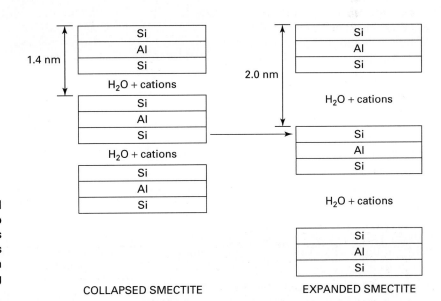

**FIGURE 16–10** Soils shrink and swell because some clays absorb water in the interlayer space. This is repeated innumerable times and is reflected in deep cracks in dry soil (Figure 16–11) and leaning telephone poles (Figure 16–12).

COLLAPSED SMECTITE            EXPANDED SMECTITE

**FIGURE 16–11** The 5- to 10-cm-wide and 70- to 120-cm-deep cracks in this Vertisol occur because the dominant clay mineral is smectite. Scale is in inches.

**FIGURE 16–12** Leaning telephone poles and fence posts are a good indication that the soil clays are mostly smectite.

Little can be done to improve high shrink-swell soils. When these soils cannot be avoided, buildings require stronger and deeper foundations and more expensive building techniques. Even in agriculture, high shrink-swell soils can be troublesome. Most often, these high-clay soils are either too wet or too dry for cultivation, and so the timing of operations becomes critical to successful farming. An alternative is to use very powerful equipment to work the soil.

A measure of the soil's shrink-swell capacity is the **coefficient of linear extensibility (COLE)**. COLE is determined by measuring a specimen in wet and dry states.

$$\frac{(\text{moist length} - \text{dry length})}{\text{dry length}} \quad (1)$$

Various moisture states have been used for moist length, but field capacity is often used. COLE is often measured from the clods' volume change rather than change of length. In such cases, Equation 2 is used to calculate COLE:

$$\text{COLE} = \sqrt[3]{\text{moist volume/dry volume}} - 1 \quad (2)$$

COLE generally varies from near 0 to 0.12. The degree of probable limitation on the soil's use is correlated with COLE (Table 16–4).

**TABLE 16–4**
Values of COLE and Interpretations of the Hazard to Engineering Use

| COLE | Interpretation |
| --- | --- |
| <0.03 | Low hazard |
| 0.03–0.06 | Moderate hazard |
| 0.06–0.10 | Severe hazard |
| >0.10 | Very severe hazard |

# 16.3  RECREATION AND CONSTRUCTION

## 16.3.1  Recreational Facilities

The *National Soils Handbook* contains guidelines for many interpretations, several pertaining to recreational facilities, including paths and trails, picnic areas, and playgrounds. Specific criteria for these uses are given in Supplement 16–2. Limitation ratings are slight, moderate, and severe. More treatment or better management is required to overcome moderate limitations than for slight limitations. Artificial drainage, control of water to reduce runoff and erosion, or other modifications are necessary with a moderate limitation rating. A severe rating is given to soils that have one or more properties unfavorable for the rated use.

## 16.3.2  Construction

Interpretations are made in soil survey reports in regard to the suitability of soils for the construction of small structures, including dwellings with and without basements, small commercial buildings, and local roads and streets. Supplement 16–3 shows several of the charts for determining suitability. They follow the same general format as the suitability charts for recreation, the major limitations being horizons that limit the ease of excavation, the strength of the material, and the shrink-swell hazard.

# 16.4  RECLAIMED AND ARTIFICIAL SOILS

## 16.4.1  Polders

In many coastal countries, there have been attempts since the time of the Roman Empire to create new land by holding back the sea. The process of building dikes and draining the land produces polders (Figure 16–13). The word *polder* is Dutch, which is appropriate since the Netherlands has over 965,000 ha of polders reclaimed from marine sediments less than 10,000 years old. Today, polders are used for all forms of agriculture, recreation, and urban development.

The first step in creating a polder is protecting the area from the sea with dikes or levees of mud, stone, sand, or concrete, depending on the available material and the size of the project. Once the polder is protected from tidal influences, it must be drained. Surface drainage by pumps and canals is the most common. The famous Dutch windmills were built to pump water as part of polder reclamation.

Polder sediments may be mineral, organic, or combinations of both. They may be calcareous if shells are present. Frequently, polder sediments are very slowly permeable clays, and draining the sediments causes great changes. Because they were saturated with water, they settle, compact, and crack when drained. These are sometimes called shrink-noswell soils because of the cracks that form as the material loses water. Often, large deep cracks form, which can help improve water penetration in clay sediments. The process of drainage and settling, or **ripening,** usually lasts from 1 to 2 years but may take many more. During this time, the polders may be cultivated to help mix and drain the soil. Salt-tolerant grasses, reeds, or sedges are planted to remove water through transpiration.

Frequently the sediments are high in sodium, and they are almost always salty. The salts and sodium must be removed before the polder can be farmed. The process is the same as that described in Chapter 11. Surface or subsoil drainage is installed, first to remove the seawater and then to remove the added leaching water. Next, adequate good-quality water is added to leach the salts from the soil. If sodium is on the exchange complex, it is replaced with Ca ions

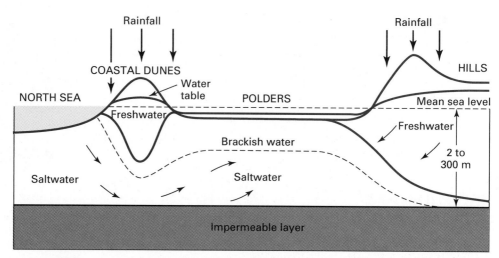

**FIGURE 16–13** Diagrammatic hydrological section across the Netherlands illustrating where a polder is created by excluding the ocean and draining the land. (From Paul Wagret. 1968. *Polderlands.* London, Methuen. With permission from Methuen and Co. Ltd.)

through the addition of gypsum. In some countries with sufficient rainfall, natural reclamation may take place after the seawater is removed.

After the polder is drained, the reduced minerals oxidize, and the soil colors change from gray, blue, or black to brown. If reduced forms of sulfur are present, they will oxidize to sulfuric acid ($H_2SO_4$). Shells ($CaCO_3$) neutralize the acid, forming calcium sulfate so that gypsum is not needed. If shells are not present, the polder can become extremely acidic (pH <4). These are called acid-sulfate soils because of the sulfuric acid generated as iron sulfides are oxidized. Equation 1 shows the reaction when the iron sulfide mineral pyrite ($FeS_2$) is oxidized.

$$FeS_2 + 3.75O_2 + 3.5H_2O \rightarrow Fe(OH)_3 + 4H^+ + 2SO_4^{2-} \quad (3)$$

Iron sulfide is a common mineral found in soils that have been waterlogged and strongly reduced. Note that 1 mol of iron sulfide releases 4 mol of hydrogen ion. The iron hydroxide does not neutralize the acidity because this mineral precipitates from solution. This is why the reaction creates extremely acidic soils. The acidity is a serious problem for growing plants in these soils, and it may be a serious off-site problem if the acid leaches into groundwater or finds its way to surface water. The strong acidity solubilizes aluminum and trace elements that are toxic to aquatic organisms and humans. Aluminum toxicity to plants

occurs at concentrations between 1 and 2 mg $L^{-1}$. Fish and other aquatic organisms are killed by Al concentrations as much as 10 times lower. Drinking water standards vary around the world but are generally lower than 1.5 mg $L^{-1}$.

The acidity is corrosive to buried concrete and steel, shortening the useful life of building foundations and pipes. Such acidity is not restricted to reclaimed ocean sediments. Similar problems occur with acidic mine tailings.

The normal microbial population of a soil does not exist in ocean sediments. Thus polders may not have sufficient microbial populations to decompose organic matter efficiently, and they are often N deficient. Legumes are sometimes planted to help reestablish microbial communities once the soil is drained and has become aerobic.

The creation of useful polders from wet, soft, impermeable clayey sediments is a triumph for the engineer and agriculturalist, though the process is expensive and slow. Although the areas involved are significant to the individual countries that have polders, the overall effect on the world food situation is small.

## 16.4.2 Mine Soils

Significant areas are being mined for coal, oil shale, and gravel by surface mining (**strip mining**). Envi-

As a rule, topsoils (A horizons) suitable for growing corn and small grains are good for turf. For example, the surface horizons of well-drained Mollisols are rich in humus, friable, and well aerated; have good water-holding capacity; are medium acid to neutral; and are naturally fertile. These characteristics are necessary for turf, because the root system of turf grass mowed at 2 to 5 cm will penetrate at least 20 to 30 cm in favorable soils.

Soils that lack these characteristics are modified by improving drainage through tile drains; by adding amendments such as fertilizer, organic matter, and lime; or, in extreme cases, by replacing the natural soil with more favorable material. Sand, organic matter, and vermiculite or perlite can be added to dilute the clay and improve drainage. Sand can be used only on small areas because of the tremendous volume required. To increase the macroporosity of fine-textured soils, the sand content must be raised to 80 percent or more. Any less sand, and the density will increase.

Grass varieties used for turf require well-aerated soil and are very sensitive to inadequate oxygen. Consequently, compaction is one of the greatest turf management problems. Sandy soils are generally good turf soils because they are well drained, and well-drained soils resist compaction. Good drainage helps ensure rapid warming of soil in spring and good early-season growth, but grass grown on sand must be watered often.

Sandy soils are often amended with manure or peat (histic material) to improve their low water-holding capacity. **Peat** is a general term that covers a wide range of organic materials. The most suitable are sphagnum or hypnum moss peats. They are fibrous, acidic materials with a high cation exchange capacity and a high water-holding capacity.

Lime is required on turf soils when the pH is below 5.5. Nitrogen is often limiting to many turf grasses, and it must be added several times during the growing season. Other essential elements must also be added as required for good grass growth. Cool-season grasses are normally fertilized in fall and early spring and warm-season grasses in early spring and midsummer.

When the natural soil cannot be used for turf, topsoil harvested from other areas is spread over the natural soil, or new soil can be created. Additional organic matter, fertilizer, and lime may be mixed with the topsoil, depending on its source and characteristics. The particle size distribution of the topsoil must be similar to that of the natural soil, or the water flow through the different layers will be slow.

If "new soil" is created, the two most important factors are adequate depth and sufficiently large pores for drainage. In addition, the soil should have good properties after compaction. The usual mixture is 90 percent sand and 10 percent clay. Sand in the range of 0.05 to 0.5 mm diameter is the preferred size fraction. Clay is used rather than silt because it has a greater cation exchange capacity. Organic matter up to 20 percent is often added to improve the mixture's water- and nutrient-holding capacity.

### 16.4.4   Horticultural Mixes

Growing plants in containers requires special soil materials. The requirements resemble those for turf, without the traffic considerations, but weight is now important—enough to keep plants upright without straining the gardener. Also, potted soil is inherently ill drained, despite the drain holes. In the field, liquid water is drawn into the underlying subsoil by matrix forces (usually much greater than gravity), but a potted soil, which is drained only by gravity, stops dripping when it is still close to saturation. (Putting a layer of gravel in the bottom, as some gardening writers still recommend, just shifts the problem a centimeter upward.) To counteract this weak drainage, potting mixes must have a stable, very open, coarse porosity, so that they can be aerated even when wet. Natural soils, potted, tend to become sodden bricks.

Most container soils or potting mixes are based on sand and organic material. The ratio of sand to organic matter is lower than in turf soils, typically 40:60 or 50:50. Mixes for propagating seedlings and cuttings, which need less nutrition and more firmness and density, have more sand and less organic matter, typically 75:25 or 70:30.

The sand for potting mixes is preferably coarse and sharp (not round) for better mixing and porosity. Sandy soil or small amounts of loamy soil can substitute for some of the sand. (Clays seldom mix satisfactorily.) Other optional substitutes include perlite (a granulated bubbly glass) and vermiculite (a heat-expanded mica, not the clay mineral). These are coarser, less dense, and more water retentive than sand. Ceramic pellets, made of clay fired at a low temperature, combine water capacity with moderately high density and expense.

Favored organic materials include leaf mold, shredded or ground bark, shredded coconut husk or other fiber, and composted sawdust, leaf litter, or manures. These materials are tending to replace increasingly expensive peat and sphagnum, which are harvested from limited wetland areas (Histosols). Careful, infrequent sphagnum harvesting might be sustainable in some cases, but peat cutting is not.

Nutrient retention capacity and some nutrients are supplied in peat, sphagnum, leaf mold, manure, soil, and some composts. Clay and vermiculite can contribute cationic nutrients and exchange capacity. Sand and perlite contribute nothing to nutrition. Coconut fiber and sawdust, due to their low nutrient content, have the potential to detract by immobilizing N and S. The kind of organic matter determines its decomposition rate and the long-term supply of available N to microbes and plants. Well-decomposed materials generally have sufficient nitrogen, but raw materials such as uncomposted sawdust, fiber, or wood shavings have very high C/N ratios that can cause N to become unavailable.

For incompletely understood reasons, some composted materials (such as pine bark) are better than others (such as peat) in suppressing root-rot fungi (such as *Pythium* and *Rhizoctonia* spp.), perhaps by stimulating growth of bacteria that parasitize these fungi.

Most potting mixes have nutrients added as manure, synthetic slow-release fertilizers, and/or a generous addition of commercial inorganic fertilizers (e.g., 10 kg each of ammonium nitrate, superphosphate, and dolomite per cubic meter of mix, plus 5 kg potassium sulfate and 0.5 kg ferrous sulfate; dolomite supplies Ca and Mg and neutralizes acidity that arises largely from the iron's precipitation as ferric hydroxide).

Mixes are commonly steam sterilized to discourage weeds, insects, and pathogens. This can induce manganese toxicity if large quantities of soil are included in the mix. It also kills mycorrhizal fungi. Replacing them with an inoculant of appropriate fungal spores or mycelium frequently improves the performance of nursery-raised tree seedlings for orchards and reforestation.

It is expensive to make container soils, so these techniques are reserved for avid gardeners, commercial flower growers, and nurseries that produce young plants for gardeners, foresters, and orchardists. Farmers and ranchers must effectively manage what they have, because creating "soils" on an agronomic scale is not economic.

## 16.5  SUMMARY

Soil is suitable for the disposal of waste if its properties allow sufficient capacity and treatment of the waste to protect surface water and groundwater supplies from contamination. Soils useful for on-site septic systems and off-site waste disposal must combine moderate permeability and sufficient depth along with suitable slope and enough active surface area for effective waste treatment.

The suitability of soils for many nonagricultural uses is determined by engineering properties. Soil strength is the soil's ability to resist stress. Stress may be applied as a weight on the surface or by shearing forces such as a plow moving through the soil. Soil strength is primarily determined by the soil's particle size distribution and water content. The kind and amount of clay strongly influence the soil's engineering properties, including the Atterberg limits and shrink-swell characteristics. Atterberg limits provide a quantitative measure of the more familiar property soil consistence. As the soil's clay content increases, the plastic and liquid limits generally increase also, and the material becomes less useful for engineering purposes. COLE, the coefficient of linear extensibility, measures the soil's shrink-swell potential. As the clay and smectite content increases, the COLE also increases.

Reclaimed and artificial soils cover a small part of the earth's surface, but they can be important to crop production, forestry, and housing, depending on the location and type of soil created. A polder is land created by dikes and drainage. Soils are also created on a smaller scale in reclaiming strip mines. Turf for golf courses and football fields requires special soil conditions, for which soils are created. Similarly, horticulture often requires created soil mixes with properties that cannot be found in nature.

## QUESTIONS

1. What is on-site disposal of waste?
2. What soil properties influence the success of on-site systems?
3. How do sewage sludge and sewage effluent differ? Do these differences influence their disposal on soil?
4. What is the perc test?
5. How does biological clogging affect on-site waste disposal? What can be done to reduce it?
6. What is soil strength? What soil fraction does it depend on?
7. How do friction and cohesion determine soil strength?
8. What is COLE?
9. What causes the shrinking and swelling of soils?
10. What is a polder?

11. What are the major characteristics that differentiate mine soils from "natural" soils?

12. What properties are most desirable in potting soils? How do they contrast with turf soils and good agricultural soils?

13. Do you think that "made soils" will ever have a major role in world food production? Explain.

## FURTHER READING

Madison, J. H. 1971. *Principles of turfgrass culture.* New York: Van Nostrand Reinhold.

National Academy of Science. 1981. *Surface mining: Soil, coal, and society.* Washington, D.C.: National Academy Press.

Poincelot, R. P. 1980. *Horticulture: Principles and practical applications.* Englewood Cliffs, N.J.: Prentice Hall.

Polders of the World. 1982. *Polder projects: Land and water management aspects.* Vol. 1. Wageningen, The Netherlands: International Institute for Land Reclamation and Improvement.

Schafer, W. M., G. A. Nielsen, D. J. Dollhopf, and K. Temple. 1979. Soil genesis, hydrological properties, root characteristics, and microbial activity of 1- to 50-year-old stripmine spoils. EPA-600/7-79-100. Cincinnati, Ohio: Industrial Environmental Research Laboratory.

Smith, S. R. 1991. Effects of sewage sludge application on soil microbial processes and soil fertility. *Advances in Soil Science* 16:191–212.

Soil Conservation Service Staff. 1984. *National soils handbook.* Washington, D.C.: U.S. Government Printing Office.

Sopper, W. E. 1992. Reclamation of mine land using municipal sludge. *Advances in Soil Science* 17:351–431.

Sprague, H. B. 1976. *Turf management handbook.* 2d ed. Danville, Ill.: Interstate Printers and Publishers.

U.S. Environmental Protection Agency. 1983. *Process design manual for land application of municipal sludge.* EPA 625/1-83-016. Cincinnati, Ohio: Municipal Environmental Research Lab.

Wagret, P. 1968. *Polderlands.* London: Methuen.

## WEB RESOURCES

The U.S. Department of the Interior has a Web site about surface mining and reclamation. It has photos and links to state and federal regulations and to many other government, industry, and nonprofit organization sites. It is at *http://www.osmre.gov/osm.htm*

The U.S. Environmental Protection Agency has a Web site with information on every environmental topic at *http://www.epa.gov/*

## SUPPLEMENT 16–1  Perc Test

The percolation test is used to determine a soil's suitability for a septic leach field. The standard method for determining a perc rate is as follows: First, a 7- to 10-cm-diameter hole is drilled to the depth of the proposed leach field. Next, the sides and bottom are cleaned and roughened to eliminate any sealing caused by the digging. The bottom of the hole is covered with 5 cm of gravel to prevent sealing of the bottom when water is added. The soil in the area of the hole is then soaked for 24 hours to ensure that any cracks are swollen shut and the soil is near saturation. After the 24-hour saturation time, the hole is filled with water, and the water level is maintained for 4 hours. The rate at which the water enters the soil is then measured every 30 minutes for 4 hours. The final 30-minute rate is recorded as the soil's percolation rate.

The percolation rate is usually given as the number of minutes necessary for 1 in (2.5 cm) of water to enter the soil. This rate is used to calculate the number of gallons of effluent per square foot of soil that can be safely added. Equation 4 shows the calculation:

$$Q = \frac{5}{\sqrt{t}} \qquad (4)$$

Here, $Q$ is the gallons per square foot per day, and $t$ is the percolation rate in minutes per inch. If the volume of effluent that is to be treated on site ($V$) is known, the total area required for treatment can be calculated by dividing $V$ by $Q$.

## SUPPLEMENT 16–2 Suitability Criteria for Recreation

Interpretative soil classifications help in identifying wetlands, managing recreation areas and parks, and selecting sites for golf courses. There also is an interpretive grouping for off-highway vehicle use and another for paths and trails.

The following tables from the *USDA National Soils Handbook* are used to determine the suitability and limitations of soils for specific recreation uses. Read the tables from left to right, starting in the upper left. For a soil to be rated as slight, all of its properties must meet the requirements of slight; even a single property can cause a soil to be rated moderate or severe.

## Suitability Criteria for Paths and Trails

| Property | Limits | | | Restrictive feature |
| --- | --- | --- | --- | --- |
| | **Slight** | **Moderate** | **Severe** | |
| USDA texture | — | — | Ice | Permafrost |
| Fraction >7.5 cm (wt. %) surface | <25 | 25–50 | >50 | Large stones |
| Depth to high water table (cm) | >60 | 30–60 | <30 | Wetness[a] |
| USDA texture (surface) | — | — | SC, SiC, C | Too clayey |
| USDA texture (surface) | — | LCOS, VFS | COS, S, FS | Too sandy |
| Unified texture (surface) | — | — | Pt[b] | Excess humus |
| Slope (%) | <15 | 15–25 | >25 | Too steep |
| K factor[c] | — | — | >0.35[d] | Erodes easily |
| Coarse fragments (wt. %) surface | — | — | >65 | Small stones |
| Flooding | None, rare, occasional | Frequent | — | Flooding |
| USDA texture (surface)[e] | — | SiL, Si, VFSL, L | — | Dusty |
| Other | — | — | [f] | Fragile |

[a]Rate as ponded if water table is above surface layer.
[b]Pt is peat soils in this engineering classification of particle size.
[c]This is the soil erodibility K discussed in Chapter 15.
[d]Disregard this rating if slopes are less than 8 percent.
[e]Note that in this rating system, USDA surface texture appears several times. In each case, different textures have different ratings. Use the table by starting at the upper left. Read to the right. The soil is considered to have slight limitations until a criterion places it into moderate or severe. The system works like a sieve. Once a soil drops from slight to moderate or severe, it remains in the lower rating.
[f]If soil is easily damaged by use or disturbance, rate as severe–fragile.
Source: Adapted from *USDA National Soils Handbook,* Soil Survey Staff, Natural Resources Conservation Service. 1999 National Soil Survey Handbook. Title 430-VI Washington, D.C. United States Government Printing Office. Table 620–15. This and other tables can be found at http://www.statlab.iastate.edu/soils/nssh/spc.cont.htm

## Suitability for Picnic Areas

| Property | Limits | | | Restrictive feature |
|---|---|---|---|---|
| | Slight | Moderate | Severe | |
| USDA texture | — | — | Ice | Permafrost |
| Slope (%) | <8 | 8–15 | >15 | Slope |
| Flooding | None, rare, occasional | Frequent | — | Flooding |
| Depth to high water table (cm) | >75 | 30–75 | <30 | Wetness[a] |
| USDA texture (surface) | — | [b] | | Large stones |
| USDA texture (surface) | — | — | SC, SiC, C | Too clayey |
| USDA texture (surface) | — | LCOS, VFS, LFS, LS | COS, S, FS | Too sandy |
| Unified texture (surface) | — | — | Pt | Excess humus |
| Coarse fragments (% surface layer) | <25 | 25–50 | >50 | Small stones |
| Sodium adsorption ratio (0–1 m) | — | — | >12 | Excess sodium |
| Salinity (dS/m) (surface layer) | <4 | 4–8 | >8 | Excess salt |
| pH (surface layer) | — | — | <3.6 | Too acid |
| Permeability (cm/hr, 0–1 m) | >15 | 1.5–15 | <1.5 | Percs slowly |
| USDA texture (surface layer) | — | SiL, Si, VFSL, L | — | Dusty |
| Depth to bedrock (cm) | — | — | <50 | Depth to rock |
| Depth to cemented pan (cm) | — | — | <50 | Cemented pan |
| Other | — | — | [d] | Fragile |

[a]If over the surface, rate as ponding.
[b]Very stony, bouldery, cobbly, or flaggy.
[c]Extremely stony or bouldery or other similar modifiers to texture that indicate severe limitations due to amount of coarse fragments.
[d]If soil is easily damaged by use or disturbance, rate as severe–fragile.

Source: Adapted from *USDA National Soils Handbook,* Table 620–13. Soil Survey Staff, Natural Resources Conservation Service. 1999. National Soil Survey Handbook. Title 430-VI. Washington, D.C. United States Government Printing Office.

## Limitations for Playgrounds

| Property | Limits Slight | Moderate | Severe | Restrictive feature |
|---|---|---|---|---|
| USDA texture | — | — | Ice | Permafrost |
| USDA texture modifier (surface) | — | Stony | Very, ext[a] | Large stones |
| Slope (%) | <2 | 2–6 | >6 | Slope |
| Coarse fragments (% surface layer) | <10 | 10–25 | >25 | Small stones |
| USDA texture (surface) | — | — | SC, SiC, C | Too clayey |
| USDA texture (surface) | — | LCOS, VFS, LFS, LS | COS, S, FS | Too sandy |
| Unified (surface) | — | — | Pt | Excess humus |
| Depth to high water table (cm) | <75 | 45–75 | <45 | Wetness |
| Flooding | None, rare | Occasional | Frequent | Flooding |
| Depth to bedrock (cm) | >100 | 50–100 | <50 | Depth to rock[b] |
| Depth to cemented pan (cm) | >100 | 50–100 | <50 | Cemented pan[b] |
| Permeability (cm/hr) (0–1 m) | >1.5 | 0.15–1.5 | <0.15 | Percs slowly |
| USDA texture (surface layer) | — | SiL, Si, VFSL, L | — | Dusty |
| Sodium adsorption ratio (0–1 m) | — | — | >12 | Excess sodium |
| Salinity (dS/m) (surface layer) | <4 | 4–8 | >8 | Excess salt |
| Soil pH (surface) | — | — | <3.6 | Too acidic |
| Other | — | — | [c] | Fragile |

[a]Very or extremely cobbly, stony, and the like—modifiers that indicate many large rocks.
[b]Rate as slight for 50- to 100-cm depths if slope is 0 to 2 percent.
[c]If soil is easily damaged by use or disturbance, rate as severe–fragile.
Source: Adapted from *USDA National Soils Handbook,* Table 620–14. Soil Survey Staff, Natural Resources Conservation Service. 1999. National Soil Survey Handbook. Title 430-VI. Washington, D.C. United States Government Printing Office.

# SUPPLEMENT 16–3 Suitability Criteria for Construction

Two tables are given as examples of the rating scheme for nonagricultural uses.

## Limitations for Dwellings with Basements

| Property | Limits | | | Restrictive feature |
| --- | --- | --- | --- | --- |
| | **Slight** | **Moderate** | **Severe** | |
| USDA texture | — | — | Ice | Permafrost |
| Total subsidence (cm) | — | — | >30 | Subsides |
| Flooding | None | — | Rare, common | Flooding |
| Depth to high water table (m) | >2 | 0.8–2 | <0.8 | Wetness |
| Depth to bedrock (cm) | | | | |
|   Hard | >150 | 100–150 | <100 | Depth to rock |
|   Soft | >100 | 60–100 | <60 | |
| Depth to cemented pan (cm) | | | | |
|   Thick | >150 | 100–150 | <100 | Cemented pan |
|   Thin | >100 | 60–100 | <60 | |
| Slope (%) | <8 | 8–15 | >15 | Slope |
| Shrink-swell[a] | Low | Moderate | High, very high | Shrink-swell |
| Unified (bottom layer) | — | — | OL, OH, Pt | Low strength |
| Fraction >7.5 cm (wt. %)[b] | <25 | 25–50 | >50 | Large stones |
| Downslope movement | — | — | [c] | Slippage |
| Formation of pits | — | — | [d] | Pitting |
| Differential settling | — | — | [e] | Unstable fill |

[a]Rate thickest layer between 25 and 150 cm.
[b]Rate weighted average to 1 m.
[c]If soil is susceptible to movement downslope when loaded, excavated, or wet, rate as severe slippage.
[d]If soil is susceptible to formation of pits caused by melting of ground ice when ground cover is removed, rate as severe pitting.
[e]If soil is susceptible to differential settling, rate as severe–unstable fill. Note that several of these are very specialized conditions. Pitting is a problem in permafrost areas such as Alaska and other areas with frozen ground.
Source: Adapted from *USDA National Soils Handbook,* Table 620–3. Soil Survey Staff, Natural Resources Conservation Service. 1999. National Soil Survey Handbook. Title 430-VI. Washington, D.C. United States Government Printing Office.

## Limitations for Dwellings without Basements

| Property | Limits | | | Restrictive feature |
|---|---|---|---|---|
| | Slight | Moderate | Severe | |
| USDA texture | — | — | Ice | Permafrost |
| Total subsidence (cm) | — | — | >30 | Subsides |
| Flooding | None | — | Rare, common | Flooding |
| Depth to high water table (cm) | >75 | 45–75 | <45 | Wetness |
| Shrink-swell[a] | Low | Moderate | High, very high | Shrink-swell |
| Unified[a] | — | — | OL, OH, Pt | Low strength |
| Slope (%) | <8 | 8—15 | >15 | Slope |
| Depth to bedrock (cm) | | | | |
|   Hard | >100 | 50–100 | <50 | Depth to rock |
|   Soft | >50 | <50 | | |
| Depth to cemented pan (cm) | | | | |
|   Thick | >100 | 50–100 | <50 | Cemented pan |
|   Thin | >50 | >50 | | |
| Fraction >7.5 cm (wt. %)[b] | <25 | 25–50 | >50 | Large stones |
| Downslope movement | — | — | c | Slippage |
| Formation of pits | — | — | d | Pitting |
| Differential settling | — | — | e | Unstable fill |

[a]Rate thickest layer between 25 and 150 cm.
[b]Rate weighted average to 1 m.
[c]If soil is susceptible to movement downslope when loaded, excavated, or wet, rate as severe slippage.
[d]If soil is susceptible to formation of pits caused by melting of ground ice when ground cover is removed, rate as severe pitting.
[e]If soil is susceptible to differential settling, rate as severe–unstable fill.
Source: Adapted from USDA National Soils Handbook, Table 620–2. Soil Survey Staff. Natural Resources Conservation Service. 1999. National Soil Survey Handbook. Title 430-VI. Washington, D.C. United States Government Printing Office.

# GLOSSARY

The following is an alphabetical list of simple definitions of soil science jargon. The terms are also explained in the text, where they are identified in **boldface** type.

**Absorbed energy**   In remote sensing and meteorology, radiant energy that is neither reflected nor transmitted by opaque or transparent materials. Absorption converts the energy to another form, usually heating the absorbing object.

**Accelerated erosion**   Water or wind erosion at more rapid than normal or geological rates, usually associated with human activities. See *water erosion* and *geological erosion.*

**Accessory cation**   A cation paired with an anion to complete a molecule or crystal structure. In aluminosilicate minerals, a cation (usually Ca, Mg, K, or Na) not coordinated with O in tetrahedra or octahedra.

**Accumulate**   Gather together; in cells or organisms, to absorb an element so that its concentration is greater than in the outside medium.

**Acid**   A substance that releases $H^+$ ions; a condition in which the activity of $H^+$ ions exceeds that of hydroxyl.

**Acidification**   The process of making something acid; lowering pH.

**Actinomycetes**   A major group of filamentous bacteria, some of which are abundant in soils; once classified as intermediate between fungi and bacteria.

**Active remote sensing**   A method for acquiring visual images of an object (i.e., pictures) in which the energy recorded is provided by the observer. Examples are radar imagery and flash photography.

**Adenosine triphosphate (ATP)**   A nucleotide compound formed during biochemical processes to store energy in a form that can be released (with breakdown to adenosine diphosphate and phosphate ion).

**Adhesion**   The molecular attraction between surfaces that holds substances together. Water adheres to soil particles.

**Adsorption**   The process of attachment of a substance to the surface of a solid or liquid.

**Adsorption, specific**   Adsorption, usually chemical (chemisorption), that is selective for a particular ion or substance.

**Adsorptive force**   The cause (energy or power) of adsorption.

**Aeolian**   Material accumulated through wind action. Loess and sand dunes are examples. Also spelled *eolian.*

**Aerobic**   Living or active only in the presence of oxygen.

**Aggregate**   Individual sand, silt, and clay particles bound together into a larger particle. Aggregates may be spheres, blocks, plates, prisms, or columns.

**Alga**   Plural form is **algae.** An aquatic, eucaryotic, plantlike, photosynthetic organism, mostly microscopic, often single celled. The term *blue-green algae,* sometimes still used, refers to blue-green bacteria.

**Alkaline**   Containing or releasing an excess of hydroxyl over hydrogen ions; pH > 7.0.

**Allophane**   A general term describing noncrystalline clay-sized soil minerals. Also, a short-range-order aluminosilicate mineral.

**Alluvium**   Loose (unconsolidated) sediments deposited by flowing water.

**Aluminosilicates**   Minerals whose major elements are silicon (Si), aluminum (Al), and oxygen (O).

**Amendment, soil**   A substance (e.g., lime, gypsum, or peat) added to soil to improve its pH or physical properties.

**Amino acid**   An organic acid containing an amino group $(-NH^2)$; building unit of peptides and proteins.

**Amorphous**   Without crystal structure. In amorphous minerals, the atoms are not in a regular order that repeats extensively in three dimensions.

**399**

**Amphiboles**   A group of aluminosilicate minerals having a structure containing double chains of linked silica tetrahedra.

**Anaerobic**   Without molecular oxygen. The opposite of *aerobic*.

**Analysis**   Fertilizer trade jargon. The percentage of the desired nutrient element(s) in the fertilizer. Also, methods used to determine plant or soil composition.

**Angular blocky**   A structure whose units (aggregates) are nearly equidimensional, square, or rectangular and the corners are not rounded.

**Anion sorption**   The process by which an anion, such as phosphate, replaces $OH^-$ groups on mineral surfaces and edges.

**Anoxia**   Shortage of oxygen.

**Arable**   Describes land suitable for the production of cultivated crops.

**Aspect**   The direction that a slope faces. A south-facing slope has a south aspect.

**Assimilation (of nutrients)**   Metabolic use of absorbed nutrients to make cell constituents.

**Atterberg limits**   An engineering classification of soil consistence at different moisture states.

**Autotroph**   An organism that need not take in organic food to get energy. For example, photoautotrophs use light, and chemoautotrophs obtain energy from oxidation of inorganic substances. See *heterotroph*.

**Availability (of a nutrient)**   Adequacy of supply, freedom, ease of release, and mobility. The amount of a nutrient in chemical forms accessible to plant roots.

**Avalanche**   The rapid downslope movement of soil and rock under the force of gravity.

**Bacteria**   Plural of *bacterium*. A diverse group of microbes containing the procaryotic organisms. Mostly single celled; sometimes occurring in chains or packets. Some, such as actinomycetes, are filamentous. Some, such as blue-green bacteria, are capable of photosynthesis.

**Banding, band placement**   Application of fertilizer below the ground surface in a narrow strip or band.

**Base**   A substance that reacts with $H^+$ ions or releases hydroxyl ions; a substance that neutralizes acid and raises pH.

**Base, exchangeable**   In soil science, an exchangeable cation other than $H^+$ or $Al^{3+}$.

**Beneficiation**   In mining and fertilizer production, treatment to improve the quality and handling of pulverized ore by processes such as sieving and flotation to remove unwanted material.

**Biological nitrogen fixation**   Reduction and assimilation of $N_2$, a capability of certain free-living and symbiotic bacteria.

**Biological (or biochemical) oxygen demand (BOD)**   A measure of the soluble organic content in waste products.

**Biosequence**   Two or more soils in which the soil-forming factor that varies the most is the vegetation (usually kind of vegetation). Other soil-forming factors are constant or vary much less than vegetation.

**Biotic factor**   One of the five soil-forming factors. *Biotic* refers to the living organisms that influence soil formation.

**Biotite**   An important soil-forming aluminosilicate mineral; a black, platy mica found in some igneous rocks such as granite. It is trioctahedral. All possible positions in the octahedral sheet are occupied by atoms.

**Bioturbation**   Mixing (turbation) of soil by organisms (biota).

**Blocky**   A soil structure type whose vertical and horizontal axes are nearly equal. See *angular blocky* and *subangular blocky*.

**Broadcasting, broadcast application**   Scattering, dropping, or spreading fertilizer or other materials on the soil surface.

**Buffering**   Processes that constrain the shift in pH when acids or bases are added. More generally, processes that constrain shifts in the dissolved concentration of any ion when it is added to or removed from the system.

**Bulk density**   In soils, the dry mass (weight) of soil per unit bulk volume.

**Capillarity**   Forces between water and soil surfaces in the small (capillary) pores.

**Carbohydrate**   An organic substance with the general formula $(CH_2O)_n$—for example, sugars and polysaccharides.

**Carboxyl group**   $-COOH$, the most common active group of organic acids.

**Cation exchange capacity (CEC)**   The total amount of positive ions (cations) that a soil can adsorb exchangeably.

**Cellulose**   The carbohydrate most abundant in plants; a polysaccharide made up of glucose units.

**Channel**  A natural stream or an excavation constructed to carry water.

**Chelate**  A compound in which a polyvalent metal cation is firmly combined with a (usually organic) molecule by multiple bonds. The binding molecule is called a *chelating agent* or *chelon*.

**Chelation**  The process of forming a chelate.

**Chlorite**  A nonexpanding 2:1 aluminosilicate mineral with a sheet of magnesium and hydroxyl atoms in the interlayer space.

**Chlorophyll**  Green pigment, a complex chelate of Mg; traps light for photosynthesis in plants, algae, and some bacteria.

**Chlorosis**  Pale color, yellowing. Caused by lack of chlorophyll in plants.

**Chronosequence**  Two or more soils in which the soil-forming factor that varies the most is time. Other soil-forming factors are constant or vary much less than time.

**Clay**  A mineral particle size less than 0.002 mm diameter. Also, a specific group of aluminosilicate minerals with a high surface area and exchange capacity.

**Clay film**  A coating of oriented clay particles on larger particle, pore, or aggregate surfaces.

**Climosequence**  Two or more soils in which the soil-forming factor that varies the most is the climate (usually amount of precipitation). Other soil-forming factors are constant or vary much less than climate.

**Clogging**  A process that blocks pores and reduces water flow in soils. Mechanisms can be physical, chemical, and biological.

**Codominant**  In forestry, trees with crowns forming the general level of the canopy and receiving full light from above but little from the sides.

**Coefficient of linear extensibility (COLE)**  A measure of the potential one-directional change in the volume of a soil as the soil moisture level changes.

**Cohesion**  Attraction between like molecules. Water molecules cohere to one another.

**Cohesionless soils**  In engineering, sands with little clay or silt.

**Cohesive soils**  In engineering, clay- and silt-rich soils.

**Colloid**  A material made of organic or inorganic particles with diameter from 0.001 to 0.1 $\mu$m.

**Colluvium**  Unconsolidated, unsorted earth material that has moved downhill under the force of gravity and has accumulated at the base or lower slopes of the hill.

**Columnar**  An aggregate shape that is longer than it is wide, with a rounded top and edges.

**Cometabolism**  Transformation of an organic compound (the substrate) by a microorganism without the organism deriving benefit (energy, carbon, or nutrients) from the substrate.

**Compaction**  A process of rearranging soil particles to decrease pore space and increase bulk density. It produces a compacted, dense soil layer.

**Compound slide**  A combination of a fall and a rotational slide.

**Compressive strength**  In engineering, a soil's ability to resist compressive (squeezing) forces.

**Concentration (of material)**  The amount of the material per unit volume of solution, gas, or solid. Also, the process of increasing the concentration at a certain place.

**Concentration gradient**  The rate of change of concentration per unit distance.

**Conductive tissue**  In plants, tissue that freely transmits water and dissolved compounds; vascular tissue, comprising xylem and phloem.

**Conductivity**  A measure of the ease with which a material or object conducts or transmits something such as heat (thermal conductivity), water (hydraulic conductivity), or diffusing substances (diffusion coefficient).

**Conductor**  A material that conducts—allows the diffusive transmission of—(usually) either electricity or heat.

**Conservation**  Protection from loss or waste. In soil science, most often denotes efforts to reduce soil loss by wind and water erosion.

**Conservation tillage**  Any tillage sequence that leaves at least 30 percent crop cover on the soil surface, with the goal of minimizing soil and water loss. It ranges from omitting one or more tillage operations to eliminating all tillage.

**Consociation**  The most detailed soil-mapping unit, composed of delineations that show the size, shape, and location of the unit.

**Contact time**  In waste disposal, the time that the waste is in contact with soil surfaces as it flows through the soil.

**Continuity**  Length of unbroken, continuous, or coherent soil pores.

**Control section**  A defined portion of the soil profile used in determining the soil's classification.

**Convection**  Bulk mixing of fluid and of heat and materials contained in the fluid. Usually caused by density differences from uneven heating or cooling or by turbulence.

**Correlation**   In general, the degree of (numerical) relationship between associated variables. In a soil survey, comparing observed profiles with defined series to ensure proper classification.

**Cortex**   Outer layer; in plant roots and stems, the layer between the epidermis and the central conductive tissue.

**Coulombic attraction**   Electrical attraction between things (ions, surfaces, and the like) with opposite charges.

**Covalent (chemical bond)**   Caused by sharing of electrons.

**Cover crop**   Any crop that is grown to provide soil protection from erosion between periods of normal crop production or between trees in orchards and vines in vineyards.

**Creep**   In soil science, the slow movement of soil down a slope under the force of gravity. Also, the rolling or pushing of particles along the ground surface by wind. See *soil creep.*

**Crown drip**   The portion of rainfall, snowmelt, or fog that drips from vegetation. May be an important part of nutrient cycling in some ecosystems.

**Crust**   A layer of increased bulk density of the top few millimeters of a soil caused by raindrop impact, clay dispersion, and clay translocation that blocks pores. See *seal.*

**Crystalline**   An arrangement of atoms in minerals and other substances in an ordered, systematic, repeating fashion.

**Cytochromes**   Metal-organic substances used in transfer of electrons to $O_2$ during cell respiration.

**Debris flow**   Downslope movement of water-lubricated soil and rock. See *mass wasting.*

**Deflocculation (of particles, colloids)**   Dispersion, or separation, of clusters of particles into separate individual particles.

**Denitrification**   Microbial reduction of nitrate to form gaseous $N_2$ or $N_2O$.

**Depletion zone**   The narrow zone next to the root where immobile nutrient concentrations in soil become markedly lowered.

**Deprotonation**   A chemical reaction involving removal of a proton ($H^+$).

**Desorption**   Migration of an ion or molecule from a surface. The opposite of sorption.

**Diffusion**   Molecular movement along a gradient. Water diffusion occurs from wet areas to dry ones. Gas and solute diffusion occur from zones of high concentration to zones of low concentration.

**Diffusion coefficient**   A numerical measure of the ease with which something can diffuse through something else (both must be specified, along with the temperature). A proportionality constant relating diffusion rate to concentration gradient.

**Disperse**   In soil science, to cause aggregates to separate into individual soil particles.

**Dissipation**   Spreading in various directions; scattering, dispersing, wasting.

**Dissociation**   Separation of a molecule into component atoms or ions—for example, $H_2O = H^+ + OH^-$.

**Dissolution**   The process by which a solid passes into solution.

**Dominant**   In forestry, the major tree species in a canopy.

**Dosing**   In waste disposal, the process of adding waste at intervals to keep the soil aerobic and to reduce clogging.

**Drain, tile**   A subsurface tunnel drain consisting of loosely fitting ceramic pipe sections into which water can flow from saturated soil. The term is still used for new perforated plastic drainpipe.

**Drawdown**   Lowering of the water table by drain or well; a change in the shape of the water table near the point of water withdrawal.

**Drill (agricultural)**   A machine for laying seed, fertilizer, and the like below the ground surface.

**Driving force**   See *potential gradient.*

**Earth flow**   The process of saturated soil moving down a slope under the force of gravity. The term is also used to describe the results of the process.

**Effective precipitation**   The amount or proportion of precipitation that infiltrates into soil.

**Effluent**   Liquid waste from either a septic tank or a sewage treatment plant.

**Electromagnetic spectrum**   Range of energy and wavelengths of electromagnetic radiation. The range from shortwave high-energy radiation to longwave low-energy radiation includes X-ray, ultraviolet, visible light, infrared, and radio waves.

**Electron**   A small, negatively charged atomic particle.

**Electrostatic (attraction or repulsion)**   Interaction between electrically charged objects.

**Element**   Any substance that cannot be further separated except by nuclear disintegration. See the periodic table at the front of the book.

**Eluvial horizon**   A soil layer (horizon) formed by the removal of constituents such as clay or iron.

**Eluviation**   Process of removing from a soil layer any soil constituents in suspension.

**Emissivity**  Relative measure of an object's ability to emit radiant energy at a given temperature.

**Emulsion**  The light-sensitive coating on photographic film, usually containing a silver salt such as silver chloride.

**Enzymes**  Protein catalysts produced in cells of living organisms that direct and control the cells' chemical reactions.

**Epidermis**  In plants, the outside layer of cells.

**Epipedons**  Surface layers of soil with specific characteristics used in classifying soils by soil taxonomy (e.g., the mollic epipedon).

**Equilibrium**  The condition of a chemical reaction or an entire ecosystem in which there are only minor changes in conditions over time.

**Equivalent**  The unit formerly used to describe cation exchange capacity or quantities of ions. One gram atomic weight of hydrogen or the amount of any other ion that will combine with or displace this amount of hydrogen. The amount that provides 1 mol of charge. Equals 1 $mol_c$.

**Erosion**  The wearing away of the land surface by water, wind, ice, or gravity.

**Eucaryotic**  In biology, referring to the type of cells with a distinct nucleus with a nuclear membrane; characteristic of fungi, protozoa, algae, plants, and animals. See *procaryotic*.

**Eutrophication**  Pollution with unwanted nutrients.

**Evaporation**  Vapor loss from soil or free water directly into the atmosphere.

**Evapotranspiration**  Evaporation plus transpiration.

**Exchangeable bases**  Exchangeable cations other than $Al^{3+}$ and $H^+$.

**Exchangeable base saturation**  Exchangeable cations other than $Al^{3+}$ and $H^+$, expressed as a percentage of cation exchange capacity measured at neutrality.

**Exchangeable ions**  Ions (charged atoms or molecules) held by electrical attraction at charged surfaces; can be displaced by exchange with other ions.

**Exchangeable sodium percentage (ESP) (of soil)**  Amount of exchangeable Na expressed as a percentage of total exchangeable cations.

**Exchange complex**  All the materials (clay, humus) that contribute to a soil's exchange capacity.

**Extracellular**  Outside the cell. Extracellular enzymes are excreted by some bacteria and fungi.

**Extracellular enzyme**  A protein substance that acts as an organic catalyst excreted outside the bacterial or plant cell. Also called an *exoenzyme*.

**Extrusive rock**  Any rock that forms from hot molten material by solidifying at the earth's surface.

**Factorial trial**  A systematic comparison of the effects of all combinations of two or more variables.

**Fall**  Occurs when part of a cliff or steep hillside falls vertically downward under gravitational stress.

**False-color infrared**  Color film carrying an emulsion sensitive to infrared energy. The color is false because the invisible infrared records as a visible color; (e.g., green objects appear red on the resulting photograph).

**Feldspar (felspar)**  A group of primary aluminosilicate minerals with a three-dimensional lattice. Important to soil genesis and fertility as source of the K, Na, and Ca they contain as accessory cations.

**Fermentation**  A set of metabolic processes by which anaerobic organisms obtain energy by converting sugars to alcohols or acids and $CO_2$.

**Fertigation**  The addition of soluble fertilizer through an irrigation system.

**Field capacity**  The amount of water retained in soil after it has been saturated and allowed to drain freely for 2 or 3 days.

**Film speed**  A measure of the emulsion's sensitivity, which determines the exposure required to produce the desired image.

**Flocculation**  Joining of colloidal particles to form clusters (flocs). Opposite of *deflocculation, dispersion*.

**Foliar**  Pertaining to leaves (e.g., foliar intake of nutrients, directly by leaves).

**Frictional resistance**  Resistance to the movement of fluids or particles caused by the interaction (rubbing) of surfaces.

**Genotype**  The genetic constitution of an organism. A group of organisms that are genetically alike.

**Geographic data**  Information that can best be displayed on maps. The distribution of soils on the earth's surface is one example.

**Geological erosion**  As opposed to accelerated erosion; erosion at natural rates, unaffected by human activity.

**Geostatistics**  Statistics that describe the variability of properties from place to place.

**Glacial outwash**  Geological material moved by glaciers and subsequently sorted and deposited by streams flowing from melting ice. Also called *glaciofluvial deposits*.

**Glacial till**  Unsorted and unstratified geological material deposited directly by glacial ice.

**Glaciation**   The process of geological erosion by means of glacial ice.

**Gleying**   A process that produces reduction of iron and other elements under conditions of prolonged saturation.

**Glucose**   A common sugar with six carbon atoms per molecule; present in all cells. A constituent of cellulose, starch, and other polysaccharides.

**Grade**   In furrow irrigation, flood irrigation, and stream flow, the slope of the flowing water surface; the effective slope down which the water runs.

**Gradient**   The rate at which a variable changes with distance. See *concentration gradient, potential gradient,* and *temperature gradient.*

**Granular**   Spherically shaped; used to describe soil aggregates, fertilizer particles, and the like.

**Grass waterway**   A natural or constructed waterway covered with erosion-resistant grasses and used to carry water and reduce erosion.

**Gravimetric**   Pertaining to weighing. Gravimetric methods determine amounts of a substance by direct or indirect weighing.

**Gravity erosion**   The movement of soil or rock by gravity.

**Greenhouse effect**   The warming of the earth's surface and atmosphere owing to absorption of outgoing radiation by $CO_2$, $CH_4$, and $H_2O$ (like absorption by glass).

**Green manure**   Plants grown to be cultivated into soil to improve soil fertility.

**Ground truth**   In remote sensing, the actual scene or data from the site that is being remotely sensed.

**Gully**   A channel cut by the concentrated flow of water during rainstorms. A gully is deeper than a rill and cannot be removed by normal tillage.

**Gypsum**   A mineral or rock composed of $CaSO_4 \cdot 2H_2O$.

**Halophyte**   Any plant that grows in saline environments and readily takes up salts.

**Heat capacity**   The amount of heat that an object must absorb or lose to produce a 1° change in its temperature.

**Heat conductivity**   A measure of the ease with which a material or object conducts or transmits heat.

**Hemicellulose**   A group of complex carbohydrates; polysaccharides that, unlike starch and cellulose, contain other sugars besides glucose. Important to plant cell walls.

**Heterotroph**   Any organism that derives its energy from organic compounds. See *autotroph.*

**Horizon**   A soil layer approximately parallel to the land surface.

**Humus**   The stable, dark-colored organic material that accumulates as a by-product of decomposition of plant or animal residues added to soil. The term is often used synonymously with *soil organic matter.*

**Hydrated**   Having water attached or incorporated as part of a chemical substance.

**Hydration**   The chemical combination of water with another substance.

**Hydraulic conductivity**   A measure of the ease (or difficulty) with which a liquid (usually water) flows through a soil (or plant) in response to a given potential gradient. The flux of water per unit gradient of hydraulic potential.

**Hydrogen bond**   An intermolecular chemical bond between the hydrogen from one molecule and the oxygen from another. Other electronegative atoms, such as N, may also form hydrogen bonds.

**Hydrolysis**   A reaction that involves products ($H^+$ or $OH^-$) of the dissociation of water. A mineral weathering reaction that adds $H^+$ to a mineral structure.

**Hydromulching**   Technique of spraying a slurry of fiber, seed, fertilizer, and chemicals onto roadsides for erosion control.

**Hydrous oxides, hydroxyoxides**   Sesquioxides. Oxides of Fe, Al, or similar metals, with different proportions of water and hydroxyl in the structure.

**Hydroxyl**   An OH ion or group.

**Igneous rocks**   Rocks formed from the cooling and solidification of hot liquid magma.

**Illuviation**   The process of deposition in lower soil horizons of material eluviated (transported) from upper horizons.

**Impermeable**   Unable to transmit water. A term often used with dense soil horizons through which water moves extremely slowly.

**Inactivated organisms**   In waste disposal, harmful organisms made harmless by reaction with the soil and soil organisms.

**Inclusion**   One or more polypedons within a map unit, not identified by the map unit name.

**Inferred properties**   Soil properties that are not seen or measured but are assumed from other properties. Suitability for a use is an inferred property.

**Infiltration**   Entry of water into the soil.

**Infrared**   Refers to the electromagnetic radiation of wavelength longer than light but shorter than radio. Sometimes called *heat radiation.*

**Inoculant**   A living culture of an organism in a form suitable for introduction to soil or some other new environment.

**In-place parent material**   Parent material that has not been transported from its original location; for example, a bedrock is an in-place parent material.

**Insulator**   A material (or space) that transmits heat or electricity poorly.

**Interaction**   A combined effect of two or more factors or variables such that the variables modify each other's effects.

**Interaction, electrostatic**   Adsorption caused by the electrical attraction of ions to a charged surface.

**Intercrop**   Two or more crops grown together on the same piece of land at the same time.

**Interveinal**   Between veins.

**Interveinal chlorosis**   Chlorosis only between leaf veins.

**Intrusive rock**   Any rock that forms by cooling from a hot molten mass below the earth's surface. Granite is an intrusive rock.

**Ion exchange**   The interchange between an ion in solution and another ion on the surface of any surface-active material such as clay or humus.

**Isomorphous substitution**   Substitution of one atom by another of similar size in a mineral without disrupting or changing the mineral structure.

**Kandite**   A family of clay minerals composed of one silicon tetrahedral sheet for each aluminum octahedral sheet.

**Kinetic energy**   Energy resulting from motion. In equation form, $KE = 1/2$ mass $\times$ (velocity squared).

**Landslide**   A rapid downhill movement of soil and rock under the force of gravity. A term also used to described the resulting landform.

**Latent heat**   A property of a material. The amount of heat involved per unit mass when the material undergoes a change of phase (e.g., the heat of melting or the heat of vaporization).

**Layer silicates**   Another term for aluminosilicate minerals that are platelike in form and composed of layers of atoms. Mica and the aluminosilicate clay minerals are layer silicates.

**Leaching**   The removal from soil or plants of soluble materials dissolved in water.

**Lichen**   Symbiosis between fungi and algae or blue-green bacteria, commonly forming a flat, spreading growth on surfaces of rocks and tree trunks.

**Ligand exchange**   A class of surface reactions on minerals in which anions from solution take the place of some of the anions (ligands) normally bound to cations in the mineral.

**Lignin**   A noncarbohydrate organic structural constituent of woody fibers in plant tissues. This slow-to-decompose material is an important part of soil organic matter.

**Lime**   Strictly, calcium oxide and hydroxide. More commonly, ground limestones in which $CaCO_3$ with or without $MgCO_3$ is the effective constituent.

**Lipids**   Organic oils, fats, and waxes. Combinations of organic acids and glycerin.

**Liquid limit**   In engineering, the water content corresponding to the limit between a soil's liquid and plastic states of consistency.

**Lithosequence**   Two or more soils in which the soil-forming factor that varies the most is the parent material. Other soil-forming factors are constant or vary much less than parent material.

**Loam**   A soil textural class name with limits of 7 to 27 percent clay, 28 to 50 percent silt, and 23 to 52 percent sand.

**Loess**   Wind-transported and deposited material of silt and clay size.

**London forces**   Weak, close-range attraction between atoms, molecules, or colloid particles due to transitory dipole–dipole interactions between their electrons.

**Longwave radiation**   In meteorology, radiation in the infrared and radio wavelengths, emitted at the earth's surface and partly absorbed in the atmosphere. Distinct from the sun's shortwave radiation.

**Macromolecule**   A large molecule, typically of colloidal size and made of hundreds or thousands of smaller molecules polymerized (joined together).

**Macronutrients**   Plant nutrient elements needed in the largest amounts. Examples are N, K, and P.

**Macropores**   Large pores, often formed by roots and small soil animals and worms.

**Magma**   A hot liquid from within the earth that solidifies to form igneous rock.

**Map unit**   Defines the contents of an area inside a polygon on a soil map.

**Mass flow**   Movement of fluid in response to pressure. Movement of heat, gases, or solutes together with the flowing fluid in which they are contained.

**Mass wasting**   A term describing all of the kinds of movement caused by gravity (such as landslides).

**Master horizons**   Major soil horizons identified by capital letters: A, B, C, E, O, and R.

**Matric potential**   A negative potential resulting from adhesive and capillary forces due to the soil matrix. These forces attract water in soil and lower its potential energy, compared with free water not in contact with soil.

**Meniscus**   The curved surface of water caused by the affinity of water to surfaces.

**Meristem**   The region of tissue in a plant shoot or root where most of the growth by cell division occurs.

**Metamorphic rocks**   Rocks formed from the alteration of preexisting igneous, sedimentary, or other metamorphic rocks by heat and pressure within the earth.

**Mica**   A platy, nonexpanding 2:1 layer aluminosilicate mineral with one octahedral (Al, Mg, Fe), O, OH sheet between two (Si, Al), O sheets. Two common forms are biotite and muscovite.

**Micronutrients**   Plant nutrient elements needed in lesser amounts than the macronutrients. *Micro-* does not imply lesser importance.

**Micropores**   Pores that are small compared with macropores. These have sometimes been referred to as *capillary pores.*

**Mineralization (of C, N, S, P, and so on)**   Release of an element in inorganic form—usually ions—during the decay of organic matter containing the element.

**Minerals**   Any of a large number of solid nonorganic soil constituents in which Al, Si, O, and cations are the major components.

**Mine soil**   Soil formed by reclamation of land used for mining.

**Minimum tillage**   A method of cultivation that reduces the number of machinery operations to the fewest required to create the proper soil condition for seed germination.

**Miscellaneous area**   A soil map unit mainly composed of disturbed soil or highly variable soil. Urban areas, beaches, dumps, minespoil, and artificial land are examples.

**Modal**   The central or most typical soil series.

**Monocategorical**   In soil classification, a system that uses only one level of classification, in contrast with a multicategorical system.

**Monomer**   A single molecule capable of joining with identical or similar monomer molecules to form a large (macro) molecule—that is, a polymer. Glucose polymerizes to form starch and cellulose, and amino acids polymerize to form protein.

**Mottles**   Spots or blotches of different colors or shades of color interspersed with the main (or matrix) soil color. Mottles indicate wetness.

**Mulch**   Any material spread on the soil surface to protect soil from raindrops, sunshine, freezing, or evaporation.

**Muscovite**   A form of mica. A platy aluminosilicate mineral with an aluminum octahedral sheet between two silicon tetrahedral sheets. It is a dioctahedral mineral because two out of three possible octahedral positions in the structure are filled with aluminum atoms.

**Natural classification**   A type of classification in which observed properties or objects are classified.

**Necrosis**   Death of a piece or pieces of tissue.

**Neutron**   An uncharged atomic particle with a mass of 1.

**Neutron probe**   An instrument that uses radiation to measure the soil water content.

**Nitrogen fixation**   Conversion of $N_2$ from air to forms usable by plants—for example, nitrate, or, more commonly, ammonia.

**No-tillage**   A farming system in which a crop is planted in the residue from a previous crop without soil tillage (such as plowing).

**Nucleic acids**   Polymers made of nucleotide units that carry and transcribe genetic information in cells.

**Nucleotides**   Compounds consisting of a ring-structured organic base, a five-carbon sugar, and one or more phosphate groups. May polymerize to form nucleic acids. See *adenosine triphosphate (ATP).*

**Nucleus**   In chemistry, the central part of an atom, consisting of positive protons and neutral neutrons in all atoms except hydrogen. In biology, the central organ within a eucaryotic cell, with a membrane enclosing nucleic acid structures that contain the cell's genetic information.

**Nutrient cycling**   The movement of nutrients through different forms and places in an ecosystem.

**Nutrient elements**   Elements or substances that contribute to an organism's growth and health. Some nutrients are essential to completion of the life cycle.

**Oblique photo**   In remote sensing, a photograph taken at an angle to the horizon, not perpendicular. See *vertical photo*.

**Observed property**   Any property that can be seen or measured.

**Octahedron**   An eight-sided molecular unit. In mineralogy, a metal cation such as Al or Fe surrounded by six oxygen atoms or hydroxyl groups.

**Off-site disposal**   Disposal of waste in a location away from where the waste is produced.

**On-site disposal**   Disposal of waste at the site where the waste is produced.

**Organic fertilizer**   Organic material that releases or supplies useful amounts of a plant nutrient when added to soil.

**Organic soil**   Soil that contains a high percentage of organic matter throughout the solum.

**Osmotic potential**   The negative potential caused by the presence of solutes in soil water; solute potential.

**Overland flow**   Water flow on the soil surface.

**Oxidation**   The addition of oxygen. More generally, the removal of electrons from an atom, ion, or molecule during a reaction. Oxidation may increase the positive charge of an element or compound. See *reduction*.

**Paralithic**   Soft or weathered rock. Literally, "like rock." In soil classification, the solum may overlie lithic or paralithic material.

**Parallax**   In remote sensing, the apparent displacement, or difference, in the apparent direction of an object seen from two different points.

**Parent material**   The unconsolidated and more or less chemically weathered mineral or organic matter from which the soil solum develops.

**Particle detachment**   The first step in the erosion process. Wind or water detaches particles from the soil mass.

**Particle transport**   The second step in the erosion process. Wind or water carries particles away from where they were detached.

**Passive remote sensing**   Remote sensing in which the energy being sensed is not supplied by the sensor. For example, any sensing that uses the sun's energy is passive remote sensing.

**Peat**   Unconsolidated soil material consisting largely of undecomposed, recognizable plant material.

**Ped**   A natural unit of soil structure. Another term for *aggregate*.

**Pedon**   A three-dimensional body of soil with lateral dimensions large enough to permit the study of horizon shapes and relations. The soil individual.

**Pellet**   A small ball of material stuck together. Seed may be pelleted with fertilizer, inoculants, and the like. Powdered fertilizers are pelleted for efficiency and ease of handling.

**Peptides**   Amino acid polymers smaller in molecular size than proteins. Protein breakdown products.

**Percolation**   The downward movement of fluid (water or waste effluent) in soil.

**Perc test**   A test to evaluate the suitability of soils for on-site waste disposal, by measuring the rate that water enters (percs) a soil layer or horizon.

**Permanent wilting point**   The water content of soils at which indicator plants will regain turgor even if the soil water content is raised. Usually estimated as $-1.5$ MPa.

**Persistence**   The property of a material such as a pesticide or organic matter to resist microbial or chemical decomposition. The more persistent a chemical, the longer time required to decompose it.

**pH**   A measure of acid intensity. $pH = -\log[H^+]$.

**pH buffering**   See *buffering*.

**Phenol, phenolic compounds**   Organic compounds with an unsaturated carbon-ring structure with one or more OH groups attached. Components of lignin and humus.

**Phloem**   One of two main kinds of conductive (vascular) tissue in plants. Phloem carries dissolved organic materials from leaves to other parts.

**Phytoremediation**   The use of plants to remediate (clean) soil by plant uptake and accumulation of soil contaminants.

**Plane table**   A flat table used in mapping soils before the use of aerial photographs became common.

**Plant-available moisture**   Soil water held loosely enough that plants can extract it for use.

**Plastic index**   In engineering, the difference between the liquid limit and the plastic limit.

**Plastic limit**   In engineering, the minimum soil water content at which the soil consistence is soft and pliable rather than hard and brittle.

**Platy**   A soil structure type whose horizontal axis is much longer than its vertical axis.

**Point data**   Data for a specific location that are most easily presented in a table or graph. Soil color and texture of a horizon are examples of point data.

**Polar**   In chemistry, an uncharged molecule (such as water) that has an uneven distribution of charge so that it has some of the properties of a charged molecule.

**Polders**   Land once flooded that has been reclaimed by means of drainage.

**Pollute**   To make unclean; to add unwanted substances.

**Polymer**   A large molecule made up of many similar small molecules (monomers) joined together. Examples are cellulose, proteins, and nucleic acids.

**Polymerization**   The process of joining two or more similar molecules to form a larger molecule.

**Polypedon**   A group of contiguous pedons.

**Polysaccharides**   Carbohydrate macromolecules—polymers—made from sugars.

**Polyuronides**   Polymers of sugar acids (sugars with attached carboxyl groups) and related monomer units. Include pectic compounds important to plant cell walls.

**Ponding**   Deliberate or accidental buildup of water on the ground surface.

**Pores**   The space not occupied by solid particles in the bulk volume of soil. Sometimes referred to as *pore space.*

**Pore size distribution**   The volume fractions of the various sizes of pores in a soil.

**Porosity**   The volume percentage of the total bulk of soil not occupied by solid particles. The volume of pores in a sample divided by the sample volume.

**Positive pore water pressure**   Pressure exerted on soil particles by water in a saturated soil.

**Potential gradient**   The rate of change of potential with distance. The driving force of water through a medium.

**Precipitation**   In chemistry, the formation of solid material from constituents of a solution. Also, the deposition of water on the ground as rain, snow, dew, and frost or the amount of deposition.

**Primary particles**   Individual mineral particles; sand, silt, and clay are primary soil particles.

**Prismatic**   A soil structure type whose vertical axis is much longer than its horizontal axis and that has angular edges, compared with a columnar structure, which has a similar shape but curved edges.

**Procaryotic**   In biology, referring to morphologically simple cells, without a membrane-enclosed nucleus; characteristic of bacteria. See *eucaryotic.*

**Profile**   Vertical gradation or sequence of a soil property—for example, temperature profile or pH profile. The appearance of a pedon's vertical face or section.

**Proteins**   Polymers of amino acids that are essential to life. They function as enzymes and structural molecules within cells and cell membranes.

**Proton**   A positively charged primary particle found in the nucleus of atoms and released by dissociation of water and other acids. $H^+$ ions.

**Protonation**   The addition of a proton to a molecule, ion, or surface.

**Protozoan**   Plural form is **protozoa.** A microscopic, often single-celled, eucaryotic, usually nonphotosynthetic animal-like organism.

**Psychrometer**   A device to measure soil or leaf moisture or water potential as a function of relative humidity.

**Purpling**   Accumulation of anthocyanin pigments in plants, causing a purple coloration of leaves or stem.

**Pyroxenes**   A family of single-chain ferromagnesian silicates.

**Quartz**   $SiO_2$. An important rock-forming mineral, resistant to weathering and abundant in many soils.

**Radiation**   Processes in which energy is sent out as waves and particles through space from atoms and molecules as they rotate, vibrate, and undergo internal change.

**Rangeland**   Land used for free-grazing livestock.

**Recalcitrant**   The property of a material to stubbornly resist decomposition. A chemical such as a pesticide that is not decomposed by microorganisms is a recalcitrant pesticide.

**Reclamation**   In regard to soil, the treatment of soil to correct waterlogging or severe excesses such as salinity or sodicity.

**Reduction**   Addition of electron(s) to an atom or molecule. Opposite of *oxidation.*

**Reemitted energy**   Radiation of previously absorbed energy.

**Reflected energy**   Radiant energy that is thrown back from an object, with no change except direction.

**Relative humidity** The concentration of water in the air or soil atmosphere relative to the maximum concentration it can hold at the given temperature.

**Remote sensing** Identifying and observing objects at a distance.

**Reradiation** Radiant emission of energy previously absorbed.

**Residual parent material** Unconsolidated and partly weathered mineral materials accumulated by disintegration of consolidated rock in place. See *transported parent material.*

**Residue conservation** Leaving straw, stubble, trash—crop residues—to rot on or in the ground instead of removing or burning them.

**Resistance** A measure of the difficulty with which a material conducts heat, ions, electricity, or water. Reciprocal of *conductance.*

**Respiration** The set of metabolic processes in which an organism obtains energy from the oxidation of sugars to $CO_2$ and water, usually with $O_2$ as the oxidizing agent.

**Response curve** A graph relating a plant's quantitative response—changes in growth, yield, or any desired attribute—to different levels of a treatment such as fertilizer application.

**Rhizosphere** The region in the soil immediately next to the root surface (within perhaps a millimeter) where the plant most strongly affects soil properties.

**Rill** A small, intermittent water course with steep sides usually only several centimeters deep and wide.

**Ripening** The process by which reclaimed land develops characteristics suitable for farming. It includes drainage, cultivation, and neutralizing acidity.

**Root cap** Cells at the tip of the root protecting the growing cells as the root pushes through the soil.

**Root density** Number mass or length of roots within a given soil volume.

**Root distribution** Arrangement of roots in space within the soil, especially the vertical root density profile.

**Root exudate** A mixture of organic acids, sugars, and other soluble plant components that escape from roots.

**Root hair** A hairlike structure produced from a root epidermal cell. Probably helps obtain water and immobile nutrients.

**Rotational slide** Mass wasting in which the mass rotates about a point rather than sliding horizontally.

**Runoff** Water that runs off the soil surface instead of infiltrating; the process of running off. Sometimes loosely includes seepage—that is, the escape of water that has flowed through the soil.

**Saline** Containing large concentrations of soluble salts. Operationally defined as the electrical conductivity of a saturation extract of $\geq 4$ deciSiemens per meter ($dS\ m^{-1}$).

**Salinization** The process of making a soil saline; increasing the concentration of soluble salts.

**Saltation** Movement of particles by bouncing or skipping, under the influence of wind or moving water.

**Salting out** Jargon. Precipitation of salts, as when concentrated fertilizer solutions are overcooled.

**Sand** A mineral particle size ranging between 0.05 and 2.00 mm in diameter.

**Sanitary landfill** An area of land used for the disposal of (household) solid waste.

**Saturated** Generally, occupying all of a capacity. With respect to water, the condition of a soil when all pores are filled with water. With respect to a particular cation or group of cations, the condition in which cations of the specified kind occupy all the exchange capacity.

**Saturation extract** The soil solution removed for analysis from a saturated soil by means of vacuum filtration.

**Scattered energy** Radiation reflected from an object in many directions. In remote sensing, this energy is useless for identifying the object.

**Scattering (of light or other radiation)** Reflection in several directions. Scattering from fog, cloud, and haze reduces energy input to the ground.

**Seal** A thin, dense layer that develops when the soil is wet. See *crust.*

**Section–range system** A legal survey system used to accurately locate a parcel of land.

**Sedimentary rocks** Rocks formed by consolidation of materials deposited or precipitated from water or air.

**Sedimentation** A technique for determining particle size distribution. Particles fall through water at a rate proportional to their size.

**Sensible heat** Ordinary everyday heat, sensed by feel or by thermometer.

**Sesquioxides** The general term for the aluminum and iron oxides in soil. See *hydrous oxides.*

**Seventh Approximation**  The first widely distributed version of the present U.S. soil classification system.

**Shear strength**  The resistance of a soil to forces acting at right angles to the soil body. A plow produces shearing forces.

**Sheet erosion**  Water erosion that removes a uniform layer of soil from the land surface.

**Sheet flow**  A thin, relatively uniform water runoff (a few millimeters deep), in which the water is not concentrated in channels.

**Shortwave radiation**  In meteorology, high-energy radiation of the ultraviolet, visible, and near-infrared wavelengths; sunshine. Unlike longwave radiation, it passes freely through air.

**Shrinkage limit**  The maximum water content at which a reduction in water content will not cause a decrease in the soil's volume.

**Silicates**  Those minerals in which silicon and oxygen are the major elemental constituents.

**Silt**  A mineral particle size ranging between 0.002 and 0.05 mm in diameter.

**Sink**  Somewhere or something for the disposal of wastewater.

**Site index**  A quantitative evaluation of a soil's relative site productivity for forest growth under the existing or specified environment.

**Size separate**  Individual mineral particles less than 2.0 mm in diameter, ranging between specified size limits. Sand, silt, and clay are common soil size separates.

**Slake**  A process of aggregate breakdown caused by internal pressures generated as water enters aggregates and air escapes.

**Sludge**  In waste disposal, a semifluid mixture of fine solid particles with a liquid. Also referred to as *biosolids*.

**Slump**  Collapse of soil and rock due to gravity.

**Smectite**  A family of expanding 2:1 (Si tetrahedral sheets to Al octahedral sheets) clay minerals with a high surface area and exchange capacity.

**Sodic**  High in sodium; having a high exchangeable Na percentage or high Na adsorption ratio.

**Sodium adsorption ratio (SAR)**  The concentration of Na divided by the square root of the sum of Ca and Mg concentrations, both expressed as molarities and measured in the saturation extract.

**Soil acidity factors**  A set of related factors often inhibiting plant growth in acidic soils (deficiency of $Ca^{2+}$ and excess of $H^+$, $Al^{3+}$, and $Mn^{2+}$).

**Soil air**  The gas within the soil; the soil atmosphere.

**Soil association**  A group of defined and named soil units occurring together in an individual and characteristic pattern over a geographic area. The individual bodies are large enough to be delineated at a scale of 1:24,000.

**Soil complex**  A map unit consisting of two or more kinds of soil that occur in a regular, repeating pattern so intricate that its components cannot be separated at the 1:24,000 map scale.

**Soil creep**  Slow mass movement of soil and soil material down relatively steep slopes under the influence of gravity. See *creep*.

**Soil density**  A measure of the relative amount of pores and solid particles. The mass per unit volume.

**Soil drainage class**  An interpretive grouping of soils based on the level of the water table during the growing season and the rate of water flow through soil.

**Soil fabric**  The pattern resulting from the arrangement of solids and pores.

**Soil-forming factors**  The five interrelated natural factors that are active in the formation of soil: parent material, climate, organisms, topography, and time.

**Soil-forming processes**  The biological, chemical, and physical processes that act under the influence of the soil-forming factors to create soils.

**Soil genesis**  Soil formation; the process of soil formation.

**Soil individual**  The pedon; a three-dimensional body large enough to contain all the properties necessary to describe a soil completely.

**Soil matrix**  Like soil fabric, the combination of solids and pores in a soil.

**Soil order**  In soil classification, the most general level of classification. All soils fit into 12 orders.

**Soil potential**  The usefulness of a site for a specific purpose using available technology at a cost expressed in economic, social, or environmental units of value.

**Soil series**  The most detailed level in soil classification. A soil series consists of soils essentially alike in all major profile characteristics except the texture of the A horizon.

**Soil solution**  The liquid phase of the soil and its solutes.

**Soil survey**  The systematic examination, description, classification, and mapping of soils in an area.

**Soil taxonomy**   The U.S. Department of Agriculture system of soil classification.

**Soil test**   A quick, routine extraction and analysis of soil for a diagnostic purpose.

**Solid waste**   Waste other than sewage, mostly solid. Household garbage is a familiar form of solid waste.

**Solum**   The upper and most weathered part of the soil profile; the A, E, and B horizons.

**Solute**   A material dissolved in a solvent to form a solution.

**Specific adsorption**   See *adsorption, specific* and *ligand exchange.*

**Specific gravity**   The ratio of the weight or mass of a given volume of a substance to that of an equal volume of a reference substance. Water is the reference for liquids and solids.

**Spectral signature**   In remote sensing, an object's spectral reflectance pattern. The wavelength and intensity of reflectance.

**Splash erosion**   The detachment and airborne movement of small soil particles by raindrop impact.

**Spoil (overburden)**   Material overlying sought-after resources that is removed and piled to gain access to the minerals, coal, or other underground materials of value. Also, the material remaining after rock has been crushed to remove valuable minerals.

**Stele**   The central conductive tissue of a root or some kinds of stem.

**Stereo pair**   Two photographs of the same object taken from slightly different positions that when viewed together give the viewer a three-dimensional image.

**Stomates** (or **stomata**)   The controllable openings in the epidermis of leaves and other parts of a plant shoot.

**Stratified**   Arranged in layers.

**Stress**   A force exerted on a body, especially one that strains or deforms its shape. An adverse condition imposed on an organism.

**Strip-cropping**   A technique to reduce soil erosion in which bare fallow or a row crop such as corn or soybeans is grown in conjunction with grass or a legume hay crop.

**Strip mining**   A process in which rock and top soil strata overlying ore or fuel deposits are scraped away to expose the desired deposits.

**Structure**   In soil, the arrangement of primary particles into secondary units or peds. The units are characterized and classified on the basis of size, shape, and degree of distinctness. In crystals, the arrangement of atoms or molecules. In molecules, the arrangement and bonding of atoms.

**Structure grade**   A grouping of soil structure based on the cohesion or stability of the structure. Grades are *structureless, weak, moderate,* and *strong.*

**Structure strength**   A term sometimes used in place of *structure grade.*

**Subangular blocky**   A structure type whose aggregates are nearly equidimensional, square, or blocklike with rounded edges.

**Substrata**   Layers of material underlying the soil horizons—for example, sediment or rock.

**Substrate**   A substratum. More often, the reactant or starting compound for a biochemical reaction or a material that provides food for microbes.

**Subsurface diagnostic horizons**   One or more soil horizons below the surface horizon with specific properties as described in the soil taxonomy (e.g., the argillic horizon).

**Subtractive trial**   A fertility trial in which deficiencies are detected by observing plant response to the elimination of individual elements from a complete mixture.

**Sugar**   A simple carbohydrate, consisting of one or two monomer units, each usually containing five or six C atoms. Rapidly used food sources for microbes in soil. Examples are glucose, fructose, and sucrose.

**Superphosphate**   A phosphate fertilizer made by treating phosphate rock with acid to increase solubility.

**Surface chelation**   A chelation reaction that holds a cation at the surface.

**Surface tension**   The force required per unit length to separate or pull apart a liquid surface.

**Suspension**   A fluid, usually liquid, bearing small particles held suspended but not dissolved.

**Symbiosis**   An association between two organisms or populations that, in the absence of environmental change, is stable.

**Technical classification**   A soil classification of inferred rather than observed properties.

**Temperature gradient**   The rate at which temperature changes with distance; in soil, usually vertical distance.

**Tensile strength**   The resistance of a soil to forces pulling from opposite directions.

**Tensiometer**   A device for measuring the soil–water matric potential in place.

**Terrace** In soil conservation, a more or less level or horizontal strip of earth usually constructed on a contour designed to reduce erosion.

**Tetrahedron** A four-sided arrangement of atoms with a cation, such as Si, in the center surrounded by four oxygen atoms or hydroxyls.

**Texture** The relative proportion of the various soil separates—sand, silt, and clay—that make up the soil texture classes as described by the textural triangle (Figure 2–4).

**Thermal properties** See *conductivity* and *heat capacity.*

**Tile drain** See *Drain, tile.*

**Tillage** The mechanical stirring or turning of the soil profile. In agriculture, horticulture, and landscaping, tillage is used to prepare soil for seeds, to incorporate organic matter or chemicals, or to control weeds.

**Tissue test** A routine, rapid chemical analysis of leaves or other plant parts for diagnostic purposes.

**Toposequence** Two or more soils in which the soil-forming factor that varies the most is the landscape (usually elevation or position on a slope). Other soil-forming factors are constant or vary much less than topography.

**Tortuosity** A term describing the winding, twisting, and crooked nature of soil pores. A factor contributing to the rate of water movement in soil.

**Trace elements** Elements contained in plants or soil in small quantities. Sometimes used for *micronutrients.*

**Transitional horizon** A horizon with properties intermediate between the horizons above and below.

**Transitional slide** A break causing a mass to slide along the face of the supporting mass under the force of gravity.

**Transmitted energy** In remote sensing, energy that passes through an object.

**Transpiration** Evaporation from leaves; the flow of water through plants from soil to atmosphere.

**Transported parent material** The opposite of *residual parent material;* unconsolidated material carried by wind or water that, once stabilized, acts as the starting material for soil.

**Trial** A term often used to describe a field or container experiment designed to determine nutrient deficiencies. See *factorial trial* and *subtractive trial.*

**Turbulent transfer** Diffusion-like movement of heat, gases, and solutes greatly accelerated by irregular mixing motion—turbulence—in the fluid medium.

**Turgor** The normal state of turgidity in living cells produced by pressure of the cell contents on the cell walls.

**Turgor pressure** Pressure exerted on a cell wall by the cell's constituents (and vice versa).

**Unconsolidated** Loose.

**Undifferentiated group** A soil map unit consisting of two or more similar soil units not in a regular geographic association. The soil units have the same or very similar use and management.

**Unit cell** The smallest definable repeating structural unit in a crystalline material.

**Universal Soil Loss Equation (USLE)** An equation developed from experimental results that relates the average annual soil loss to rainfall, soil, cover, erosion control, and topographic factors.

**Uronide** A sugar with a COOH group. See *polyuronides.*

**van der Waal's forces** Combined polar and London attractions between molecules in close proximity.

**Variable charge** An electrical charge on clay or organic matter that changes with changes in soil pH.

**Vascular** Having to do with veins. See *conductive tissue.*

**Vermiculite** An expanding 2:1 layer aluminosilicate clay mineral, in which one sheet of Al, O, and OH in octahedral coordination lies between two sheets of Si and Al, with O in a tetrahedral coordination.

**Vertical photo** In remote sensing, a photo taken from directly above an object rather than at an angle to the object. See *oblique photo.*

**Vesicular pore** A soil pore not connected to other pores.

**Void ratio** A measure of porosity. The ratio of void volume to soil bulk volume.

**Voids** Pores.

**Volcanic ash** Fine particles of rock blown into the air by a volcano. Settles, often in layers, to become soil parent material. Often consists of short-range-order minerals.

**Washed-in zone** Formed by clay separating from the top few tenths of a millimeter of soil and being translocated into the next few tenths to clog the pores.

**Water erosion** The natural wearing away of the earth's surface by rainfall and surface runoff. See *accelerated erosion.*

**Waterlogged** Saturated with water.

**Water potential** The tendency of soil water to move; the sum of gravity, pressure, matric, and solute components.

**Water profile** The pattern of vertical variation in water content or potential with depth in the soil.

**Water retention curve or soil–water characteristic curve** The graphical relationship between soil water content (by mass or volume) and the soil–water matric potential (the energy required to remove the water).

**Water table** The level, elevation, or imaginary plane corresponding to the top of a water-saturated zone of soil, sediment, or rock.

**Weathering** The chemical or physical processes at or near the earth's surface that change rocks and minerals exposed to air and water.

**Wetting front** The advancing boundary between dry soil and wetted soil during infiltration.

**Xylem** One of two main kinds of conductive tissue in plants. Xylem has tubelike cells that conduct water and ions rapidly from roots to leaves. Wood is xylem tissue.

## REFERENCES

*Glossary of Soil Science Terms.* 1975. Madison, Wis.: Soil Science Society of America.

*Glossary of Soil Science Terms.* 1997. Madison, Wis.: Soil Science Society of America.

*Resource Conservation Glossary.* 1976. Ankeny, Iowa: Soil Conservation Society of America.

# INDEX

convenience and ease of use of, 212
determining needs, 213–20
liquid, 231
methods and timing of application of, 220–24
nitrogen as, 208–10
nutrient content, 212
nutrient cycling and control of losses, 224
organic, 211–12
phosphorus as, 210–11
potassium as, 211
release rates, 212
side effects of, 212–13
solid, 231
specifications and calculations, 231–32
sulfur as, 211
Fibrists, 328
Field capacity, 55–56
Field trials, 213
Fifth-order surveys, 313
Film speed, 307
Fine-earth fraction, 16
Flocculation, 256
effects of cation layer on, 257
Flooded soils, 167, 169
saturated flow in, 90
Flooding in correcting acidity, 236, 238
Flood irrigation, 115–16, 119–20
Flux, 73
Focal length, 305
Foliar spray, 220, 224
Foliar uptake, 187
Folists, 328
Food chains, 201
Forests, acid precipitation in cool, humid, 257–59
Forest Service maps, 299
Fourier's heat flow equation, 73, 75–76
Fourth-order surveys, 313
Fragipan, 318–19
*Frankia* spp., 145
Frictional resistance, 382–83
Fungi, 133, 140
Furrow irrigation, 116, 119–20

## G

Gases
movement of, 83–84
solubility, 66
Gelisols, 9, 315, 321

Genotype, 198
Geographic data, 294–97
handling, 307–11
Geographic information, 293
Geographic variations in soil climate, 79
Geological cycle, soils as part of, 261
Geological erosion, 344–45
Geostatistics, 300–301
Gibbsite, 252, 253
Glacial outwash, 266
Glacial till, 266
as transported parent materials, 264
Glaciations, 266
Gley colors, 270, 290
Gleying, 270
Global Operational Environmental Satellite (GOES), 305
Global positioning system (GPS), 302
Grade, 119
Graded terraces, 365
Gradient of temperature, 73
Grand Canyon, 344
Granular structure, 19
Granulation, 135–36
Grass waterway, 365
Gravel, 4
Gravimetric method, 88
Gravitation, infiltration and, 111–12
Gravity, specific, 51
Gravity erosion, 352–54
processes, 352–54
Greenhouse effect, 70
Greenhouses, soil temperature in, 82
Green manures, 165
Ground covers, effects of, on soil climate, 79, 80
Ground truth, 304
Groundwater pollution, 230
Growth per unit of water applied, 126
Gullies, 344
erosion, 346
Gypsum, 246

## H

Half-life of radioactive nucleus, 46
Halloysite, 31–32
Halphytes, 243
Head, 87